MW00422598

FIXED BROADBAND WIRELESS

SYSTEM DESIGN

FIXED BROADBAND WIRELESS
SYSTEM DESIGN

HARRY R. ANDERSON, Ph.D., P.E.

CONSULTING ENGINEER
USA

Telephone (+44) 1243 779777

Email (for orders and customer service enquiries): cs-books@wiley.co.uk
Visit our Home Page on www.wileyeurope.com or www.wiley.com

Other Wiley Editorial Offices

John Wiley & Sons Inc., 111 River Street, Hoboken, NJ 07030, USA

Jossey-Bass, 989 Market Street, San Francisco, CA 94103-1741, USA

Wiley-VCH Verlag GmbH, Boschstr. 12, D-69469 Weinheim, Germany

John Wiley & Sons Australia Ltd, 33 Park Road, Milton, Queensland 4064, Australia

John Wiley & Sons (Asia) Pte Ltd, 2 Clementi Loop #02-01, Jin Xing Distripark, Singapore 129809

John Wiley & Sons Canada Ltd, 22 Worcester Road, Etobicoke, Ontario, Canada M9W 1L1

Wiley also publishes its books in a variety of electronic formats. Some content that appears in print may not be available in electronic books.

Library of Congress Cataloging-in-Publication Data

Anderson, Harry R.
Fixed broadband wireless system design / Harry R. Anderson.
p. cm.
Includes bibliographical references and index.
ISBN 0-470-84438-8 (alk. paper)
1. Wireless communication systems – Design and construction. 2. Cellular telephone systems – Design and construction. 3. Broadband communication systems. I. Title.

TK5103.4 .A53 2003
621.3845′6 – dc21

2002033360

British Library Cataloguing in Publication Data

A catalogue record for this book is available from the British Library

ISBN 0-470-84438-8

Typeset in 10/12pt Times by Laserwords Private Limited, Chennai, India
Printed and bound in Great Britain by Biddles Ltd, Guildford, Surrey
This book is printed on acid-free paper responsibly manufactured from sustainable forestry in which at least two trees are planted for each one used for paper production.

Contents

Preface

The growing demand for high-speed data connections to serve a variety of business and personal uses has driven an explosive growth in telecommunications technologies of all sorts including optical fiber, coaxial cable, twisted-pair telephone cables, and wireless. Nations have recognized that telecommunications infrastructure is as significant as roads, water systems, and electrical distribution in supporting economic growth. In developing countries it is not particularly unusual to see cell phone service in a town or village that does not yet have a water or sewer system. In the United States, recent government initiatives have recognized the importance of broadband telecommunications to economic growth.

This book focuses on fixed broadband wireless communications – a particular sector of the communication industry that holds great promise for delivering high-speed data to homes and businesses in a flexible and efficient way. The concept of 'broadband' communications is a relative one. Compared to the 1200-baud modems commonly used 20 years ago, today's dial-up phone connections with 56-kbps modems are 'broadband'. The demands and ambitions of the communication applications and their users have expanded, and will continue to expand, on what is meant by 'broadband'. The term is evolving, as is the technology that is classified as broadband. Nevertheless, for the purposes of this book I will use the somewhat arbitrary definition that broadband wireless systems are those designed for, and capable of handling baseband information data rates of 1 Mbps or higher, knowing that future developments may well move this threshold to 5 or 10 Mbps and beyond. The term 'broadband' also has an engineering significance that will be discussed in some detail in this book. Broadband wireless channels, as distinguished from narrowband channels, are those whose transfer characteristics must be dealt with in a particular way, depending on the information transmission speed and the physical characteristics of the environment where the service is deployed.

The term 'fixed' has also become somewhat nebulous with the technological developments of the past few years. Whereas fixed and mobile were previously well-understood differentiators for system types, we now have intermediate types of network terminals including fixed, portable, nomadic, and mobile, among others. Recent system standards such as those for 3G UMTS W-CDMA define different service levels and data rates depending on whether the user is in a fixed location, walking, or moving at high speed. This trend portends a convergence of fixed and mobile system types whose operation and availability are largely transparent to the application users. As will be shown, whether the system user is at a fixed location or in motion affects several decisions about the system design, the most appropriate technology, and the quality and performance that can be expected from a wireless application.

Although there have been a few books recently written on broadband, and specifically wireless broadband, in general they have been intended for non-technical audiences.

This book is intended for engineers who are faced with designing and deploying fixed broadband wireless systems, and who must also have sufficient understanding of the theory and principles on which the designs are based to formulate creative solutions to special engineering problems that they will eventually face. Along with generally accepted design assumptions and simplifications, the underlying theory and requisite mathematics are included where necessary to provide this foundation knowledge.

In addition to design engineers who deal with fixed broadband wireless systems on a daily basis, this book is also well suited to graduate and post-graduate level courses that are focused on wireless communications engineering. Wireless communication system design and planning is an increasingly important area that warrants serious academic treatment.

This book also covers some areas that have not classically fallen in the domain of wireless RF engineers; in particular, traffic modeling, environment databases, and mapping. Wireless system design is driven by the commercial requirements of the system operators who ultimately build viable businesses by successfully serving the traffic demands of the customers in their service areas. Detailed statistical modeling of packet-based traffic for a variety of applications (e-mail, web-browsing, voice, video streaming) is an essential consideration in fixed broadband system design if the operator's capacity and quality of service objectives are to be achieved.

The chapters in this book are organized with the fundamentals of electromagnetic propagation, channel and fading models, antenna systems, modulation, equalizers and coding treated first since they are the building blocks on which all wireless system designs are based. Chapters on multiple access methods and traffic modeling follow. The remaining chapters set forth the specific details of many types of line-of-sight (LOS) and non-line-of-sight (NLOS) systems, including elemental point-to-point links as well as point-to-multipoint, consecutive point, and mesh networks. Because of their importance, a separate chapter is devoted to designing both LOS and NLOS point-to-multipoint networks. The final chapter deals with the important subject of channel assignment strategies where the capacity and service quality of the wireless network is ultimately established.

Fixed wireless design relies on a number of published sources for data and algorithms. For convenience, the essential data, such as rain rate tables and maps, is included in the Appendices. In general, the referenced publications chosen throughout are currently available books or journal papers which are readily accessible in academic libraries or on-line. For the most recent or unique work, technical conference papers are also utilized.

A book of this type is clearly not a solo effort. I would like to thank several people who offered valuable comments, including Tim Wilkinson for reviewing Chapters 7 and 8, George Tsoulos for reviewing Chapter 6, and Jody Kirtner for reviewing Chapter 5, and for her efforts in proofreading the entire manuscript. Creating and refining a technical work such as this book is an evolutionary process where comments, suggestions, and corrections from those using it are most welcome and encouraged. I hope and anticipate that this book will prove to be a worthwhile addition to the engineering libraries of those who design, deploy, and manage fixed broadband wireless systems.

Harry R. Anderson
Eugene, Oregon, USA
January, 2003.

1

Fixed broadband wireless systems

1.1 INTRODUCTION

The theoretical origin of communications between two points using electromagnetic (EM) waves propagating through space can be traced to James Maxwell's treatise on electromagnetism, published in 1873, and later to the experimental laboratory work of Heinrich Hertz, who in 1888 produced the first radio wave communication. Following Hertz's developments at the end of the nineteenth century, several researchers in various countries were experimenting with controlled excitation and propagation of such waves. The first transmitters were of the 'spark-gap' type. A spark-gap transmitter essentially worked by producing a large energy impulse into a resonant antenna by way of a voltage spark across a gap. The resulting wave at the resonant frequency of the antenna would propagate in all directions with the intention that a corresponding signal current would be induced in the antenna apparatus of the desired receiving stations for detection there. Early researchers include Marconi, who while working in England in 1896 demonstrated communication across 16 km using a spark-gap transmitter, and Reginald Fassenden, who while working in the United States achieved the first modulated continuous wave transmission. The invention of the 'audion' by Lee DeForest in 1906 led to the development of the more robust and reliable vacuum tube. Vacuum tubes made possible the creation of powerful and efficient carrier wave oscillators that could be modulated to transmit with voice and music over wide areas. In the 1910s, transmitters and receivers using vacuum tubes ultimately replaced spark and arc transmitters that were difficult to modulate. Modulated carrier wave transmissions opened the door to the vast frequency-partitioned EM spectrum that is used today for wireless communications.

Radio communications differed from the predominate means of electrical communication, which at the time was the telegraph and fledgling telephone services. Because the new radio communications did not require a wire connection from the transmitter to the receiver as the telegraph and telephone services did, they were initially called *wireless communications*, a term that would continue in use in various parts of the world for several

Fixed Broadband Wireless System Design Harry R. Anderson
 ISBN: 0-470-84438-8

decades. The universal use of the term *wireless* rather than *radio* has now seen a marked resurgence to describe a wide variety of services in which communication technology using EM energy propagating through space is replacing traditional wired technologies.

1.2 EVOLUTION OF WIRELESS SYSTEMS

As the demand for new and different communication services increased, more radio spectrum space at higher frequencies was required. New services in the Very High Frequency (VHF) (30–300 MHz), Ultra High Frequency (UHF) (300–3,000 MHz), and Super High Frequency (SHF) (3–30 GHz) bands emerged. Table 1.1 shows the common international naming conventions for frequency bands. Propagation at these higher frequencies is dominated by different mechanisms as compared to propagation at lower frequencies. At low frequency (LF) and Mediumwave Frequency (MF), reliable communication is achieved via EM waves propagating along the earth–atmosphere boundary – the so-called *groundwaves*. At VHF and higher frequencies, groundwaves emanating from the transmitter still exist, of course, but their attenuation is so rapid that communication at useful distances is not possible. The dominant propagation mechanism at these frequencies is by space waves, or waves propagating through the atmosphere. One of the challenges to designing successful and reliable communication systems is accurately modeling this space-wave propagation and its effects on the performance of the system.

The systems that were developed through the twentieth century were designed to serve a variety of commercial and military uses. Wireless communication to ships at sea was one of the first applications as there was no other 'wired' way to accomplish this important task. World War I also saw the increasing use of the wireless for military communication. The 1920s saw wireless communications used for the general public with the establishment of the first licensed mediumwave broadcast station KDKA in East Pittsburgh, Pennsylvania, in the United States using amplitude modulation (AM) transmissions. The 1920s also saw the first use of land-based mobile communications by the police and fire departments where the urgent dispatch of personnel was required.

From that point the growth in commercial wireless communication was relentless. Mediumwave AM broadcasting was supplemented (and now largely supplanted) by

Table 1.1 Wireless frequency bands

Frequency band	Frequency range	Wavelength range
Extremely low frequency (ELF)	<3 kHz	$>100,000$ m
Very low frequency (VLF)	3–30 kHz	100,000–10,000 m
Low frequency (LF)	30–300 kHz	10,000–1,000 m
Mediumwave frequency (MF)	300–3,000 kHz	1,000–100 m
High frequency (HF)	3–30 MHz	100–10 m
Very high frequency (VHF)	30–300 MHz	10–1.0 m
Ultra high frequency (UHF)	300–3,000 MHz	1.0–0.1 m
Super high frequency (SHF)	3–30 GHz	10–1.0 cm
Extra high frequency (EHF)	30–300 GHz	1.0–0.1 cm

frequency modulation (FM) broadcasting in the VHF band (88–108 MHz). Television appeared on the scene in demonstration form at the 1936 World Fair in New York and began widespread commercial deployment after World War II. Satellite communication began with the launch of the first Russian and American satellites in the late 1950s, ultimately followed by the extensive deployment of geostationary Earth orbit satellites that provide worldwide relay of wireless communications including voice, video, and data.

Perhaps the most apparent and ubiquitous form of wireless communication today are cellular telephones, which in the year 2002 are used by an estimated one billion people worldwide. The cellular phone concept was invented at Bell Labs in the United States in the late 1960s, with the first deployments of cell systems occurring in the late 1970s and early 1980s. The so-called third generation (3G) systems that can support both voice and data communications are now on the verge of being deployed.

Fixed wireless systems were originally designed to provide communication from one fixed-point terminal to another, often for the purpose of high reliability or secure communication. Such systems are commonly referred to as 'point-to-point (PTP)' systems. As technology improved over the decades, higher frequency bands could be successfully employed for fixed communications. Simple PTP telemetry systems to monitor electrical power and water distribution systems, for example, still use frequencies in the 150- and 450-MHz bands. Even early radio broadcast systems were fixed systems, with one terminal being the transmitting station using one or more large towers and the other terminal the receiver in the listener's home. Such a system could be regarded as a 'Point-to-Multipoint (PMP)' system. Similarly, modern-day television is a PMP system with a fixed transmitting station (by regulatory requirement) and fixed receive locations (in general). Television can also be regarded as 'broadband' using a 6-MHz channel bandwidth in the United States (and as much as 8 MHz in other parts of the world), which can support transmitted data rates of 20 Mbps or more.

The invention of the magnetron in the 1920s, the 'acorn' tube in the 1930s, the klystron in 1937, and the traveling wave tube (TWT) in 1943 made possible efficient ground and airborne radar, which saw widespread deployment during World War II. These devices made practical and accessible a vast new range of higher frequencies and greater bandwidths in the UHF and SHF bands. These frequencies were generically grouped together and called *microwaves* because of the short EM wavelength. The common band designations are shown in Table 1.2. Telephone engineers took advantage of the fact that

Table 1.2 Microwave frequency bands

Microwave band name	Frequency range (GHz)
L-band	1–2
S-band	2–4
C-band	4–8
X-band	8–12
Ku band	12–18
K-band	18–27
Ka band	27–40

PTP microwave links used in consecutive fashion could provide much lower signal loss and consequently higher quality communication than coaxial cables when spanning long distances. Although buried coaxial cables had been widely deployed for long-range transmission, the fixed microwave link proved to be less expensive and much easier to deploy. In 1951, AT&T completed the first transcontinental microwave system from New York to San Francisco using 107 hops of an average length of about 48 km [1]. The TD-2 equipment used in this system were multichannel radios manufactured by Western Electric operating on carrier frequencies of around 4 GHz. Multihop microwave systems for long-distance telephone systems soon connected the entire country and for many years represented the primary mechanism for long-distance telecommunication for both telephone voice and video. The higher frequencies meant that greater signal bandwidths were possible – microwave radio links carrying up to 1800 three-kilohertz voice channels and six-megahertz video channels were commonplace.

On the regulatory front, the Federal Communications Commission (FCC) recognized the value of microwave frequencies and accordingly established frequency bands and licensing procedures for fixed broadband wireless systems at 2, 4, and 11 GHz for common carrier operations. Allocations for other services such as private industrial radio, broadcast studio-transmitter links (STLs), utilities, transportation companies, and so on were also made in other microwave bands.

Today, these long-distance multihop microwave routes have largely been replaced by optical fiber, which provides much lower loss and much higher communication traffic capacity. Satellite communication also plays a role, although for two-way voice and video communication, optical fiber is a preferred routing since it does not suffer from the roughly 1/4 s round-trip time delay when relayed through a satellite in a geostationary orbit 35,700 km above the Earth's equator.

Today, frequencies up to 42 GHz are accessible using commonly available technology, with active and increasingly successful research being carried out at higher frequencies. The fixed broadband wireless systems discussed in this book operate at frequencies in this range. However, it is apparent from the foregoing discussion of wireless system evolution that new semiconductor and other microwave technology continues to expand the range at which commercially viable wireless communication hardware can be built and deployed. Frequencies up to 350 GHz are the subject of focused research and, to some extent, are being used for limited military and commercial deployments.

The term *wireless* has generally applied only to those systems using radio EM wavelengths below the infrared and visible light wavelengths that are several orders of magnitude shorter (frequencies several orders of magnitude higher). However, free space optic (FSO) systems using laser beams operating at wavelengths of 900 and 1100 nanometers have taken on a growing importance in the mix of technologies used for fixed broadband wireless communications. Accordingly, FSO systems will be covered in some detail in this book.

1.3 MODELS FOR WIRELESS SYSTEM DESIGN

The process of designing a fixed broadband wireless communications system inherently makes use of many, sometimes complex, calculations to predict how the system will perform before it is actually built. These models may be based on highly accurate measurements, as in the case of the directional radiation patterns for the antennas used in the system, or on the sometimes imprecise prediction of the levels and other characteristics of the wireless signals as they arrive at a receiver. All numerical or mathematical models are intended to predict or simulate the system operation before the system is actually built. If the modeling process shows that the system performance is inadequate, then the design can be adjusted until the predicted performance meets the service objects (if possible). This design and modeling sequence make take several iterations and may continue after some or all of the system is built and deployed in an effort to further refine the system performance and respond to new and more widespread service requirements.

The ability to communicate from one point to another using EM waves propagating in a physical environment is fundamentally dependent on the transmission properties of that environment. How far a wireless signal travels before it becomes too weak to be useful is directly a function of the environment and the nature of the signal. Attempts to model these environmental properties are essential to being able to design reliable communication systems and adequate transmitting and receiving apparatus that will meet the service objectives of the system operator. Early radio communication used the LF portion of the radio spectrum, or the so-called long waves, in which the wavelength was several hundred meters and the propagation mechanism was primarily via groundwaves as mentioned earlier. Through theoretical investigation starting as early as 1907 [2], an understanding and a model of the propagation effects at these low frequencies was developed. The early propagation models simply predicted the electric field strength as a function of frequency, distance from the transmitter, and the physical characteristics (conductivity and permittivity) of the Earth along the path between the transmitter and receiver. The models themselves were embodied in equations or on graphs and charts showing attenuation of electric field strength versus distance. Such graphs are still used today to predict propagation at mediumwave frequencies (up to 3000 kHz), although computerized versions of the graphs and the associated calculation methods were developed some years ago [3].

All wireless communication systems can be modeled using a few basic blocks as shown in Figure 1.1. Communication starts with an information source that can be audio, video, e-mail, image files, or data in many forms. The transmitter converts the information into a signaling format (coding and modulation) and amplifies it to a power level that is needed to achieve successful reception at the receiver. The transmitting antenna converts the transmitter's power to EM waves that propagate in the directions determined by the design and orientation of the antenna. The propagation channel shown in Figure 1.1 is not a physical device but rather represents the attenuation, variations, and any other distortions that affect the EM waves as they propagate from the transmitting antenna to the receiving antenna.

By using EM waves in space as the transmission medium, the system is necessarily exposed to sources of interference and noise, which are often beyond the control of the system operator. Interference generally refers to identifiable man-made transmissions. Some systems such as cellular phone systems reuse frequencies in such a way that interference transmitters are within the same system and therefore can be controlled. Cellular system design is largely a process of balancing the ratio of signal and interference levels to achieve the best overall system performance.

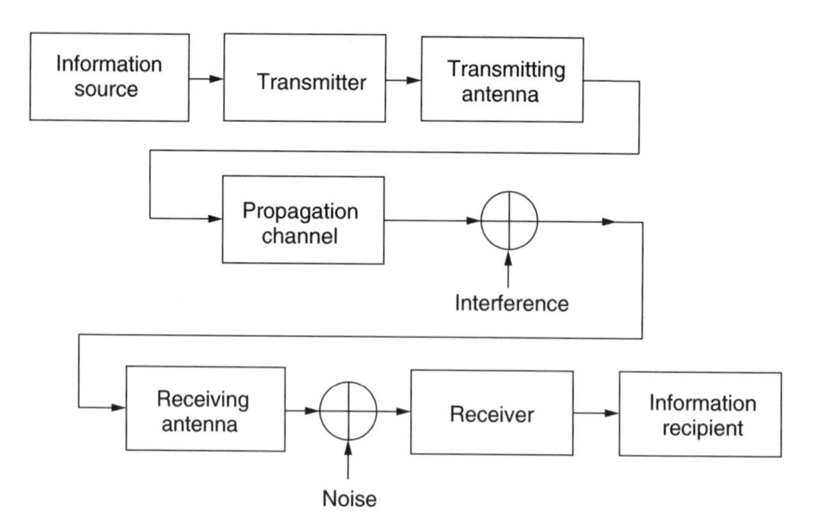

Figure 1.1 Block diagram of a basic wireless communications system.

External noise sources may be artificial or natural, but are usually differentiated from interference in that they may not be identifiable to a given source and do not carry any useful information. Artificial noise sources include ignition noise from automobiles, noise from all sorts of electrical appliances, and electrical noise from industrial machinery among others. Natural external noise includes atmospheric noise from the sun's heating of the atmosphere and background cosmic noise. The noise power from these various sources is very much a function of frequency, so depending on the frequency band in use, these noise sources may be important or irrelevant to the system design.

At the receiver, the receiving antenna is immersed in the EM field created by the transmitting antenna. The receiving antenna converts the EM fields into power at the terminals of the receiving antenna. The design and orientation of the receiving antenna compared to the characteristics of the transmitted field in which it is immersed, determine the amount of power that is present at the receiving antenna terminals. Besides the transmitted field, the EM fields from the interference and noise sources are also converted to power at the receiving antenna terminals, again depending on the design and orientation of the receiving antenna. The so-called smart or adaptive antennas, to be discussed later in this book, can actually change their characteristics over time to optimize signal reception and interference rejection. The power at the receiving antenna terminals is coupled to the receiver that processes the power in an effort to recover exactly the source information that was originally transmitting. For some systems this process can be quite complex, with methods for decoding signals, correcting data errors, mitigating or exploiting signal variations, and rejecting interference being part of modern fixed broadband receiving systems. Ultimately after processing, the received information is presented to the system user in the form of audio, video, images, or data. The accuracy and fidelity of the received signal when compared to originally transmitted source information is a broad general measure of the quality of the communication system and the success of the system design.

1.4 DEMAND FOR COMMUNICATION SERVICES

The creation of any wireless communication system is driven by a need for services by individuals, businesses, governments, or other entities. Government and military demand for services is an ongoing requirement that is largely accommodated first when spectrum resources are allotted. The remaining spectrum is divided into blocks or bands that generally are intended to be best suited to particular service objectives. Within these bands, regulatory authorities over the years have in many cases established rigid technical standards so that equipment manufacturers, system operators, and the buyers (consumers) of telecommunications equipment could rely on the equipment being compatible and working correctly together. Over the past two decades there has been a trend by the FCC to simply assign frequency bands for various services and let the wireless industry choose the appropriate technology through marketplace competition or standards-setting processes conducted by private organizations. The debate between government-mandated standards and marketplace forces setting standards continues today with valid arguments for both regulatory and marketplace approaches.

Ultimately standards are intended to achieve reliable service to the target market. The type and nature of the services that wireless communication systems must provide is constantly changing, which perhaps has become the greatest stress on the standards-setting process. Whereas 5 decades ago nationwide standards for AM, FM, and TV broadcasting could be established and work effectively for several decades, the rapidly changing services that must be delivered have lead to standards being revised and replaced every 10 years. The cellular telephone industry is a perfect example. The early, so-called 1G, standards established in the 1980s were quickly recognized as inadequate because the demand for capacity was much greater than expected. The 2G standards established in the late 1980s and early 1990s are now being replaced by 3G standards, with 4G standards in the planning stages. The need to replace standards in such a short time has been entirely driven by the demand for services and the type of services demanded. A ubiquitous mobile cell phone service that offered simple voice calls was a significant achievement in the 1980s, but now demand for a wide range of data services at increasing data rates is considered essential to having a competitive wireless service offering.

For the fixed broadband wireless system, the digital service demand can be broken down into two basic classes – Internet access for the public and businesses and general private high-speed data communications for small, medium, and large businesses. The explosive growth in Internet usage over the past decade has made it the new community connection that everyone feels compelled to have available – as were telephones 50 years ago. Some of the services or applications that are most commonly used on the Internet are

- E-mail
- Web-browsing
- File and image download and general file transfer via file transfer protocol (FTP)
- Streaming audio files for 'real-time' audio connections
- Streaming video files for 'real-time' video connections
- Voice over Internet protocol (VoIP), also a 'real-time' service.

As discussed in some detail later in this book, each of these applications has particular characteristics in terms of data rate, the statistical distribution of the data flow, and user expectations that affect the way a fixed wireless system must be designed to successfully support them. From a simple inspection, it is clear that some of these services are much more demanding on the communication system than others. Whether the system operator considers the additional cost of deploying a system that can support some or all of these applications a worthwhile expenditure in light of anticipated revenue is a business decision that may be difficult to make. The cost of deployment in turn is controlled by the technology utilized and the efficiency of the system design. The savings in deployment costs that can be achieved through intelligent and accurate system design often far outweigh the cost savings achieved by choosing one technology over another.

The other major service requirement for fixed broadband wireless systems is private high-speed data connections for business, military, and government. This type of service can be regarded as the 'traditional' domain of PTP fixed wireless networks such as the transcontinental microwave systems carrying telephone and video traffic described earlier. Besides telephone companies, many organizations used fixed microwave for internal business communication, among them

- Utilities that used such links to connect dams, power generating stations, substations, pumping stations, and so on.
- FM and TV broadcasters who need to connect studio facilities with often remote mountaintop transmitting facilities, and to relay signals to remote auxiliary repeater or translator transmitting stations, or remote electronic news gathering (ENG).
- Businesses that need to connect various offices, plants or other facilities with broadband services including data and internal computer networks such as local area networks (LAN) or wide area networks (WAN).
- Educational institutions that must connect various campus facilities or remote campuses for high-speed data and video transmissions including teleconferencing.
- Backhaul links that connect cellular base transmitting stations (BTS) to mobile switching centers (MSCs) carrying all the voice and data traffic to the public switched telephone network (PSTN).

As with Internet services, the types and carriage requirements of such services continues to expand, thus placing growing demands on the technology, the system design techniques, and on spectrum regulators to provide adequate spectrum to accommodate these requirements. As discussed in the next section, the current international spectrum allocations have a significant impact on how fixed broadband wireless systems can be built to meet the described service requirements.

1.5 LICENSED FREQUENCY BANDS

The use of radio spectrum worldwide is regulated by the International Telecommunications Union (ITU), which operates with the participation of all member nations under the auspices of the United Nations. The ITU serves to address the needs of all countries during the World radio communications Conference (WRC, formerly WARC) held every three years; the next WRC will be held in Geneva, Switzerland in 2003. At the WRC, the delegations must juggle and resolve the often conflicting demands of member nations and of different service operators that require spectrum allocations for mobile, fixed, and satellite technologies. Within the bands set by the WRC, the Radio Regulation Board (RRB, formerly International Frequency Registration Board or IFRB) established rules for how actual assignments and sharing are to be handled in the band assignments made at the WRC.

The spectrum available for the construction of fixed broadband wireless systems can be divided into licensed and license-exempt frequency bands. In general, licensed spectrum provides for some degree of interference protection because each new licensee must demonstrate compliance with certain standards for limiting interference to other existing nearby licensed systems. There are also radiated transmitter power level and other parameter limitations that each licensee must observe. License-exempt bands do not require individual transmitters to be licensed in order to operate, but there are still radiated power restrictions that usually keep power at low levels as a *de facto* way of limiting interference. There may also be a rudimentary channelization scheme and modulation standard; again, to make possible as many successful operations as possible without destructive interference. Some cooperation and coordination may sometimes be necessary to make the most of these measures. Cordless telephones, remote control toys, and IEEE802.11b/802.11a wireless LAN devices (to be discussed in this book) are examples of license-exempt systems.

There are a number of frequency bands that have been allocated throughout the world for use by licensed fixed broadband services. Within the general ITU band designations, individual countries may elect to implement or not implement polices that allow those frequencies to be licensed and used within their country boundaries. This is especially true for fixed broadband wireless services. Because of these country-specific differences, it is not useful in the context of this book to present a comprehensive tabulation of all these frequency bands. However, Tables 1.3 and 1.4 provide a convenient summary for the United States and most European countries, respectively. The frequency bands listed are intended as examples of the variety of services that have access to the microwave spectrum for fixed services. The tables include the major bands used for newer PTP and PMP broadband services such as Local Multipoint Distribution Service (LMDS). The information in Table 1.3 was extracted from [4,5] while the information in Table 1.4 was extracted from [6,7].

In addition to requirements to obtain a license for systems operating in these bands, each band also has a number of technical criteria that each system must satisfy. In general, these criteria are established to reduce or minimize interference among systems that share the same spectrum, and to ensure that the spectrum efficiency (information transmitted) is sufficiently high to justify occupying the spectrum. In a given band, there may be requirements for minimum and maximum radiated power levels, particular efficient modulation types, and even standards for the radiation patterns of directional antennas to reduce interference to other operators in the band. These technical standards can be detailed and complex, and may vary from country to country. Designing and deploying

Table 1.3 Examples of US licensed fixed wireless bands

Frequency band (GHz)	Service name	Notes
2.150–2.156	MDS1	Single 6-MHz channel for MMDS services
2.156–2.162	MDS2	Single 6-MHz channel for MMDS services
2.156–2.160	MDS2A	Narrow 4-MHz MMDS channel
2.500–2.690	MMDS/ITFS	Thirty-one 6-MHz channels that are shared between ITFS and MMDS operators
3.8–4.2	—	Common carrier band for PTP link systems
5.9–7.1	—	Common carrier band for PTP link systems
10.7–11.7	—	Common carrier band for PTP link systems
12.7–13.25	—	CARS band for cable television relay services
17.7–18.820	—	Shared use for broadcast auxiliary, common carrier, CARS, private operational fixed PTP systems
24.25–25.25	DEMS	DEMS = digital electronic messaging service. The band includes 5 × 40-MHz FDD channels with 800-MHz spacing
28	LMDS	LMDS = local multipoint distribution service. Block A is 1,150 MHz in three parts: 27.5–28.35 GHz, 29.10–20.25 GHz, and 31.075–31.225 GHz, Block B is 150 MHz in two parts: 31.0–31.075 and 31.225–31.3 GHz
38	—	50-MHz FDD channels 38.6–38.95 GHz with channel pairs at 39.3–39.65 GHz

Note: MMDS = Multipoint Multi-channel Distribution Service.
ITFS = Instructional Television Fixed Service.
CARS = Cable Television Relay Service.

a fixed wireless system in any particular country requires a careful review and functional understanding of the administrative rules that govern the use of the intended licensed spectrum space.

1.6 LICENSE-EXEMPT BANDS

As mentioned above, there is a growing interest in using the so-called license-exempt bands. One of the primary reasons is that it allows users of the wireless service to purchase off-the-shelf wireless modems for connecting to a system. In the United States, the 11-Mbps IEEE 802.11b standard that specifies Direct Sequence Spread Spectrum (DSSS) technology operating in the 2.4-GHz band is the best current example of self-deployed license-exempt technology. However, license-exempt bands offer no regulatory

Table 1.4 Examples of European licensed fixed wireless bands

Frequency band (GHz)	Service name	Notes
3.4–3.6	—	Duplex spacings of 50 or 100 MHz are employed. 3.7 GHz is the upper limit of this band is in some countries
3.8–4.2	—	High capacity public operator band for PTP link systems
5.9–7.1	—	High capacity public operator band for PTP link systems
7.1–8.5	—	Medium and high capacity public operator band for long haul PTP systems
10.15–10.65	—	5 × 30-MHz channels with duplex spacings of 350 MHz
10.7–11.7	—	High capacity public operator band for PTP link systems
12.7–13.3	—	Low and medium capacity public operator band
14.4–15.4	—	Fixed link operations of various types
17.7–19.7	—	Public operator band for low and medium capacities
21.2–23.6	—	Public operator band for PTP link systems of various types
24.5–26.5	—	ETSI 26-GHz band. 3.5- to 112-MHz FDD channels with 1008-MHz duplex spacing. Channel widths vary from country to country
37–39.5	—	Common carrier band for PTP link systems

interference protection except that afforded by the interference immunity designed into the technology itself. With relatively modest penetration of these systems to date, the robustness of the design for providing the expected quality of service in the presence of widespread interference and many contending users has yet to be fully tested. As the number of people using license-exempt equipment increases in a given area, the ultimate viability of having a multitude of people using a limited set of frequencies will be tested. Table 1.5 shows the license-exempt bands currently used in the United States for fixed broadband communications.

The license-exempt spectrum has been designated in Europe, though the uptake of the technology has been slower than in the United States. As discussed in the next section, several long-running standard-setting efforts designed for this purpose did not bear fruit in a timely fashion, resulting in many of these efforts being suspended or abandoned in favor US standards already in place. Table 1.6 shows the license-exempt bands currently available for use in Europe. At the time of this being written, the IEEE 802.11a high-speed network standard has not been certified for use in Europe, although this is expected to happen in the year 2002.

Table 1.5 US license-exempt fixed wireless bands

Frequency band (GHz)	Service name	Notes
2.4–2.483	ISM	ISM = industrial, scientific, and medical. This band is where IEEE 802.11b DSSS networks operate
5.15–5.35	U-NII	U-NII = unlicensed national information infrastructure. This band is where IEEE 802.11a orthogonal frequency division multiplexing (OFDM) systems operate among several other proprietary standards. Channel widths are 20 MHz. Particular power limits apply for segments of this band intended for indoor and outdoor applications
5.725–5.825	U-NII	Same as 5.15–5.35-GHz U-NII band except this band is intended only for outdoor applications with radiated power levels up to 4 W

Table 1.6 European license-exempt fixed wireless bands

Frequency band (GHz)	Service name	Notes
2.4–2.483	ISM	ISM = industrial, scientific, and medical. This is the same band where IEEE 802.11b DSSS networks operate in the United States
5.15–5.35	HiperLAN	HiperLAN is the fast wireless network standard for Europe, which uses an OFDM transmission standard similar to IEEE 802.11a. This band is intended for indoor operations with radiated powers limited to 200 mW
5.470–5.725 GHz	HiperLAN/2	Proposed frequency band for outdoor operations with radiated power levels limited to 1 W

1.7 TECHNICAL STANDARDS

Many fixed broadband wireless systems, especially private PTP microwave systems, use technology and engineering methods and technology that comply with a minimum regulatory framework but otherwise are proprietary methods that have been developed to achieve an advantage over their commercial competition. Since communication is intended only among nodes or terminals within of the same network, there is no need for public standards that would facilitate a manufacturer developing and marketing equipment. Over the years, this approach has lead to considerable innovation in fixed-link equipment with new power devices, receivers, coding and decoding schemes, and very spectrum-efficient high-level modulation types being successfully developed and deployed.

As noted above, there has been a trend in regulatory agencies, especially the FCC, to set the minimum technical standards necessary to control interference among different system operators, with the details of the transmission methods left to individual operators. This is the case with the LMDS bands in the United States, for example, where operators with licenses to use these bands in different cities can chose any technology they wish to employ. The pivotal question here is 'Is the system intended to serve a large customer base that needs low-cost terminal devices or is the system intended to serve a narrow set of customers who sufficiently value the service to pay higher prices for terminal equipment capable of greater performance?'

Even in this context there is still considerable motivation to establish standards, especially for systems that expect to provide service to vast numbers of users in businesses and residences randomly dispersed throughout a service area. With detailed transmission standards, two particular benefits may be achieved

- Competing companies will manufacture large quantities of standards-compliant devices, thus drastically reducing the price of individual devices.
- Operators will more willingly deploy systems that comply with standards because they can expect a large quantity of inexpensive terminal devices available for use by their customers, thus enlarging their customer base.

Included here is a brief summary of the standard-settings efforts and organizations that are focused on the fixed broadband wireless systems for widespread deployment. The actual details of the standards are not discussed here since they are extremely detailed, usually requiring several hundred pages to document. Interested readers can consult the references for more specific information on these standards. Moreover, except in limited ways, the details of standards, especially many aspects of the medium access control (MAC) layer, are not germane to the wireless network design process.

1.7.1 IEEE 802.11 standards

The IEEE 802.11 Working Group is part of the IEEE 802 LAN/MAN Standards Committee (LMSC), which operates under the auspices of the IEEE, the largest professional organization in the world. The committee participants representing equipment manufactures, operators, academics, and consultants from around the world are responsible for establishing these standards.

The original IEEE 802.11 standard provided for wireless networks in the ISM band that provide data rates of only 1–2 Mbps. These rates were substantially less than inexpensive wired Ethernets that routinely ran at 10 or 100 Mbps speeds and could be readily deployed with inexpensive equipment. To improve the capability of these wireless networks, two additional projects were started.

The IEEE 802.11b project was actually started in late 1997 after 802.11a project (hence, a 'b' suffix instead of an 'a'). The standard was completed and published in 1999 to provide for wireless networks operating at speeds up to 11 Mbps using the unlicensed 2.4-GHz ISM band in the United States and other parts of the world. With this standard, the 2.4-MHz band is divided into six channels, each 15 MHz wide. Power levels of

802.11b devices are limited to mW, and use of spread spectrum transmission technology is required to reduce the potential of harmful interference to other users. To manage access by multiple users, it provides for the collision sense multiple access (CSMA) approach for sharing the channels.

The IEEE 802.11a standard was also completed and published in 1999. It provides for operation of the 5-GHz U-NII bands (see Table 1.5) using OFDM modulation. Using 20-MHz channels, it provides for data rates up to 54 Mbps. IEEE 802.11a also specifies CSMA as the multiple access technology.

The most recent standard from this committee is 802.11g, which is intended to provide better data rates than 802.11b but still use the 2.4-GHz band. As of this writing, this standard is not well defined although it likely will use OFDM of some sort.

Further information can be found at the 802.11 Working Group web site [8], or through IEEE.

1.7.2 IEEE 802.16 standards

The IEEE 802.16 Working Group on Broadband Wireless Access is also part of IEEE 802 LMSC. It was originally organized to establish standards for fixed broadband systems operating above 11 GHz, especially the 24-GHz DEMS, 28-GHz LMDS, and 38-GHz bands. The purpose was to speed deployment of systems through the benefits of mass marketing of standard terminal devices. Since then, the committee work has expanded to include systems operating on frequencies from 2 to 11 GHz; this standards effort is now designated IEEE 802.16a.

The 802.16 WirelessMAN™ Standard ('Air Interface for Fixed Wireless Access Systems') covering 10 to 66 GHz was approved for publication in December 2001. This followed the publication in September 2001, of IEEE Standard 802.16.2, a Recommended Practice document entitled 'Coexistence of Fixed Broadband Wireless Access Systems', also covering 10 to 66 GHz. The corresponding standard for 2 to 11 GHz will be designated IEEE 802.16.2.a.

The 802.16a standard for the 2- to 11-GHz standard uses the same MAC layer as 802.16, but necessarily has different components in the physical layer. Balloting on the 802.16a air interface standard is expected to be completed and the 802.16a standard approved and published in mid to late 2002.

Further information can be found at the 802.16 Working Group web site [9], or through IEEE.

1.7.3 ETSI BRAN standards

The European Telecommunications Standards Institute (ETSI) and its committee for Broadband Radio Access Networks (BRAN) has worked on several standards for wireless networking for a number of years. These include

- *HIPERLAN/2*: This is a standard that has a PHY (Physical) layer essentially the same as 802.11a but a different MAC layer using time division multiple access (TDMA) rather than CSMA. Like 802.11a, it is intended to operate in the 5-GHz band and provide data rates up to 54 Mbps. The first release of the HIPERLAN/2 standard was published in April 2000. There is also ongoing work to develop bridge standards to IEEE networks and IMT-2000 3G cell phone systems.

- *HIPERACCESS*: This is intended as a long-range variant of HIPERLAN/2 intended for PMP operation at data rates up to 25 Mbps in various kinds of networks. HIPERACCESS is intended to operate in the 40.5- to 43.5-GHz band, although these spectrum allocations have not yet been made.
- *HIPERMAN*: This standard is design for interoperable fixed broadband wireless access in the 2- to 11-GHz frequency range, with the air interface designed primarily for PMP. According to [10], the HIPERMAN standard uses the 802.16a standard as a baseline starting point.
- *HIPERLINK*: This standards effort is designed for short range (<150 meters), high-speed (up to 155 Mbps) links that would connect HIPERMAN and HIPERACCESS networks. Work on this standard has not yet started.

At this time there is some contention between the IEEE 802.11a and HIPERLAN/2 standards for high-speed wireless access in the 5-GHz spectrum. Attempts are currently being made to bridge the differences and provide certification for the IEEE 802.11a standard in Europe, along with coexistence rules for neighboring systems.

1.8 FIXED, PORTABLE, AND MOBILE TERMINALS

As mentioned in the Preface, the differences that distinguish fixed and mobile systems have become somewhat blurred. The term *fixed* is clear – the transmitting and receiving terminals of the wireless transmission circuit are physically fixed in place. A microwave link system with the transmitting and receiving antennas mounted on towers attached to the ground, a rooftop or some other structure is a reference example of a fixed system. In fact, any system that incorporates a high-gain fixed pattern antenna such as a parabolic dish or horn antenna is necessarily a fixed system since precise alignment of the antennas so that they point in the proper directions is required for the system to work properly. MMDS–type antennas mounted on the outside wall or roof of a residence to receive signals from a transmitter toward which the antenna is pointed is also a good example of a fixed system, even though the antennas may have less directionality than those used in PTP microwave link systems. Even television broadcast systems are fixed systems in this sense. The transmitting and receiving antennas are fixed, as is the TV itself while it is being watched. An exception to this are high-gain antennas receiving satellite signals on board ships where sophisticated gimbaling systems are required to keep the antenna pointed in the correct orientation regardless of the movements of the ship.

The IEEE 802.11 standards are primarily designed for fixed and 'portable' terminal devices. Also referred to as *nomadic* in ITU and other European documents, a portable terminal is one that stays in one place while it is being used, but can readily be picked up and moved to another location. A notebook computer with an 802.11b wireless access

PC card is a good example of such a portable device. Another portable device would be a desktop wireless modem that is connected to a computer via a USB port, but while operating, the modem is expected to be in more or less one place – on a desktop, for example. Moving it across the desk or to another room makes it portable, but while in use, it is stationary. The concept of such an indoor portable wireless modem with a data rate capability of 5 to 10 Mbps currently dominates much of the leading system design and manufacturer developments in fixed broadband wireless access in licensed bands below 11 GHz.

The classification of fixed systems can be further refined by recognizing that for some networks, one terminal of the transmission link is at an *ad hoc* location rather than an 'engineered' location; that is, no engineering knowledge or effort has been used to determine a good location for the terminal device. Instead, the terminal location has been chosen by the user to be the place that is most convenient. A notebook computer with a wireless modem of some sort placed on an arbitrarily located desk is an example of such a system. This contrasts to a system in which an antenna is mounted on the outside or roof of a structure and carefully pointed by a technician to achieve a certain performance level. *Ad hoc* fixed systems present new challenges to system design since the problem of analyzing coverage and interference, and ultimately performance, is quite similar to that of mobile radio or cellular systems in which the design must provide for terminals located essentially anywhere in the system service area.

The differences between engineered and *ad hoc* fixed wireless systems have a dramatic impact on the commercial success of the system. An engineered system requires the expensive step of sending a trained technician to every terminal location at least once to complete a successful installation. The value to the operator of this customer's business must be significant enough to justify the cost of this 'truck roll'. Certainly for some customers such as large businesses that require microwave links carrying hundreds of megabits of data, this may very well be the case. However, for systems designed to serve thousands of more casual communication users such as homes, home offices, and small businesses, a system design that can work effectively with *ad hoc* 'self-installed' terminals without the necessity of truck roll is needed for that system to be commercially viable.

Finally, the term *mobile* can be distinguished as applying to those systems designed to support terminals that are in motion when being used. The recent 3G UMTS standards even differentiate the level of service that should be provided on the basis of the speed of the mobile terminal. The 3G specification identifies three levels of mobile speed

- 0 km/hr, where data rates up to 2.048 Mbps can be provided.
- 3 km/hr (pedestrian), where data rates up to 384 kbps can be provided.
- 30 km/hr (vehicular), where data rates up to 144 kbps can be provided.
- 150 km/hr (fast train), where data rates up to 64 kbps can be provided.

While 2.5G and 3G cellular systems of all types (GPRS, EDGE, UMTS W-CDMA and CDMA2000) will be important mechanisms for providing voice and data communication worldwide, they are not fixed systems as defined here. However, the engineering methods they employ and the data rates that are possible make the extension of these technologies to fixed broadband wireless scenario a logical step. Several system hardware developers are currently pursuing exactly this course in an effort to provide the high-speed, *ad hoc* wireless broadband service described earlier. A good example is the TDMA TDD (time division duplex) version of the UMTS W-CDMA standard. Although primarily designed for mobile services, these technologies will be treated in this book to the extent that they are applicable to the fixed broadband network deployment.

1.9 TYPES OF FIXED WIRELESS NETWORKS

The types of fixed wireless network topologies that will be treated in this book fall into four broad categories. Each is briefly introduced in the following sections with more detailed technical design and analysis to follow in later chapters.

1.9.1 Point-to-point (PTP) networks

Point-to-point (PTP) networks consist of one or more fixed PTP links, usually employing highly directional transmitting and receiving antennas, as illustrated in Figure 1.2. Networks of such links connected end to end can span great distances as in the case of the original AT&T 4-GHz link network that crossed the United States in 1951. Links connected end to end are often referred to as tandem systems, and the analysis for the end-to-end reliability or availability of the whole network must be calculated separately from the availability of individual links.

1.9.2 Consecutive point and mesh networks

Consecutive point networks (CPN) are similar to PTP networks in that they consist of a number of links connected end to end. However, as illustrated in Figure 1.3, CPN

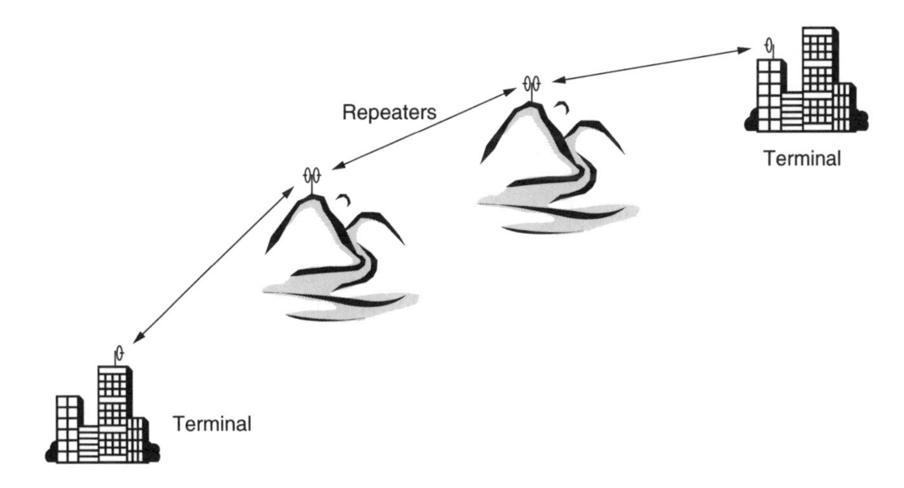

Figure 1.2 Point-to-point (PTP) network connecting two cities through mountaintop repeaters.

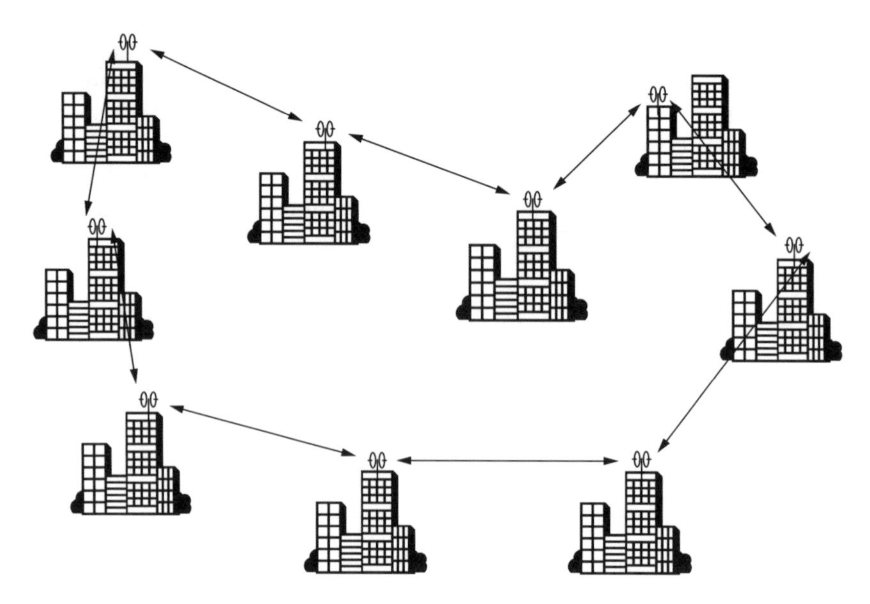

Figure 1.3 Consecutive point network (CPN) connecting buildings within a city.

are configured as rings usually attached to an optical fiber node at some point along the ring that ultimately connects into worldwide optical fiber networks. The data traffic travels in both directions around the ring. The main advantage of such a ring system is that if a problem develops at some point along the ring such that the traffic flow is interrupted, the traffic can be automatically rerouted in the other direction around the ring. The disadvantage is that the traffic originated by and destined for customers on the ring must share the same radio link data transmission capacity. Depending on the capacity of the CPN and the number of customers on the ring (and their data rate requirements), this can be an important limitation that must be considered in link dimensioning.

Mesh networks are a variation of CPNs that are generally configured as links connected in both rings and branching structures. The main advantage is that mesh networks provide alternate paths for connected customers who might otherwise lack line-of-sight (LOS) visibility to the network, increasing the potential number of connected customers. However, like CPNs, the traffic for any given customer must sometimes be routed through several nodes, possibly straining the data capacity of the links connecting those nodes and possibly introducing data delivery delays (latency) that can affect the quality of service for services that require real-time response (such as VoIP). The additional nodes involved in achieving an end-to-end connection can result in lower reliability than multipoint networks in which only one wireless link is needed to connect to the network.

1.9.3 Point-to-multipoint (PMP) networks

PMP networks used a 'hub and spoke' approach to deliver data services as illustrated in Figure 1.4. The hub is analogous to the base station in a cellular system. It consists of

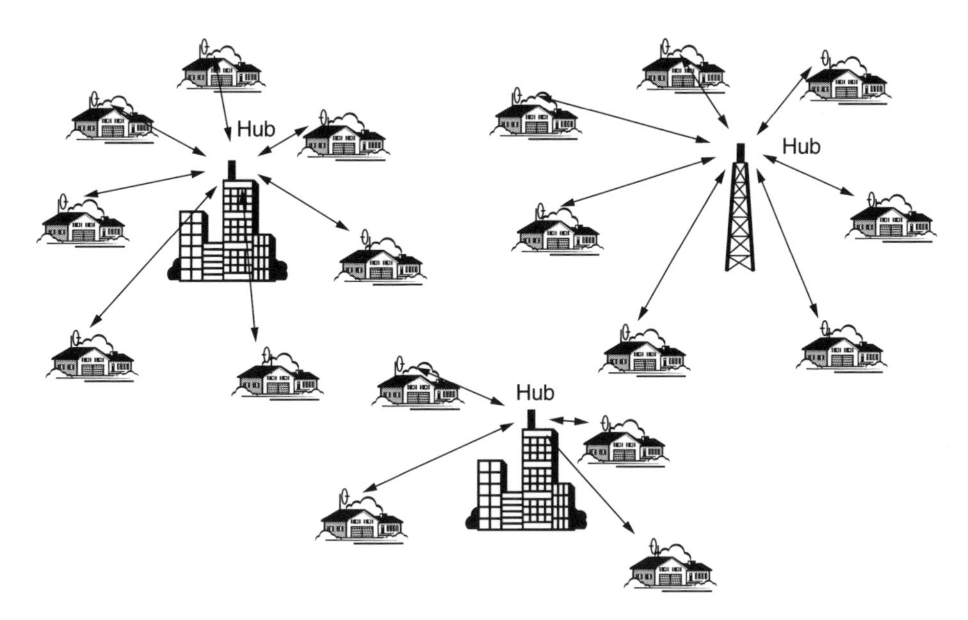

Figure 1.4 Point-to-multipoint (PMP) network.

one or more broad-beam antennas that are designed to radiate toward multiple end-user terminals. Depending on the frequency band employed, and the data rates to be provided to end users, normally several hubs are needed to achieve ubiquitous service to a city. The remote end-user terminals are engineered installations in which directional antennas have been installed in locations that are in the LOS to the hub and oriented by a technician to point at the hub location. In some cases this may require extensive work at each terminal location.

PMP network architecture is by far the most popular approach to fixed broadband wireless construction. It mimics the network topology successfully used for decades in wired telephone networks, cable television networks, and even electrical, gas, and water utilities of all sorts. For wireless, the major drawback is the cost of the infrastructure to construct the hubs needed to achieve comprehensive LOS visibility to a large percentage of the service.

1.9.4 NLOS point-to-multipoint networks

Non-line-of-sight (NLOS) PMP networks are identical in topology to the PMP networks described above. The difference lies in the nature of the remote terminals. Instead of the remote terminals being engineered and professionally installed to achieve successful performance using an outside antenna, the terminals are arbitrarily positioned at the convenience of the end user inside a house or office. In most cases, the location of these terminals will be places that do not have a clear, obstruction-free view of a network hub and are thus called non-line-of-sight. The signal attenuation and amplitude variability that occurs along the wireless signal path from the network hub to NLOS location present new challenges to system designers in their efforts to provide a reliable high-speed data service to every terminal. The engineering problem is similar to the problem of providing service to mobile phones; however, as explained in later chapters, the fixed wireless engineer can exploit some advanced techniques in the terminal and network that are not yet practical for cellular system engineers.

1.10 ORGANIZATION OF THIS BOOK

This book is organized into several chapters that provide detailed discussions of the engineering principles on which the design of fixed broadband wireless systems are based, followed by several chapters that discuss on a more pragmatic level methods and techniques that can be utilized in designing real fixed wireless systems. The methods described are intended to be applied to generic system types rather than any particular manufacturer's equipment or approach to network construction. While different manufacturers may tout their products as being uniquely better than those of their competitors, in reality all are based on the same engineering principles described in this book.

Chapters 2 through 6 provide an engineering foundation for the physical mechanisms and current technology for fixed broadband wireless systems. Chapter 2 deals with the theoretical fundamentals of EM wave and wave propagation and, in particular, the impact that the physical environment has on EM wave propagation. This discussion includes both effects from the natural environment such as terrain, rain scattering, atmospheric refraction, fog, and so on, and effects from the artificial environment including shadowing, reflection, diffraction, and scattering from buildings and other structures.

Chapter 3 uses the theoretical propagation mechanisms described in Chapter 2 to construct propagation or channel models. As the word implies, a 'model' is a close representation for the real thing; it is not the real thing. Since the models will be used to design fixed wireless systems, the closer the model is to the 'real thing', the better. Over the decades, considerable effort has been made into making the models better. Distinctive classifications and subclassifications of models have emerged that fall into three general areas: theoretical, empirical, and physical models. Propagation models themselves traditionally have been used to describe the median EM field strength at distance from the transmitter. A more comprehensive approach is a channel model that attempts to describe not only the field strength at some point away from the transmitter but also the variations in the field as a function of time, frequency, and location. The performance of many modern fixed broadband wireless systems, especially NLOS PMP systems, depends on information provided by a channel model.

Chapter 4 deals with models of signal fading. Signal fading occurs due to changes in atmospheric conditions, the presence of rain and other precipitation, and changes in the locations of objects in the propagation environment. Fading phenomena are described in statistical terms. For high reliability digital links, these fading model statistical descriptions have a primary impact on the impact of predicted availability and reliability of the link.

Chapter 5 also discusses the important topic of terrain, clutter, and building (structure) models that are used in conjunction with propagation and channel models. As discussed in this chapter, the accuracy of the propagation model is often limited by the accuracy of the physical databases rather than the engineering methods in the model itself.

Chapter 6 discusses the important subject of antennas for fixed wireless systems. This chapter includes descriptions of traditional PTP fixed antennas such as parabolic dishes and horns, as well as 'smart' or adaptive and MIMO (multiple input, multiple output) antenna systems that are emerging as important techniques for achieving high capacity NLOS links.

Chapter 7 discusses the basic principles of digital modulation, equalizers, and coding. Every fixed wireless system uses a modulation method of some sort, whether it is a multiplex analog FM as in the original telephone carrier systems, or digital modulation that ranges from simple lower efficiency methods such as BPSK (binary phase shift keying) and QPSK (quadrature phase shift keying) to more elaborate and efficient methods such as 64QAM or 256QAM (quadrature amplitude modulation). Even simple OOK (on–off keying like a telegraph) is used in FSO systems.

Chapter 8 also deals with the important subject of multiple access and duplexing techniques such as

- FDMA (frequency division multiple access)
- TDMA (time division multiple access)
- CDMA (code division multiple access)
- SDMA (space division multiple access)
- OFDMA (orthogonal frequency division multiple access)
- FDD (frequency division duplexing)
- TDD (time division duplexing).

All multiple-access techniques are fundamentally trying to increase the number of users that can simultaneously access the network while maintaining a certain level of service quality (data rate, throughput, delay, and so on). The ability of a multiple-access scheme to achieve this objective can have the largest impact on the network's commercial success.

Chapter 9 lays the groundwork for the fixed wireless system design by discussing traffic and service models, the physical distribution of traffic and various traffic types, and service application models. This chapter also includes new traffic simulation results that provide a convenient approach for dimensioning the capacity of a wireless network hub to achieve a given service quality to a projected population of end users.

Chapter 10 describes traditional PTP fixed link design methods that have been in use for many years, with a focus on developing and using link budgets to assess performance and availability. This process includes choosing tower locations and heights, path clearance analysis, rain and fade outage analysis, and ultimately link availability. Link budgets and fading criteria for NLOS links are also discussed. The link design methods and analyses presented in this chapter are basic building blocks that are also used to design consecutive point and mesh networks, as well as PMP networks.

Chapter 11 provides the steps to designing both LOS and NLOS PMP networks, including identifying traffic sources and choosing hub locations that provide adequate coverage based on the link budgets developed in Chapter 10. Dimensioning the hub cell service areas, the multiple-access channel data rates and the hub sector capacities to meet the projected traffic load is also discussed. The use of Monte Carlo simulations to evaluate the quality of a design is also dealt with in this chapter. The objective of a design is to produce a system plan that can be subject to the iterative process of refining and updating the network plan as needed.

Chapter 12 deals with the important subject of channel planning. The efficient assignment of frequencies, time slot, and codes to meet traffic demands that vary with time and location is critical to realizing the highest capacity and commercial benefit of the available spectrum. The channel assignment techniques discussed include static and dynamic methods, including dynamic packet assignment (DPA). The integration of newer technologies that enhance network capacity is also discussed. Chapter 12 also discusses some ideas that generalized the concept of spectrum space as a Euclidean space, with path loss, time, and frequency as the dimensions of space rather than the traditional three physical dimensions as well as time and frequency. This approach to viewing the wireless spectrum offers some insights into efficient system planning that would not otherwise be as apparent.

1.11 FUTURE DIRECTIONS IN FIXED BROADBAND WIRELESS

Fixed broadband wireless has a history that spans several decades. The progress in the technology, especially hardware innovations for transmitters and receivers, has opened the door to effectively exploiting higher and higher frequency bands. Even more dramatic developments in high density very large scale integration (VLSI) has led to highly efficient signal processing, coding, and multiple-access techniques that were not feasible in mass-produced devices even a few years ago.

The upward innovation ramp will continue with access to frequencies above 60 GHz, which is now imminent. The administrative process to allow commercial deployment in these frequencies has already begun. Developments in higher power transmitting devices and lower noise receiving devices, along with adaptive spatial signal processing, will also permit more extensive use of existing bands beyond their currently perceived limitations. Employing 28-GHz frequencies for NLOS networks, currently considered infeasible, will probably become possible with upcoming hardware improvements.

New approaches to utilizing spectrum will also assume a more prominent role. Ultra wideband (UWB) technology is one example that has recently gained FCC approval in the United States. UWB uses very narrow pulse technology that spreads the signal power over a very wide bandwidth resulting in average power levels that are intended to be sufficiently low to preclude interference to other spectrum occupants. A similar innovation is MIMO technology that can potentially increase the data rate capacity between link terminals by as much as an order of magnitude over conventional approaches.

The availability of low-cost, accessible license-exempt technologies led by IEEE 802.11b (Wi-Fi) in the 2.4-GHz band has captured the imagination of millions worldwide by unveiling the exciting possibilities of high-speed wireless data connections to the Internet. The energy in the enterprises pursuing business opportunities using these networks will also flow into more sophisticated and capable fixed broadband wireless technologies discussed in this book. The rapidly growing number of wireless Internet service providers (WISPs) is a harbinger of the business opportunities that will abound for service providers, equipment vendors, and application developers.

Worldwide, fixed wireless networks offer the best opportunity for bringing electronic communication to the majority of the Earth's population still lacking even basic telephone service. Wireless networks can rapidly extend service into remote areas at much lower cost when compared to installing poles or digging trenches to accommodate wired networks.

The technical innovations and potential applications have moved wireless closer to the mythical concept of a secure, private, high capacity, dimensionless, low-cost, tetherless communications connection that is available everywhere at low cost and minimal environmental impact. The current progress in fixed broadband wireless technology is an incremental step in this human evolutionary pursuit.

1.12 CONCLUSIONS

This chapter has presented a brief history of wireless technology and in particular fixed wireless technology as it originally saw wide-scale deployment with multilink networks for cross-continent telephone traffic. The growing demand for various types of high-rate data services were also discussed along with the frequency bands for licensed and license-exempt fixed broadband networks that regulators have allotted to meet the recognized demand for data services. Several network topologies include PTP, consecutive point (CPN), PMP, and NLOS PMP, which are presented in schematic form. The chapters that follow provide the substance and the details required to build on this framework, and extend these concepts to creating successful, high-performance fixed broadband wireless systems.

Compared to wired alternatives, wireless technologies offer substantially more flexibility in choosing how and where services are deployed and the types of applications that can be supported. The ability to more readily modify, reconfigure and enhance wireless networks compared to wired networks ensures a significant and growing future for fixed broadband wireless communications technology.

1.13 REFERENCES

[1] William Alberts, "Bell System Opens Transcontinental Radio-Relay," *Radio and Television News*, October, 1951.

[2] A.D. Watt. VLF Radio Engineering. Oxford: Pergamon Press. 1967. pp. 180 ff.

[3] H.R. Anderson, "A computer program system for prediction and plotting mediumwave groundwave field strength contours." *IEEE Transactions on Broadcasting*, pages 53–62, Sept. 1980.

[4] Rules of the Federal Communication Commission. Part 101.101. CFR Title 47. United States Government Printing Office.

[5] Rules of the Federal Communication Commission. Part 74. CFR Title 47. United States Government Printing Office.

[6] International Telecommunications Union (ITU).
[7] European Telecommunication Standardization Institute (ETSI).
[8] IEEE 802.11 Working Group web site. http://ieee802.org/11/.
[9] IEEE 802.16 Working Group web site. http://WirelessMAN.org.
[10] Roger B. Marks. "Status of 802.16 Efforts: Successes, New Products & Next Steps," *Wireless Communication Association 8th Annual Technical Symposium, January, 2002, verbal presentation.*

2

Electromagnetic wave propagation

2.1 INTRODUCTION

Wireless communication from one point to another requires a transmitter and an antenna to create electromagnetic (EM) waves that are modified in some way in response to the information being communicated. The amplitude, phase, and frequency (wavelength) of a wave can all be modified to represent the information. On the other end of the link, the receiving antenna and receiver detect these amplitude, phase, or frequency variations of the wave and convert them into a form that the recipient can use. Understanding EM waves and how they get from one place to another is fundamental in determining how a wireless communication link will perform.

While the chapter begins with a restatement of Maxwell's equations and wave equations for convenience, it is not necessary to understand or find solutions to wave equations to design fixed broadband wireless systems. The task of working with fundamental EM wave principles and calculating convenient representations of how waves propagate in a wide variety of physical circumstances mechanisms is the primary subject of this chapter. In Chapter 3, the representation of these basic mechanisms will be combined to form the propagation models that are used in wireless systems design.

2.2 MAXWELL'S EQUATIONS AND WAVE EQUATIONS

Early researchers originally regarded electricity and magnetism as separate physical phenomena. In the eighteenth and the nineteenth centuries, through laboratory experiments researchers found that an electric current produces a magnetic field and that a changing

Fixed Broadband Wireless System Design Harry R. Anderson
© 2003 John Wiley & Sons, Ltd ISBN: 0-470-84438-8

magnetic field produces an electric field. There was a clear inter-relationship between the electric and magnetic effects that Maxwell set down in an elegant and concise form:

$$\nabla \times \mathbf{E} = -\mathbf{M}_i - \frac{\partial \mathbf{B}}{\partial t} \tag{2.1}$$

$$\nabla \times \mathbf{H} = \mathbf{J}_i + \mathbf{J}_c + \frac{\partial \mathbf{D}}{\partial t} \tag{2.2}$$

$$\nabla \times \mathbf{D} = q_{ev} \tag{2.3}$$

$$\nabla \times \mathbf{B} = q_{mv} \tag{2.4}$$

Where:

$\mathbf{E}$ = electric field intensity (volts/meter)
$\mathbf{H}$ = magnetic field intensity (amperes/meter)
$\mathbf{D}$ = electric flux density (coulombs/meter)
$\mathbf{B}$ = magnetic flux intensity (webers/meter)
$\mathbf{J}_i$ = source electric current density (amperes/squaremeter)
$\mathbf{J}_c$ = conduction electric current density field intensity (amperes/square meter)
$\mathbf{M}_i$ = source magnetic current density (volts/square meter)
q_{ev} = electric charge density (coulombs/cubic meter)
q_{mv} = magnetic charge density (webers/cubic meter)

The concept of source magnetic current density and magnetic charge density are inserted to make the equations symmetrical – they are not physically realizable. For example, (2.4) is often written as $\nabla \times \mathbf{B} = 0$, so there is no single unique form to express Maxwell's equations. Equations (2.1) through (2.4) are complete forms as found in [1].

Even though Maxwell's equations are not particularly useful or accessible to those not fluent in fields and vector calculus, the important concepts can still be recognized. Equation (2.1) says that a time-varying magnetic field produces an electric field ($\partial \mathbf{B}/\partial t \neq 0$). Equation (2.2) says that a time-varying electric field or a current produces a magnetic field ($\partial \mathbf{D}/\partial t \neq 0$, $\mathbf{J}_i \neq 0$, or $\mathbf{J}_c \neq 0$). The validity of these assertions can be appreciated by starting with the simple concept of two charged objects. Depending on whether the charges are the same or opposite, a directional force exists that will cause the objects to move away from or towards each other along the line that joins them. An electric field is *by definition* the magnitude and direction of this force per unit charge. If the charges are now put in motion, the electric field will obviously change in a corresponding way, which in turn creates a magnetic field and a magnetic flux density. From a rudimentary perspective that is what (2.1) and (2.2) state.

Although Maxwell's equations can be used to find electric and magnetic fields directly, they are typically not used this way because they are coupled differential equations; that is, (2.1) and (2.2) contain variables for both electric and magnetic fields. To uncouple the equations so that (2.1), for example, has electric field as the only variable requires some manipulation of the equation and the substitution of (2.2) into (2.1) or vice versa, to arrive at the second-order differential *wave equations*. Interested readers can refer to [1] for this derivation.

Solving either Maxwell's equations or the wave equations requires finding functions for the electric field **E** and the magnetic field **H** that make the equalities true. The task of solving these equations (finding the functions **E** and **H**) for innumerable physical circumstances has been ongoing for the past 100 years. The results of these efforts will be drawn upon to show how EM waves propagate in real wireless system environments.

2.3 PLANE AND SPHERICAL WAVES

An EM plane wave traveling or propagating through free space in the z direction is shown in Figure 2.1. The electric field is oriented along the x-axis, while the magnetic field is oriented along the y-axis or orthogonal to the electric field orientation. Free space strictly means in a vacuum. For short propagation paths through a stable, dry atmosphere, the free space assumption can be used. However, as will be discussed later in this chapter, for many paths through the atmosphere, through rain, over the earth, or around constructed objects, the free-space assumption can only be used in limited ways.

Figure 2.1 shows the EM wave traveling in a straight line along the z-axis. The planes perpendicular to the z-axis are known as the wave fronts. If the wave is moving in a straight line and all the wave fronts are parallel, the wave is called *a uniform plane wave*. Again, when the wave is not in free space or encounters objects, it can change amplitude, phase, and direction and may no longer be a uniform plane. In short, the plane wave characterization is just a temporary description of the wave at a given time and at a given location. This characterization is often invoked in analyzing the fields around objects because it simplifies the mathematical analysis. The vector in the direction of travel of the wave is called *the Poynting vector*.

The EM wave is shown as sinusoidal in Figure 2.1, although the wave can consist of any time-varying electric and magnetic fields. Clearly, through Fourier analysis any arbitrary wave can be resolved into sinusoidal components, so describing the EM wave as having sinusoidal variations is sufficiently general.

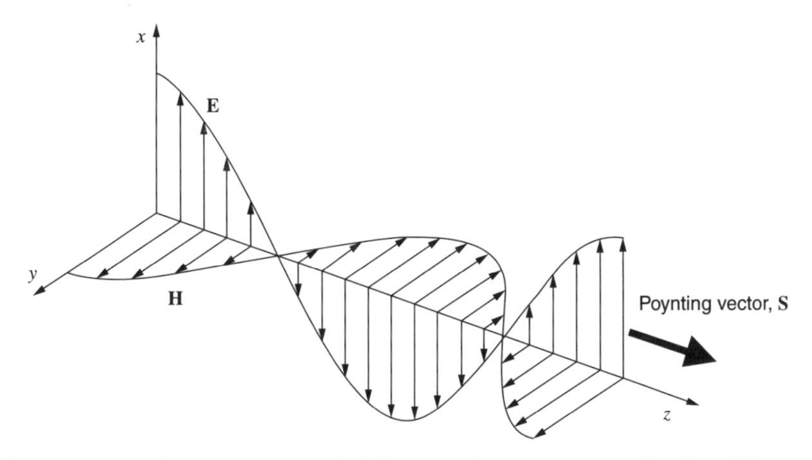

Figure 2.1 Electric (**E**) field, magnetic (**H**) field, and Poynting vector, **S**, of an EM wave propagating in direction z.

With a sinusoidal wave the following relationships can be written for the fields. The electric field is

$$\mathbf{E} = E_0 \cos(\omega t - kz)\hat{\mathbf{x}} \tag{2.5}$$

where E_0 is the peak amplitude of the wave, ω is the angular frequency ($\omega = 2\pi f$), t is the elapsed time, k is the phase constant, z is the distance along the z-axis, and $\hat{\mathbf{x}}$ is the unit vector in the positive x direction in Figure 2.1. The wavelength λ in meters is inversely proportional to the frequency ($\lambda = c/f$) where c is the speed of light in meters per second. The phase constant simply relates the change in phase over a given distance and is directly proportional to wavelength λ.

$$k = \frac{2\pi}{\lambda} \tag{2.6}$$

The equation for the magnetic field can be written in a similar way:

$$\mathbf{H} = H_0 \cos(\omega t - kz)\hat{\mathbf{y}} \tag{2.7}$$

where $\hat{\mathbf{y}}$ is the unit vector in the y-axis direction in Figure 2.1.

If the medium through which the wave is traveling is lossless, the amplitude of the wave does not diminish with distance. The equations for the electric and magnetic fields can also be written in complex notation:

$$\mathbf{E} = E_0 e^{j(\omega t - kz)\hat{\mathbf{x}}} \tag{2.8}$$

$$\mathbf{H} = H_0 e^{j(\omega t - kz)\hat{\mathbf{y}}} \tag{2.9}$$

2.3.1 Impedance of free space and other transmission media

The amplitudes of the electric and magnetic fields are related by constants of proportionality that are in general different for each medium the wave may pass through. The constants are known as permeability μ in henrys per meter and permittivity ε in farads per meter. Usually, these are given in terms relative to the values for free space as follows:

$$\mu = \mu_0 \mu_r \tag{2.10}$$

$$\varepsilon = \varepsilon_0 \varepsilon_r \tag{2.11}$$

The values μ_0 and ε_0 are the permeability and permittivity of free space, respectively, with $\mu_0 = 4\pi \times 10^{-7}$H/m and $\varepsilon_0 = 8.854 \times 10^{-12}$F/m. Since the reference values are for free space, in this case $\mu_r = 1$ and $\varepsilon_r = 1$.

The amplitudes of the electric and magnetic fields are thus related:

$$\frac{|\mathbf{E}|}{|\mathbf{H}|} = \frac{E_0}{H_0} = \sqrt{\frac{\mu}{\varepsilon}} = Z \tag{2.12}$$

Since the electric field is given in volts per meter (V/m) and the magnetic field is given in amperes per meter (A/m), the ratio in (2.12) is described as an impedance following Ohm's law and given the units of ohms. This impedance can be found in this way for any medium using the appropriate values of μ and ε. For free space, using the values of μ and ε given above, the resulting impedance is about 377 ohms.

2.3.2 Power in a plane wave

The instantaneous power in the wave is calculated in a way analogous to Ohm's Law for circuits; that is, voltage in the electric field component multiplied by amperes in the magnetic field component. However, since these fields are vectors, this multiplication is done with the cross product:

$$\mathbf{S} = \mathbf{E} \times \mathbf{H}^* \tag{2.13}$$

where * denotes the complex conjugate. Since $\mathbf{S}$ is a vector, this power relationship in fact defines the magnitude and direction of the Poynting vector. The average power in the Poynting vector is found by integrating the instantaneous Poynting vector $\mathbf{S}$ over one period:

$$\mathbf{S}_{\text{AVE}} = \frac{1}{2}\text{Re}(\mathbf{E} \times \mathbf{H}^*) = \frac{1}{2}E_0 H_0 \hat{\mathbf{z}} \cos\theta \quad \text{W/m}^2 \tag{2.14}$$

where θ is the angle between the electric and the magnetic field (0 degrees in free space) and 'Re' indicates the real part of the vector product.

2.3.3 Spherical waves

For most antennas used in wireless communications, the field strength of interest is generally at a distance that is large compared to the size of the antenna. This is known as the *far field* (see Chapter 6). In the far field the antenna appears essentially as a point source with EM waves (of different amplitudes) spreading out in all directions from the source. The spreading waves from a point source are known as *spherical waves* since the wave fronts follow the curved surface of the sphere. As shown in Figure 2.2, the power in the spherical wave is the power passing through a solid angle δA. As the distance from the source is increased, the area of the sphere across this solid angle increases. Doubling the distance from r to $2r$ quadruples the area on the surface of the sphere across

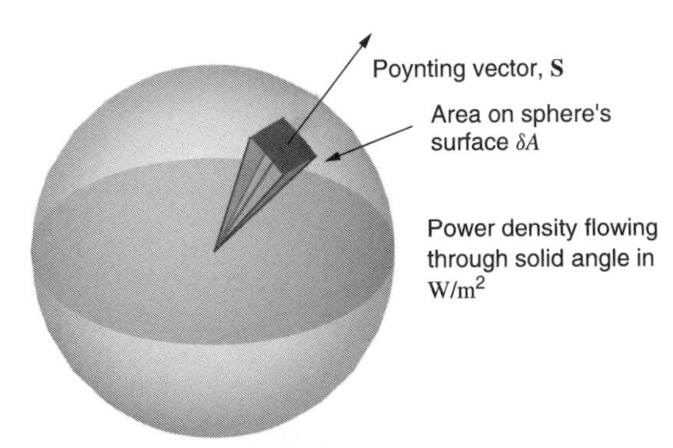

Figure 2.2 Constant power flowing through solid angle from a spherical radiator. Power density (W/m^2) decreases with increasing sphere radius.

this solid angle, while the power flowing through the solid angle remains the same. The *power density* through the surface is thus divided by 4 so the average power density per square meter at distance $2r$ is 1/4 of what it was at distance r. From (2.13) the electric and magnetic filed amplitude are each reduced by a factor of 2.

Viewing the radiation from an antenna as a spherical wave is important to understanding why the field amplitudes diminish proportional to $1/r$. However, at distances that are practical for wireless communication links, the radius of the sphere is large and the solid angle over which the fields need to be found is small. At such distances, over the small area represented by the aperture of the receiving antenna the surface of the sphere is essentially flat so that approximating it as a uniform plane wave is generally justified.

2.4 LINEAR, CIRCULAR, ELLIPTICAL, AND ORTHOGONAL POLARIZATIONS

As shown in Figure 2.1 for a uniform plane wave, the electric field vector has a particular orientation. By convention, the polarization of the wave is taken to be the direction that is parallel to the electric field vector. For example, for a dipole antenna (see Chapter 6), the electric field vector is parallel to the axis of the dipole. If the dipole is vertical (perpendicular to the earth's surface), the electric field vector also has a vertical orientation and the wave is referred to as *vertically polarized* (VP). In a similar way, if the dipole is oriented parallel to the earth's surface, the electric field vector is also horizontal and the wave is referred to as *horizontally polarized* (HP). These different polarizations are illustrated in Figure 2.3.

If two dipoles are used, one vertically oriented and the other horizontally oriented, and the phase of the currents in the dipole are adjusted such that they are 90 degrees out a phase, then a *circularly polarized* wave is created, which is also illustrated in Figure 2.3. Depending on whether the phase of the vertical field is 90 degrees ahead or behind the horizontal field, the circularly polarized can have a right-hand or left-hand rotation.

Circular polarization is actually just a special case of elliptical polarization in which the amplitudes of the linear electric field vectors are equal. If they are not equal, then the wave is *elliptically polarized.*

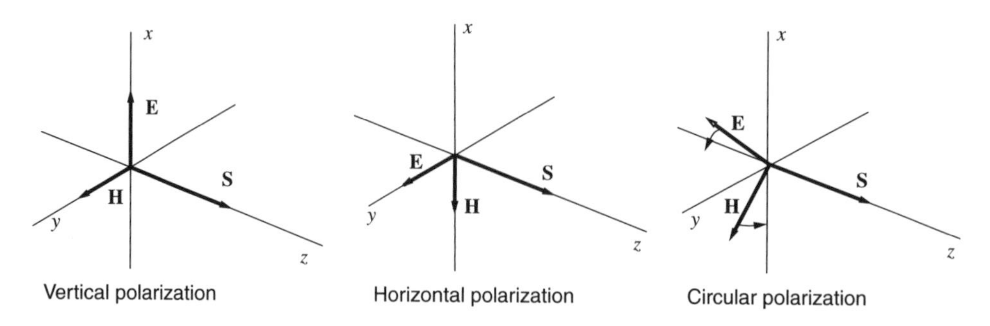

Figure 2.3 Wave polarizations.

The polarization of the wave is important in calculating how the wave will interact with natural and constructed objects, including gas molecules and rain drops in the atmosphere. The calculations that follow in this chapter for wave reflection and diffraction, for example, depend on the polarization of the wave.

The examples of vertical and horizontal polarizations are simply special cases of the more general principle of orthogonal polarization. If the transmitted polarizations are orthogonal, then antennas can be used at the receiver to independently detect both polarizations. This idea can be exploited to send different signals in each polarization, and in essence double the amount of information that can be transmitted along the link. The practicality of this approach is limited by interactions in the environment that cause the polarization of the wave to change. This polarization change is called *depolarization*, and results to some extent from most interactions that the wave will have in transit from the transmitter to the receiver. Depolarization not only affects the strength of the desired signal arriving at the receiver but also affects the amplitude of the interference from the other links. In fixed broadband wireless systems with a high density of links such as Local Multipoint Distribution Service (LMDS), a signal-to-interference (S/I) ratio that is adequate in clear weather may be inadequate in rainy weather because of the depolarization resulting in a failed link. Such factors are important in designing LMDS and other systems and will therefore be treated quantitatively in later sections of this book.

2.5 FREE-SPACE PROPAGATION

Free-space transmission is a primary consideration in essentially all fixed broadband wireless communication systems. Free space strictly means a vacuum, but can be successfully applied to dealing with short-range space-wave paths between elevated terminals. However, it should be kept in mind that all useful terrestrial links are by definition close to the earth's surface (the terminals are mounted on buildings or towers) and thus it will always be necessary to consider potential interactions with terrain features and constructed objects in any system design.

Depending on path length, propagation effects through the atmosphere must also be carefully considered; so even if the link path is clear of terrain or other objects, and is passing through the atmosphere, there are many effects that can substantially impact the communication link performance. Atmospheric effects will be discussed later in this chapter.

Two calculations of free-space propagation are most commonly used.

- The path loss or attenuation between the transmitting and receiving terminals
- The electric field strength at some distance from the transmitter.

Each calculation is presented below.

2.5.1 Path attenuation between two antennas

As illustrated in Figure 2.2, the radiated power at some distance from a transmitting antenna is inversely proportional to the square of the distance from the transmitter.

$$W = \frac{P_T G_T}{4\pi r^2} \tag{2.15}$$

where P_T is the transmitter power, G_T is the transmitting antenna gain in the direction of the receiver, and $4\pi r^2$ is the surface area of the sphere at radius r.

A receive antenna placed at distance r intercepts the radiated power (is immersed in the field created by the transmitter). The amount of power the receive antenna intercepts depends on the antenna's effective aperture A_e.

$$P_R = \frac{P_T G_T}{4\pi r^2} A_e \tag{2.16}$$

The effective aperture of the antenna is given by [2] as:

$$A_e = \left(\frac{\lambda^2 G_R}{4\pi}\right) \tag{2.17}$$

where G_R is the gain of the receive antenna in the direction of the transmitter and λ is the wavelength. The free-space path loss or attenuation between the transmitter and the receiver is simply the ratio of the received power to the transmitted power.

$$\text{Free space pathloss} = L_f = \frac{P_R}{P_T} = G_T G_R \left(\frac{\lambda}{4\pi r}\right)^2 \tag{2.18}$$

The first presentation of this simple formula for path loss is attributed to Friis [3]. This ratio formula is most commonly used in decibel (dB) units.

$$L_f = 32.44 - 10\log G_T - 10\log G_R + 20\log f + 20\log d \quad \text{dB} \tag{2.19}$$

where f is the frequency in MHz and d is the distance in kilometers (km). This convenient relationship is the most common starting point to designing wireless communications systems and as such, will be used extensively in this book.

2.5.2 Field strength at a distance

Considering a sphere around a radiating source, all the power P_T from the source must pass through the surface of the sphere. Since the surface area of the sphere is $4\pi r^2$, the average power density S_{ave} of the power passing through a square meter of surface area at distance d is $P_T/4\pi r^2$, assuming that the radiating source is an isotropic antenna. The corresponding free-space real microwave systems (RMS) field strength is given by

$$E_{\text{RMS}} = \sqrt{Z_o S_{\text{ave}}} = \frac{1}{r}\sqrt{\frac{Z_o P_T}{4\pi}} \tag{2.20}$$

where Z_o is the plane wave free-space impedance.

Field strength is often given in terms of dB relative to one microvolt per meter (dBμV/m). Converting (2.20) and using a more convenient distance unit gives

$$E_r = 74.77 + 10\log P_T - 20\log d \quad \text{dBuV/m} \tag{2.21}$$

where the distance d is in kilometers. Unlike the path loss in (2.19), field strength does not depend on frequency. The path loss in (2.19) is the path loss between the two antennas with particular gains. For a given physical effective antenna aperture from (2.17), the gain increases with frequency (λ^2), which is then inversely accounted for by the frequency term in (2.19).

2.6 REFLECTION

Reflection is one of the most significant wave propagation mechanisms involved in almost every type of fixed wireless system, even point-to-point (PTP) links with high-gain antennas. Three basic reflection types that will be considered here are

- specular reflection from smooth surfaces
- reflections (scattering) from rough surfaces
- physical optic reflections

The normal handling of specular reflection with a single specular point is simplification of the more complete physical optics (PO) approach to reflection, which is briefly discussed here. As shown in Chapter 3, calculating the magnitude and phase of reflection coefficients plays an important role in all physical propagation and channel models.

2.6.1 Specular reflection

When a transmitting signal intersects the ground, a wall or any other surface (without edges or discontinuities), the magnitude and phase of the reflected and transmitted fields can be represented by the standard Fresnel reflection and the transmission coefficients multiplied by the incident field. The reflection occurs from the 'specular point', that is, the point where the angle of incidence from the source equals the angle of reflection to the point where the reflected field is being computed. The two-dimensional geometry for the reflection is shown in Figure 2.4. The general reflection coefficient R is

$$R = R_s g \tag{2.22}$$

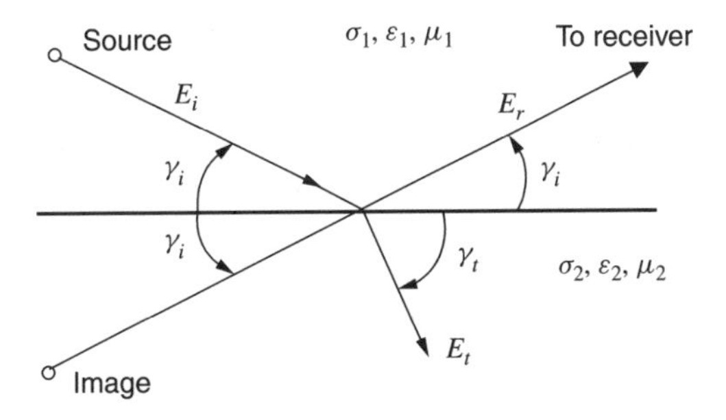

Figure 2.4 Specular reflection with source image.

where R_s is the smooth surface reflection coefficient and g is the surface roughness attenuation factor (a scalar quantity).

For parallel and perpendicular polarizations, respectively, the smooth surface reflection coefficients are [4]

$$R_{\parallel} = \frac{\eta_2 \sin \gamma_t - \eta_1 \sin \gamma_i}{\eta_2 \sin \gamma_t + \eta_1 \sin \gamma_i} \quad \text{(parallel polarization)} \tag{2.23a}$$

$$R_{\parallel} = \frac{-\varepsilon_r \sin \gamma_i + \sqrt{\varepsilon_r - \cos^2 \gamma_i}}{\varepsilon_r \sin \gamma_i + \sqrt{\varepsilon_r - \cos^2 \gamma_i}} \quad (\mu_1 \cong \mu_2 \cong \mu_0, \sigma = 0) \tag{2.23b}$$

$$R_{\perp} = \frac{\eta_2 \sin \gamma_i - \eta_1 \sin \gamma_t}{\eta_2 \sin \gamma_i + \eta_1 \sin \gamma_t} \quad \text{(perpendicular polarization)} \tag{2.24a}$$

$$R_{\perp} = \frac{\sin \gamma_i - \sqrt{\varepsilon_r - \cos^2 \gamma_i}}{\sin \gamma_t + \sqrt{\varepsilon_r - \cos^2 \gamma_i}} \quad (\mu_1 \cong \mu_2 \cong \mu_0, \sigma = 0) \tag{2.24b}$$

where γ_i is the angle of incidence relative to a plane tangent at the point of reflection and $\eta_{1,2}$ is the complex permittivity given by

$$\eta_{0,1} = \sqrt{\frac{j\omega\mu_{1,2}}{\sigma_{1,2} + j\omega\varepsilon_{1,2}}} \tag{2.25}$$

where σ, ε, and μ are the conductivity, permittivity, and permeability of the air and the reflecting material and ω is the frequency of the incident radiation in radians. For lossless materials, $\sigma = 0$ and (2.25) reduces to (2.12). Equations (2.23b) and (2.24b) are simplified forms that apply under the listed conditions. The relative dielectric constant ε_r in these equations equals $\varepsilon_r = \varepsilon_2/\varepsilon_1$.

The roughness attenuation factor g is a quantity given approximately by [5] as

$$g^2 = e^{-2\Delta\rho} \tag{2.26}$$

$$\Delta\rho = \frac{4\pi \Delta h}{\lambda} \sin \gamma_i \tag{2.27}$$

where Δh is the standard deviation of the normal (Gaussian) distribution used to describe the surface roughness variations. From (2.22), the effect of the roughness factor is to decrease the amplitude of the specular reflection field. This occurs because the rough surface tends to scatter energy rather than reflect it in a single specular direction. This is an approximate way of treating the scattering from rough surfaces. Rough surface scattering is addressed in more detail in Section 2.6.3.

Regarding polarization, the $R_{\parallel}$ coefficient is for the incident radiation in which the E field is oriented parallel to the plane of incidence and not to the plane of the reflection surface. Likewise, $R_{\perp}$ is for E field polarization perpendicular to the plane of incidence. With vertical wall reflections in an urban environment, for example, horizontal polarization is *parallel* to the plane of the incidence, while vertical polarization is *perpendicular* to

Table 2.1 Material constants for reflection coefficient curves shown in Figure 2.5

Curve	Polarization	σ(S/m)	ε_r	Material
1	Perpendicular	1×10^{-12}	6	Glass
2	Perpendicular	1×10^{-3}	3.4	Dry sandy soil
3	Perpendicular	1×10^{-1}	10	Poor insulator (wood)
4	Perpendicular	5	15	Seawater
5	Perpendicular	2×10^{6}	1	Iron
6	Parallel	1×10^{-12}	6	Glass
7	Parallel	1×10^{-3}	3.4	Dry sandy soil
8	Parallel	1×10^{-1}	10	Poor insulator (wood)
9	Parallel	5	15	Seawater
10	Parallel	2×10^{6}	1	Iron

the plane of incidence. Conversely, for horizontal ground surface reflections, horizontal polarization is *perpendicular* and vertical polarization is *parallel*. The separate horizontal and vertical polarization effects will play an important role in polarization diversity, which is examined in later chapters.

Figures 2.5(a) and 2.5(b) show the reflection coefficients (magnitude and phase) for a source frequency of 2 GHz and several material types from [6] as identified by key numbers in Table 2.1. Equations (2.23) and (2.24) were used to calculate the numbers in the graphs. From these figures it can be seen that perpendicular incidence results in an almost constant 180° phase shift and a coefficient magnitude of nearly one for most angles of incidence. It is also apparent that for low angles of incidence ($\gamma_i < 10$ degrees) the conductivity and permittivity of the reflection boundary are relatively unimportant to the magnitude and phase of the reflection coefficient. It is also useful to note that at shallow angles of incidence ($\gamma_i < 10$ degrees) that are commonly encountered in wireless system design, the magnitude of the reflection coefficient for perpendicular polarization is greater than 0.8 for a wide range of material types; so for many situations exact information on material conductivity and permittivity is not necessarily critical for accurate reflection calculations.

2.6.2 Physical optics

By taking into account the phase of the EM field, the PO approach provides a more realistic representation of the reflection field as shown in Figure 2.6. The incident field induces currents over the entire reflecting surface, but the only areas of the surface that significantly contribute to the reflection field at the field point are those that are in the first few Fresnel zones around the specular reflection point. Additional minor reflection energy lobes also exist in nonspecular directions as shown. In the general case with $\gamma_i < 90°$, the Fresnel zone areas are elliptical (Figure 2.6). As the angle of incidence gets smaller, the ellipse in which significant surface currents contribute to the reflected field becomes more elongated to the point where, for a finite surface (a real building wall, for example), some portion of the ellipse may include the surface edge. At such oblique incidence, there

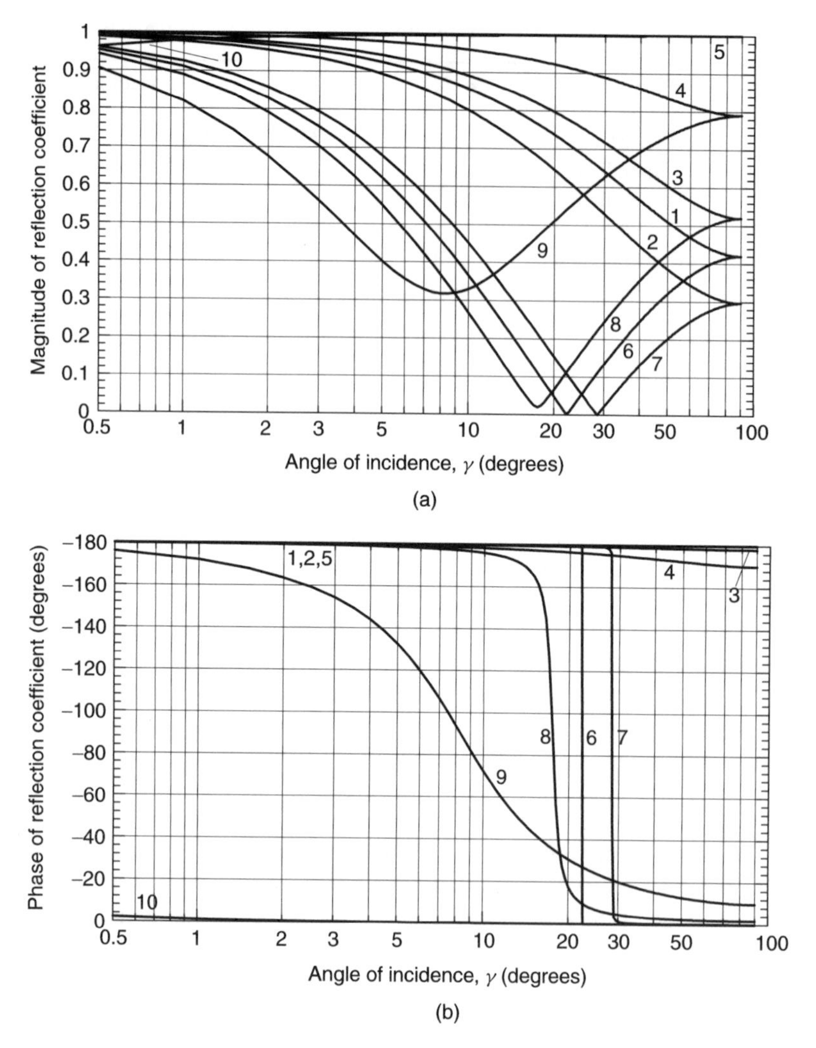

Figure 2.5 (a) Magnitude of reflection coefficient at 2 GHz for various material types shown in Table 2.1; (b) Phase of reflection coefficient at 2 GHz for various material types shown in Table 2.1.

is also a surface wave created that causes a backscatter field due to the edge impedance discontinuity creating a reflected surface current [7]. The area of the ellipse described by the first Fresnel zone intersection with the boundary will depend on the distance to the source and the observation point, the angle of incidence, and the frequency. The total area in the first Fresnel zone is

$$A = \frac{\pi d\sqrt{\lambda d}}{4} \frac{3/2}{[1 + (h_1 + h_2)^2/\lambda d]^{3/2}} \tag{2.28}$$

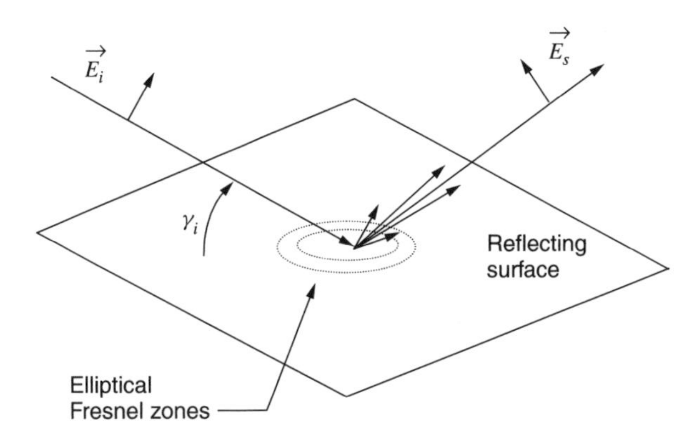

Figure 2.6 Physical optics result for a specular reflection.

where h_1 = the transmitter height of the distance from the reflecting surface, h_2 is the receive point distance from the reflecting surface, and d is the direct distance from the transmitter to the receiver.

The more valid PO description of the reflected field is discussed here to demonstrate that the specular reflection model is not a complete picture. Nonetheless, the specular reflection calculations will be used for propagation modeling and system design bearing in mind that in some limited circumstances the accuracy of the results may have some corresponding limitation.

In practice, the geometry of the reflection from a surface is conveniently used as an 'image' of the source. As shown in Figure 2.4, this image is a 'mirror' position on the opposite side of the reflection boundary. The total ray path length from the image to the observation point is the same as the path through the reflection. This reflection ray path length is used to calculate the delay time for the signal to travel from the transmitter to the receiver. The delay time along with the reflection magnitude is one component used for the dispersion (frequency–selective) channel models as described in Chapter 3.

2.6.3 Reflections from rough surfaces

A smooth reflecting surface is an idealized surface that is only occasionally encountered in real propagation environments. Typically encountered surfaces have random variations as in the earth's surface or have systematic variations such as in the walls and roofs of artificial structures. Depending on the wavelength of the wireless signal, the height of these variations may or may not be significant in terms of how reflection amplitude is calculated. In extreme cases, the surface may appear to be a pure scatterer. The degree of roughness, or the criterion on which roughness warrants considerations, is often given by [8] which is also known as the Rayleigh criterion:

$$h_R \geq \frac{\lambda}{8 \sin \gamma_0} \tag{2.29}$$

where h_R is the difference in the maximum and minimum surface variations as illustrated in Figure 2.7. For a frequency of 2 GHz and an incident angle of 20 degrees, $h_R = 5.5$ cm. Terrain and the outside surfaces of buildings can easily exhibit surface variations greater than this.

Theoretical analysis and scale modeling have been used to assess the degree of scattering from random rough surfaces including Monte Carlo simulation techniques, as described in [9]. In this work, scattering from a rough surface with a random Gaussian distribution of surface variations was analyzed using finite difference time domain (FDTD) and integral equation (IE) methods. An example of the results is redrawn in Figure 2.8 for $\gamma_0 = 45°$. The specular reflection component at 45 degrees is clear, but along with it the graph shows that the energy of varying magnitudes is reradiated in all directions from the surface. The RMS variation of the surface in this example is the equivalent of about 2.6 cm at 2 GHz.

In addition to random rough surfaces, systematic rough surfaces are also encountered. An analysis of the reflection and scattering from several systematic rough surfaces using ray-tracing was reported in [10]. An example of a structure representative of variations

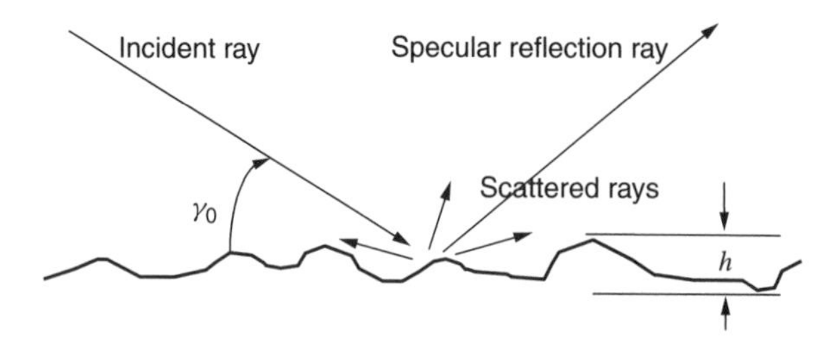

Figure 2.7 Reflection and scattering from a rough surface.

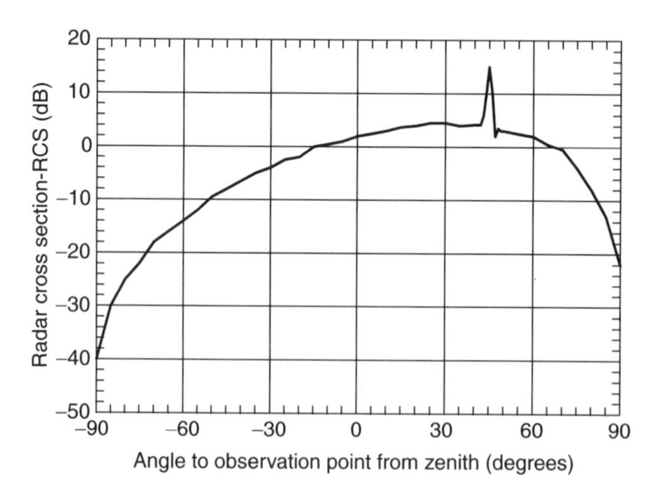

Figure 2.8 Radar cross section (RCS) for a rough surface (data from [9]).

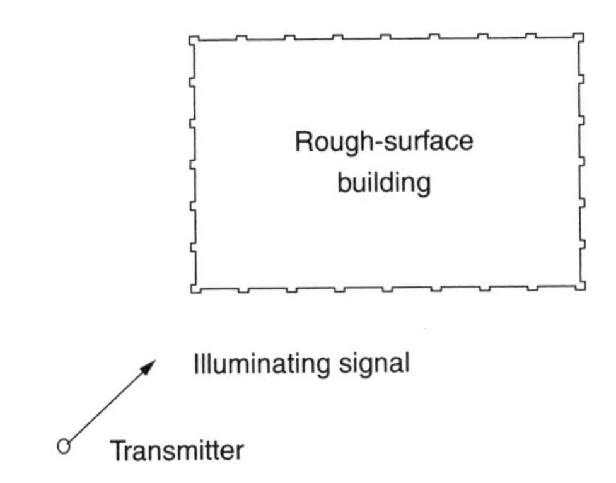

Figure 2.9 Model of the periodic rough surface of a building.

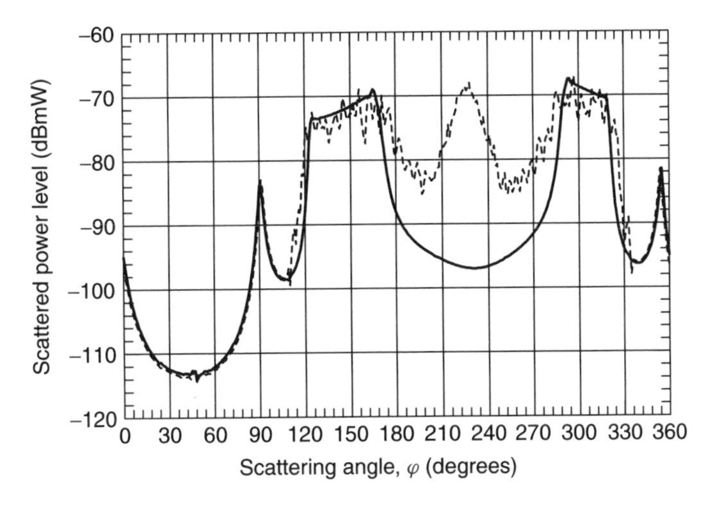

Figure 2.10 Scattered power levels at scattering angles around the structure shown in Figure 2.9.

along the exterior wall of the building with columns and recessed windows in between is shown in Figure 2.9.

The reflection/scattering results along a circular path around this rough structure are shown in Figure 2.10, with the transmitter located to the lower left (at an azimuth of 225 degrees) from the center of the structure in Figure 2.9. For comparison purposes, the scattered power levels of the same structure with smooth walls (solid line) and with the wall variations shown in Figure 2.9 (dashed line) are included in Figure 2.10. Of particular interest is the spike in the reflected signal in the direction of the transmitter. This is a backscatter component resulting from the dihedral corners formed by the intersection of the wall or window with the protruding column.

These examples serve to illustrate the wide variations that can occur with rough and systematic rough surfaces. However, 'rough' really serves to characterize those features of the surface that are unknown or otherwise not explicitly treated by the analysis. Using higher resolution methods such as ray-tracing, FDTD, and others, roughness simply becomes a set of small-scale features that the analysis method can treat exactly. The backscatter calculated using ray-tracing for the building wall is such a case.

2.7 DIFFRACTION

Diffraction is an important wave propagation phenomenon that occurs with any propagating wave including EM waves used for wireless communications, light waves, sound waves, and water waves. Diffraction occurs when there is a partial blocking of a portion of the wave front by an object of some kind. The object can have a sharp or round edge; the important feature is that it partially blocks the propagating wave and thus it is disguised from a reflecting surface that is usually presumed to be infinite in extent as discussed in Section 2.6. In conceptual terms, the collision of the wave with an object results in some of the wave proceeding past the object and part of the wave reflecting or scattering in other directions.

Propagation models for wireless system design commonly use two approaches to diffraction:

- Wedge diffraction
- Knife-edge diffraction

Each approach is discussed in detail below. For the propagation modeling process, each has its advantages and weaknesses.

2.7.1 Wedge Diffraction

An important feature in city propagation environments is diffracting wedges. Diffracting wedges occur at the corners of buildings, at the edge of walls where they intersect roofs, and at the junction of walls with the ground or street (inside wedges). The problem of computing the field scattered from a wedge discontinuity has been addressed by a number of researchers, notably in [11,12]. Keller's geometric theory of diffraction (GTD) [11] extended geometric optics (GO) so that it could treat edge effects, although it still would predict incorrect fields at the singularities occurring at the shadow and reflection boundaries, as shown in Figure 2.11. The singularities resulted in infinite diffracted fields. Kouyoumjian and Pathak [12] resolved the singularity problem by adding a coefficient based on a Fresnel integral. This correction term in effect approaches zero as the GTD terms becomes infinite. The net result is a finite and useful result at the reflection and shadow boundaries. The Kouyoumjian and Pathak approach is called *the uniform theory of diffraction* (UTD). The formulas of Keller, and Kouyoumjian and Pathak, dealt only with perfectly conducting wedge materials. This was extended by Luebbers [13] who used a heuristic approach in which the reflection coefficients were included to the UTD

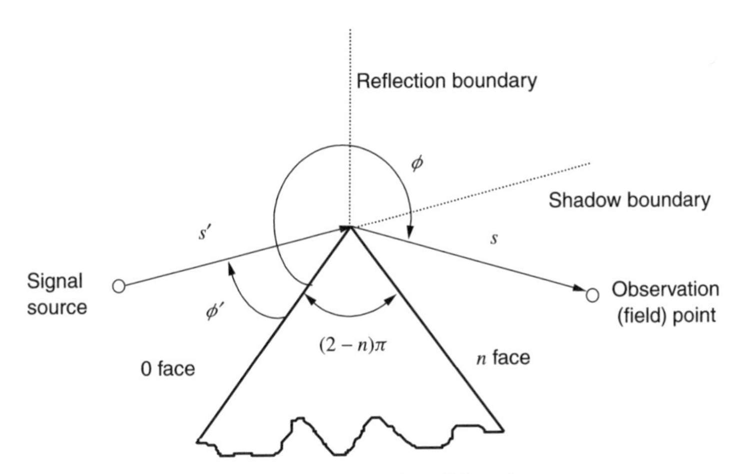

Figure 2.11 2D of view wedge diffraction geometry.

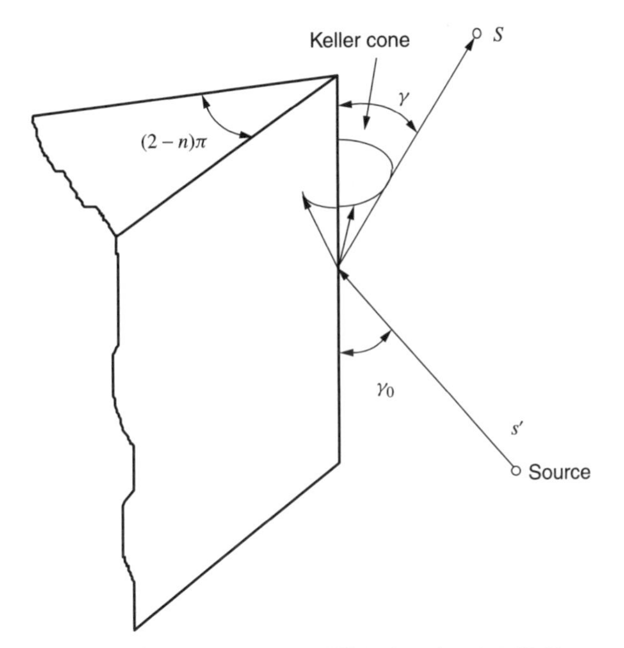

Figure 2.12 3D view of wedge diffraction showing Keller cone.

diffraction coefficients so that the so-called 'lossy' or dielectric wedges could be treated. Luebber's diffraction equations in [13] are used directly here.

The development of the diffraction coefficients from EM field theory is quite involved and will not be repeated here. Using the geometry and notation in Figure 2.11 and Figure 2.12, the diffraction coefficients are

$$D_{\parallel}^{\perp} = \frac{-e^{-j(\pi/4)}}{2n\sqrt{2\pi k}\sin\gamma_0}\left[\begin{array}{l}\cot\left(\dfrac{\pi+(\phi-\phi')}{2n}\right)F(kLa^{+}(\phi-\phi')) \\ +\cot\left(\dfrac{\pi-(\phi-\phi')}{2n}\right)F(kLa^{-}(\phi-\phi')) \\ +R_{\parallel,0}^{\perp}\cot\left(\dfrac{\pi-(\phi+\phi')}{2n}\right)F(kLa^{-}(\phi+\phi')) \\ +R_{\parallel,n}^{\perp}\cot\left(\dfrac{\pi+(\phi+\phi')}{2n}\right)F(kLa^{+}(\phi+\phi'))\end{array}\right] \tag{2.30}$$

The Fresnel integral to correct for the singularities at the shadow boundaries is given by

$$F(X) = 2j\sqrt{X}e^{jX}\int_{\sqrt{X}}^{\infty} e^{-j\tau^2}\,d\tau \tag{2.31}$$

where X represents the various possible arguments of $F(\cdot)$ in (2.30). The distance term L is

$$L = \frac{ss'\sin^2\gamma_0}{s+s'} \tag{2.32}$$

where s is the distance from diffracting edge to the field point, and s' is the distance from the edge to the illuminating source, as shown in Figure 2.11. The parameter a in (2.30) is given by

$$a^{\pm}(\phi-\phi') = 2\cos^2\left(\frac{2n\pi N^{\pm}-(\phi-\phi')}{2}\right) \tag{2.33}$$

In (2.33), the parameters $N^{\pm}$ are integers, which most closely satisfy the equations

$$2\pi nN^{+} - (\phi\pm\phi') = \pi \tag{2.34}$$

$$2\pi nN^{-} - (\phi-\phi') = -\pi \tag{2.35}$$

The terms R_0 and R_n refer to the reflection coefficients of the incidence wedge face (0 face) and the opposite wedge face (n face) as shown in Figure 2.11. They are computed for parallel and perpendicular polarizations using (2.23) and (2.24), respectively.

Recently it was determined that the angle arguments of the reflection coefficients R in Luebbers diffraction coefficients in (2.30) were incorrect for certain geometries [14]. On the basis of [14], the angles γ_0 for the 0 and n used in the reflection coefficient (2.23) and (2.24) should be those shown in Table 2.2.

The diffracted field from the wedge at some distance s is found by multiplying the incident field strength by the diffraction coefficient in (2.30) and the spatial attenuation factor $A(s',s)$ as follows:

$$\mathrm{E}^{d} = \mathrm{E}^{i}A(s',s)\overline{\overline{D}}e^{-jks} \tag{2.36}$$

Table 2.2 Revised arguments for reflection coefficients in (2.30) from [14]

Region	γ_0 for 0 face	γ_0 for n face
$\phi' < \pi$, illum.	$-\phi'$	$-(\phi + \phi')$
$\phi' < \pi$, shadow	ϕ'	$n\pi - (\phi + \phi')$
$\phi' > \pi$, illum.	ϕ'	$n\pi - (\phi + \phi')$
$\phi' > \pi$, shadow	$n\pi - \phi'$	ϕ

The spatial or distance attenuation factor depends on the incident wave:

$$A(s', s) = \frac{1}{\sqrt{s}} \tag{2.37}$$

for plane and cylindrical incident waves.

$$A(s', s) = \left[\frac{s'}{s(s' + s)}\right]^{1/2} \tag{2.38}$$

for spherical incident waves. For $s' \gg s$, (2.38) is essentially the same as (2.37) as expected.

Figures 2.13(a–c) show the magnitude of the diffracted field for a 90° wedge, the wedge shape appropriate to modeling the corners of most artificial structures. The direction ϕ' to the illuminating source shown by the bold triangle illumination source varies in each case showing the impact on the diffraction field. The illuminating field for Figure 2.13 is VP (electric field vector parallel to the edge of the wedge). Figures 2.14(a–c) are the same as Figure 2.13 except that the illuminating field is horizontally polarized.

For Figures 2.13 and 2.14, it was assumed that the wedge material is a perfect electrical conductor (PEC) with a 90° interior angle as shown by the blank area in the polar plots. The figures show the magnitude of the diffraction field relative to the incident field at a distance of 10λ from the edge. For example, Figure 2.13(a) shows that the magnitude of the diffraction field at azimuth 45° is 6 dB below the incident field magnitude when the incident illumination is from azimuth direction 225°. The peaks in the diffraction field at the shadow and reflection boundaries are clear, especially in Figure 2.13(a).

The PEC wedges used in Figures 2.13 and 2.14 are idealized wedges where the reflection coefficient for the wedge faces is essentially 1. A more realistic wedge is one with conductivity and permittivity values similar to what might be encountered on real propagation paths. Figures 2.15(a–b) show the diffraction coefficient as a function of the azimuth for vertical and horizontal polarizations, respectively, using a dielectric or 'lossy' wedge with a conductivity of 1×10^{-3} Siemens/m and a relative permittivity of 3.4. The azimuth of illumination is 225 degrees; so comparisons can be made between Figures 2.13(a) and 2.15(a), and Figures 2.14(a) and 2.15(b). Depending on the azimuth, the 'lossy' wedge can have substantially different diffraction properties when compared to the PEC wedge.

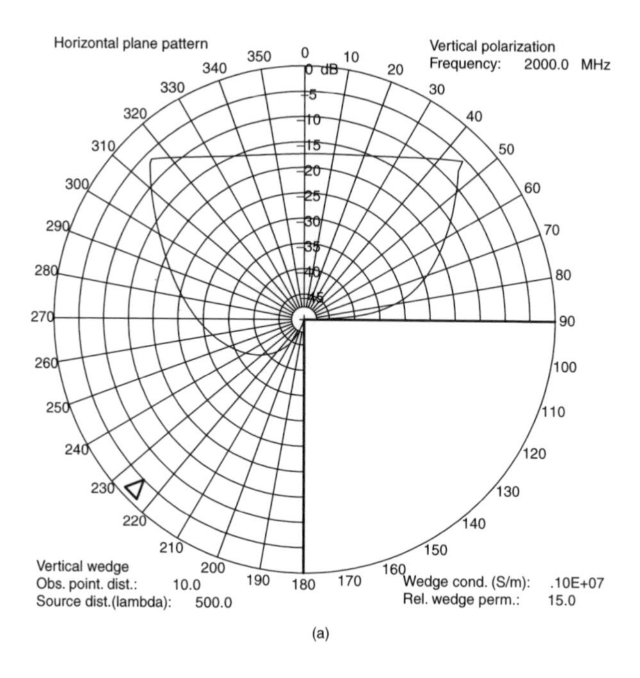

(a)

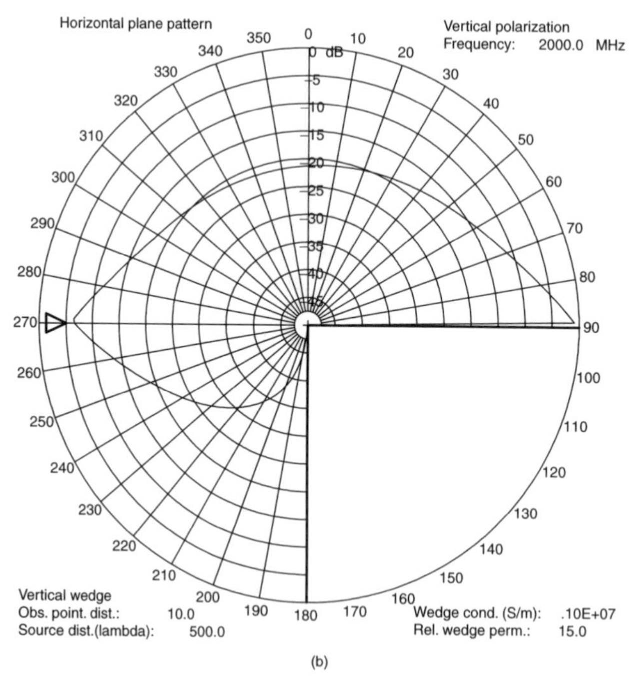

(b)

Figure 2.13 (a) VP diffraction coefficient, PEC wedge, illumination at 225 degrees; (b) VP diffraction coefficient, PEC wedge, illumination at 270 degrees; (c) VP diffraction coefficient, PEC wedge, illumination at 315 degrees.

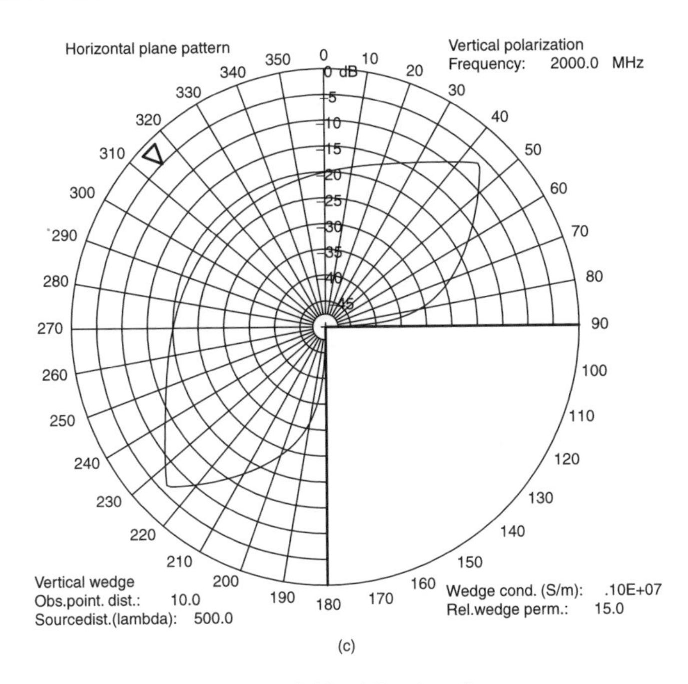

Figure 2.13 (*Continued*).

The diffraction coefficients in (2.30) include a term $\sin\gamma_0$, which accounts for incidence energy at angles that are not perpendicular to the edge. As with the reflection case, the diffracted rays leave the edge at an angle that is equal to γ_0, creating a cone of diffracted rays, which is called the Keller cone (see Figure 2.12). The diffraction coefficients only apply to finding the fields at points on the Keller cone, not elsewhere. For the purely two-dimensional case, $\gamma_0 = 90^\circ$ and the Keller cone degenerates into a circular disk in the same plane as the incident energy ray. It should also be noted that perpendicular polarization on wedges is often called *soft polarization*, while parallel polarization is referred to as *hard polarization*.

Although many fixed broadband wireless systems in the high microwave frequencies are designed for line-of-sight (LOS) paths from the transmitter to the receiver, the ability to calculate path loss on obstructed paths is still important in high-density systems in which the effects of interference on obstructed paths must be considered. The wedge diffraction coefficients described here can be used to find the diffraction attenuation for an obstructed interference path over a rooftop edge or the parapet of a building, for example. Recent experimental results have demonstrated the validity of this wedge diffraction calculation to predicting signal levels on a path obstructed by a building corner [15].

2.7.2 Knife-edge diffraction

If the interior angle of the wedge is assumed to be zero degrees, the wedge becomes a so-called 'knife-edge' and diffraction effects can be analyzed using a simplified and

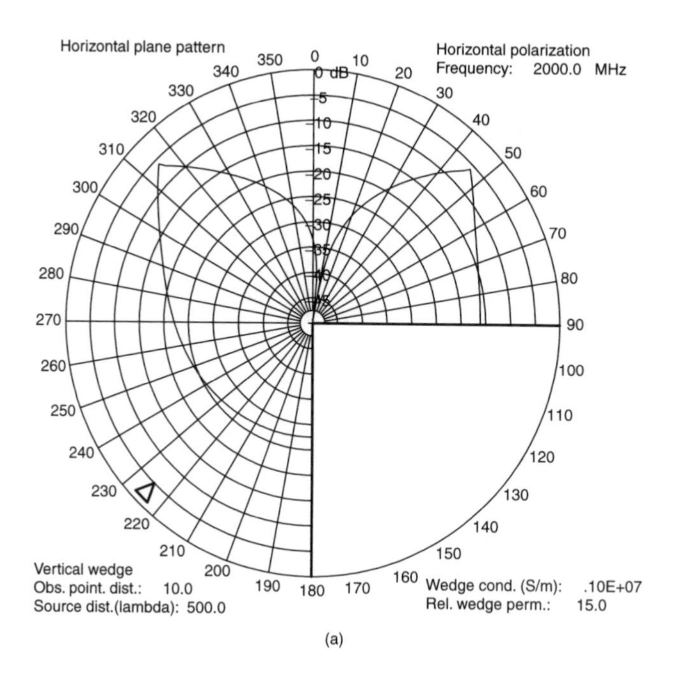

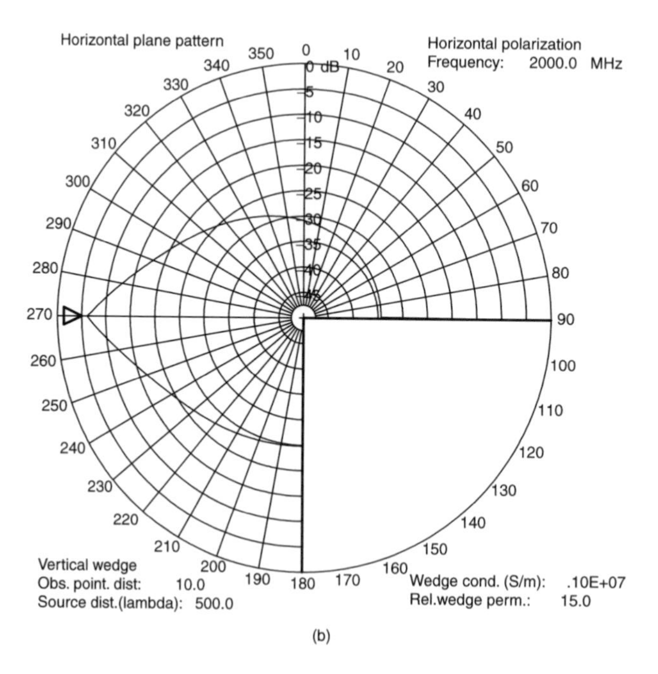

Figure 2.14 (a) HP diffraction coefficient, PEC wedge, illumination at 225 degrees; (b) HP diffraction coefficient, PEC wedge, illumination at 270 degrees; (c) HP diffraction coefficient, PEC wedge, illumination at 315 degrees.

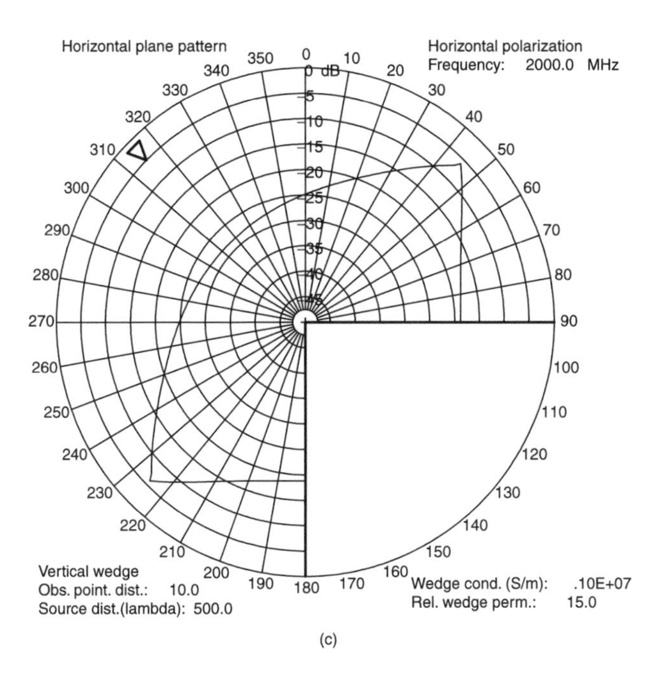

Figure 2.14 (*Continued*).

physically accessible approach compared to the wedge diffraction in Section 2.7.1. As discussed in Chapter 3, the knife-edge diffraction approach is also used in many propagation models because of its simplicity and speed of calculation efficiency, even though in many cases its use represents a very approximate treatment of the diffraction over an obstacle.

To analyze the effect of a knife-edge obstacle on path loss, Huygen's principle will be applied. Huygen's principle states that each point on a primary wave front can be considered as a secondary emitting source, with the secondary wave front represented by the envelope of the waves emanating from each of these sources. In three dimensions, the sources on the primary wave front are point sources and the waves they radiate are spherical waves. In two dimensions, the sources are line sources and the waves they radiate are cylindrical waves.

Figure 2.16 shows a plane wave striking a knife-edge obstacle from the left. The two-dimensional case is considered here; so the knife-edge is actually a half plane extending infinitely in the y direction (into and out of the page). The secondary cylindrical sources are shown positioned along the x-axis above the knife-edge half plane. The field at the observation point is the vector sum of the fields resulting from each of the secondary sources. This field summation takes into account the relative amplitudes and phase of the fields from each of the secondary sources, which in turn depend on the distance from each secondary source to the observation point. The magnitude of the radiation from each secondary source is assumed to be the same since it represents a uniform plane wave front.

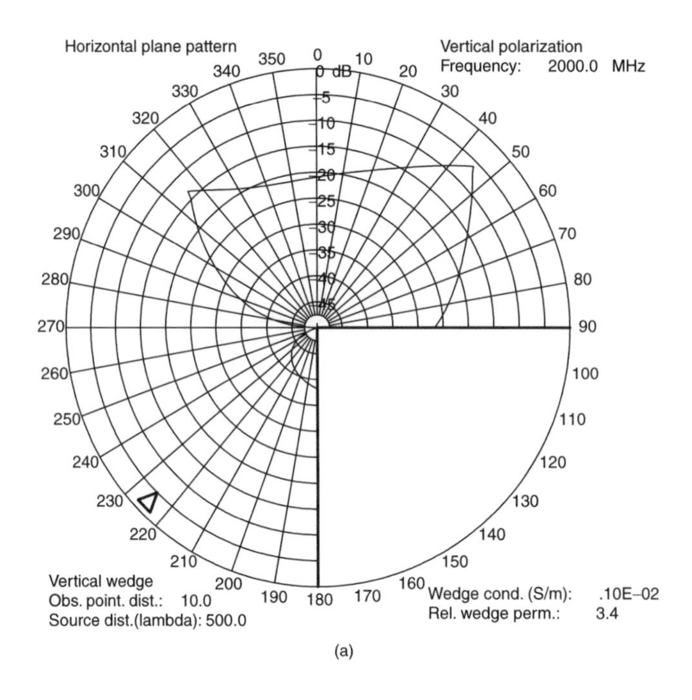

(a)

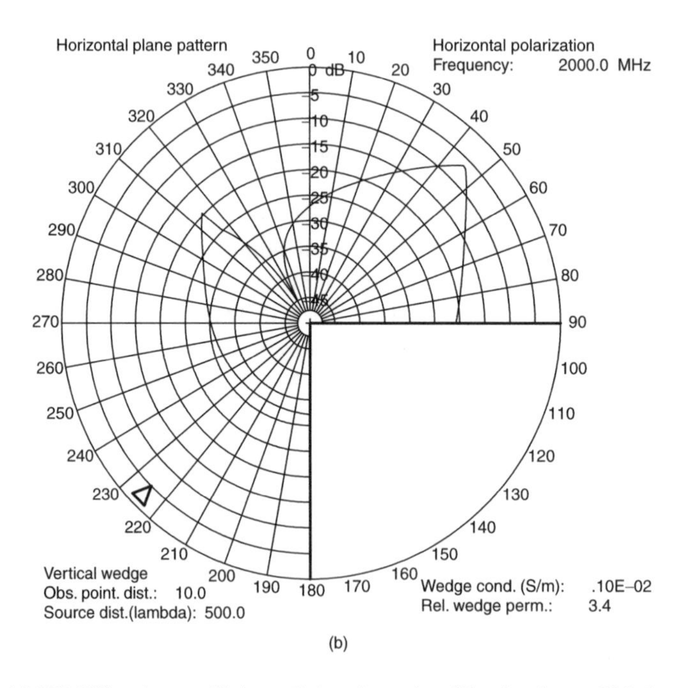

(b)

Figure 2.15 (a) VP diffraction coefficient, dielectric wedge, illumination at 225 degrees and (b) HP diffraction coefficient, dielectric wedge, illumination at 225 degrees.

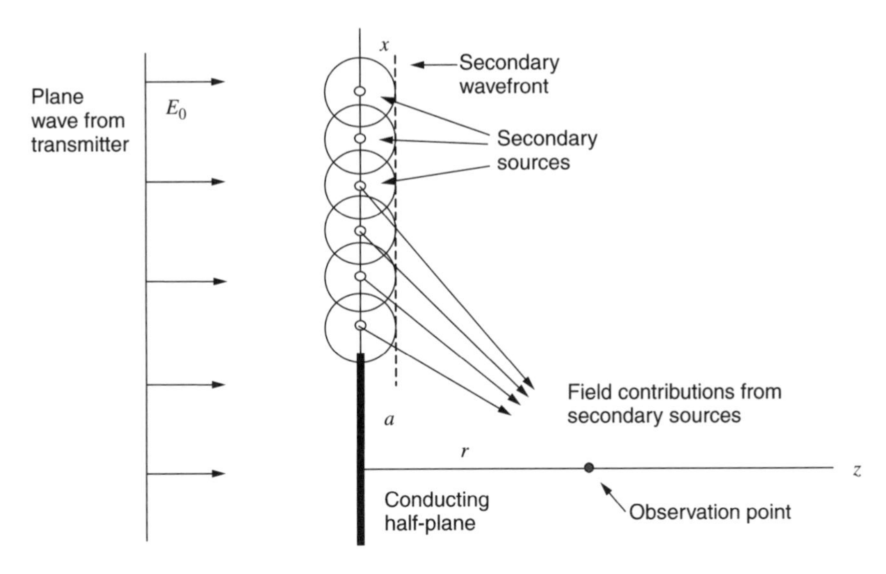

Figure 2.16 Knife-edge diffraction geometry.

The field strength at the observation point can thus be found by summing the fields from each source or integrating along the infinite line of sources from the top of the knife-edge to infinity in the positive x direction. The details and approximations used for this integration can be found in [16]. The resulting integral is actually in the form of a Fresnel integral as follows:

$$E_{OP} = E_0 \frac{(1+j)}{2} \int_{\nu}^{\infty} e^{-j(\pi/2)u^2}\, du \tag{2.39}$$

where E_0 = incident field strength, and ν is the Fresnel diffraction parameter defined from Figure 2.17 as

$$\nu = h\sqrt{\frac{2(d_1 + d_2)}{\lambda(d_1 d_2)}} \tag{2.40}$$

As with the wedge, if $\nu = 0$, the observation point is on the shadow boundary. Figure 2.17 shows how the parameters used to calculate ν are related to a and r in Figure 2.16.

The integral in (2.39) can be solved by recognizing that

$$\int_{\nu}^{\infty} e^{-j(\pi/2)u^2}\, du = \int_{\nu}^{\infty} \cos\left(\frac{\pi}{2}u^2\right) du - j\int_{\nu}^{\infty} \sin\left(\frac{\pi}{2}u^2\right) du \tag{2.41}$$

Since integration of the Fresnel integral from 0 to ∞ equals 1/2, the two terms on the right side of (2.41) can be written as

$$\int_{\nu}^{\infty} \cos\left(\frac{\pi}{2}u^2\right) du = \frac{1}{2} - \int_{0}^{\nu} \cos\left(\frac{\pi}{2}u^2\right) du \tag{2.42a}$$

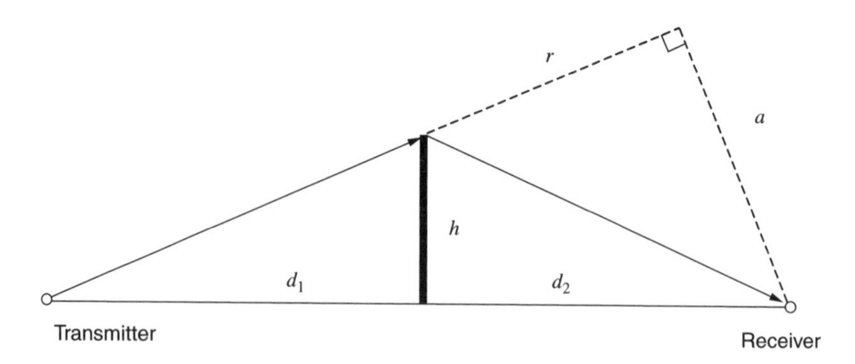

Figure 2.17 Simple geometry for diffraction over a knife-edge obstacle.

$$\int_{\nu}^{\infty} \sin\left(\frac{\pi}{2}u^2\right) du = \frac{1}{2} - \int_{0}^{\nu} \sin\left(\frac{\pi}{2}u^2\right) du \tag{2.42b}$$

The terms on the right side of the two equations in (2.42a) and (2.42b) are the Fresnel cosine and sine integrals, respectively, and are usually written as $C(\nu)$ and $S(\nu)$. The attenuation of the knife-edge (diffraction coefficient of the field strength by the knife-edge) is the ratio of the field at the observation point and the incident field or E_{OP}/E_0. Using (2.41) and (2.42) to represent the Fresnel integral and squaring the terms to form a power ratio, the diffraction coefficient $D_{KE}(\nu)$ becomes:

$$D_{KE}(\nu) = \frac{1}{2}\left[\left(\frac{1}{2} - C(\nu)\right)^2 + \left(\frac{1}{2} - S(\nu)\right)^2\right] \tag{2.43}$$

The Fresnel cosine and sine integrals $C(\nu)$ and $S(\nu)$ can be evaluated to adequate accuracy using the equations and approximations found in Section 7.3 of [17]. Figure 2.18 shows a plot of the resulting ratio in dB as a function of ν. The oscillatory effect for values in the nonshadowed regions results from constructive and destructive vector addition of the diffraction field from the edge of the half plane and the field from the incident plane wave. The phase shift is due to the longer path length off the edge as compared to the direct path. If the observation point is on the shadow boundary ($\nu = 0$), the field strength is reduced by 6 dB from the incident field strength value. From there the field strength attenuation increases as ν increases. As described in Chapter 3, approximate equations to fit the curve in Figure 2.18 have been developed that are applicable in the shadow region and just above. For modeling purposes, these are computationally more efficient than evaluating (2.43) directly.

As shown in Chapter 3, the knife-edge diffraction mechanism described here is used as a model for many obstructed path circumstances including paths with terrain obstructions where those obstructions may only be gently rolling hills that have very little resemblance to a knife-edge. The knife-edge diffractions formulas can be modified to apply to rounded obstacles instead of knife-edge obstacles. With this enhancement, the predicted signal level in the shadowed region is always less than that for the case of the knife-edge

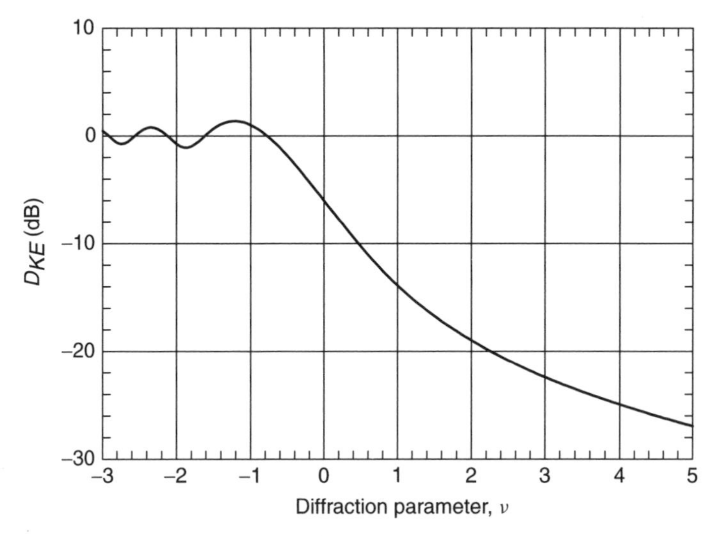

Figure 2.18 Attenuation behind a diffracting knife-edge as a function of parameter ν.

diffraction obstacle, and is usually larger than reasonable on the basis of field experience. Section 3.5.2 discusses modifications to the knife-edge diffraction loss equation to account for rounded obstacles.

2.8 FRESNEL ZONES AND PATH CLEARANCE

In fixed wireless link design, a fundamental design objective is to achieve adequate path clearance for the link. Adequate path clearance means that the transmit and the receive antennas are placed sufficiently high above their respective sites (the ground, a rooftop, etc.) such that every point on the path between them has a certain distance or spacing from any obstacles along the path. As will be discussed later, the 'path' may in fact not be a straight line and may change with the changing refractivity of the atmosphere. Evaluating clearance to achieve a reliable link, therefore, may require evaluating path clearance under a variety of atmospheric conditions.

Figure 2.18 in the preceding section showed that when the path between the transmitter and the receiver was unobstructed, the field varied in an oscillatory fashion, with 0 dB diffraction loss occurring at a particular value of ν. The oscillatory pattern was due to the diffraction field and the direct field adding constructively and destructively as the path lengths change and the phase of the two fields are in-phase or out-of-phase. The locus of the points where the diffracted path length is multiples of 180 degrees ($\pi/2$) different from the direct path length are called the Fresnel zones. As shown in Figure 2.19, the Fresnel zones form elliptically shaped solids of revolution around the transmit–receive propagation path. The Fresnel zone locus of points exist where

$$d_1 + d_2 + \frac{n\lambda}{2} = a + b \tag{2.44}$$

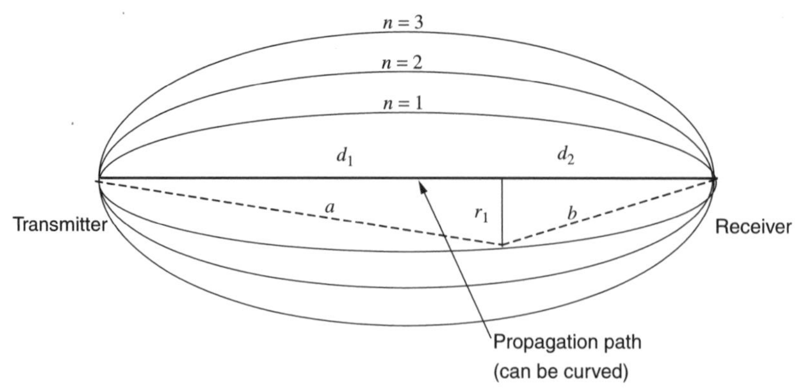

Figure 2.19 Fresnel zones around a propagation path shown in 2 dimensions.

where n is an integer. If the path length is much greater than the Fresnel zone radius, r, (the usual case) then the Fresnel zone radius can be approximated as

$$r_n = \sqrt{\frac{n\lambda d_1 d_2}{d_1 + d_2}} \tag{2.45}$$

The units of (2.45) are all in meters. Equation (2.45) can be converted to the more useful form for fixed broadband wireless systems operating at microwave frequencies:

$$r_n = 17.3\sqrt{\frac{n d_1 d_2}{f(d_1 + d_2)}} \tag{2.46}$$

where distances d_1 and d_2 are in kilometers, the frequency f is in GHz, and the radius r_n is in meters. For a short 3 km path in an LMDS system operating at 28 GHz, the first Fresnel zone radius is 2.83 m at the center of the path. For any given link, the largest radius of a given Fresnel zone occurs in the center of the path.

Conceptually, the first Fresnel zone can be thought of as the region where the significant power is transmitted. If the first Fresnel zone is significantly obstructed or blocked, the power available at the receiver will be diminished. A general criterion for link system design is to set the path clearance so that a radius equal to 60% of the first Fresnel zone is unobstructed – the so-called *0.6 first Fresnel zone* criterion. From Figure 2.18, the 0.6 first Fresnel zone occurs at a value $\nu = -0.8$, which corresponds to the diffraction attenuation of 0.0 dB.

While path clearance is usually considered for terrain, foliage, and structure obstacles underneath the path, the Fresnel zones are in fact three-dimensional, so the path clearance must be considered in all directions around the path. In most cases, obstacles below the path are the primary concern, but with high-density fixed broadband links in urban areas, link paths could certainly exist in close proximity to the sides and corners of tall buildings that extend to heights above the subject path. Full three-dimensional Fresnel zone clearance is thus a consideration, although as pointed out above, a worst-case radius of 2.83 m for a 3-km LMDS path means that there is very little 'gray area' of partial obstruction. The path is either obstructed or has adequate clearance, at least as can be determined within the limitations of the practical resolutions of modern urban building databases. This subject will be addressed again in Chapter 5.

2.9 MATERIAL TRANSMISSION

So far the EM wave propagation has been discussed in terms of free-space waves encountering reflecting surfaces or diffracting edges while in transit from the transmitter to the receiver. Until recently, these mechanisms would suffice to support fixed broadband wireless system design at microwave frequencies. As noted in Chapter 1, fixed broadband wireless systems now include systems operating in the 2.0 to 2.7 GHz, 3.5 to 3.7 GHz, and 5.1 to 5.8 GHz frequency ranges that are specifically intended to work when the transmitter and receiver are non-line-of-sight (NLOS). The path is obstructed by foliage or more likely by intervening structures or the walls of the structure where the receiver is located. In fact, NLOS systems of this type are currently one of the most prominent areas of commercial interest because they would allow a simple desktop wireless modem device that could be installed by the customer (placed on the desk), thus saving the substantial expense of having a technician perform the installation. For this reason, the subject of how EM waves propagate through various materials is now particularly relevant to fixed broadband wireless systems at these frequencies.

When an EM wave encounters a wall or actually any boundary between one medium (material) and another, part of the energy is reflected and part of it is transmitted into the second medium. As shown in Figure 2.20, the transmitted wave changes amplitude and direction as it continues into the second material. The transmission coefficients take a form similar to the reflection coefficients in (2.23) and (2.24), as follows:

$$T_{\parallel} = \frac{2\eta_2 \sin\gamma_i}{\eta_2 \sin\gamma_t + \eta_1 \sin\gamma_i} \tag{2.47}$$

$$T_{\perp} = \frac{2\eta_2 \sin\gamma_i}{\eta_2 \sin\gamma_i + \eta_1 \sin\gamma_t} \tag{2.48}$$

where $T_{\parallel}$ and $T_{\perp}$ are the transmission coefficients for parallel and perpendicular incident waves (see Section 2.6.1) and $\eta_{1,2}$ is given by (2.25).

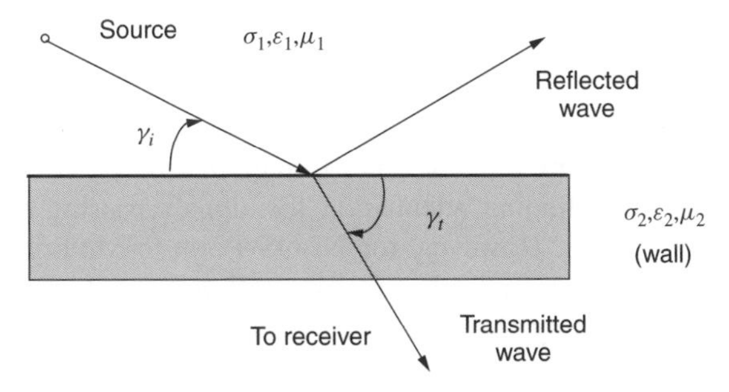

Figure 2.20 Wave transmission from one material (σ_1, ε_1, μ_1) through a second material (σ_2, ε_2, μ_2) such as a wall.

2.9.1 Transmission into structures

The expressions in (2.47) and (2.48) are exact for transmission from one medium into another, but in practice a signal is normally passing from free-space through a material like a wall and back into free-space where the receiver is located. These equations must therefore be applied twice, once for each transition. For typical walls, the construction is usually not a single homogeneous material with well-known properties. A typical residence in the United States will have walls constructed of a collection of materials such as wood framing, fiberglass insulation (often with aluminum foil backing), sheetrock interior cladding, and wood or even metal exterior cladding. Others may be made of brick. Concrete and glass are the typical modern building materials used to construct office buildings. Elsewhere in the world, the materials may be quite different. A substantial number of buildings in Europe are old constructions in which very thick stonewalls are commonplace. The relative proportion of windows to opaque walls adds to the nonhomogeneity of the material that a signal must pass through in transit from an external transmitting antenna to a receiver located inside a building.

Because of the complexities of the materials used in real structures, (2.47) and (2.48) are rarely used for the purpose of finding wall transmission loss. Instead, measurements are made for a variety of structures and material types and approximate net attenuation values at different frequencies are derived from those measurements. Such measurements have been made by many researchers for cellular and personal communication service (PCS) frequencies below 2 GHz [18]. In [19], a variety of measurements at widely spaced frequencies of 9.6 GHz and above are reported. For fixed broadband wireless systems currently being designed, the bands from 2 to 2.7 GHz, 3.5 to 2.7 GHz, and 5.1 to 5.8 GHz are of primary importance. Unlike cellular frequencies, there is relatively little published data for building penetration losses for these frequencies. Some attenuation results taken from currently available published work are shown in Table 2.3.

The loss values in Table 2.3 are intended to be added to the total path loss as calculated along the path from the transmitter to the outside of the structure in which the receiver is located, as discussed further in Chapter 3. It should also be noted that these paths are reciprocal; that is, the locations of the transmitter and receiver can be interchanged and the path loss is still the same. These data are therefore applicable to two-way wireless systems.

2.9.2 Transmission through foliage

PTP microwave systems are typically designed so that adequate 0.6 first Fresnel zone clearance is achieved above any trees or other foliage that may exist along the path. For this reason, the problem of finding attenuation for signals passing through foliage has not been of great importance. However, for NLOS Point-to-Multipoint (PMP) systems as described in Chapter 1, trees and other foliage will very likely be present on a wireless transmission. Consequently, for residential, small office buildings, and other low profile structures where trees may extend above the height of the receiver antenna location, the attenuation or path loss due to foliage must be considered along with building penetration loss.

Table 2.3 Measured building penetration loss, frequencies >2 GHz

Frequency (GHz)	Material	Loss, L (dB)	Source
2.30	Stone university and office buildings	12.8	20
2.40	University buildings	20.0	21
2.57	Suburban houses	9.1	22
5.85	Brick house wall, paper-backed insulation	12.5	23
5.85	Brick house wall, foil-backed insulation	16.3	23
5.85	Wood house siding	8.8	23
5.85	Concrete wall of house	22.0	23
5.85	Basement (subterranean) rooms	31.0	23
5.85	Interior plaster walls	4.7	23
9.60	2 sheets of sheetrock	2.0	19
9.60	2 sheets dry 3/4″ plywood	4.0	19
9.60	2 sheets wet 3/4″ plywood	39.0	19
9.60	Aluminum, 1 sheet of $1/8^{th}$ thick	47.0	19
28.8	2 sheets of sheetrock	2.0	19
28.8	2 sheets dry 3/4″ plywood	6.0	19
28.8	2 sheets wet 3/4″ plywood	46.0	19
28.8	Aluminum, 1 sheet of $1/8^{\text{th}}$ thick	46.0	19

Foliage transmission loss is actually more difficult to treat theoretically than building penetration loss because the structure of foliage is essentially random. In fact, fractal models are the most successful at synthesized foliage growth that appears realistic. Depending on the frequency, the leaves on the trees can directly block and absorb energy from the EM wave. The branches in their various orientations and configurations tend to act as scattering elements. Depending on the length, they can also be close to a resonant frequency of the incident EM wave, thus acting like small parasitic antenna elements that act differently depending on whether they are wet or dry. The problem is further complicated by the wide variety of trees that may be encountered, and the fact that deciduous trees lose their leaves in the fall and in winter, introducing a seasonal variation in the foliage loss that a signal will experience. Trees also move in the wind, so the attenuation and scattering effects will have a temporal variation proportional to the wind speed and directions. In spite of these difficulties, some attempts at analytically treating the impact of trees on a signal have been made [24]. The scattering approach was also attempted to assess wind variations in foliage loss as reported in [25].

All of the issues associated with attenuation and scattering through trees quickly lead to the conclusion that an empirical measurement approach is currently the only viable way to gain a useful understanding of how the signal will be affected. Table 2.4 is a tabulation of some of the measurement work done for frequencies above 2 GHz. Figure 2.21 shows a graph of the data taken from [26] showing foliage attenuation (dB/m) as a function frequency. The data in Figure 2.21 are the composite of many different measurements

Table 2.4 Foliage loss, frequencies >2 GHz

Frequency (GHz)	Foliage configuration	Loss, L (dB)	Source
5.85	Small deciduous tree	3.5	27
5.85	Large deciduous tree	10.7	27
5.85	Large conifer tree	13.7	27
9.60	Single conifer tree	15.0	28
9.60	6 conifer trees	21.0	28
28.8	Single conifer trees	15.9	28
28.8	6 conifer trees	33.0	28
28.8	Single deciduous tree no leaves	5.0	29
28.8	Single deciduous tree with leaves	7.0	29
28.8	8 deciduous trees, no leaves	15.0	29
28.8	8 deciduous trees, with leaves	30.0	29
38.0	Various trees	2.0/m	30
38.0	Single oak tree with leaves	17.0	31

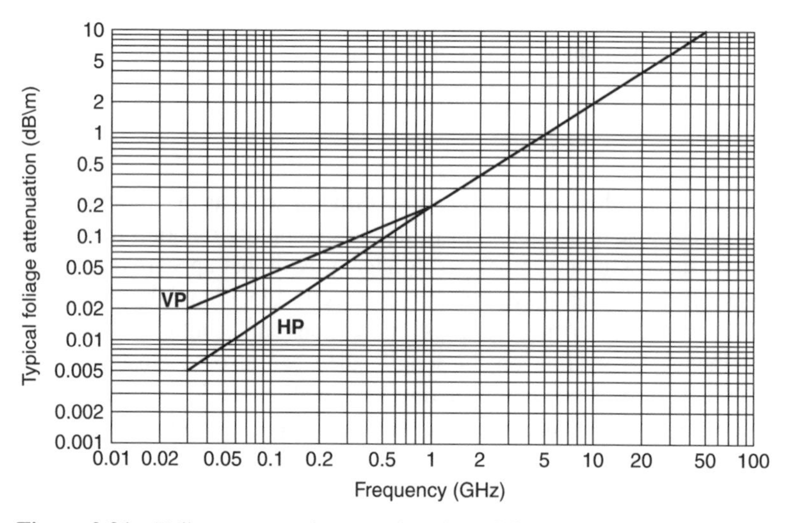

Figure 2.21 Foliage attenuation as a function of frequency taken from [26].

and should be regarded as 'typical' because it does not take into account the many foliage variations as noted above.

2.10 ATMOSPHERIC REFRACTION

The EM waves for wireless systems generally are propagating through the air or the atmosphere, only occasionally encountering reflecting, diffracting, or attenuating objects as discussed earlier in this chapter. The atmosphere, of course, is not a vacuum but a collection of gases, primarily oxygen, nitrogen, and carbon dioxide. The atmosphere also contains water vapor resulting from evaporating bodies of water or wet ground.

For terrestrial wireless systems (those between two points on the earth's surface) that are the subject of this discussion, the EM waves are traveling through a layer of atmosphere called the troposphere. Above the troposphere is the stratosphere. The troposphere begins at the earth's surface and ranges up to a height of about 10 to 17 km, depending on the location on the earth and the prevailing meteorological conditions.

In general, the temperature of the air in the troposphere decreases about 2°C for every 300 m increase in height above ground. Because the temperature is decreasing as a function of height, the phase velocity of the wave front is slightly higher at higher altitudes (lower at lower altitudes) resulting in the EM wave from tilting versus altitude. The net effect is that the wave front propagation direction bends as the wave travels through the atmosphere. This bending process is known as *refraction*. It is essentially the same effect as the change in direction of a wave transmitted from one medium into another as illustrated in Figure 2.20.

The refractivity of the atmosphere is described by a refractive index n. The value of the refractive index differs from one by a small amount – about 315 parts per million ($n = 1.000315$). The refractive index decreases approximately exponentially with height above ground. To avoid dealing with the small decimal fractions close to 1, a refractivity value in N-units has been defined:

$$N = (n - 1) \times 10^6 \tag{2.49}$$

The refractivity N can also be computed using the temperature, total pressure, and water vapor pressure as follows [32]:

$$N = \frac{77.6}{T}\left(P + 4810\frac{e}{T}\right) \tag{2.50}$$

where:

P = total atmospheric pressure in millibars (mb)
e = water vapor pressure in millibars (mb)
T = temperature in degrees kelvin (K)

Since the temperature and pressure are falling exponentially with height h, the value N for any height up to a reference height H can be written in terms of a reference refractivity value N_s at the earth's surface:

$$N(h) = N_s \exp\left(\frac{-h}{H}\right) \tag{2.51}$$

where $N_s \cong 315$ and $H = 7.35$ km, the standard reference values set forth in [32]. These expressions are derived for a standard atmosphere with stable characteristics. In reality, real atmospheres and their refractivity values are highly variable, a variability that must be taken into account for successful wireless link design.

For the purposes of link design, it is useful to think in terms of the EM waves leaving the transmitter as rays whose trajectories are bent or otherwise modified by refraction in the atmosphere. For a standard atmosphere, the rays will bend because of refraction in such a way that they tend to curve along the surface of the earth. The radius of this curvature can be found by considering the refractivity gradient. Near the earth's surface the exponential change in refractivity can also be approximated by a linear function such that:

$$N = N_s - \frac{N_s}{H}h \tag{2.52}$$

with this approximation, a linear change in refractivity with height (refractivity gradient) results, which in turn implies that the radius of curvature of the wireless ray is a constant, thus forming a circle. The radius of this circle can be found from the refractivity gradient. A commonly used median value for the refractivity gradient of standard atmosphere is -39 N-units/km, which equals a gradient $dn/dh = -39 \times 10^{-6}$ from (2.49). From [33], the effective earth radius k is given by

$$k = \frac{1}{1 + R(dn/dh)} \tag{2.53}$$

where R is the real earth radius. The earth is actually not a true sphere but more closely resembles an oblate spheroid with a major and minor axis. For the purposes here, an earth radius value of 6735 km is adequate since there are other approximations in this derivation. For a refractivity gradient of -39 N-units/km, (2.53) yields an effective earth radius of $k = 1.35$ or about 4/3.

A graph of a PTP link system along a path over a terrain profile is shown in Figure 2.22. The profile is a display of the terrain elevations along the path under the link. In this case the profile has been curved to correspond to a 4/3-earth radius. The radio path between the transmitter and receiver is therefore drawn as a straight line. Also shown on this graph is the 0.6 first Fresnel zone under the radio path. As will be discussed later, this type of diagram is one of the most widely used for link profile design. Figure 2.23 shows the same terrain profile, but in this case the terrain is shown as flat and the radio paths are curved. Three different curvatures are shown corresponding to effective radii (k factors) of 4/3, 1, and 2/3. A graph such as that shown in Figure 2.23 is useful because a single picture can show path clearance with a range of refractivity gradients that encompass those values that are likely to be encountered from 5% to 95% of the time, for example. The different shades of gray under the terrain profiles in Figures 2.22 and 2.23 correspond to different land use or morphology types along the path.

Worldwide maps showing mean temperatures, reference surface refractivity, refractivity gradients, and percentage time the refractivity gradient is less than -100 N-units for different times in the year can be found in [34]. Some of these refractivity maps are included in Appendix A.

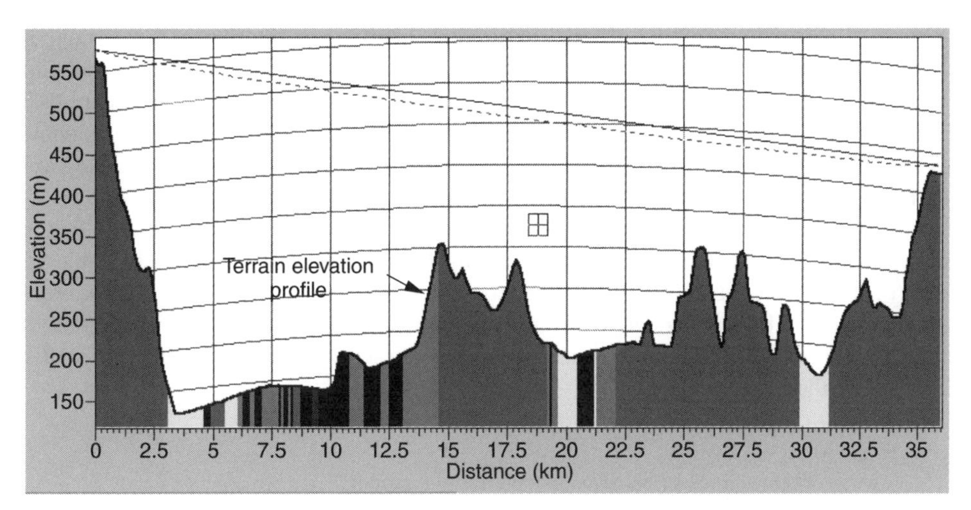

Figure 2.22 Link profile over terrain with the terrain elevations adjusted for 4/3 effective earth radius curvature. The dashed line is the 0.6 Fresnel zone.

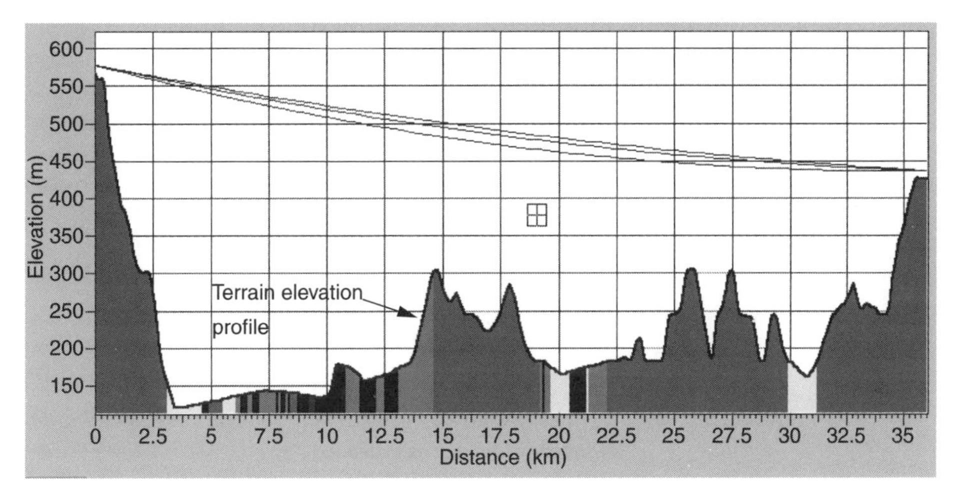

Figure 2.23 Link profile showing flat terrain with three curved radio paths corresponding to 4/3, 1 and 2/3 effective earth radius k.

2.10.1 Statistics of varying refractivity gradients

The typical value of 4/3 effective earth radius is a widely used starting point to characterize atmosphere refractivity for the purposes of wireless system planning. In reality, the refractivity gradient is continuously changing. The refractivity gradient is commonly assigned the symbol G, such that $G = dN/dh$. A value of $G = -39$ N-units/km, corresponding to $k \cong 4/3$, is a median value found in the continental temperate climate conditions in the Northern Hemisphere. The median value is the value that exceeds 50% of the time.

Figure 2.24 shows an example of the measured probability distribution of the refractivity gradient G for Buenos Aires. It is clear that the refractivity gradient can vary substantially over time. In this case, for 10% of the time the refractivity gradient is less than -100 N-units/km, corresponding to a k factor of 1.56. Graphs of measurements such as shown in Figure 2.24 have been developed over many years for several locations on the earth. Reference [35] is one of many documents that include compilations of this data.

By using the equations given above it is possible to find some key values of G that are used to provide broad descriptions of the atmospheric conditions as shown in Table 2.5. The probability that sub-refraction and super-refraction conditions will exist along a radio path must be carefully considered when designing high reliability links where reliability values exceeding 99.999% may be required. Sub-refraction and super-refraction will be discussed in the next sections.

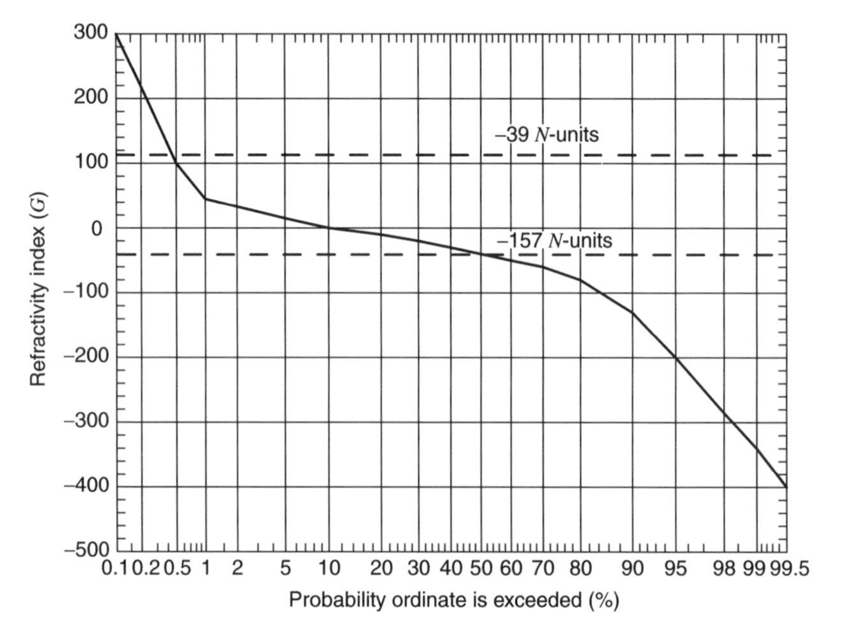

Figure 2.24 Probability of refractivity gradient occurring.

Table 2.5 Refractivity gradients and k factor

Refractivity gradient, G	k factor	Refractivity description
−39	4/3	Normal
0	1	Normal
>0	<1	Sub-refraction
−157	∞	Super-refraction
<−157	–	Super-refraction (ducting)

2.10.2 Sub-refraction

Sub-refraction means that the refractivity gradient is less than zero and that the effective earth radius is less than the real earth radius. The result is that the earth in effect 'bulges' up toward the radio path, thus increasing the possibility that terrain, building or foliage obstacles will intrude into the 0.6 first Fresnel zone. A sketch of an example of the refractivity gradient profile as a function of height is shown in Figure 2.25.

Sub-refraction conditions can exist for a variety of reasons, including temperature inversions in which the temperature decreases with height more rapidly than normal, or the water vapor pressure actually increases with height instead of decreasing as it normally does. This condition is often found in coastal regions where very moist air masses can move in from the ocean over warm dry land with the result that the humidity is higher at higher heights above ground.

2.10.3 Super-refraction and ducting

Super-refraction has a profile in which the refractivity gradient is decreasing with height as drawn in Figure 2.26. For a value of $G = -157$, the radio wave is actually following the curvature of the earth and can consequently travel long distances over flat terrain or bodies of water. When G decreases below -157, it often does so at a specific height or at a very narrow range of heights. Such severe refraction bends the radio waves

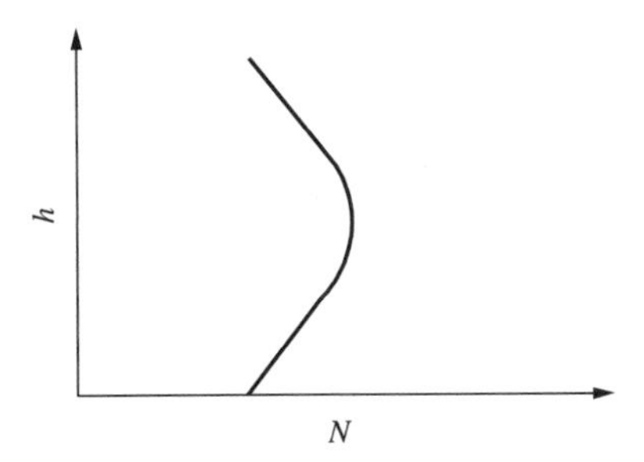

Figure 2.25 Refractivity gradient profile for sub-refraction.

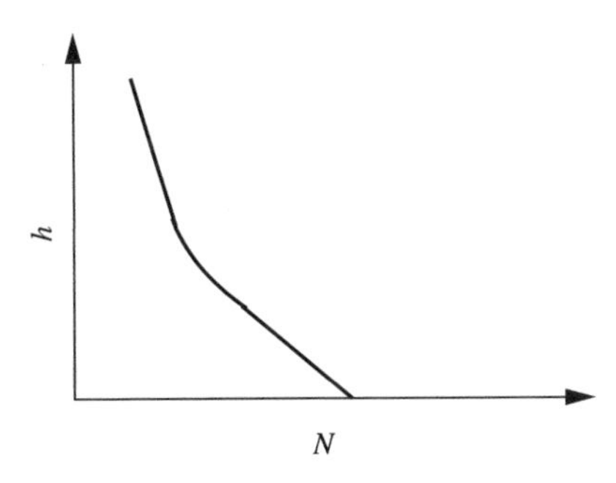

Figure 2.26 Refractivity gradient profile for super-refraction.

back to the earth where they can reflect off the surface and back into the atmosphere in which it can be refracted again. This 'ping-pong' effect essentially traps the wave in this layer, a phenomenon called *ducting*. The duct can exist between the earth's surface and an atmospheric layer (surface duct), or between two layers up to about 1,500 m in the atmosphere (elevated duct) in which the duct layers may be spaced only a few meters apart. By deliberately launching a signal into the duct (which may require the antenna height be adjusted to tract the duct height), very long range communication can be achieved. Such ducting conditions often exist in hot dry desert areas along bodies of water such as those found around the Persian Gulf and the Arabian Gulf. For ducts to exist, the atmosphere must also be unusually calm and stable. For that reason, ducts essentially never occur in mountainous or hilly areas where the air naturally becomes more turbulent as it moves over the undulating surface of the earth.

Worldwide maps showing statistics about the existence and elevation of elevated ducts can be found in [34].

2.11 ATMOSPHERIC ABSORPTION

As mentioned earlier, the atmosphere is made up of a collection of gases. The temperature and pressure of these gases result in refraction as discussed in the preceding section. At certain frequencies, the gases can also directly attenuate the EM wave. This occurs for two primary reasons:

- Polar molecules like water vapor (H_2O) will tend to align themselves with the variations in the changing electric field vector in the EM wave. This process necessarily takes energy that is lost by the wave. The loss in dB/km tends to increase as frequency increases because of the more rapid field oscillations.
- Magnetic moments in molecules like oxygen (O_2) and others will try to align with the incident fields, which also results in energy loss.

The amount of energy in the wave, which is lost to these processes, depends on the frequency, with several resonant peaks in the loss at various frequencies. Figure 2.27 shows a graph of the atmospheric absorption loss as a function of frequency. The values in the graph may also be calculated using the formulas found in [36].

The clear peaks in the attenuation in Figure 2.27 have resulted in certain frequency bands being considered less desirable than others for wireless communication purposes. If the path is very long, the accumulated attenuation can be quite large. The decision to allocate frequencies at about 30 GHz for LMDS-type services was to some extent determined by the dip in the gaseous attenuation curves at that frequency band.

2.12 RAIN ATTENUATION AND DEPOLARIZATION

Rain is one of the more challenging phenomena to be considered in EM wave propagation for fixed broadband wireless systems. In fact, in many systems and in many locations,

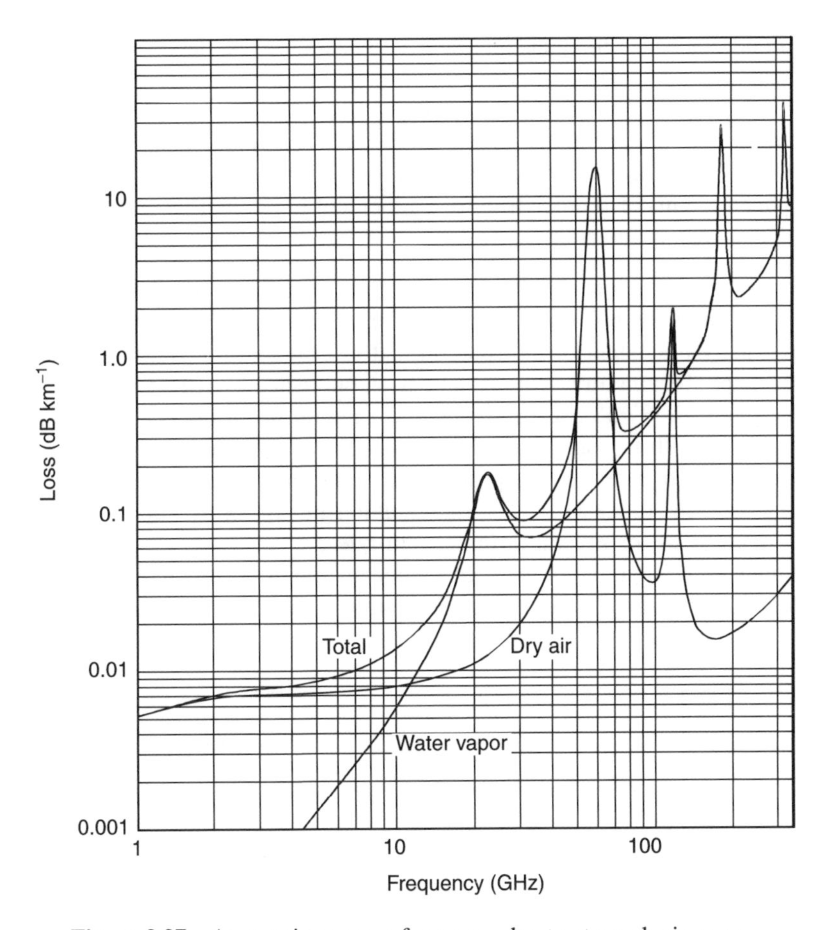

Figure 2.27 Attenuation versus frequency due to atmospheric gases.

attenuation due to rain is the single factor limiting the range of high-reliability microwave links. Signal path attenuation due to rain is a flat-fading phenomenon rather than a spatial-fading phenomenon (like multipath fading) so it cannot be mitigated using space or polarization diversity techniques.

Rain affects radio wave propagation in three important ways:

- Raindrops smaller than a wavelength will absorb the energy in the incident wave through a basic heating effect, causing the energy in the signal to be reduced. Depending on frequency, this will be the most significant source of rain attenuation.
- Larger raindrops on the order of wavelength in size will scatter the incident EM wave resulting in a reduction of the wave amplitude at the desired receive location. The amount of scattering depends on the size distribution, orientation, and especially the intensity of the raindrops.

- The raindrops cause the incident wave polarization to be depolarized, again, largely via scattering. Surface currents will be created on the drops by the illuminating EM wave that in turn causes new radiated fields. The new radiated fields will be a function of the raindrop size distribution, intensity, and orientation. If the incident wave is VP, the waves scattered by the rain will also have horizontally polarized field components. This also holds true for an incident horizontal wave that is scattered so that VP waves are created.

Usable information about the distribution of raindrop sizes and their shape and orientation is usually unknown for any particular area. Simple rain intensity measurements in mm/h are usually the extent of information that is available for system design purposes. Worldwide maps showing rain regions have been published by R.K. Crane and the ITU-R and can be found in [37,38], respectively. Each region has a rain rate probability distribution associated with it from which the additional path attenuation and path availability can be calculated. For reference purposes, these maps and tables of rain rate probabilities are also included in Appendix A. Rain attenuation in dB/km as a function frequency is shown in Figure 2.28 for several different rain rates.

The attenuation on a radio path also depends on how much of the path is subject to rain. Rain, especially high-intensity rain, is normally associated with rain cells

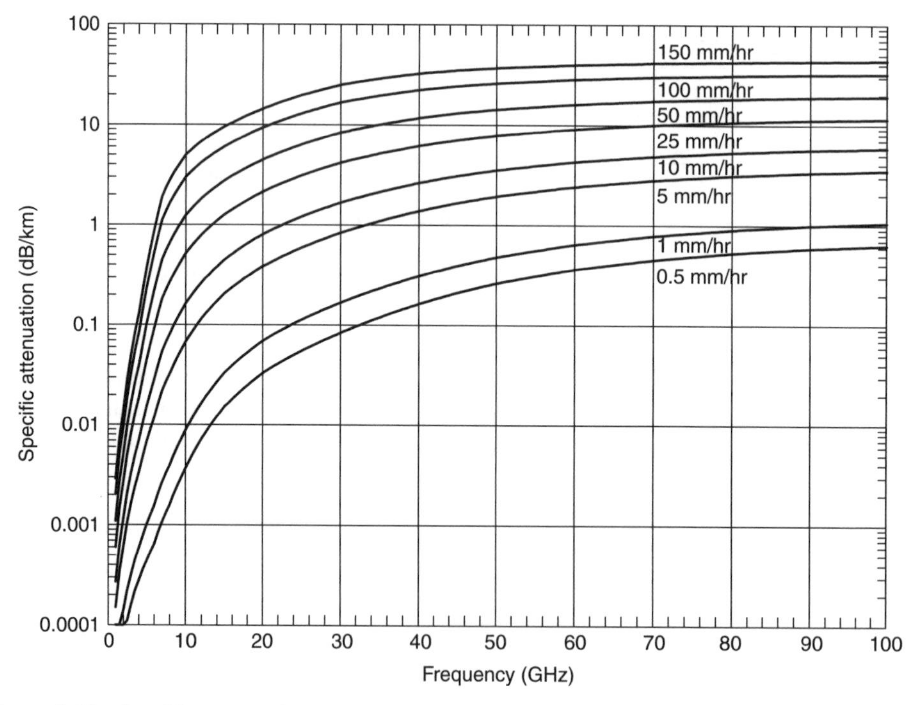

Figure 2.28 Specific attenuation due to rain in dB/km versus frequency for vertical polarization ($\tau = 90$ degrees) and a horizontal propagation path ($\Theta = 0$ degrees) as calculated from (2.54).

that vary in size and shape and move as a storm or weather front moves through an area. For a metropolitan area with microwave links widely deployed, the movement and size of the rain cells can actually be tracked on the basis of which microwave links are experiencing abnormal amounts of attenuation as a storm moves through the area. The movement, size, and shape of the rain cells are as difficult to predict as the weather itself.

Rain attenuation in dB per km is found from the following formula:

$$\gamma_R = kR^\alpha \quad \text{dB/km} \tag{2.54}$$

where k and α are coefficients found from values in Table A.1 (Appendix A) for the relevant frequency, and R is the rain rate in mm/h for a given percentage of time for a given rain region of the world. The values in Table A.1 are used to find k and α using the following equations:

$$k = \frac{[k_H + k_V + (k_H - k_V)\cos^2\Theta\cos 2\tau]}{2} \tag{2.55}$$

$$\alpha = \frac{[k_H\alpha_H + k_V\alpha_V + (k_H\alpha_H - k_V\alpha_V)\cos^2\Theta\cos 2\tau]}{2k} \tag{2.56}$$

where Θ is the path elevation angle and τ is the polarization tilt angle relative to horizontal. For linear horizontal or vertical polarization used here, the polarization tilt angle τ is either zero or 90 degrees, respectively. Likewise, for terrestrial radio paths, the path elevation angle is typically very small, especially for path distances where rain attenuation might be important. From (2.55) and (2.56), it is clear that horizontal and vertical polarizations have somewhat different attenuation values. Some values of attenuation versus frequency for various rain rates and vertical polarization are shown in Figure 2.28.

The use of orthogonal polarizations is an effective tool for increasing the capacity of LOS fixed broadband systems. Consequently, it is important to consider the changes in cross-polarization discrimination (XPD) due to the depolarizing effects during rain. The system design models for handling the changes in XPD due to rain are described in Section 4.3.5.

2.13 FREE-SPACE OPTICS (FSO) PROPAGATION

Free-space optics (FSO) links use lasers for short range, high capacity communications. The basic technique has actually been around for several decades, having first been used for short-range wireless video connections for temporary broadcast interconnections. Today, the wireless industry is envisioning high-density systems with many buildings connected in a mesh of links. The main advantages of FSO are the high data rate capacity ($<$1 Gbps) and the fact that EM spectrum at optical frequencies well above 350 GHz is unregulated. Free-space optics systems therefore can be set up very rapidly without the need for any licenses or regulatory approval.

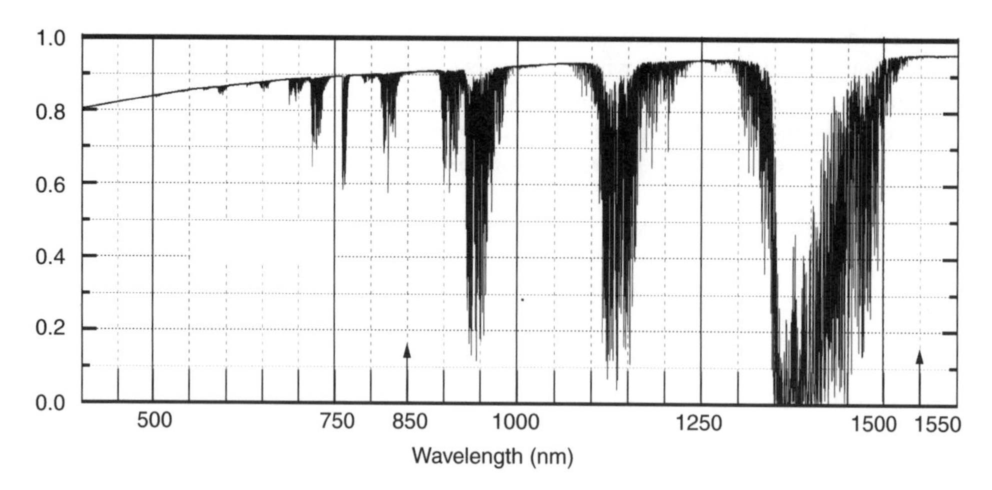

Figure 2.29 Clear sky atmospheric transmission along a 200-m path calculated using MODTRAN. Data adapted from [39].

The lasers used in FSO all operate at light wave frequencies, normally at one of two wavelengths –850 nm (352,941 GHz) or 1,550 nm (193,548 GHz). At the longer wavelength, 1,550 nm lasers have been found to present less danger to the human eye than 850 nm lasers because the longer wavelengths are effectively scattered by the cornea before reaching the retina. Because of this, the allowed power for 1,550 nm systems is higher and longer ranges are possible. Beyond this difference, the propagation mechanisms affect both laser wavelengths in much the same way. The power used in FSO laser transmitters is typically from 10 mW to 80 mW.

Figure 2.29 shows the relative transmission levels for part of the visible spectrum. The high transmission at the typically used laser wavelengths of 850 and 1,550 nm are evident.

2.13.1 Beam divergence

The laser beam diverges as the distance from the transmitter increases, thus causing the signal intercepted at the receiver to be weaker at greater distances. The amount of divergence in the beam is a function of the transmitter design and can be chosen to be more focused or more dispersed to accommodate various link design considerations. The beam can be thought of as a cone with the cone diameter given by:

$$r = d\tan(\theta/2) \tag{2.57}$$

where θ = beam divergence angle, and d = distance from the transmitter in meters. If θ equals a typical value of 6 milli-radians (0.343 degrees), at a distance of 500 m the radius of the cone is about 1.5 m (diameter = 3 m). This geometry is shown in Figure 2.30.

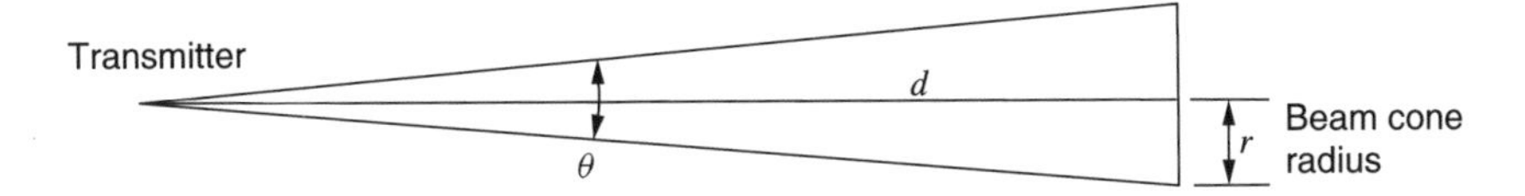

Figure 2.30 Divergence beam of free-space optical transmitter.

2.13.2 Fog, snow, and rain attenuation

The small wavelength of the laser light beam means that very small particulates in the atmosphere can cause the beam to scatter and thus be weaker at its intended receive location. This scattering is usually regarded as Mie scattering. The degree of forward scatter and reverse scatter is a function of the size of the scattering particles and the wavelength, much the same as that at microwave frequency propagation through the atmosphere. For this reason, raindrops and snowflakes have relatively little impact on FSO system attenuation because of their size and low density. Similarly, haze that is made up of particles a few microns in size also is relatively insignificant. The limiting factor on FSO path length and link margins is fog.

Fog is composed of water vapor drops on the order of a few microns to 100 microns in size. Because of the variability of sizes, the scattering effects are fairly uniform with frequency through the infrared and visible light range, although lasers operating in the far-infrared range offer somewhat better penetration. Because of fog scattering, the range of a successful FSO is limited by the length of time the fog of a given density will be present along the path. As will be discussed in Chapter 5, some meteorological databases of fog exist though their resolution is inadequate for planning links that only span a few hundred meters. In general terms, high-reliability FSO system link ranges may be limited to the distances from 300 m to as much as 1,000 m depending on the fog characteristics in the area where the link is deployed.

2.13.3 Atmospheric scintillation

Atmospheric scintillation is actually a refractivity effect as described in Section 2.10. It is caused by rapid changes in temperature or humidity. Scintillation is the term given to highly localized, dramatic changes in refractivity in the atmosphere. The wavy nature of visible light rising from a hot desert and twinkling stars are two examples of scintillation affecting the propagation of EM waves at optical frequencies. As with microwave frequencies, the refractivity index boundaries can cause bending and defocusing of the laser beam in an FSO system.

Scintillation pockets or events can occur in a variety of sizes. Pockets that are much larger than the beam can to some extent be mitigated by automatically resteering the beam. Scintillation pockets that are much smaller than the beam will spread the beam, increasing the radius of the beam cone and decreasing the amplitude of the signal reaching the receiver.

Scintillation events that are about the same size as the beam will defocus the beam and cause a sudden reduction or fade of the optical signal level at the receiver that can range from 7 to 10 dB. Such fades may be deep enough to cause the received signal level to fall below the threshold required for adequate signal-to-noise ratio (SNR) and associated bit error rate (BER). These types of fades can be partially combated using path diversity. Path or space diversity is a commonly used technique for improving the reliability of all types of wireless communication links at all frequencies. Path diversity makes use of the fact that separate paths may not experience fades at the same time (the fades are not perfectly correlated) and by using two or more paths, and combining them in a constructive fashion, the overall net fading effect is reduced. Path diversity with FSO systems is typically achieved using multiple laser transmitters spaced approximately 200 mm apart mounted in the same transmitter housing. Further details on the construction of FSO transmitters and receivers can be found in [40].

2.14 CONCLUSIONS

Electromagnetic wave propagation is a complex and multidimensional topic, which has been the subject of much research and extensive publications. The elements of EM wave propagation presented here are a selection of this mechanism that is most relevant and most commonly employed in propagation models used for designing fixed broadband wireless systems at frequencies above 2 GHz. Those models are discussed in detail in Chapter 3.

2.15 REFERENCES

[1] C.A. Balanis. *Advanced Engineering Electromagnetics*. New York: John Wiley & Sons. 1989, pp. 104 ff.
[2] W.L. Stutzman and G.L. Thiele. *Antenna Theory and Design*. New York: John Wiley & Sons, 1981, page 61.
[3] H.T. Friis, "A note on a simple transmission formula", *Proceedings of the IRE*, vol. 34, pp. 254–6. May, 1946.
[4] J.D. Kraus. *Electromagnetics*. New York: McGraw-Hill, 1984.
[5] K.A. Chamberlain and R.J. Luebbers, "An evaluation of Longley-Rice and GTD propagation models," *IEEE Transactions on Antennas and Propagation*, vol. AP-30, no. 6, pp. 1093–1098, Nov. 1982.
[6] J.D. Kraus. *Electromagnetics*. New York: McGraw-Hill, 1984. page 123.
[7] E.F. Knott, J.F. Shaeffer, and M.T. Tuley. *Radar Cross Section*. Dedham, MA: Artech House, 1985, pp. 116 ff.
[8] J.D. Parsons. *Mobile Radio Propagation Channel*. 2nd Edition. Chichester, U.K.: John Wiley & Sons, 2000, pp. 24–26.
[9] F.D. Hastings, J.B. Schneider, and S.L. Broscaht, "A monte-carlo FDTD technique for rough surface scattering," *IEEE Transactions Antennas and Propagation*, vol. AP-43, no. 11, pp. 1183–1191, Nov. 1995.
[10] H.R. Anderson, "A second generation 3-D ray-tracing model using rough surface scattering," *Proceedings of the 1996 Vehicular Technology Society Conference*, Atlanta, pp. 46–50, April, 1996.
[11] J.B. Keller, "Geometric Theory of Diffraction," *Journal of the Optical Society of America*, vol. 52, pp. 116–130, 1962.

[12] R.G. Kouyoumjian and P.H. Pathak, "A uniform theory of geometric diffraction for an edge in a perfectly conducting surface," *Proceedings of the IEEE*, vol. 62, pp. 1448–1461, Nov. 1974.
[13] R.J. Luebbers, "Finite conductivity uniform GTD versus knife-edge diffraction on predicting propagation path loss," *IEEE Transactions Antennas and Propagation*, vol. AP-32, no. 1, pp. 70–76, January 1984.
[14] K.A. Remley, H.R. Anderson, and A. Weisshaar, "Improving the accuracy of ray-tracing techniques for indoor propagation modeling," *IEEE Transaction on Vehicular Technology*, vol. vt-49, no. 6, pp. 2350–2356, November, 2000.
[15] H.R. Anderson, "Building corner diffraction measurements and prediction using UTD," *IEEE Trans. Antennas and Propagation*, vol. AP-46, no. 2, pp. 292–293, Feb. 1998.
[16] W.L. Stutzman and G.L. Thiele. *Antenna Theory and Design*. New York: John Wiley & Sons, 1981, pp. 460–461.
[17] M. Abramowitz and I.A. Stegun. *Handbook of Mathematical Functions*. Washington: National Bureau of Standards, Tenth printing, December, 1972.
[18] T.S. Rappaport. *Wireless Communications – Principles & Practice*. Upper Saddle River, New Jersey: Prentice Hall PTR, 1996, pp. 122–127.
[19] E. Violette, R. Espeland, and K.C. Allen, "Millimeter-wave propagation characteristics and channel performance for urban-suburban environments," NTIA Report 88-239, December, 1988.
[20] A.F. de Toledo, A.M.D. Turkmani, and J.D. Parsons, "Estimating coverage of radio transmission into and within buildings at 900, 1800 and 2300 MHz," *IEEE Personal Communications*, pp. 40–47, April, 1998.
[21] T.S. Rappaport, "The wireless revolution," *IEEE Communications Magazine*, vol. 35, pp. 52–71, November, 1991.
[22] P.I. Wells, "The attenuation of UHF radio signals by houses," *IEEE Transactions on Vehicular Technology*, vol. VT-26, no. 4, pp. 358–362, November 1977.
[23] G. Durgin, T.S. Rappaport, and H. Xu, "Measurements and models for radio path loss and penetration loss in and around homes and trees at 5.85 GHz," *IEEE Transactions on Communications*, vol. 46, no. 11, pp. 1484–1496, November, 1998.
[24] S.A. Torrico, H.L. Bertoni, and R.H. Lang, "Modeling tree effects on pathloss in a residential environment," *IEEE Transactions on Antennas and Propagation*, vol. 46, no. 6, pp. 872–880, June, 1998.
[25] International Telecommunications Union, "Propagation data and prediction methods required for the design of terrestrial broadband millimeter radio access systems operating in a frequency range of about 20–50 GHz," Recommendation ITU-R P.1410-1, 2001.
[26] International Telecommunications Union, "Attenuation in Vegetation," Recommendation ITU-R P.833-3, 2001.
[27] G. Durgin, T.S. Rappaport, and H. Xu, "Measurements and models for radio path loss and penetration loss in and around homes and trees at 5.85 GHz," *IEEE Transactions on Communications*, vol. 46, no. 11, pp. 1484–1496, November, 1998.
[28] D.L. Jones, R.H. Espeland, and E.J. Violette, "Vegetation loss measurements at 9.6 28.8, 57.6 and 96.1 GHz through a conifer orchard in Washington state," Propagation NTIA Report 89-251. October, 1989.
[29] P.B. Papazian, D.L. Jones, and R.H. Espeland, "Wideband Propagation Measurements at 30.3 GHz through a Pecan Orchard in Texas," NTIA Report 92–287. September 1992.
[30] A. Seville, U. Yilmaz, P. Charriere, N. Powel, and K.H. Craig, "Building scatter and vegetation attenuation measurements at 38 GHz," *Proceedings of ICAP, IEE Conference Publication* 407, pp. 46–50, April, 1995.
[31] H. Xu, T.S. Rappaport, R.J. Boyle, and J.H. Schaffner, "38 GHz wideband point-to-multipoint radio wave propagation study for a campus environment," *Proceedings of the 1999 Vehicular Technology Society Conference*, Houston, CD-ROM, May, 1999.
[32] International Telecommunications Union, "The radio refractivity index: its formula and refractivity data," Recommendation ITU-R P.453-8, 2001.

[33] International Telecommunications Union, "Definition of terms relating to propagation in non-ionized media," Recommendation ITU-R P.310-9, 1994.
[34] International Telecommunications Union, "The radio refractivity index: its formula and refractivity data," Recommendation ITU-R P.453-8, 2001.
[35] C.A. Samson, "Refractivity Gradients in the Northern Hemisphere," U.S. Department of Commerce, Office of Telecommunications, April, 1975, Access number AS-A009503.
[36] International Telecommunications Union, "Attenuation by atmospheric gases," Recommendation ITU-R P.676.5, 2001.
[37] R.K. Crane. *Electromagnetic Wave Propagation Through Rain*. New York: John Wiley & Sons, 1996.
[38] International Telecommunications Union, "Specific attenuation model for rain for use in prediction methods," Recommendation ITU-R P.383-1, 1999.
[39] E.L. Woodbridge and W.S. Hartley, "Weather, climate and FSO link availability," *8th Annual WCA Technical Symposium*, San Jose, CA. CD-ROM, January, 2002.
[40] H. Willebrand and B.S. Ghuman. *Free-space Optics: Enabling Optical Connectivity in Today's Networks*. Indianapolis: SAMS Publishing, 2002.

3

Propagation and channel models

3.1 INTRODUCTION

Propagation models are fundamental tools for designing any fixed broadband wireless communication system. A propagation model basically predicts what will happen to the transmitted signal while in transit to the receiver. In general, the signal is weakened and distorted in particular ways and the receiver must be able to accommodate the changes, if the transmitted information is to be successfully delivered to the recipient. The design of the transmitting and receiving equipment and the type of communication service that is being provided will be affected by these signal impairments and distortions. The role of propagation modeling is to predict the system performance with these distortions and to determine whether it will successfully meet its performance goals and service objectives. If the performance is inadequate, the system design can be modified accordingly before the system is built.

Traditionally, 'propagation models' is the term applied to those algorithms and methods used to predict the median signal level at the receiver. Early communication systems were narrowband systems in which median signal level prediction along with some description of signal level variability (fading) statistics were the only models needed to adequately predict system performance. Modern communication systems achieve higher capacity (higher data rates) by using a wider band of frequencies. For such systems, narrowband prediction of signal levels and fading alone does not provide enough information to predict system performance. As a consequence, the concept of a propagation model has been broadened to include models of the entire transfer function of the channel. These models are intended to represent all the modifications the transmitted signal undergoes in traveling from the transmitter to the receiver. Such models include signal level information, signal time dispersion information and, in the case of mobile systems, models of Doppler shift distortions arising from the motion of the mobile. Appropriately, such models that provide this additional information are called *channel models*.

Fixed Broadband Wireless System Design Harry R. Anderson
 ISBN: 0-470-84438-8

Propagation and channel modeling is a very pragmatic endeavor. A model is developed so that it adequately provides the information necessary for the system performance prediction task at hand. A model is chosen by a system designer to be appropriate to the design problem being addressed. For example, for the preliminary step of dimensioning an Local Multipoint Distribution Service (LMDS) system at 28 GHz, a simple model that predicts the service radius of a hub is all that is required to estimate the number of hubs needed to cover the intended service area. However, when detailed system design is undertaken, a comprehensive point-to-point model, which can determine whether a path is line-of-sight, is needed. Such models make use of detailed terrain and building databases and the best available methods for predicting the availability of the links with multipath and rain fading conditions.

System designers sometimes make inappropriate propagation model choices. In the early days of cellular system deployment, the Hata–Okumura model was very widely used for predicting the coverage of cell sites. Unfortunately, this model was developed in relatively flat areas so it did not explicitly take into account the mountains or tall structures that can create low signal levels, a process commonly called *shadowing*. Because the Hata–Okumura model could not predict shadowing, it failed rather badly in predicting system performance in hilly or mountainous areas like the San Francisco Bay area, or cities with large office buildings like New York. The result was poor coverage, impaired system performance, and dissatisfied customers. While much more advanced and sophisticated models are now available, to some extent this type of problem still afflicts modern cellular system design. Choosing and applying the appropriate propagation model is an important aspect of wireless system design.

3.1.1 Model classifications

Given the significance of propagation and channel models in designing and building successful communication systems, a considerable amount of effort has been devoted by the industry to developing such models. In this chapter, propagation and channel models will be divided into three basic classifications:

- Theoretical
- empirical
- physical

These model classifications will be discussed in more detail later with some specific examples. Theoretical models have relatively little application to fixed broadband wireless systems except as they may be applied to predict rain attenuation, as discussed in Section 2.12. Empirical models are based on measurement or observations of signal performance in real propagation environments. Their use for dimensioning non-line-of-sight (NLOS) point-to-multipoint systems is becoming more widespread. Physical models make use of the physical mechanisms of electromagnetic (EM) wave propagation, as discussed in Chapter 2, to predict signal attenuation and channel response. Such models are the most widely used for fixed broadband wireless systems. Physical models are sometimes described as *deterministic* models; a term borrowed from probability analysis that is more appropriately applied to channel modeling in which the channel response characteristics may be divided between deterministic and random processes.

3.1.2 Fading models

Models of multipath and other fading are often appropriately included with a discussion of wireless propagation and channel modeling because the fading characteristics of the channel have a very significant impact on the system performance. For fixed broadband systems, the design objective for communications reliability can be 99.9% or higher – much higher than the typical design objective for cellular or other mobile radio communications. Accordingly, specialized fading models have been developed for multipath and rain fades for such systems. Given the importance of these models, a separate chapter (Chapter 4) has been devoted to the discussion of these specialized fading models as well as standard mobile radio fading models that are best applied to the NLOS system design.

3.2 THEORETICAL, EMPIRICAL, AND PHYSICAL MODELS

Modeling the radio channel is fundamental to predicting the performance of a wireless radio system. Consequently, there have been many researchers over the past several decades who have created channel models of one kind or another. The purpose of this section is to briefly review representative examples of this work as a means of creating a context for the propagation models that are used for fixed broadband link design.

Channel models are often classified as *narrowband* or *wideband*. A narrowband model is one based on the predicted mean signal level and some assumptions about the signal envelope fading statistics around this predicted mean level. The assumption is that because of the limited bandwidth of the signal being sent through the channel, more detailed information is not required to provide a useful prediction of the channel effect on the signal. Because the signal bandwidth is narrow, the fading mechanism will affect all frequencies in the signal passband equally. For this reason, a narrowband channel is often referred to as a *flat-fading* channel.

For wideband channel models, the assumption is that the bandwidth of the signal sent through the channel is such that time dispersion information, in addition to mean signal level information, is required. Time dispersion causes the signal fading to vary as a function of frequency, so wideband channels are often called *frequency-selective fading* channels.

It is clear that the usual distinction between narrowband and wideband channels depends both on the signal bandwidth and the nature of the propagation environment. It would be more useful to describe a channel based solely on the nature of the propagation environment and not on the signal bandwidth. A proper channel model definition should be insensitive to the signal transmission since the physical reality of how radio energy gets from the transmitter to the receiver has nothing to do with modulation type, data rate, or occupied signal bandwidth. With a physical channel model based on EM propagation mechanisms as described in Chapter 2, the distinction between narrowband and

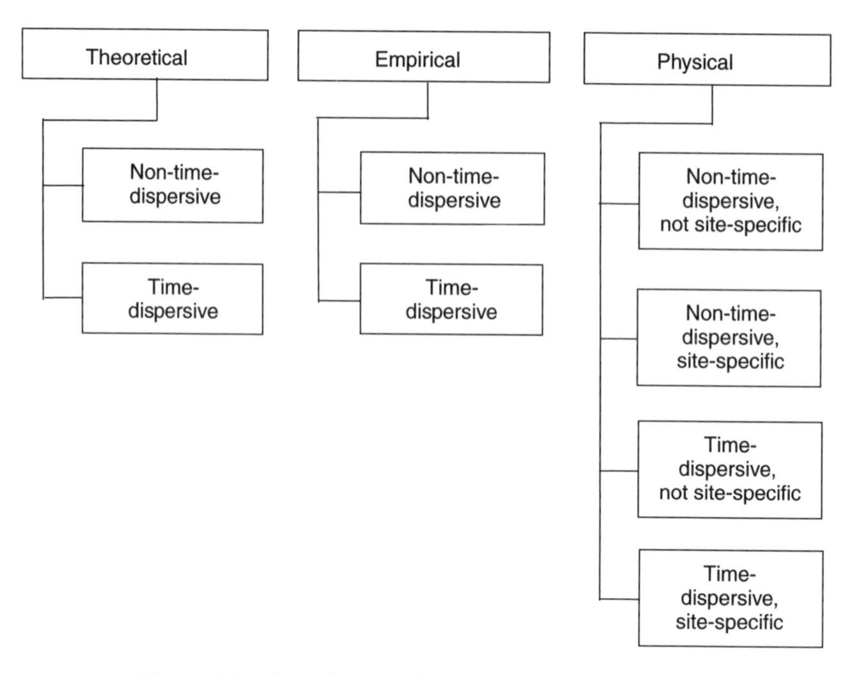

Figure 3.1 Classification of propagation/channel models.

wideband is unnecessary. Even so, the shorthand notation of narrowband and wideband is convenient and will occasionally be used in this book.

Instead of narrowband and wideband, previous channel modeling work will be discussed here according to the classifications shown in Figure 3.1. This outline defines channel models in terms of how they work and the information they provide, rather than the bandwidth of the signal that can successfully be used with them. The three main model categories are theoretical, empirical, and physical, with non-time-dispersive and time-dispersive models as the primary subcategories. A time-dispersive model is one designed to provide information about the time delay experienced by a transmitted signal and its multipath replicas in reaching the receiver. Each of the categories in Figure 3.1 is briefly summarized along with citations of representative published examples of the models.

3.2.1 Theoretical channel models

Models in this category are based on some theoretical assumptions about the propagation environment. They do not directly use information about any specific environment, although the assumptions may be based on measurement data or physical laws. Theoretical models are useful for analytical studies of the behavior of communication systems under a wide variety of channel response circumstances, but because they do not deal with any specific propagation information, they are not suitable for planning communication systems to serve a particular area. With this objective, they usually rely on assumptions that lead to mathematically tractable formulations.

3.2.1.1 Theoretical, non-time-dispersive

The classic channel models by Bello [1], Turin [2], and Clarke [3] are examples of theoretical models. Of these, the Clarke model is essentially not time-dispersive. It is based on the assumption that the signal arriving at the receiver consists of a collection of scattered waves in the horizontal plane that are uniformly distributed in angle and independent. From this, first and second-order statistics of a fading envelope can be developed along with characteristics of the channel spectrum. Because it only considers envelope-fading statistics, models of this type are classified here as not time-dispersive.

3.2.1.2 Theoretical, time-dispersive

Examples of theoretical time-dispersive models are those of Bello [1] and Turin [2] and those by two of Turin's coworkers – Suzuki [4] and Hashemi [5]. They both used measurement data to infer the characteristics of theoretical models of channel response, usually by seeking common descriptions for the fading behavior in time bins of arriving energy. A similar example is represented by the work of Saleh and Valenzuela [6] who fitted Poisson distributions to arrival times and amplitudes based on signals they measured for indoor environments.

Another theoretical, time-dispersive model is the so-called *tapped delay line* model in which densely spaced tap delays and multiplying constants and tap-to-tap correlation coefficients are determined on the basis of measurements or some other theoretical interpretation of how the propagation environment affects the signal. Examples of the tapped delay approach can be found in [7,8]. The tapped delay line channel model is described in more detail in Section 3.3.1.2.

Again, the distinction here is that theoretical models are not intended to be applied to any real propagation circumstance. As such, there is no way to relate the parameters of these models to physical parameters of any particular propagation environment.

3.2.2 Empirical channel models

In the VHF/UHF (Very High Frequency/Ultra High Frequency) frequency bands, a classic example of an empirical propagation model is the Federal Communications Commission (FCC) model [9]. The FCC model is actually composed of equations that were fitted to a set of signal strength measurements done at several locations in the United States. The propagation model itself is represented as a set of curves for different frequency bands showing field strength versus distance for a range of height above average terrain(HAAT) values. The ITU-R have similar curves based on HAAT, which are set forth in Rec. 370-5 [10]. Both methods also provide for corrections to take into account 'terrain roughness' or Δh, the 10% to 90% interdecile terrain variation over the path. No element of these models makes use of electromagnetic principles to assist in refining the prediction. As such, the FCC and ITU-R models are classic examples of purely empirical models.

Empirical models use what are known as *predictors* or *specifiers* in general statistical modeling theory. Predictors are parameters, which have been found through statistical analysis, to bear a relationship to (are correlated with) the quantity that is to be predicted. In econometric models, the objective may be to predict gross national product (GNP). In doing so the model may use values for unemployment, disposable income, balance of trade, and so on as predictors. All of these factors may have been found to be correlated with GNP, but none of them directly *causes* GNP to go up or down. There are other mechanisms at work. The classic textbook axiom is *"Correlation does not prove causality."*

In the case of the FCC model, through statistical analysis a correlation was found between antenna HAAT and signal strength. But this was only a correlation, not a causal relationship. Indeed, one could not conceive of an electromagnetic propagation mechanism in which the simple average elevation value directly changes the magnitude of an electric or magnetic field. The consequence of this approach is easily understood by considering, for example, two terrain profiles along a 25 km path separating the transmitter and receiver where the terrain from 3 to 15 km is the same but beyond 15 km, one profile has a hill. Behind the hill, the signal will be greatly weakened, yet the empirical model based on 3 to 15 km average terrain will predict the same signal level for both paths.

A similar example could be constructed for Δh in which a valley and a mountain along two paths have the same interdecile elevation variation, yet the field strength at the receiver on the path with the mountain will be much lower than on the path with the valley. The inability to explicitly account for particular features of the propagation environment is perhaps the greatest limitation of empirical, measurement-based models.

Propagation models developed by communication engineers often lack basic statistical significance testing. For example, field measurements of signal levels for two different system configurations may be made and the results separately analyzed by some curve-fitting technique using ordinary least squared error (OLSE) minimization. The curve from one set of data may be 2 dB different from the other curve, and a claim of such a difference is offered as a conclusion. However, given the variance of the measurement data, a standard test for statistical significance (such as the Student T test) would show the difference between the two is not statistically significant. In statistical terms, the hypothesis that the two system configurations produce different results must be rejected. This notion can be extended to an empirical model such as the FCC model. The FCC model will surely predict a higher field strength for a transmit antenna HAAT value of 250 m when compared to a HAAT of 225 m, but is such a prediction difference justified given the variance in the original measurement data? The lack of proper statistical significance testing in propagation measurement analysis, and the failure to include such information in the resulting propagation model description, is a weakness of many empirical propagation models that have been put forward for communications engineering.

The accuracy and usefulness of such empirical models also depend on the environment in which the original data for the model were taken and how universally applicable that environment is. A common problem is trying to use empirical models in areas where the propagation environment is widely different from the environment in which the data were gathered. An example is the Hata model [11], based on the work of Okumura [12], in which propagation path loss is defined for urban, suburban, and open environments. These correction factors in Okumura's work are an effort to mitigate their limitations, but unless the characteristics of 'urban', 'suburban', and 'open are reasonably similar to those in Japan where the measurement data were taken, these finer-grained classifications may not be of much use. In spite of their limitations, empirical models such as the FCC, ITU-R, and Hata models are still widely used because they are simple and allow rapid computer calculation. They also have a certain 'comfort' factor in that people using them in certain circumstances over time have come to know what to expect and make their own localized 'corrections' to the predicted values provided by the model.

As pointed out above, new communication systems require a wider variety of information from propagation models to predict performance than just EM field strength. With empirical models, each new category of information represents another set of measurements that have to be done. For example, real microwave systems (RMS) delay spread (defined later in this chapter) has recently become a routinely used factor for predicting the performance of wideband digital communication systems. For an empirical model to be useful for such systems, another set of measurement data would have to be acquired and appropriate statistical analysis done to determine statistically significant predictors of RMS delay spread. All the same limitations of empirical modeling pointed out above would still apply, but when field strength and RMS delay-spread predictions are both considered as separate dimensions in the prediction problem, the inadequacies of the empirical approach multiply. This problem is increasingly aggravated as other information types such as signal fading statistics are added, and the attraction of the empirical modeling approach diminishes.

Empirical models fundamentally use experimental measurement data to deduce a relationship between the propagation circumstances and expected field strength or time dispersion results. Because they use statistical specifiers that have no direct physical relationship to the quantity being predicted, they are inherently noncausal. They are also inherently not site-specific since they do not explicitly take into account the unique features of a given propagation environment along a path from a transmitter to a receiver.

3.2.2.1 Empirical, non-time-dispersive

For the non-time-dispersive case, there are a large number of models that have been developed over the years. The FCC [9], ITU-R [10], and Hata [11] models mentioned earlier are examples of empirical propagation models. There are several others, including Egli [13], Edward–Durkin [14], Blomquist–Ladell [15], and Allsebrook–Parsons [16]. All of these models use some measurement sets and a statistical analysis of the data to construct a curve through the data. The objective is to simply predict mean path loss as a function of antenna heights, height above average terrain, terrain roughness, or other parameters such as local clutter (foliage and buildings). The Allsebrook–Parsons model is interesting in that it attempts to predict path loss in urban areas using the orientation and widths of streets. Good summary references for the algorithms used in such irregular terrain models and their relative prediction success can be found in [17,18].

3.2.2.2 Empirical, time-dispersive

Empirical models that provide time dispersion information are rare. Like predicting mean path loss based on a measurement set, a corresponding time dispersion model would use measured time delay profiles from which statistically derived curves, or simple look-up tables, yield a way of predicting time dispersion in some given environment based on its characteristics.

Models in this classification were developed by the Institute of Electrical and Electronic Engineers (IEEE) 802.16 Working Group, for use in planning fixed broadband systems in the 2 to 11 GHz frequency range [19]. Also known as the Stanford University Interim (SUI) models, they are discussed in detail in Section 3.4.1.

3.2.3 Physical channel models

As pointed out earlier, physical models rely on the basic principles of physics rather than statistical outcomes from experiments to find the EM field at a point. Physical models are causal by design. Depending on whether they consider the particular elements of the propagation environment between a transmitter and receiver, physical models may or may not be site-specific. Also, they may or may not provide time dispersion information. The various categories of physical models as shown in Figure 3.1 are discussed below.

One aspect that affects the capabilities and success of a physical model is the kind of information about the propagation environment it can use and what it does with it. This is an important point about physical propagation modeling. The quality of the model's predictions is a direct consequence of how the model maps the *real* propagation environment into the *model* propagation environment. For a channel model to be a physical, site-specific model, it not only must use the physical laws of EM wave propagation but it must also have a systematic technique for mapping the real propagation environment into the model propagation environment.

3.2.3.1 Physical, non-time-dispersive, not site-specific

A physical, not site-specific model uses physical principles of EM wave propagation to predict signal levels in a generic environment in order to develop some simple relationships between the characteristics of that environment and propagation. An apt example of this approach is the Walfisch–Ikegami model for mobile radio systems in urban areas where roof edges are considered as a series of diffracting screens with a final edge diffraction from building roof to the street level being included. This composite model is based on work published in [20,21]. The diffracting screen model of clutter attenuation is described in detail in Section 3.5.6.1.

3.2.3.2 Physical, non-time-dispersive, site-specific

A physical, not time-dispersive model is one in which the EM field at the receiver is predicted using physical laws governing wave propagation, but no signal time delay information is available from the model. Models in this category include a collection of propagation algorithms to predict signal attenuation over terrain. An early example is by Bullington [22] who actually prepared nomographs to make his model more directly useful to practicing engineers. Other examples of models are the Longley–Rice [23], TIREM [24], Free Space+RMD[25], and the Anderson 2D [26]. Most of these models differ in how they interpret physical data describing the propagation path. In particular, the issue of how best to treat multiple terrain obstacles has spawned particular algorithms for this purpose. Examples are the Epstein–Peterson model [27], which is used in TIREM, Free Space+RMD, and the Anderson 2D models. Other approaches to calculating composite signal attenuation over multiple obstacles include those by Deygout [28] and Vogler [29]. The rigorous method of Vogler is computationally demanding but the most accurate of these approaches.

These physical models listed here are specifically applicable to propagation prediction in fixed broadband wireless systems and as such they will be discussed in some detail later in this chapter.

3.2.3.3 *Physical, time-dispersive, site-specific*

Examples of site-specific, time-dispersive models are found in [30,31]. Physical laws are used along with one-to-one mappings from the real environment to the model environment. These models are basically known as *ray-tracing*, a high-frequency approximation method that tracks the trajectory of EM waves leaving the transmitter as they interact with objects in the propagation environment. Since a ray-tracing model tracks ray trajectories, it not only provides time dispersion information but also angle-of-arrival information that is of great interest in assessing the operation of adaptive or 'smart' antennas.

Since the early pioneering efforts with ray-tracing models, the technique has grown in acceptance; several papers and articles have been published suggesting enhancements to the fundamental concept that improve its accuracy and computation speed. Ray-tracing will be discussed in Section 3.5.7.

Interestingly, perhaps the earliest example of ray-tracing models comes from broadcasting rather than from the mobile radio. The NTSC television signal used in the United States and elsewhere employs a 6 MHz bandwidth to transmit a vestigial sideband picture signal and an frequency modulation (FM) sound signal. Starting in the late 1940s, it is the first example of an areawide linear modulation system in which channel time dispersion was important to understand and control. Of course, channel time dispersion in analog television manifests itself as ghosting – replicated transmitted images that appear on the viewer's screens. While mobile radio was focused on narrowband FM channel models and associated Rayleigh fading, broadcast engineers throughout the 1960s and 1970s were using simple ray techniques to model the amplitude and delay of time dispersion from mountains, buildings, and other features of the propagation environment. Some of the early published examples using ray techniques for such time dispersion prediction of broadcast signals are by Ikegami and his coworkers [32,33].

3.3 GENERIC WIDEBAND CHANNEL MODEL

As mentioned above, channel models are usually divided into narrowband and wideband categories, wideband generally indicating those channels in which time dispersion and frequency-selective fading have a significant impact on the signal being transmitted. From this definition, defining a wideband channel depends on the propagation environment and the signal. In this book all channels will be considered wideband; that is, the generic channel model will comprehensively include all those elements that describe how the physical propagation channel responds to a transmitted signal regardless of what that signal is. For signals whose occupied bandwidth is narrow compared to the correlation bandwidth of the channel, the generic wideband model simplifies to a narrowband model in which many elements of the wideband model may be discarded.

Although this book is focused on fixed wireless systems, the generic wideband model presented here includes terms that take into account motion in the channel. This could be motion of the transmitter or receiver terminals, or motion of some other element of the propagation environment that causes time-variability in reflected waves arriving at the receiver.

The generic wideband model is also a site-specific channel model; that is, the channel response the signal experiences, while in transit, from a specific transmitter terminal location or site to a specific receiver terminal location or site. However, a site-specific model can still be applied to the prediction of wireless system performance over a wide area, such as the coverage area of an multipoint, multi-channel distribution service (MMDS) or LMDS system hub. A site-specific model breaks down the modeling problem into point-to-point elements which, when taken as an ensemble, provide a much more detailed understanding of the areawide results. This will then lead to transmitter and receiver design and overall system design, which are more finely optimized for real system deployment circumstances.

The concept of the channel model is shown schematically in Figure 3.2. The propagation of EM energy from the transmit antenna to the receive antenna is the radio channel to be modeled. Under ideal circumstances, the channel would have no effect on the signal, that is, the signal electric field at the receive antenna terminal, E_r, would be the same as the signal at the transmitter. This can be written as

$$E_r = AE_t \exp(-j\omega t + \theta) \tag{3.1}$$

where E_r is the (complex) electric field voltage or magnetic field current at the receive antenna, E_t is the magnitude of the transmitted signal (voltage or current), and ω is the carrier frequency in radians. For simplicity only the electric field will be represented in the following equations with the understanding that there is an associated magnetic field. The multiplicative factor A is the propagation attenuation while θ is some phase delay or phase shift introduced by the channel.

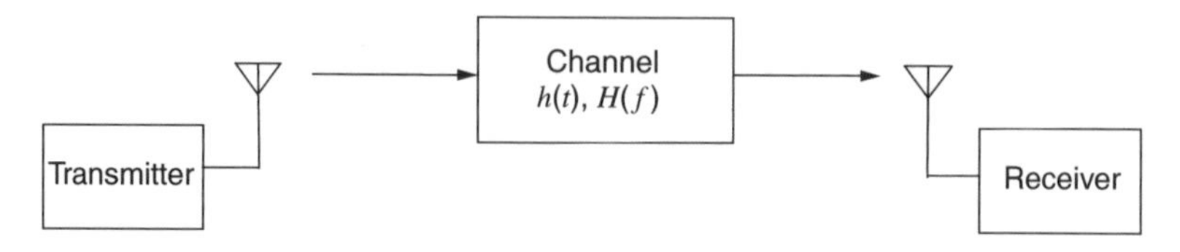

Figure 3.2 Schematic diagram of generic wideband channel model.

If the channel is now considered as a filter with some low-pass impulse response, then that impulse response would be given by

$$h(t) = A\delta(t - \tau)\exp(-j\theta) \tag{3.2}$$

A sinewave signal at frequency ω leaving the transmitting antenna would arrive at the receiver reduced in amplitude by factor A, shifted in phase by θ, and delayed by τ seconds. Such a model of the transmission channel is applicable for free-space propagation conditions in which signal energy arrives at the receiver directly (via one path) from the transmitter.

If the channel consisted of two transmission paths for the transmitted energy to arrive at the receiver (for example, with the addition of a single ground reflection), the channel impulse response would be the sum of the effect of the two paths:

$$h(t) = A_1\delta(t - \tau_1)\exp(-j\theta_1) + A_2\delta(t - \tau_2)\exp(-j\theta_2) \tag{3.3}$$

This is the impulse response of the so-called 'two-ray' channel model. Generalizing (3.3) to N possible transmission paths, $h(t)$ becomes

$$h(t) = \sum_{n=1}^{N} A_n\delta(t - \tau_n)\exp(-j\theta_n) \tag{3.4}$$

This is the channel impulse response to a particular point $p_2(x_2, y_2, z_2)$ from a transmitter located at point $p_1(x_1, y_1, z_1)$. The more general impulse response can thus be written as

$$h(t, p_1, p_2) = \sum_{n=1}^{N(p_1,p_2)} A_n(p_1, p_2)\delta[t - \tau_n(p_1, p_2)]\exp[-j\theta_n(p_1, p_2)] \tag{3.5}$$

where the number, amplitude, phase, and time delay of the components of the summation are a function of the location of the transmit and receive antenna points in the propagation space.

The channel impulse response given by (3.5) is for a single static point in space. For mobile communication, the receiver is often moving and that motion can affect the phase relationship of the components of (3.5) in a way that may be important to data symbols being transmitted over the channel. This motion will result in a frequency or Doppler shift of the received signal, which will be a function of speed and direction of motion, and the angle of arrival (AOA) of the signal energy. Equation (3.5) can be modified to include the Doppler shift and thus account for this motion as follows:

$$h(t, p_1, p_2) = \sum_{n=1}^{N(p_1,p_2)} A_n(p_1, p_2)\delta[t - \tau_n(p_1, p_2)]\exp[-j\theta_n(p_1, p_2) + \Delta\theta_n(p_1, p_2)] \tag{3.6}$$

where $\Delta\theta_n$ is a phase displacement due to the motion. It should be kept in mind that this is motion of either the transmitter or receiver *relative to every other element in the propagation environment.* The transmitter or receiver may be fixed, but a signal from a moving reflection source (such as a moving bus) may result in a nonzero $\Delta\theta_n$ for a particular component of (3.6). For a mobile receiver, $\Delta\theta_n = (2\pi vt/\lambda)\cos(\varphi_n - \varphi_v)$ where φ_n is the arrival angle of the nth component, v is the speed of motion, φ_v is the direction of motion, and λ is the wavelength.

By an inspection of (3.6), it is clear that taking the Fourier transform of the output of a channel described by this equation, where the input to the channel is a sinewave, for example, will result in a frequency or Doppler shift of that sinewave. That Doppler shift will be a function of $\Delta\theta_n$.

In general, the amplitudes A_n and phase shifts θ_n, will be functions of the carrier frequency ω because they are controlled by the interaction of the transmitted energy with impedance boundaries and other features of the propagation environment. Inserting the frequency dependence into (3.5) gives

$$h(t, p_1, p_2, \omega) = \sum_{n=1}^{N(p_1,p_2)} A_n(p_1, p_2, \omega)\delta[t - \tau_n(p_1, p_2)]\exp[-j\theta_n(p_1, p_2, \omega)] \qquad (3.7)$$

Modeling the channel response of a propagation link as a sum of discrete impulses, as shown in (3.4), was proposed by Turin [2] after analyzing mobile measurements of channel response in an urban area of San Francisco. Turin intended the impulses to represent a multitude of generally unspecified reflections from buildings and other sources in the propagation environment. A drawing of how this might be depicted for a real environment is shown in Figure 3.3. Equations (3.4 and 3.5) can actually be interpreted more broadly. As $N \to \infty$, the values of A_n, τ_n and θ_n could be chosen to represent any arbitrary filter function with an arbitrary resulting signal voltage at the receiver. As such, this model could represent *any* EM propagation transfer function, not just those based on Turin's original physical rationalization.

The concept of physical channel modeling seeks to define the overall propagation mechanism as an ensemble of known physical elemental constructs which, when taken together, can explain and model the statistical behavior of signal levels and time-dispersive

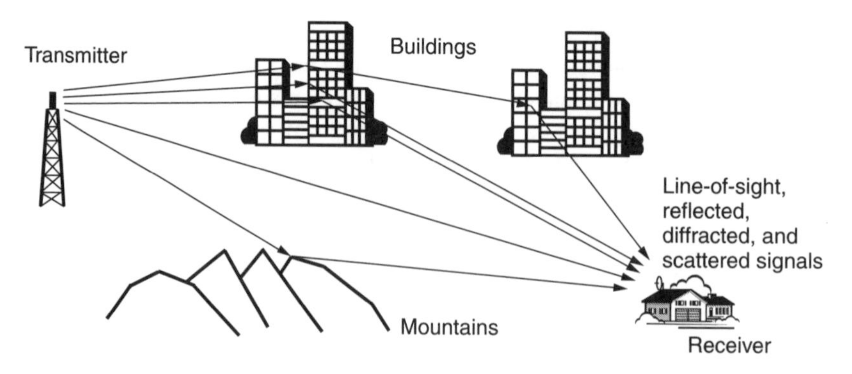

Figure 3.3 Multipath propagation environment.

characteristics commonly observed for signals measured in the field. When considered with physical models, these elemental propagation constructs are called *propagation primitives* here and include fundamental EM field-boundary interactions such as free space (no boundary), reflection, edge diffraction, diffuse scattering, and boundary (wall) transmission. These propagation mechanisms were explained in Chapter 2. The important point is that by starting with purely physical EM field propagation laws, it is possible to construct site-specific descriptions of both the *stochastic* nature of signal level variations, and the fading-dependent time-dispersive (frequency-selective) properties of the channel.

The notion of a 'site' is a heuristic concept of a local region where propagation conditions are uniform and where a system user (a desktop wireless modem, for example) might occupy. The notion of regions for channel modeling purposes has been previously addressed by Bello [1] who proposed a two-stage model in which the first stage is a small area defined as the region where the mean signal level is constant and prominent features of the propagation environment remain unchanged. In this way the statistics of the signal in this small-scale region can be considered wide-sense stationary (WSS). The second stage of the Bello model is constructed by considering the small-scale statistics over larger areas. The notion of a site as used here coincides with the physical region in space where Bello's small-scale definition applies. A site is thus the region where the signal level as a random process is WSS.

Normally a propagation model will predict EM fields at a dimensionless point in space. Knowing fields with this precision is not useful here because of variations in the locations of objects in the propagation environments and changing atmospheric conditions. As a practical matter, a site could be thought of as a circle (2D) or sphere (3D) with perhaps a 1-m radius. A site can also be interpreted as a given wavelength displacement around a dimensionless point in space where an EM field propagation study has been done. The variations of the channel characteristics (impulse response) within the site because of the vector addition of the components of (3.6) over the wavelength displacement range, can then be explored to yield sampling descriptions of the channel response at the site. The definition of a site as that region where the signal is wide-sense stationary is automatically satisfied by this approach since the site is defined by a single fixed collection of signals A_n given by the summation in (3.6), which has only one mean value and one variance.

3.3.1 Wideband channel response

From (3.5), the impulse response of the channel can be considered as a series of impulses of different amplitudes A_n occurring at times τ_n relative to some arbitrary starting time $\tau = 0$. Figure 3.4 shows a graph of what the impulse response for a point in a typical urban environment might look like as calculated using a physical ray-tracing propagation model [31]. This impulse response is also called a power delay profile. A typical measured power delay profile taken from [34] is shown in Figure 3.5. Clearly it is not possible to represent ideal delta functions as in Figure 3.4. The bandwidth of any real filter cannot be infinitely broad and thus an ideal impulse response is averaged by the practical limitations of the response of the measurement equipment used for the profile in Figure 3.5.

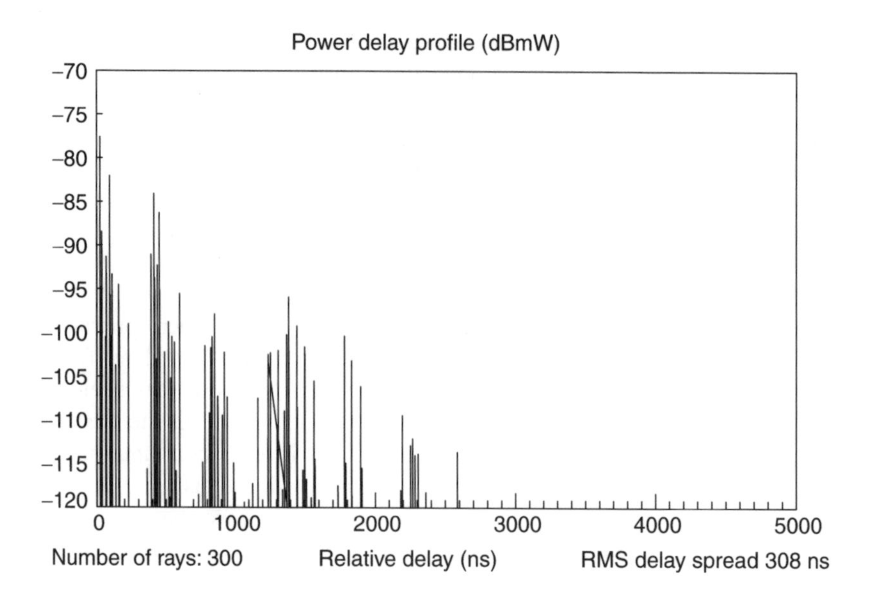

Figure 3.4 Channel response $h(t, \tau)$ (the power delay profile) predicted using a ray-tracing propagation model.

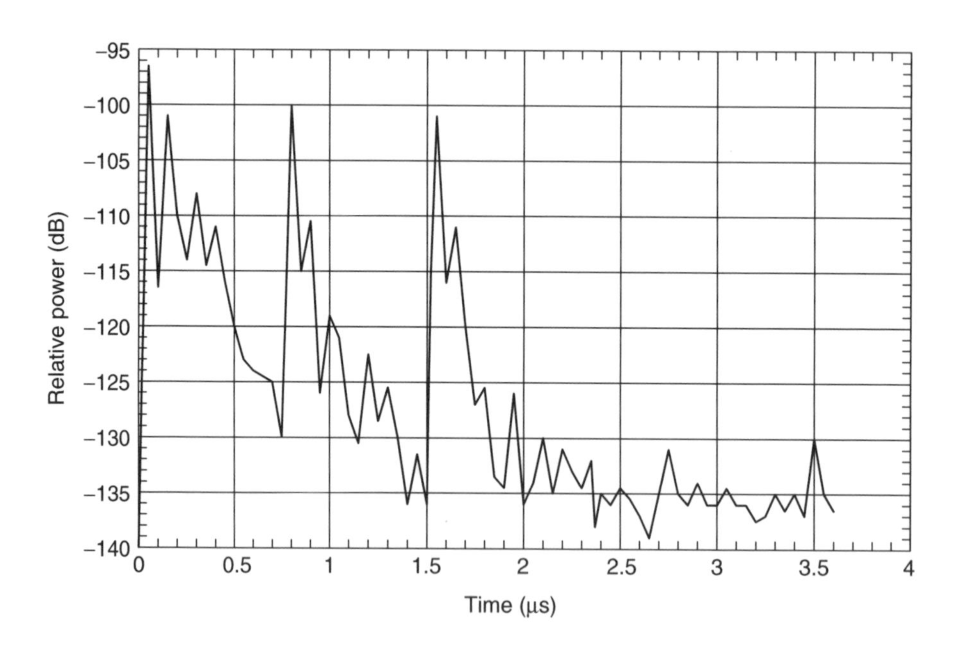

Figure 3.5 Measured power delay profile. (Data from [34].)

A common measure of the amount of time dispersion in a channel is the RMS delay spread σ_τ. The RMS delay spread can be found from the power delay profile as follows:

$$\sigma_\tau = \left[\sum_{n=1}^{N} (\tau_n - \overline{\tau})^2 p(\tau_n) \right]^{\frac{1}{2}} \tag{3.8}$$

where the mean value of the power delay profile is

$$\overline{\tau} = \sum_{n=1}^{N} (\tau_n) p(\tau_n) \tag{3.9}$$

and

$$p(\tau_n) = \frac{A_n^2}{\sum_{n=1}^{N} A_n^2} \tag{3.10}$$

The amplitudes and time delays are those found in (3.5), for example, if the amplitudes of the individual arriving waves are known or predicted. They can also be derived from measurements if time windows or bins are created to segment the data for use in the summation. This approach is the basis for the tapped delay channel model discussed in Section 3.3.1.2.

The appeal of the RMS delay spread is that it is a single number that is indicative of amount of time dispersion in the channel. However, as a statistical result it averages out much of the important details about the specific distribution of the multipath delayed power. RMS delay spread is also not unique for a given channel. Clearly, a channel with a low amplitude wave arriving at the receiver with a long delay could have the same RMS delay spread as a channel with a much higher amplitude wave arriving at a shorter delay time. Even though the RMS delay spreads in these two channels are the same, the behavior of the channels in terms of introducing intersymbol interference (ISI) is quite different. This is quantitatively demonstrated in [35].

3.3.1.1 Time–variant and static channels

The impulse response in (3.5) is clearly a function of time and of the delay τ. If a very narrow signal pulse is transmitted into the channel described by (3.5), the output of the channel at the receiver will be a series of pulses arriving at delay times τ_n with amplitudes A_n, and carrier phase shifts θ_n. If the transmitter and receiver are fixed in place, and nothing in the propagation environment moves or changes (not even the atmosphere), then the impulse response in (3.5) will be static (although still a function of time). The output of the channel $y(t)$ at the receiver is still a function of time but will be the same regardless of when the pulse is transmitted. This is a static channel. It could be assumed that because fixed systems are being modeled, this assumption could be applied here. For fixed broadband system types, this assumption can be made, or at least it can be assumed that the channel is quasi-static – that a given set of conditions will exist for relatively long periods of time compared to the length of a particular message transmission. Long-term ducting conditions are an example of such an aberrant refractivity condition that can exist for long periods of time.

In the general case, however, it is necessary to assume the propagation environment will change with time and hence a time-variant channel model is needed. For mobile communication systems, the terminals are inherently assumed to be in motion so no other approach is really appropriate.

To deal with a time-variant channel, Bello [1] defined the *input delay-spread function* $h(t, \tau)$. This is the low-pass response of the channel at some time t to a unit impulse function input at some previous time τ seconds earlier. Note that this delay τ is different from the discrete $\tau_1, \tau_2 \ldots \tau_n$ arrival delay times in (3.5) that define when the signal waves reach the receiver.

The output of the channel $y(t)$ can then be found by the convolution of the input signal $u_m(t)$ with $h(t, \tau)$ integrated over the delay variable τ

$$y(t) = \int_{-\infty}^{\infty} u_m(t - \tau)\, h(t, \tau)\, d\tau \tag{3.11}$$

which can also be written as

$$y(t) = u_m(t)^* h(t, \tau) \tag{3.12}$$

A time-variant channel can vary for many reasons and movement of the transmitting or receiving terminals is the most obvious. Changing atmospheric refractivity is another. For mobile systems, a simple way to simulate the time variations is to move the mobile along a short distance of several wavelengths and calculate the changes in the channel output as given by (3.5).

In a practical communication system, the bandwidth of the system is designed to accommodate the transmitted signal without significantly attenuating its power so as to maximize the detected signal-to-noise ratio. The time averaging effect apparent in Figure 3.5 will thus depend on the designed system bandwidth rather than the bandwidth of the measuring equipment. The channel output $y(t)$ for a transmitted signal is represented by low-pass function $u_m(t)$ into the channel with site-specific response $h(t, \tau)$.

For digital signals, the input $u_m(t)$ will often be in the form of a string of pulses or channel symbols, each symbol representing some portion of the information data bits being communicated. In the case of spread spectrum signals, several transmitted channel symbols may be used to represent one bit of information.

Using an impulse response from a ray-tracing propagation model, and convolving it with a single raised cosine symbol pulse, the result is a *time signature* of the channel as depicted in Figure 3.6. This figure shows several time signatures as a function of wavelength position around an analysis (field) point. The variations in signatures are due to the relative phase changes in the terms of (3.5), not changing amplitudes. These variations show that the channel is a *dynamic* time-dispersive channel at this particular site as defined earlier.

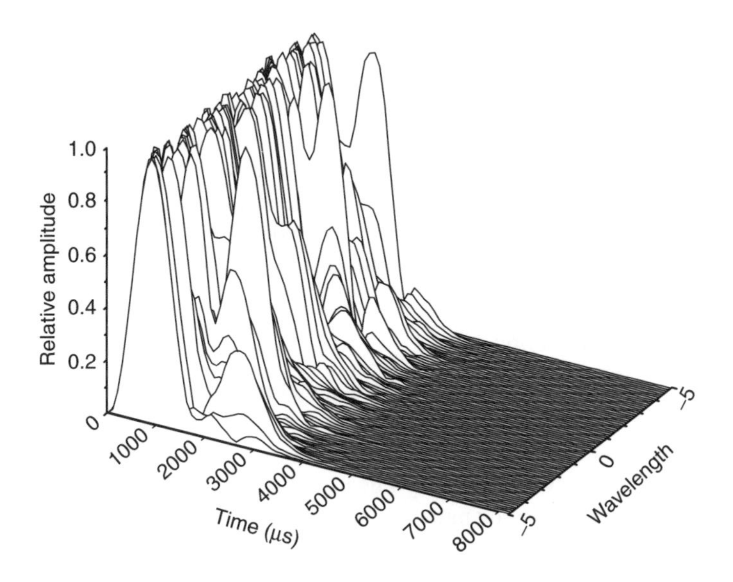

Figure 3.6 Time signature of channel response.

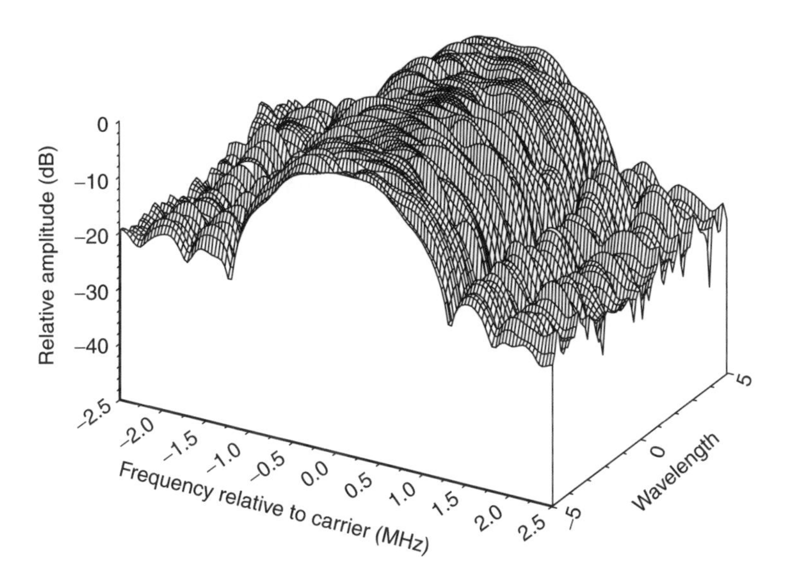

Figure 3.7 Spectrum signature of channel response.

By taking the Fourier Transform of the time signatures in Figure 3.6, it is possible to created *spectrum signatures* as shown in Figure 3.7. The nonuniform frequency response shown in Figure 3.7 (not the ideal spectrum of an undistorted raised cosine pulse) represents the frequency-selective nature of the channel. The changing spectrum response as a function of wavelength position shows that the channel is a *dynamic* frequency-selective channel at this particular site as defined above.

It should be emphasized that the nature of the time and spectrum signatures like those shown in Figures 3.6 and 3.7 are a direct result of the time duration and bandwidth of the raised cosine pulse used to create these examples. A different raised cosine pulse should yield different signatures. Accordingly, whether the dynamic characteristics of the time and frequency response are significant to successful detection of the pulse by the receiver depends on the pulse itself and the bandwidth of the system.

3.3.1.2 Tapped delay line model

A common way to model the response of the channel is with a tapped delay line as shown in Figure 3.8. In this model the output signal is the sum of the input signal delayed by incremental amounts and multiplied by a time-varying weighting function. The choice of delay times $\Delta\tau$ is somewhat arbitrary but usually chosen to be related to the symbol duration time of the symbols transmitted in the channel or the inverse of the bandwidth of the transmission system. In Figure 3.8 the delay times are shown as being equal but this is not strictly necessary for this model.

The significant features of this model are the multiplicative functions $r(t)$. Each of the delay taps essentially creates a time bin. All of the waves arriving during the time bin are vectorially added together to create the function $r(t)$. If there were only one wave arriving during the time bin, then the amplitude of $r(t)$ would be a constant with some phase shift dependent on the propagation path length of that wave. If more than one wave arrives during a time bin, the $r(t)$ is the vector sum of those waves. As those waves move in and out of phase, the envelope amplitude of the $r(t)$ functions varies accordingly. The result is that the $r(t)$ functions are described as fading functions that modify the signal amplitude as the fading envelope varies. The sources in the propagation environment that cause the multipath waves can generally be classified as *scatterers*. The common assumption is that these fading functions are uncorrelated, giving rise to the tapped channel model being described as *wide-sense stationary uncorrelated scattering* (WSSUS).

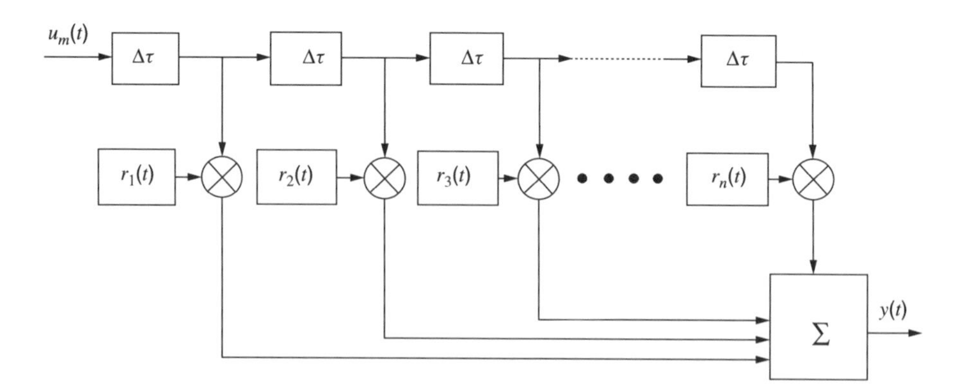

Figure 3.8 Tapped delay line model for channel response.

If the tap delay times in this model were made arbitrarily short, of variable length, and centered on the arrival time of a single arriving wave, the tapped delay line model would yield the generic wideband model in (3.5). In that case, the $r(t)$ functions would be constants with envelope amplitudes equal to A_n.

3.3.1.3 Frequency domain representations

The discussion of the wideband channel model thus far has focused on the time domain representation. The Fourier transform of the time domain channel response $h(t, \tau)$ taken with respect to delay τ is often called the time-variant transfer function shown by

$$T(f, t) = \int u(f)h(t, \tau)\,d\tau \tag{3.13}$$

The frequency response for the output signal as a function of time can then be found as

$$Y(f, t) = U_m(f)T(f, t) \tag{3.14}$$

The frequency domain representations are a useful approach to finding the *delay-Doppler domain* representation of the channel, which is important in assessing channel effect for a mobile terminal. For fixed broadband wireless systems, the time domain representations in the preceding sections are the most common way to characterize the channel.

Depending on the system type and equipment used, the time-varying nature of the channel response could be a significant design consideration. As discussed in Chapter 4, for fixed link systems carrying high-speed digital signals, the time-varying response of the channel can impact the detected bit error rate (BER). Modern high data rate radios include a specification for dispersive fade margin (DFM), which in general terms is a measure of the equipment's ability to perform in time-dispersive (frequency-selective) channels.

3.4 EMPIRICAL MODELS

Empirical models, by definition, are based on observations or measurements. For propagation models, these measurements are typically done in the field to measure path loss, delay spread, or other channel characteristics. Empirical models can also be developed from measurements made in a lab or with scale models of the propagation environment.

Empirical models are widely used in mobile radio and cellular system engineering. In fact, many cellular operators have ongoing measurement or *drive-test* programs that collect measurements of signal levels, call quality, and network performance. The measurements are then used to refine empirical propagation models used in the system-planning tool. These models are referred to here as *feedback* models, as compared to *blind* models whose predictions are not refined with measurements on an ongoing basis.

For traditional point-to-point fixed microwave systems, empirical models of path loss were of no use since the terminals are at fixed locations and simple physical models proved highly accurate at predicting median received signal level. However, empirical models of multipath and rain fading events are important to link performance predictions. Such models are discussed in Chapter 4.

With the growing interest in fixed broadband systems for business and residential use, empirical propagation models have taken on new importance since they can offer simple predictions for these environments without the need for detailed propagation environment databases as required by site-specific physical models. Their simplicity also limits empirical models to *system dimensioning* (an approximate count of the cell sites or hubs needed to serve an area). They cannot be used for detailed system planning where specific site information, shadowing effects, and so on must be considered.

The empirical models currently being applied to fixed broadband systems above 2 GHz, particularly the MMDS band at 2.5 to 2.7 GHz, have been adapted from measurements made at personal communications service (PCS) frequencies (1.8 to 1.9 GHz). The validity of extending measurement results from one frequency band to another is one issue associated with any empirical model.

The models discussed in the following sections are those currently being used for system dimensioning or proposed as part of standards for this purpose.

3.4.1 IEEE 802.16 (SUI) models

As mentioned in Chapter 1, the IEEE Working Group 802.16 has been one of the primary groups developing technical standards for fixed broadband systems. Their first task was developing technical and interoperability standards for bands above 11 GHz. Those were adopted and published by the IEEE in the year 2001 [36,37]. For frequencies below 11 GHz, the proposed standard includes channel models informally know as the Stanford University Interim (SUI) models because of the close participation of Stanford University in their development. The basic path loss model and categorization of propagation environments was taken from [38]. This basic path loss equation is as follows:

$$L = A + 10\gamma \log\left(\frac{d}{d_0}\right) + X_f + X_h + s \qquad \text{for } d > d_0 \tag{3.15}$$

where d is the distance in meters, $d_0 = 100$ meters, h_b is the base station height above ground in meters ($10\,\text{m} < h_b < 80\,\text{m}$), with

$$A = 20\log\left(\frac{4\pi d_0}{\lambda}\right) \tag{3.16}$$

and

$$\gamma = \frac{a - bh_b + c}{h_b} \tag{3.17}$$

Table 3.1 Model constants for IEEE 802.16 model for 2.5–2.7 GHz band

Model constant	Terrain Type A	Terrain Type B	Terrain Type C
a	4.6	4.0	3.6
b	0.0075	0.0065	0.005
c	12.6	17.1	20

The constants a, b and c are chosen on the basis of one of three environments designated as A, B or C as shown in Table 3.1.

The terms X_f and X_h are correction factors for frequency and receiver (remote terminal) antenna height above ground, respectively. These corrections are defined as

$$X_f = 6.0 \log\left(\frac{f}{2000}\right) \tag{3.18}$$

and

$$X_h = -10.8 \log\left(\frac{h_m}{2.0}\right) \quad \text{for Terrain Types A and B} \tag{3.19}$$

$$X_h = -20.0 - \log\left(\frac{h_m}{2.0}\right) \quad \text{for Terrain Type C} \tag{3.20}$$

where f is the frequency in MHz and h_m is the receiver (remote terminal) height above ground in meters.

The term s is a lognormal-distributed path loss factor that takes into account shadow fading from trees and structures. From [39], the standard deviation of s is typically 8.2 to 10.6 dB, depending on the terrain type.

The RMS delay spread information found in [39] proposes a model for delay spread found in [40], which was developed from a variety of measurement data taken at cellular frequencies. The approach in [40] conjectures a relationship of delay spread to distance and shadow fading and uses a collection of data to formulate some parameters for the model. The equation of delay spread from this approach is

$$\tau_{\text{RMS}} = T_1 \, d^{\alpha} u \tag{3.21}$$

where

- T_1 the mean value of τ_{RMS} at 1 km that equals 400 ns for urban microcells, 400 to 1000 ns for urban macrocells, 300 ns for suburban areas, 100 ns for rural areas, and >500 ns for mountainous areas.
- α equal to 0.5 for urban, suburban and rural areas, and 1.0 for mountainous areas.
- u is a zero mean lognormal variate with a standard deviation falling in the range of 2 to 6 dB.

Table 3.2 SUI model classifications

SUI model	Terrain type	Delay spread	Rice k factor	Doppler
SUI-1	C	Low	High	Low
SUI-2	C	Low	High	Low
SUI-3	B	Low	Low	Low
SUI-4	B	Moderate	Low	High
SUI-5	A	High	Low	Low
SUI-6	A	High	Low	High

The model in (3.21) was on the basis of measurements for mobile terminals in cellular systems in which omnidirectional antennas are used. Given the rather wide parameter ranges allowed for in (3.21), the MMDS measurements described in Section 3.4.3 are consistent with this model for the omnidirectional case. For directional antennas, τ_{RMS} is reduced by a factor that can be derived from the ratio of the omnidirectional/directional measurements in [41]. The description in [39] suggests that the delay spread from (3.21) be reduced by a factor of 2.3 and 2.6 for antenna beamwidths of 32 and 10 degrees, respectively.

The IEEE 802.16 SUI models define two models for each of the three terrain types making up a total of six classifications, as shown in Table 3.2. In addition to the basic path loss equations for each environment, the SUI models include time dispersion information in the form of the amplitude, time delay, and Rician k factor of the signals from taps as represented by the tapped delay model discussed in Section 3.3.1.2. For the SUI models, three taps have been defined with tap values provided for both an omnidirectional receive antenna and 30-degree beamwidth antenna.

The detailed tap parameters for each of the six SUI models can be found in [39]. Figure 3.9 is a graph showing path loss as a function of distance for the three terrain types using a base station antenna height of 30 m.

The SUI models were specifically developed for use in the US MMDS frequency band from 2.5 to 2.7 GHz. Reference [39] asserts the model should perform adequately in the 2 to 4 GHz range, although tests of its performance in the European 3.5 GHz band have apparently not been carried out or published.

The SUI path loss equation as shown in (3.15) was basically derived from measurements in suburban areas. No correction factors are included for urban or heavily built-up areas, or for rural areas. Also, there is no way to relate the three terrain type categories (A, B and C) to commonly available clutter or terrain databases, so the method for selecting which category to apply for any particular system deployment scenario is not systematic.

The measurements were done at distances extending out to approximately 7 km, which are adequate for coverage and service assessment. However, for multicell fixed broadband networks in which frequency reuse will be required, interference signal levels from cells that may be several cell radii away (20–30 km) are needed. The SUI models offer

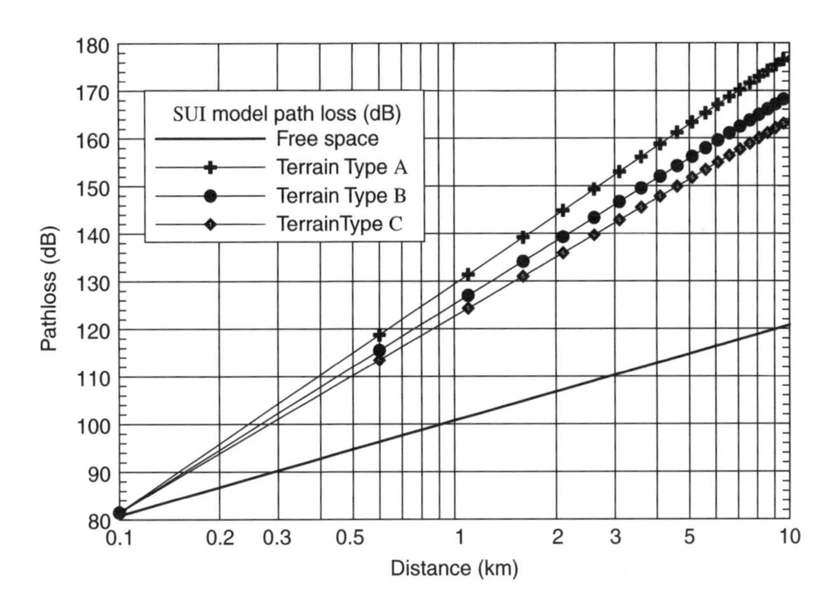

Figure 3.9 Path loss for SUI models with $h_b = 30\,\text{m}$, $h_m = 2\,\text{m}$ and frequency $= 2.6\,\text{GHz}$.

no guidance on path loss at such distances. From the slope of the path loss curves in Figure 3.9, it can be expected that extrapolated SUI path loss values at such distances would almost certainly be larger than those experienced with real systems.

Because of these limitations, the SUI models are best suited for system dimensioning or equipment development purposes rather than detailed, location-specific, system planning. For planning purposes, physical models that can exploit terrain, clutter, and building database information are a more appropriate choice. Such models are discussed in Section 3.5.

3.4.2 COST-231 Hata model

The COST-231 Hata model [42] was devised as an extension to the Hata–Okumura model [11, 12]. The Hata–Okumura model is an empirical model developed for the 500 to 1500 MHz frequency range using measurements done by Okumura [12] and equations fitting to the path loss curves by Hata [11]. The COST-231 model also has correction for urban, suburban, and open areas. Further extensions to these models could perhaps adapt them to the MMDS band. For these reasons, the COST-231 Hata is included here. The basic path loss equation for urban areas is

$$L_u = 46.3 + 33.9 \log_{10} f - 13.82 \log_{10} h_b - ah_m + (44.9 - 6.55 \log_{10} h_b) \log_{10} d + c_m \text{dB} \tag{3.22}$$

$$ah_m = (1.1 \log_{10} f - 0.7) h_m - (1.56 \log_{10} f - 0.8) \tag{3.23}$$

where

c_m = 0 dB for medium sized city and suburban centers with moderate tree density. This correction is used when the open or suburban categories are selected.
c_m = 3 dB for metropolitan centers. The correction is used when the urban category is selected.
f = frequency in MHz.
d = distance from the base station to the receiver (remote terminal) in kilometers
h_b = height of the base station (hub) above ground in meters
h_m = height of the receiver (remote terminal) above ground in meters

For urban (large city) areas, the term ah_m is defined as follows:

$$ah_m = 8.29(\log(1.54h_m))^2 - 1.1 \qquad \text{for } f \leq 200\,\text{MHz} \tag{3.24}$$

$$ah_m = 3.20(\log(11.75h_m))^2 - 4.97 \qquad \text{for } f > 400\,\text{MHz} \tag{3.25}$$

For suburban (medium–small city) or open (rural) areas the term ah_m is defined as follows:

$$ah_m = (1.1\log f - 0.7)h_m - (1.56\log f - 0.8) \tag{3.26}$$

The COST-231 Hata model has many of the same limitations for detailed system planning as the SUI models. Nonetheless, because of their simplicity, they are widely used for system dimensioning and other generic system concept formulations.

3.4.3 MMDS band empirical path loss

A comprehensive set of measurements of path loss and time dispersion (delay spread) at about 2.5 GHz for the MMDS band was reported in [41]. The measurements were made in the suburban Chicago area at locations within 10 km of the base station transmitter using both directional and omnidirectional receive antennas and three different receive antenna heights of 16.5, 10.4, and 5.2 m.

The basic path loss equation applied to these measurements takes the standard form

$$L(d) = L(d_0) + 10n\log\left(\frac{d}{d_0}\right) + X_\sigma\ \text{dB} \tag{3.27}$$

where $L(d_0)$ is the average close-in path loss, n is a path loss exponent that determines the slope of the path loss curve, and X_σ is a lognormal-distributed random variable with zero mean and standard deviation of σ. The various parameter values for

Table 3.3 Parameter values for MMDS empirical path loss model

	LOS		NLOS	
Receive antenna	n	σ (dB)	n	σ (dB)
5.2 m, directional	1.7	3.1	4.1	12.6
5.2 m, omnidirectional	2.1	3.5	4.2	10.5
10.4 m, directional	2.9	4.7	2.9	11.4
10.4 m, omni-directional	2.6	4.6	3.3	10.1
16.5 m, directional	2.4	3.8	2.1	10.5
16.5 m, omnidirectional	2.4	3.4	2.7	9.4

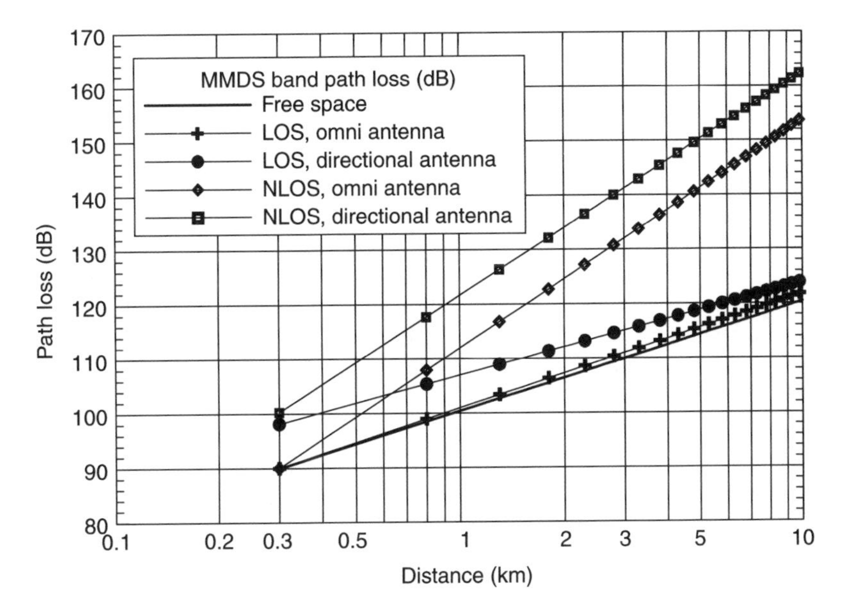

Figure 3.10 Regression curves for MMDS path loss measurement data reported in [41].

the different antenna types and heights, and both LOS and NLOS paths, is shown in Table 3.3.

A weakness of the data is that the receive antenna heights are larger than the heights of an indoor remote terminal, which would normally be located in a residence for an NLOS network. Even the lowest height of 5.2 m is near the roof peak for many single-story family residences. Nonetheless, the measured values are instructive. Figure 3.10 shows path loss values calculated using (3.27) for the 5.2-m antenna height, directional and omnidirectional antennas, and for both LOS and NLOS paths.

The measurements in [41] also include delay spread using a channel sounder with a 50 nanosecond (ns) resolution. The results of these measurements show that RMS delay spread for the LOS paths, using the directional antenna, ranged from 20 to 40 ns with a mean of 20 ns. For LOS paths with the omnidirectional antenna, the RMS delay spreads ranged from 20 ns to 2390 ns with a mean of 130 ns. For NLOS paths using the directional antenna, the RMS delay-spread values ranged from 20 ns to 5260 ns, while with the omnidirectional antenna the values ranged from 20 ns to 7060 ns with a mean of 370 ns. The use of the directional antenna clearly suppresses multipath components resulting in lower RMS delay-spread values – a concept that has been demonstrated using a ray-tracing propagation model for an LMDS system in an urban area [43].

3.4.4 3D path loss surface models

The empirical models described thus far are two-dimensional (2D) models in that a single equation has been used to describe pathloss as a function of distance from the transmitter. Applying this model in all directions from the transmitter, and assuming the same antenna height and environment are used, would result in predicted coverage area that is circular. In reality, for systems in nonhomogenous propagation environments, the coverage area for a given base station or hub transmitted is highly noncircular; in fact, it may be discontiguous.

The basic concept of an empirical model, however, is not limited to one or more 2D curves: it can be extended to three dimensions in which the path loss values form a surface for the area around each transmitter site. Of course, the path loss surface for each transmitter site will be different, as different as the propagation environment itself around each transmitter site. For cellular systems with hundreds of sites, this will require hundreds of path loss surfaces.

As noted above, collecting measurement data on signal levels has become much more streamlined in recent years as a result of small, inexpensive GPS-equipped units that accurately establish position, and measurement equipment attached to notebook computers that can conveniently store and analyze measurement data. The analysis can include BER, messaging integrity, and even call quality in the case of cellular systems. A vast amount of measurement point data is the key to constructing a path loss surface.

Once collected, the path loss surface is formed by creating a regular grid of path loss values. Since it is unlikely that path loss measurements have been taken in a regular grid, it is necessary to use a 'gridding' algorithm of some sort that essentially interpolates path loss values among known values to fill in the surface grid. An overview describing several such algorithms along with primary references can be found in [44].

With the 3D path loss surface in place for a given site, finding the signal level at some point is a simple matter of looking up that point in the path loss surface grid and using that value along with transmitter effective radiated power (ERP) and receive antenna gain to find the received power level. As such, the predictions are easily scalable for different power levels, antenna pattern types, and antenna orientation and beamtilts. If the coverage of the surface is extensive enough, it can also be used to predict coverage as well as interference at points that are distant from the cell or hub site. Figure 3.11 shows an example of a path loss surface or a mountaintop transmitting site.

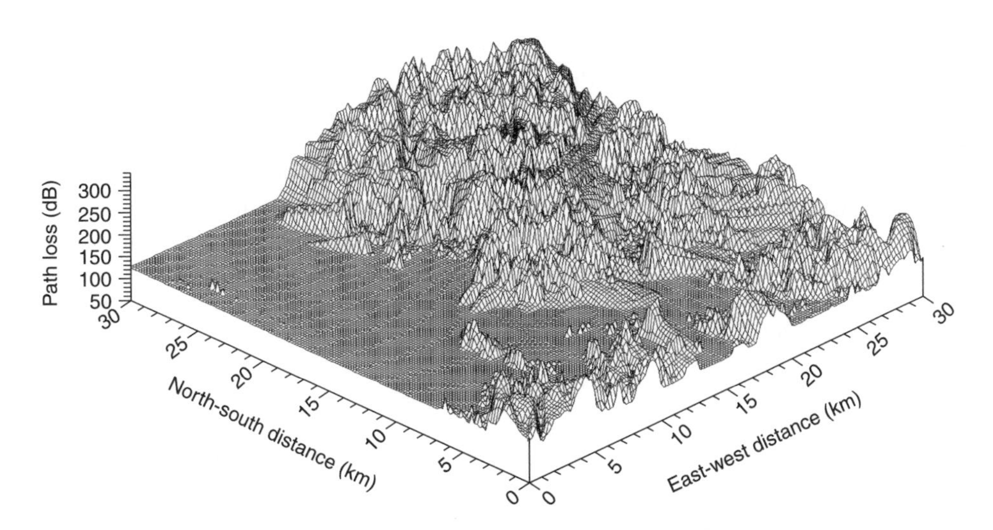

Figure 3.11 Path loss surface propagation model for a mountaintop 2.6 GHz transmitting site. High path loss to the northeast is because of mountains in this area.

The drawback to this approach, of course, is the extensive measurement campaign required. The gridding process also necessarily introduces approximations from the interpolation. However, in the case of a fixed broadband system, where the locations of the remote terminals are known and signal strength (path loss) is measured in the normal course of operating the system, the measurement campaign can essentially be replaced with real operational data. As more customers are added on the system, and the path loss surface is regridded using this information, the more refined and accurate future signal level predictions can use the path loss surface approach for that site.

3.5 PHYSICAL MODELS

The most widely used propagation models for fixed broadband wireless systems are physical models. The tens of thousands of traditional point-to-point microwave links that have been constructed over the past several decades have been designed using simple free-space path loss equations and perhaps some consideration of reflections from intervening flat surfaces or bodies of water that fall along the path. The free-space path loss model is combined with models of multipath fading and rain fading (Chapter 4) to provide a link design with the required reliability. Databases of terrain elevations, clutter (tree and foliage) heights, atmospheric refractivity conditions, and rain intensity rates are all used in the design process. Such traditional links have typically spanned many tens of kilometers, however, the same principles and the same propagation and fade modeling techniques can be applied to short microwave links for the LMDS systems, mesh networks, cellular backhaul connections, or other short-range systems in which the link lengths may only be a few kilometers.

The free-space path loss mechanism is a fundamental ingredient to all physical propagation models. The various models differ primarily in how they treat terrain and building obstacles, reflections, and clutter, building penetration, and foliage obstruction attenuation. Over the years, many approaches for handling these factors have been proposed, each with at least some validation derived from field measurements. The physical propagation modeling techniques discussed in the following sections represent a sampling of such methods that are particularly relevant to fixed broadband wireless design, either because of their strong technical capabilities, their current widespread use in the industry, or because they are mandated by regulatory authorities like the FCC in the United States for use in fixed broadband analysis.

3.5.1 Free space + RMD

The modeling technique known as $Freespace + RMD$ was developed at EDX Engineering in the late 1980s [25]. The 'RMD' stands for 'Reflection/Multiple Diffraction'. While designed for propagation analysis from 30 MHz to 40 GHz, it is particularly relevant to fixed broadband analysis because it was selected by the FCC for performing interference and coverage analysis for construction permit applications to build two-way MMDS systems. The details of the model are described in Appendix D of the FCC Rules pertaining to such two-way operations [45]. The model consists of three basic parts:

- Free-space propagation and partial Fresnel zone obstruction for LOS paths,
- Single reflection contribution for LOS paths and
- Single and multiple diffraction loss for NLOS paths.

Each of these model elements is described in some detail below because they also constitute some fundamental algorithms that are found in many physical models. Given the fundamental nature of its operation, the Free space + RMD model can also be used for basic path loss calculations for most point-to-point microwave link systems with engineered transmit and receive terminals.

3.5.1.1 Line-of-sight assessment

The first calculation that is done with this model is a determination of whether a path is LOS. This is done by determining the depression angle for the path from the transmitter to the receiver and to each point along the terrain profile in between the transmitter and receiver, as well as each point. The depression angle is found as

$$\theta_{t-r} = \frac{h_r - h_t}{d_r}\frac{d_r}{2a} \text{ radians} \tag{3.28}$$

where

θ_{t-r} is the depression angle in radians for the ray path between the transmitter and receiver.

h_t is the transmit antenna center of radiation height above mean sea level (AMSL) in meters.

h_r is the receive antenna center of radiation height AMSL in meters.

d_r is the great circle distance from the transmitter to the receiver in meters.

a is the effective Earth radius in meters taking into account atmospheric refractivity.

As discussed in Chapter 2, the effective Earth radius or k factor is used to incorporate, in a simple way, the path trajectory given the Earth refractivity. While any appropriate k factor could be used for finding a in (3.28), for the FCC MMDS Rules this value has been set at the nominal value of 1.333, resulting in a value $a = 8{,}451{,}000\,\text{m}$.

Using an equation similar to (3.28), the depression angle to every elevation point along the terrain/clutter profile between the transmitter and receiver is found as

$$\theta_{t-r} = \frac{h_p - h_t}{d_p} \frac{d_p}{2a} \text{ radians} \tag{3.29}$$

where h_p is the elevation AMSL of the terrain profile point and d_p is the distance to that point from the transmitter along the great circle path. The depression angles in (3.29) are determined at each point along the path. If at any point $\theta_{t-p} > \theta_{t-r}$, the path is NLOS; otherwise, the path is LOS.

3.5.1.2 LOS path analysis

If the path is LOS, then a simple two-ray propagation model is used in which one ray represents the free-space energy arriving at the receiver and the other ray represents the energy arriving at the receiver from a single reflection point. As presented in Chapter 2, the free-space field strength E_r at a distance is given by

$$E_r = \frac{1}{d_r}\sqrt{\frac{P_t G_t Z_0}{4\pi}} \tag{3.30}$$

where P_t is the transmit power in Watts, G_t is the transmit antenna gain in the direction of the receiver, and Z_0 is the plane wave free-space impedance (377 ohms). Written in decibel terms, (3.30) becomes

$$E_r = 76.92 - 20.0\log(d_r) + P_T \text{ dB uV/m} \tag{3.31}$$

where P_T is the transmitted effective radiated power relative to a dipole (ERP_d) in dBW.

The magnitude and phase of the ground reflected ray are found by first calculating the complex reflection coefficient using equations similar to those in Section 2.6.1.

$$R_{\|} = \frac{-\varepsilon_r \sin\gamma_i + \sqrt{\varepsilon_r - \cos^2\gamma_i}}{\varepsilon_r \sin\gamma_i + \sqrt{\varepsilon_r - \cos^2\gamma_i}} \tag{3.32}$$

$$R_{\perp} = \frac{\sin\gamma_i - \sqrt{\varepsilon_r - \cos^2\gamma_i}}{\sin\gamma_t + \sqrt{\varepsilon_r - \cos^2\gamma_i}} \tag{3.33}$$

where γ_i is the angle of incidence relative to a plane tangent at the point of reflection and ε_r is the complex permittivity given by

$$\varepsilon_r = \varepsilon_1 - j60\sigma_1\lambda \tag{3.34}$$

with ε_1 is the relative dielectric constant and σ_1 is the conductivity of the reflecting surface material in S/m. In the case of a ground reflection, a vertically polarized transmit signal is at parallel polarization relative to the reflecting surface and a horizontally polarized transit signal is at perpendicular polarization relative to the reflecting surface. While any values may be used for the conductivity and permittivity to represent the characteristics of the reflecting surface material, for the purposes of the FCC MMDS analysis, $\sigma_1 = 0.008$ S/m and $\varepsilon_1 = 15$. For general use of this model, using appropriate material constants would improve the reflection calculation. For example, for a microwave path over the sea, using a conductivity value $\sigma_1 = 5$ S/m would be appropriate.

Over typical long rage paths where the Free Space+RMD model is used, the lengths of the direct ray path and the ground reflected ray path are essentially the same (differing by perhaps several wavelengths) so the spatial attenuation due to path length is essentially the same for both rays. The reflected ray, however, is multiplied by the appropriate reflection coefficient as given by (3.32 or 3.33). The vector addition of the two rays is thus given by

$$E_r = E_d \sin(\omega t) + E_d R \sin(\omega t + \Delta\varphi) \tag{3.35}$$

where

- E_d is the magnitude of the direct ray
- ω is the signal carrier frequency in radians
- R is the complex reflection coefficient as given in (3.32 and 3.33)
- $\Delta\varphi$ is the phase delay of reflected ray in radians

The carrier frequency term is usually suppressed so (3.35) becomes

$$|E_r| = E_d|1 - Re^{j(\varphi_r + \Delta\varphi)}| \tag{3.36}$$

$$|E_r| = E_d\sqrt{(1 + |R|\cos(\varphi_r + \Delta\varphi))^2 + (|R|\sin(\varphi_r + \Delta\varphi))^2} \tag{3.37}$$

where φ_r is the phase angle of the reflection coefficient. The term $\Delta\varphi$ is the phase delay found from the difference between the direct ray path length and the reflected ray path length. For a two-ray path geometry over a curved Earth path, the approximate path length difference is given by [46] as

$$\Delta r \cong \frac{2h'_t h'_r}{d_r} \tag{3.38}$$

where h'_t and h'_r are the heights in meters of the transmit and receive antenna above the reflecting plane, respectively, and $d_r \gg h'_t, h'_r$. For plane surface reflection using the source image approach shown in Figure 2.5, the path length difference is given by

$$\Delta r = \sqrt{d_r^2 + (h'_t + h'_r)^2} - \sqrt{d_r^2 + (h'_t - h'_r)^2} \tag{3.39}$$

The phase delay $\Delta\varphi$ is then given by

$$\Delta\varphi = \frac{2\pi\,\Delta r}{\lambda} \quad \text{(modulo } 2\pi \text{ radians)} \tag{3.40}$$

For an irregular terrain profile between the transmitter and receiver, the usual difficulty is determining the location of the reflection point. The terrain profile itself is usually a set of equally spaced samples drawn from a terrain digital elevation model (DEM) database (see Chapter 5). The set of samples forms a piecewise linear approximation to the actual terrain. Each piecewise linear segment of the approximation is a potential reflecting surface. The typical procedure for finding a reflection point is to proceed along the path from transmitter to receiver and at each segment evaluate whether a specular reflection can exist from that segment in which the angle of incident from the receiver is equal to the angle of reflection to the receiver. There could be no reflection points or multiple reflection points along the path. As a practical matter, for irregular terrain profiles it is unlikely that the angle of incidence from the transmitter will ever exactly equal the angle of reflection to the receiver, so some window or range of acceptance must be employed to reasonably assess the presence and location of the reflection point.

For very low receiver heights, the ground reflection point is very near to the receiver, the length of the reflected ray path is nearly equal to the direct ray path length. For very small angles of incidence, the reflection component is shifted in phase by 180 degrees (see Figure 2.5b). The amplitude of the reflected ray will be nearly equal to 1 because the incident angle is small (see Figure 2.5a). With this phase reversal, very little phase shift occurs due to path length differences, and a reflection amplitude nearly equal to the direct signal; at very low antenna heights the reflected signal reduces the amplitude of the directly received ray because the two add out of phase. As the receive antenna height is raised, the net signal magnitude from this vector addition increases as the out-of-phase condition is reduced. This physical mechanism is the primary reason for the so-called *height-gain* effect, which is sometimes incorporated in nonphysical propagation models as the antenna height dependent path loss equation term.

From the preceding discussion, the path loss attenuation term A_r resulting from the reflection can be written as

$$A_r = -20\log\sqrt{(1+|R|\cos(\varphi_r+\Delta\varphi))^2+(|R|\sin(\varphi_r+\Delta\varphi))^2}\ \text{dB} \tag{3.41}$$

Figures 3.12(a and b) show graphs of the two-ray signal attenuation A_r as a function of path length for both vertical and horizontal polarizations, respectively.

The final component included in the LOS analysis is partial Fresnel zone obstruction. As explained in Section 2.8, the simple (infinitesimally thin) ray angle geometry to assess LOS does not take into account that part of the Fresnel zone, which may be obstructed. At the point where the ray just grazes the obstruction, the attenuation is 6 dB as shown in Figure 2.18. As the ray clearance increases above grazing, the loss is reduced from 6 dB to 0 dB. This effect can be modeled with a simple linear relationship

$$A_{fr} = 6.0\left(1.0 - \frac{C_{obs}(d_p)}{R_{FR}(d_p)}\right)\ \text{dB} \tag{3.42}$$

with $C_{obs}(d_p)$ as the height difference in meters between the ray path height at distance d_p and the obstacle height at distance d_p. The term $R_{FR}(d_p)$ is the 0.6 first Fresnel zone radius in meters at distance d_p along the path. This radius can be found from (2.46). The validity of (3.42) is limited to $0 \leq C_{obs}(d_p) \leq R_{FR}(d_p)$.

Gathering the various loss terms, the field strength for this model under LOS path conditions is

$$E_r = 76.92 - 20\log(d_r) + P_T - A_r - A_{fr}\ \text{dBuV/m} \tag{3.43}$$

The total path loss in dB for LOS path conditions is then

$$L_{LOS} = 32.45 + 20\log f + 20\log d_r + A_r + A_{fr}\ \text{dB} \tag{3.44}$$

As Figures 3.12(a and b) illustrate, the out-of-phase contribution of the reflected signal can result in very deep fades or attenuation in the received signal level. These fades can only occur if the geometry to the reflection point is precisely known. For that reason, the reflection contribution is only explicitly considered in calculating mean signal strength when the reflection surface geometry is well known. Otherwise, the reflection ray is ignored and the reflection contributions are instead accounted for using a fading model (see Chapter 4). For the particular purpose of FCC-required analysis of interference and coverage of MMDS systems in the US, the reflection analysis has been excluded for this same reason.

3.5.1.3 NLOS path analysis

If the path is obstructed on the basis of the calculation in Section 3.5.1.1, the diffraction loss over the obstacles are taken into account and no reflection analysis is done. Over the years, several approaches have been developed for calculating signal attenuation over signal and multiple path obstructions. These methods are described in the next section. For the Free Space+RMD model used by the FCC, the chosen multiple obstacle attenuation calculation method is the Epstein–Peterson method. The total obstacle loss is simple added to the free-space path loss as follows:

$$L_{NLOS} = 32.45 + 20\log f + 20\log d_r + A_{diff}\ \text{dB} \tag{3.45}$$

where A_{diff} is the total multiple obstacle loss in dB as described in the next section.

3.5.2 Multiple obstacle analysis

Obstacles on a propagation path can take the form of terrain features, foliage (trees), buildings or other structures. These obstacles are generally modeled as diffracting edges of some sort. The wedge and knife-edge diffraction propagation mechanisms described in Section 2.7 are normally used to represent these obstacles. In the case of buildings, a 90-degree wedge can be a very close model to vertical corners or horizontal parapet edges of buildings. In the case of terrain obstacles, the shape is not a simple geometry

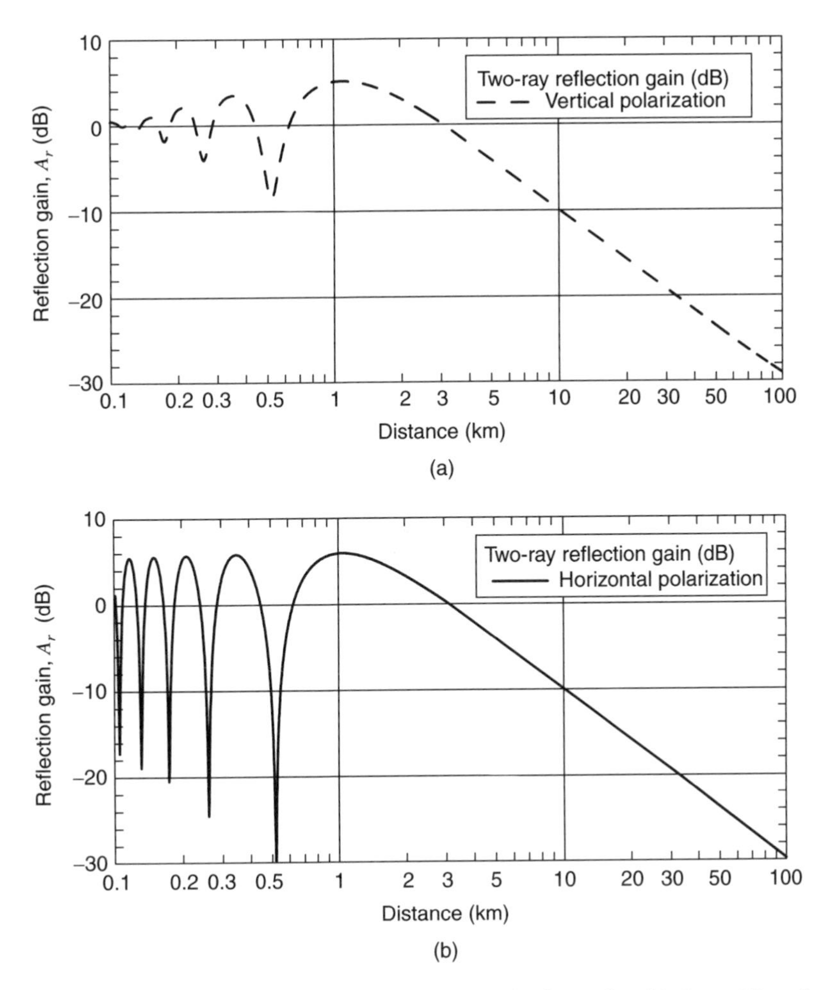

Figure 3.12 (a) Vertically polarized ground reflection gain for path with $h_m = 30\,\text{m}$, $h_m = 1\,\text{m}$, $\varepsilon_1 = 15$, $\sigma_1 = 0.008\,\text{S/m}$ and frequency $= 2.6\,\text{GHz}$. (b) Horizontally polarized ground reflection gain for path with $h_t = 30\,\text{m}$, $h_m = 1\,\text{m}$, $\varepsilon_1 = 15$, $\sigma_1 = 0.008\,\text{S/m}$ and frequency $= 2.6\,\text{GHz}$.

like a wedge but nonetheless, wedges and knife-edges have been used with reasonable success to model the attenuation over terrain obstacles.

Reference [47] is an example of the use of wedges to model terrain obstacles. The difficulty encountered in this approach is finding a systematic, mechanical approach to 'linearizing' the terrain so that it may be handled as a series of reflecting plates and diffracting wedges. When considered over a distance of several tens of kilometers as might be relevant to microwave propagation, most hills, ridges, and mountains are actually very flat so that wedges with very large interior angles close to 180 degrees are needed to approximately fit the shape of the terrain. Calculating the wedge diffraction coefficients as

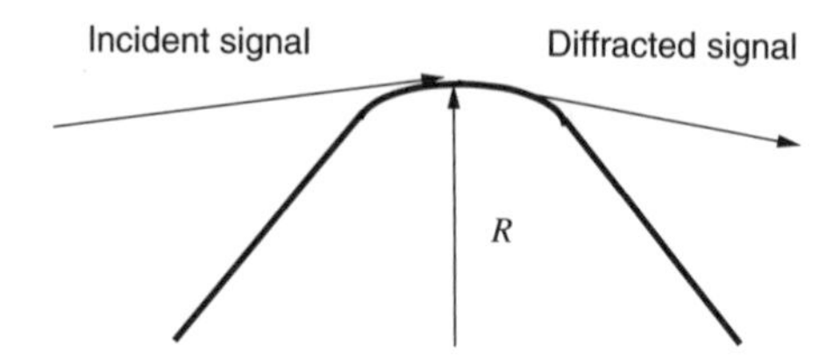

Figure 3.13 Diffraction geometry over a round terrain obstacle.

set forth in Section 2.7 is also computationally more demanding than finding diffraction loss over simple knife-edge obstacles.

Knife-edge approximations to terrain obstacles, of course, represent far different shapes than the actual terrain obstacles. Attempts have been made to lessen the disparity by using rounded obstacles (wedges with rounded edges like the side of cylinder) as illustrated in Figure 3.13. In general, the rounded obstacle approach in which the radius is chosen to fit the shape of the terrain has resulted in diffraction losses much higher than those observed with real terrain. For these reasons, most propagation models employed for point-to-point microwave link system planning use simple knife-edge diffraction models for terrain obstacles rather than wedges or rounded cylinder obstacles.

With a single obstacle, the additional path loss due to the obstacle is a straightforward use of the equations in Section 2.7.2, or a simplified version of those equations that have been presented in [48]. It is primarily a function of the parameter ν from (2.40) that is related to the path clearance over the obstacle. The total diffraction loss, $A(v, \rho)$, in dB is the sum of three parts: $-A(v, 0)$, $A(0, \rho)$, $A(0, \rho)$, and $U(v, \rho)$. The equations to calculate each part are given below:

$$A(v, \rho) = A(v, 0) + A(0, \rho) + U(v, \rho) \tag{3.46}$$

$$A(v, \rho) = 6.02 + 9.0\nu + 1.65\nu^2 \quad \text{for } -0.8 \leq v \leq 0 \tag{3.47}$$

$$A(v, 0) = 6.02 + 9.11v - 1.27v^2 \quad \text{for } 0 < v \leq 2.4 \tag{3.48}$$

$$A(v, 0) = 12.593 + 20\log_{10}(v) \quad \text{for } v > 2.4 \tag{3.49}$$

$$A(0, \rho) = 6.02 + 5.556\rho + 3.418\rho^2 + 0.256\rho^3 \tag{3.50}$$

$$U(v, \rho) = 11.45v\rho + 2.19(v\rho)^2 - 0.206(v\rho)^3 \quad \text{for } v\rho \leq 3 \tag{3.51}$$

$$U(v, \rho) = 13.47v\rho + 1.058(v\rho)^2 - 0.048(v\rho)^3 - 6.02 \quad \text{for } 3 < v\rho \leq 5 \tag{3.52}$$

$$U(v, \rho) = 13.47v\rho + 1.058(v\rho)^2 - 0.048(v\rho)^3 - 6.02 \quad \text{for } 3 < v\rho \leq 5 \tag{3.53}$$

where the curvature factor is

$$\rho = 0.676R^{0.333}f^{-0.1667}\sqrt{\frac{d}{d_1 d_2}} \tag{3.54}$$

The obstacle radius R is in *kilometers*, and the frequency f is in MHz. For knife-edge diffraction, the terms depending on ρ ($A(0, \rho)$ and $U(v, \rho)$) are zero so the diffraction loss in dB is given by $A(v, 0)$ alone.

While the diffraction attenuation over a single obstacle can be calculated using detailed solutions such as those in Section 2.7 or the simplified approach mentioned earlier, the further question is how to calculate loss over paths that have multiple obstacles. Two common approaches are discussed below.

3.5.2.1 Epstein–Peterson method

The Epstein–Peterson approach [27] to handling multiple obstacle diffraction loss is incorporated in the Free space + RMD, TIREM, Anderson 2D, and many other propagation models used for fixed broadband and mobile wireless communication system design. The geometry for this method is shown in Figure 3.14. The determination of the obstacles on the path is done by considering the depression angles to each point on the terrain profile as described in Section 3.5.1.1. At the point with the maximum depression angle (which may in fact be above the horizon) the first obstacle is identified. This point is then considered the new transmitting location and the depression angles calculated to all remaining points along the terrain profile. As each maximum is encountered, it is considered as a new transmitting location and the process repeated. The net result of this procedure is to identify those points that touch a line or 'stretched string' along the profile from the transmitter to the receiver.

The parameter ν is calculated for each obstacle in turn with the preceding obstacle (or transmitter for the first obstacle) as the transmitter and the next obstacle (or receiver for the last obstacle) as the receiver. With ν found for each obstacle, the obstacle loss $A(\nu, 0)$ is calculated using (3.47 to 3.49). The total diffraction loss is then just the sum of the individual obstacle losses in dB.

$$A_{diff} = \sum_{n=1}^{N} A_n(\nu, \rho) \text{ dB} \tag{3.55}$$

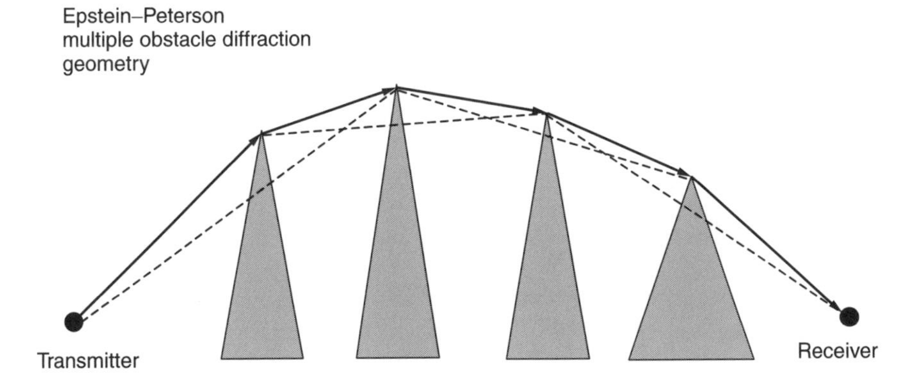

Figure 3.14 Epstein–Peterson multiple obstacle diffraction loss geometry.

The Epstein–Peterson method is straightforward to implement but tends to underestimate the total loss when the path has two or more closely spaced obstacles.

3.5.2.2 Deygout method

The Deygout method [28] attempts to rationalize the obstacles on the path into 'dominant' obstacles and subordinate obstacles, essentially ordering the weight of the diffraction loss contributions based on the magnitude of the parameter ν.

The first step in the Deygout method is to identify all the edges using the 'stretched string' approach as with the Epstein–Peterson method. The parameter ν is then calculated for every edge assuming no other edges are present on the path. The edge with the largest value of ν is deemed the 'main edge'. The loss over the main edge is then calculated for this value of ν. The ν values for the secondary edges are then found by considering the path from the transmitter to the main edge and from the main edge to the receiver. The total loss is the sum of the loss for the main edge and for the secondary edges. The losses themselves are calculated using the knife-edge model and the rigorous formulas for knife-edge loss as set forth in Section 2.7.2 or the approximations given above.

The Deygout method is often applied to only three edges as illustrated in Figure 3.15, however, the method can be generalized to more than three edges. This is done by identifying a 'subsidiary main edge' on the path from the transmitter to the primary main edge, and from the primary main edge to receiver. As before, the subsidiary main edge is identified as the edge having the greatest value of ν along this subsection of the profile. This procedure may be applied recursively to consider any number of obstacles on the path.

The Deygout method can achieve reasonably good agreement with more rigorous approaches but generally overestimates the diffraction loss in increasingly greater amounts as the path length increases. Given the end-to-end geometry used to find ν for the main edge, the overprediction of diffraction loss is understandable since it does not take into account that diffraction fields from preceding obstacles in essence redirect the propagation direction of the EM wave.

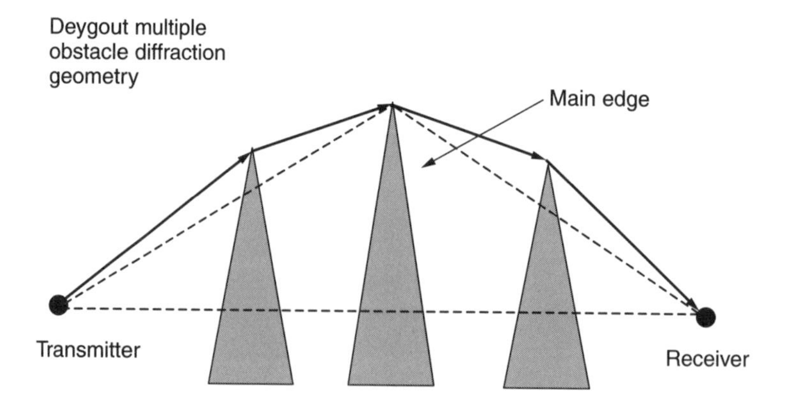

Figure 3.15 Deygout multiple obstacle diffraction loss geometry.

3.5.3 Longley–Rice model

The Longley–Rice model was developed in the late 1960s and reported in an ESSA technical report [23] as a computerized method that could take into account the detailed terrain features along a path from transmitter to receiver. The model is included here as a physical model because it uses the two-ray reflection analysis and the knife-edge diffraction loss calculations for terrain obstacles. However, it uses these fundamental calculations to construct path loss slope lines for three main regions: from the transmitter to the horizon, beyond the horizon to the point at distance d_x where forward scatter loss and diffraction loss are equal, and beyond distance d_x where forward scatter loss calculations are used. At distances from the transmitter to the horizon, the path loss actually includes a weighted portion of the diffraction loss *beyond* the horizon. Having the path loss to a receiver location be affected by terrain obstacles beyond that receiver location is clearly noncausal and violates physical reasoning for a single path two-dimensional model.

Since its original publication, the Longley–Rice model has been updated with revisions and corrections as described in [49]. In spite of its shortcomings, this model has been widely used for many years, including for DTV channel allocation studies in the United States. Because of its nonphysical elements, and because its stated upper frequency range limit is 20 GHz, the Longley – Rice model is not normally used for point-to-point microwave propagation studies, or studies in the MMDS or U-NII bands at 2.6 GHz and 5.8 GHz, respectively.

3.5.4 TIREM model

The Terrain Integrated Rough Earth Model (TIREM) was first developed by the US Electromagnetic Compatibility Analysis Center to be included as part of a larger Master Propagation System (MPS11). MPS11 is a suite of program modules that included modules for propagation at low frequency (LF) long waves, high frequency (HF) short waves, and scatter effects. The TIREM module was intended for 20 MHz to 20 GHz.

Like the Free space + RMD model, TIREM uses a rigorous application of the depression angle calculation to determine whether a path is LOS or NLOS. For LOS paths, TIREM uses either the Longley–Rice two-ray model or the Longley–Reasoner approximation, depending on the frequency. For NLOS paths, it uses the Epstein–Peterson method for calculating multiple diffraction loss. Further details of the TIREM can be found in [24].

3.5.5 Anderson 2D model

The Anderson 2D model [26] is very similar to the Free space + RMD model described above in that it includes the two-ray reflection analysis for LOS paths and the Epstein–Peterson multiple diffraction loss analysis with knife-edge approximations for NLOS paths. However, the Anderson 2D model contains some important enhancements to the Free space + RMD model, specifically:

- For obstructed paths, it includes two-way reflection analysis for the path between the transmitter and first obstacle and the receiver and the last obstacle. The caveat on using the reflection analysis only in circumstances when the reflection surface geometry is well known also applies to the use of these reflection contributions.

- It includes an algorithm to detect and deal with 'plateau' features along a terrain profile where there is a succession of terrain elevation points that are approximately the same value. Such plateaus may be real terrain features but often they are falsely created by approximations in the terrain elevation model (DEM). Using the standard approach for finding obstacles along a path, the series of elevation points on a plateau could all be considered obstacles, greatly overestimating the total path loss. The Anderson 2D model detects such anomalies and handles them by considering only the leading and trailing edges of the plateau as diffracting edges.

The validity of the Anderson 2D model was assessed in [50] in which it was found to outperform the Longley–Rice, TIREM, and several other models when compared to measurements at frequencies ranging from 150 MHz to 9.6 GHz. The average standard deviation of the prediction errors across this wide range of frequencies was approximately 10 dB. Because of its accuracy, the Anderson 2D model was adopted as the definitive model by the Telecommunications Industry Association (TIA) for Technical Services Bulletin TSB-88A [26] for predicting coverage and interference for mobile radio systems in the United States. The Anderson 2D model is designed to be used at frequencies from 30 MHz to 40 GHz.

3.5.6 NLOS dominant ray path loss model

The basic elements used in the preceding physical propagation models can be combined with additional path loss factors for local clutter to create a general purpose path loss model that is applicable to NLOS paths as shown in Figure 3.16. This model can be used

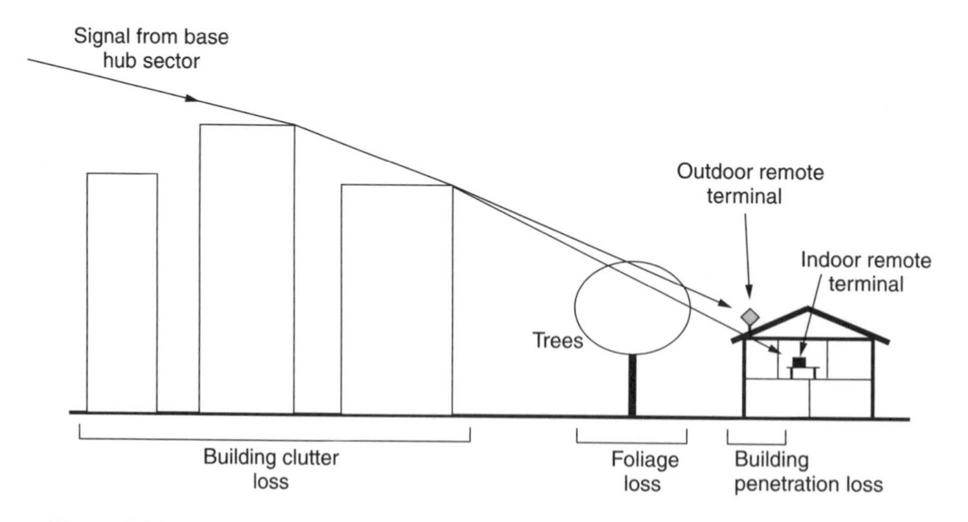

Figure 3.16 Path loss mechanisms for NLOS path links in urban and residential areas.

to evaluate various paths from one terminal to another so that the lowest loss path (the dominant, strongest ray path), can be chosen. Signals from other paths are accounted for in the fading model. This model has the following form:

$$L_T = L_b(p_T, p_R) + C + F + B \tag{3.56}$$

$L_b(p_T, p_R)$ is the basic free-space path loss plus diffraction loss given by

$$L_b(p_T, p_R) = 32.44 + 20\log f + 20\log d + \sum_{n=1}^{N} A_n(\nu, \rho) \tag{3.57}$$

The diffraction term $A_n(\nu, \rho)$ includes diffraction loss over buildings that can be explicitly identified from a building database. The term C is the loss associated with structures along the path that cannot be explicitly identified as obstacles from a database so it must be regarded as 'clutter'. Calculating the mean and standard deviation of the building clutter loss distribution is discussed in more detail in Section 3.5.6.1.

The term F is the loss associated with foliage and modeled as a lognormal-distributed random variable. Estimates of the mean value for the foliage loss distribution can be found from the information in Section 2.9.2.

Finally, the term B is the loss associated with building penetration and also modeled as a lognormal random variable. Estimates of the mean value of the building penetration loss distribution can be found from the information in Section 2.9.1.

As stated, the building clutter, foliage, and building penetration losses in the model in (3.56) are all taken as random variables with lognormal distributions. The mean and standard deviation of the associated distributions will depend on the locations of the transmitter and receiver. Ideally, the values for the mean and standard deviation would be chosen on the basis of information in a propagation environment database. Unfortunately, such specific relationships with any known database have not been established so only some very broad classifications for these values can be used. For LOS paths, the mean values and standard deviations for these three loss terms will be zero (by definition). For NLOS paths, the mean values will assume values that are found as described above, with estimated standard deviations that accounts for observed variation in the loss values.

The general-purpose model given in (3.56) has the following attributes:

- It provides the correct path loss predictions for LOS links including traditional point-to-point microwave paths.
- It provides useful path loss predictions for urban LOS links, including NLOS signals from interferers that are obstructed by building obstacles.
- It provides useful path loss predictions for general NLOS links where the remote terminal is obstructed by local building clutter, foliage, and penetration loss from the outside to the inside of a building.

The model in (3.56) will be used in Chapter 10 for urban/suburban NLOS link design.

For multipath conditions, it is possible that the strongest signal arriving at the remote terminal is not arriving via a direct path but rather via a reflected, diffracted, or locally scattered path. The strongest signal path, of course, corresponds to the path with the lowest loss. The total power is the (noncoherent) power sum of all arriving rays. Full recognition of all the arriving rays (rather than just the dominant ray) leads to the general ray-tracing model described in Section 3.5.7.

3.5.6.1 Building clutter loss

While the foliage and building penetration losses in the path loss model given above are derived from measurement data found in Chapter 2, the losses associated with intervening structures along the path (building clutter) can be estimated using some physical modeling techniques. For this approach, the signal level *outside* the house or building will be found assuming (for this calculation) that there is no intervening foliage. The currently active assumption is that a commercially attractive system deployment has remote terminals that are 'user-installed' desktop devices located inside a building. The path loss to such points can be found through the sum of the path loss to points outside the building plus the path loss due to foliage and building penetration. The path loss to a location outside the building is therefore the relevant term C for use in (3.56).

A common model for building clutter loss is the 'row house' model [51] as illustrated in Figure 3.17. The assumption is that the buildings along the propagation path from the hub to the remote terminal can be thought of as successive rows of houses. This model of structures was largely formulated from the way houses are laid out in older sections of East coast cities in the United States and some European cities. However, in modern urban planning, especial in suburban areas, such uniformity is specifically avoided; instead, irregular street layouts and detached residential structures are much more common. Even for relatively closely spaced structures organized in rows along a street, the spaces between houses are still several wavelengths, especially at fixed broadband frequencies above 2 GHz (see the aerial photograph in Figure 9.4). Significant energy passing between houses is not accounted for by the row house model. Instead, a 'shadow fading' statistic is added into the path loss calculation to account for this omission.

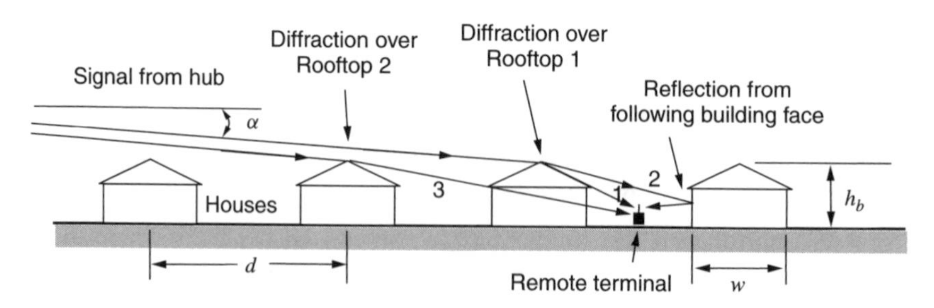

Figure 3.17 Geometry of conventional row house model for the building clutter loss to mobile locations between rows.

Nonetheless, the construction of the row house path loss model is instructive and will be described here.

As Figure 3.17 shows, the propagation paths from the hub to a (shadowed) outdoor remote terminal consists of diffracted rays over the rooftops between the remote and the hub, and also reflection of the diffracted ray from the face of the house beyond the remote terminal location. Additional rays for diffraction from the peak of the house beyond the mobile could also be included. However, for typical hub antenna heights of about 50 m, the angle α will be only about 0.6 degrees at a range of 5 km. The result is that the diffracting angle from the roof peak of the following house is highly acute and the diffraction field corresponding low.

From (3.56) the total path loss consists of the free-space path loss plus the building clutter loss (as well as the other losses). From Figure 3.17 the building clutter loss term can be divided into two parts:

- The path loss L_1 or attenuation that occurs because of the rows of houses preceding the rooftops where diffraction for paths 1, 2, and 3 in Figure 3.17 take place.
- The diffraction/reflection/transmission attenuation L_2 that occurs from the rooftop diffraction points along paths 1, 2, and 3 to the remote terminal.

Taking the L_2 term first, the amplitude attenuation of the signal at the remote terminal from the three propagation paths shown in Figure 3.18 is given as [51] (in terms of dB)

$$L_2 = -10\log\left[\frac{1}{2\pi k}\frac{1}{(h_b - h_r)^2}[\rho_1 + \rho_2|R|^2 + \rho_3|T|^2]\right] \text{dB} \tag{3.58}$$

where h_b is the height of the roof peak above ground, h_r is the height of the remote unit above ground, R is the reflection coefficient for path 2, and T is the transmission coefficient through the house for path 3, and $k = 2\pi/\lambda$ is the standard wave constant. The attenuation in (3.58) makes use of simple approximations for the Felsen's diffraction coefficients [52]. The path geometry factors ρ are given by

$$\begin{aligned}
\rho_1 &= \sqrt{(h_b - h_r)^2 + (0.5d)^2}\\
\rho_2 &= \sqrt{(h_b - h_r)^2 + (1.5d - w)^2}\\
\rho_3 &= \sqrt{(h_b - h_r)^2 + (1.5d)^2}
\end{aligned} \tag{3.59}$$

A typical residential street geometry derived from the aerial photo in Figure 9.4 has an average spacing between rows d of about 40 m and a house row width w of 10 m. Using a building height $h_b = 3.8$ m (a single-story dwelling), a remote height $h_r = 1.5$ m, a reflection coefficient of 0.5 (see Figure 2.5a for wood and $\gamma_i \cong 90$ degrees), a transmission coefficient of 0.1, a hub height of 40 m, and path length of 5 km, and positioning the remote

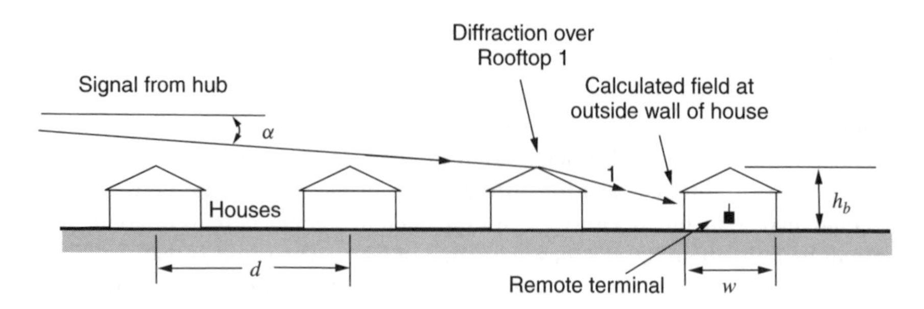

Figure 3.18 Geometry for calculating building clutter loss at the outside wall of a building.

terminal halfway between rows, the loss calculated from (3.58) is 17.4 dB at 2.6 GHz and 20.9 dB at 5.8 GHz.

If the remote terminal is located inside a house instead of in the middle of the street as in Figure 3.17, the path loss to the outside wall (prior to the building penetration loss) is the calculation objective as shown in Figure 3.18. For this geometry path 2 is irrelevant and path 3 is omitted, since at the frequencies above 2 GHz the transmission loss through the house/building will be substantial. The resulting contribution to the received power from path 3 will be much less than that of path 1. With these adjustments, L_2 in (3.58) reduces to

$$L_2 = -10\log\left[\frac{1}{2\pi k}\frac{\rho_1}{(h_b - h_r)^2}\right]\text{dB} \tag{3.60}$$

where

$$\rho_1 = \sqrt{(h_b - h_r)^2 + \left(d - \frac{w}{2}\right)^2} \tag{3.61}$$

For this geometry, using (3.60) results in $L_2 = 17.1$ dB at 2.6 GHz using the same assumptions cited above. A further calculation using the same geometry but simply calculating the diffraction loss using the knife-edge approximations in (3.47) through (3.49) results in $L_2 = 16.7$ dB. All of these local loss values are within 1 dB of each other. A similar result is found for 5.8 GHz. From these results, simplifying the problem to a single knife-edge obstacle representing the preceding obstacle appears justified.

Using this approach, a simulation was constructed to examine the statistics of the 2.6 GHz path loss for a set of 10,000 random locations inside a 5 km service radius around a hub site at 40 m above ground level (AGL). Three structure heights were used with two height distributions as shown in the legend of Figure 3.22. Oblique paths across the rows of houses were taken into account by adjusting the spacing d to the preceding diffraction obstacle as a function of the oblique path-crossing angle. The cumulative distribution function (CDF) results are shown in Figure 3.19. From this simulation, the mean path loss value for L_2 is 14.9 dB with a standard deviation of 6.5 dB. The cdf curve shows a clear dependence on the relative population of 1-story, 2-story and 3-story obstructions.

The other component of the building clutter path loss, L_1, is the EM field attenuation that occurs on the path between the hub transmitter and the rooftops where the diffraction

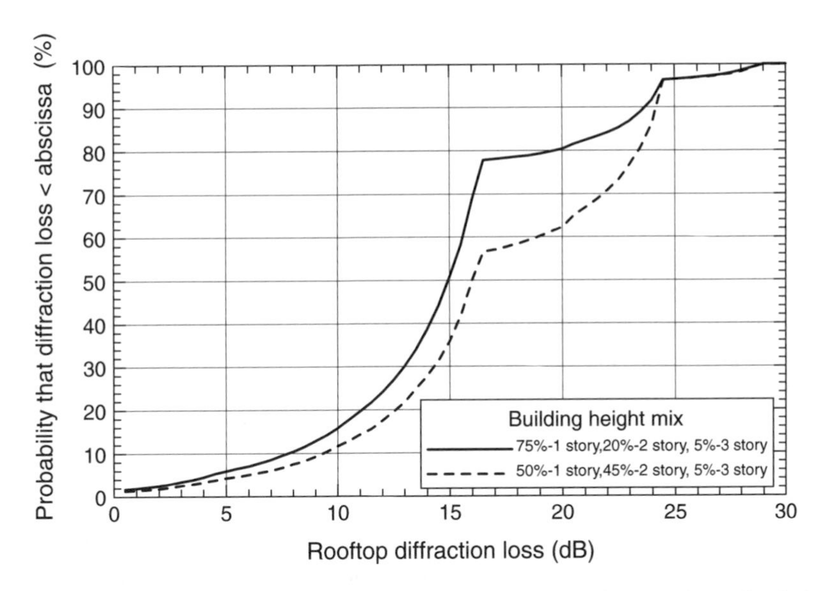

Figure 3.19 Cumulative distribution function (cdf) for rooftop diffraction from simulation using knife-edge diffraction model.

takes place for path 1 in Figure 3.18. With the row house model, this attenuation is calculated by modeling the rows of houses as a series of absorbing knife-edge screens, as shown in Figure 3.20. Normally the assumption is made that the row spacing is uniform and the house row heights are the same. This limitation is not strictly required, but it results in a simpler formulation of the composite diffraction loss over the series of edges.

A common solution to this multiple edge diffraction problem is to employ physical optics in which the field in the region above each screen is considered as the source of the fields beyond that screen. This is a very common approach that is also used to analyze the radiation patterns of horn and other open aperture antennas as described in Chapter 6. Following the analysis in [51], the integral for the field above screen $n + 1$ can be written

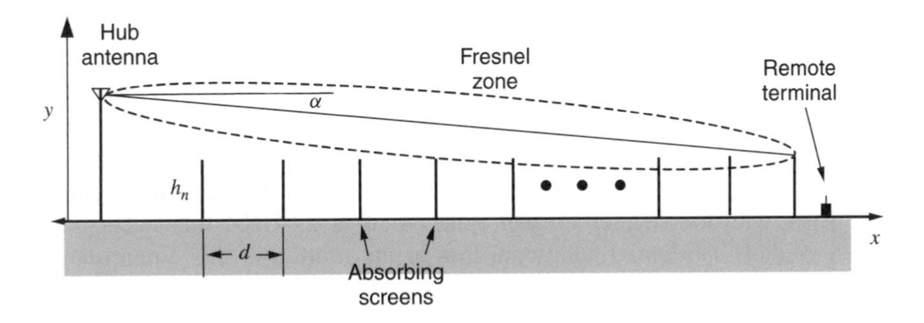

Figure 3.20 Model of row houses using uniformly spaced absorbing knife-edge screens.

in terms of the field in the aperture above screen n

$$E(x_{n+1}, y_{n+1}) = e^{j\pi/4}\sqrt{\frac{k}{2\pi}}\int_{h_n}^{\infty} E(x_n, y_n)\frac{e^{-jk\rho}}{\sqrt{\rho}}dy_n \tag{3.62}$$

where the distance between points in the aperture above screen n to the aperture above screen $n+1$ is

$$\rho = \sqrt{(x_{n+1} - x_n)^2 + (y_{n+1} - y_n)^2} \tag{3.63}$$

and the other variables are defined in Figure 3.23. The integral in (3.62) is a viable but inefficient solution. For the special case of uniform screen heights and spacings, the geometric parameter g_p can be defined as

$$g_p = \sin\alpha\sqrt{\frac{d}{\lambda}} \tag{3.64}$$

A reduction of the integral in (3.62) leads to a set of values for field strengths based on g_p and the number of rows crossed. Those field strengths oscillate up to the distance where the first row is inside the Fresnel zone. Beyond that point the field settles to a relatively constant value. The settling field value can be used to develop a relationship for field strength versus distance across the row of screens, and thus a distance path loss function. Fitting a polynomial to the field attenuation curves yields

$$Q(g_p) = 3.502g_p - 3.327g_p^2 + 0.962g_p^3 \tag{3.65}$$

The path loss using this field attenuation is then

$$L_1 = -20\log(Q(g_p)) \tag{3.66}$$

The total building clutter loss for the row house model is the sum of the two losses from (3.58 or 3.60) and (3.66)

$$C = L_1 + L_2 \tag{3.67}$$

The described row house model for calculating building clutter loss is interesting in that it produces path loss values that increase as approximately the fourth power of the distance from the transmitter to the receiver. This result has been found empirically from many measurement campaigns in the cellular and PCS frequency bands below 2 GHz. However, in at least two situations the row house model produces results that are not consistent with measurements. The first is the MMDS measurements in Figure 3.10, which show a path loss exponent of about 2 ($n \cong 2$ or free space) for the LOS case and an antenna height of 5.2 m. Even with the smaller Fresnel zone radius at 2.6 GHz, the successive screen model still predicts substantial additional loss to the rooftop in this circumstance, or a path loss exponent $n > 2$. This suggests that the underlying theory of successive screens representing rows of houses is not a valid model for the field strength attenuation that is occurring at the rooftop level.

The second situation in which the row house model has difficulties is the path loss along oblique paths crossing the rows of houses. The row house model is based on the link path from the transmitter to the receiver being perpendicular to the orientation of the rows – a relatively rare situation. Extending the calculation to paths that cross the rows at oblique angles can be done with the expected result that the path loss decreases as the angle gets more oblique. Unfortunately, the oblique angle corrections to the formulations given above do not produce results that are supported by measurements for paths across rows at oblique angles. Instead, the measured path loss is largely insensitive to the crossing angle except when the signal path is parallel to the rows, in which case the remote terminal could be near line-of-sight to the hub transmitter. There currently is no theory to account for this lack of agreement in the oblique case [51].

From the measured results in [41] shown in Figure 3.10, the path loss to the rooftop can be modeled using free-space loss calculations. From the rooftop to the outside wall of the structure, the loss over the preceding obstacle can be modeled as knife-edge diffraction. The simulation results from Figure 3.19 can then be used as representative losses for the term C in (3.56).

3.5.7 Ray-tracing

In recent years, a propagation modeling approach known as *ray-tracing* has seen considerable interest. Ray-tracing itself is a not a single cohesive mathematical technique but a collection of methods based on geometric optics (GO), the uniform theory of diffraction (UTD), and other scattering mechanisms, which can predict EM scattering from objects in the propagation environment as described in Chapter 2. This collection of field calculation methods are drawn upon mainly because none alone can successfully deal with all the geometric features of propagation environments likely to be encountered in broadband communication systems.

The notion of a 'ray' is fundamental to ray-tracing. It arises in GO where EM energy is considered to be flowing outward from a radiating source in ray tubes. While this simple concept has certain limitations, it is an important concept when considering the arrival time and angle of a particular pulse of EM energy at the receiver. Characterizing time dispersion of the received energy is an essential attribute of a site-specific physical communication channel model, and the concept of energy flowing along ray paths to the receiver, with the associated time delay proportional to the total ray path length, provides this time dispersion and angle-of-arrival information.

Ray-tracing itself is a long-used technique that can be traced back to Huygens in 1690 and Fresnel in 1818 in their studies of optical phenomena. Maxwell established a connection between optics and electromagnetism in 1873. These early treatments dealt only with rays as they traveled in free space or encountered smooth reflecting surfaces. Summerfeld [53] developed mathematical expressions for the diffracted field around a diffracting knife-edge half plane in terms of Fresnel integrals. A significant extension in geometric ray-tracing theory was made by Keller [54] who devised a way to handle edge diffraction effects using the geometric ray concept. This approach is called the geometric theory of diffraction (GTD). However, GTD had limitations in that it contained singularities, and predicted infinite diffraction fields, at the shadow and reflection boundaries around the edge. This problem was overcome by Kouyoumjian and Pathak [55] who devised the UTD, which applied a Fresnel integral to correct for the singularities at the shadow and reflection boundaries. With GO and UTD, a useful tool set was available for dealing with a wide variety of high-frequency electromagnetic problems in which the wavelength is much smaller than the dimensions of the physical features involved.

The application of ray-tracing methods to propagation prediction for communication systems has also been around for several decades. As explained above, the notion of predicting path loss over terrain obstructions by a series of diffracting edges was used by Epstein and Peterson [27] in the earlier 1950s. Throughout the 1950s and 1960s the technique of modeling path loss over irregular terrain obstructions using GO and ray diffraction became the commonly used approach with several different methods proposed to treat collections of terrain obstacles. In [32] ray-tracing was proposed as a way to predict multipath time delay, which had been observed and measured for television broadcast signals. The concept of using rays to model propagation in mobile radio, cellular, and indoor applications was extended to comprehensive ray ensembles as described in the earliest ray-tracing publications [30,31,56].

This background information is important in understanding the limitations of the ray-tracing approach. Ray-tracing does not provide a complete and accurate calculation of the field at all locations in the environment. As will be discussed, there are certainly circumstances where ray-tracing is not applicable, or where there are uncertainties in the results. In the latter case, the uncertainties are largely due to an incomplete or insufficiently refined description of the propagation environment. With this in mind, ray-tracing is a 'model' in the classic sense – it provides useful results, which are not otherwise conveniently available with empirical models, but it is not a full wave EM field solution – it is not accurate everywhere.

The mechanisms involved in ray-tracing models are in the form of 'propagation primitives' which, when used singly or in combination, describe the amplitude, phase, and time delay of ray energy arriving at the receiver. The five propagation primitives usually included in a ray-tracing model are:

- free-space propagation
- specular reflection
- diffraction
- diffuse wall scattering
- wall transmission

As a ray encounters an element in the propagation environment, one of these primitives is used to describe that interaction. As rays continue to propagate, additional environment elements will be encountered which in turn are modeled by other propagation primitives. The result is that each ray upon reaching the receiver has undergone a cascade of interactions that determine its amplitude, phase, and time delay. The details of each of the propagation primitives were discussed in Chapter 2.

An important assumption in the use of ray-tracing, GO, and UTD, is that the physical dimensions of the scattering objects are large compared to the wavelength. With this assumption, it is reasonable from a physical standpoint to regard the scattered field from one point on the scattering object to any other point on the scattering object to be insignificant compared to the incident field. Using this assumption, the interaction of the propagation primitives with the rays can be considered separately.

Although ray-tracing is founded on GO and UTD EM field calculations, it can be viewed as a model concept on a more fundamental level. A 'ray' can be thought of as an elemental, time-invariant construct of the propagation model. As such, it is hypothetical and approximate. The ray has the same relationship to the propagation model as the elements have to finite element analysis. The ray is an element that is formulated so that its properties are predictable and mathematically tractable, just as element size and shape are chosen in finite element analysis so that differential equations describing the field in an element are approximated by simple difference equations. As with finite element analysis, the approximation of the ray-tracing propagation model becomes better as the number of rays increases, and the description of their interactions with the environment boundaries becomes more refined.

As noted above, the amplitude and phase of the EM field at the receiver represented by an arriving signal ray can be found by considering the impact on the ray of all the interactions with the propagation environment it has had in transit from the transmitter to the receiver. The field at the receiver is thus given by

$$E_r = \frac{1}{s'_f}\sqrt{\frac{P_T G_T Z_0}{4\pi}}\left[\prod_i \mathbf{R}_i\right]\left[\prod_n A(s'_n, s_n)\overline{\overline{\mathbf{D}_n}}\right]\left[\prod_l A_{scat,l}\right]\left[\prod_k A_{tran,k}\right] \tag{3.68}$$

where:

s'_f is the total ray trajectory length

$\sqrt{\frac{P_T G_T Z_0}{4\pi}}$ is the free-space attenuation component from Section 2.5.2

$\mathbf{R}_i$ is the reflection coefficient from Section 2.6.1 for the ith reflection on the ray path (a complex number)

$\overline{\overline{\mathbf{D}_n}}$ is the diffraction coefficient from Section 2.7.1 for the nth diffraction wedge on the ray path (a complex number)

$A(s'_n, s_n)$ the spatial attenuation factor diffraction coefficient from Section 2.7.1

$A_{scat,l}$ is the scattering coefficient if scattering is included for objects in the model

$A_{tran,k}$ is the wall transmission coefficient if the model includes this feature.

Note that the diffraction coefficient and diffraction spatial attenuation factor are specific to the path lengths into and away from the nth wedge from wherever the illuminating source and onward receiving point or surface are located. If diffuse wall scattering or wall transmission occur along the ray path, the additional attenuation factors shown as terms 4 and 5 of (3.68) are included.

In the computer implementation of a ray-tracing model in [31], all of the rays are handled as complex electric field voltages that are affected by the magnitude and phase of the reflection and diffraction coefficients. Using (3.68) and a database describing the propagation environment, the resulting ray trajectories from this model can be found as shown by the example in Figure 3.21. In this two-dimensional (2D) model, the multitude of reflections and diffractions can easily be seen. Figure 3.22 shows a much simplified view of ray trajectories from a 3D enhancement to the model in [31]. These enhancements include diffraction over rooftops and diffraction from vertical and horizontal edges using the Keller cone geometry shown in Figure 2.12.

Several studies have been done over the past decade to validate ray-tracing, generally with good results when comparing ray-tracing signal level predictions with measurement values. Less success has been achieved when comparing predicted RMS delay-spread values with delay-spread values predicted by the ray-tracing model. A few examples comparing the predictions of the model in [31] with measurements done by others are shown in Figures 3.23 and 3.24. Figure 3.23 shows a comparison with measurements reported in [57]. The standard deviation of the prediction error 3.9 dB, a figure that is significantly better than what would be obtained with non-site-specific models such as Walfisch – Ikegami model. This is further illustrated in Figure 3.24 where ray-tracing predictions along a more complex urban route are compared with measurements from [58]

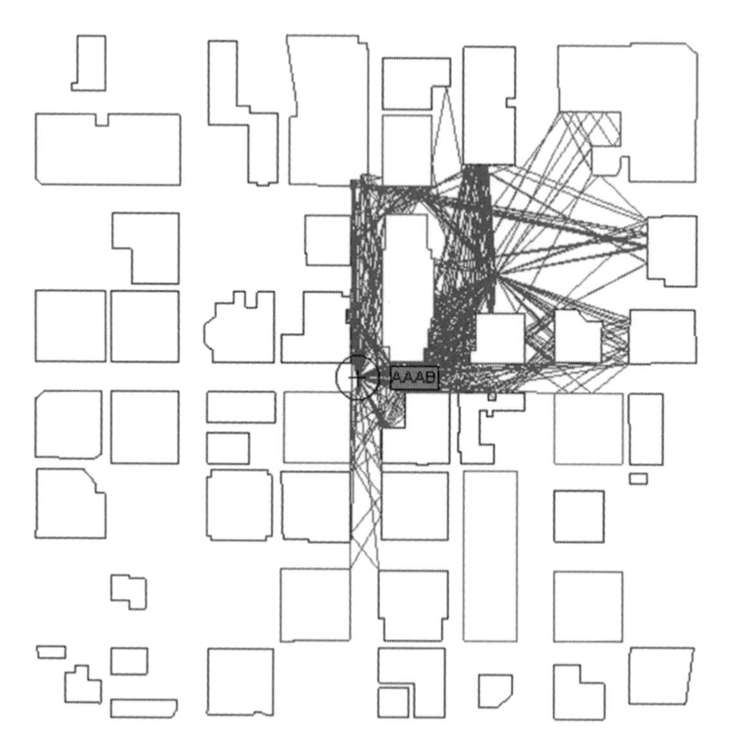

Figure 3.21 2D view of ray trajectories for ray-tracing propagation model.

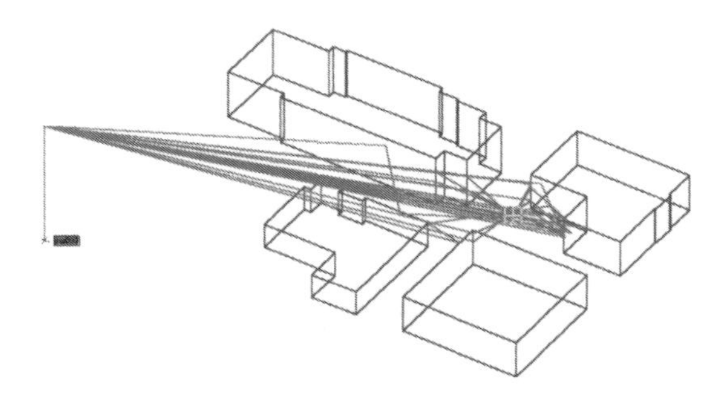

Figure 3.22 Simplified 3D ray trajectories include oblique diffraction from horizontal and vertical edges.

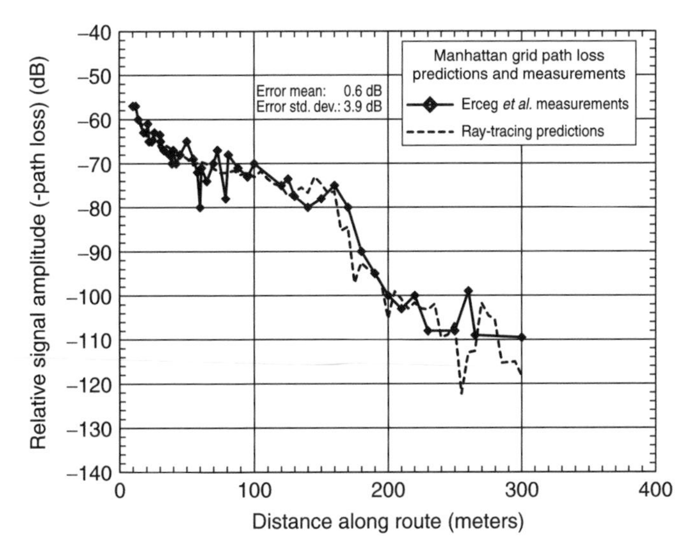

Figure 3.23 Comparison of ray-tracing predictions with measurements in Manhattan microcell using the model in [31].

and with predictions done using the Walfisch–Ikegami and Hata models. The improved accuracy that ray-tracing provides is clear from Figures 3.23 and 3.24.

For traditional LOS microwave systems through urban areas, ray-tracing has not been employed simply because it is not needed. Such links employ very high gain antennas with beamwidths of only a few degrees. For directions away from the maximum radiation directions (often called the *boresight*), very weak illumination of the off-boresight propagation environment results at the transmitting terminal and very strong rejection of off-boresight multipath components occurs at the receiving terminal. Consequently, multipath can be successfully ignored in such systems except perhaps for ground or structure reflections occurring directly along the antenna boresight path.

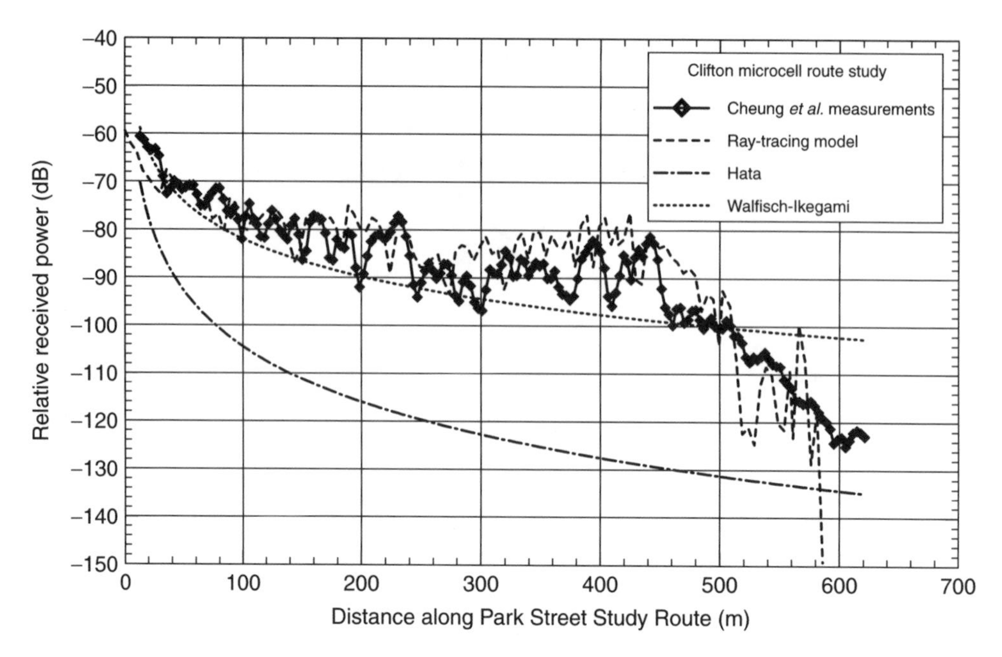

Figure 3.24 Comparison of measurements and predictions for signal levels along a route on a street in Clifton, Bristol, UK. The ray-tracing model is from [31].

For a point-to-multipoint system in which the hub is using a broadbeam antenna with a beamwidth of 30, 45 or 90 degrees, the illumination of the off-boresight environment can be substantial, indicating a potentially greater application for ray-tracing propagation models. If the receiving antenna is also low gain, or adaptive, ray-tracing models can potentially provide angle-of-arrival information that could be applied to characterize the angular distribution of energy arriving at the receiver.

The primary drawback to ray-tracing models is that the computations can take some time to complete. Since the pioneering work done in ray-tracing in the early 1990s, most of the effort with these models has been to improve the computation speed, either through simplification of the geometry of elimination of ray and ray calculations that are too weak to be relevant to the received signal. While these methods do greatly improve speed, they also sacrifice fine details that should not necessarily be excluded. The result is increased standard deviation of predictions to 7 dB or more as compared to the nearly 4 dB achieved in Figures 3.23 and 3.24 using a full-featured ray-tracing implementation. Increased processor speed has helped this situation to some extent, but generally speaking, ray-tracing is still primarily a research tool rather than a modeling approach that is widely used for day-to-day wireless system planning and deployment.

3.5.8 Simplified indoor model

The Simplified Indoor Model (SIM) uses basic elements of a physical propagation model and ray-tracing in an effort to provide a fast and simple model that predicts signal levels but does not provide time dispersion or arrival angle information. Its application to fixed broadband wireless systems is mainly in the area of license-exempt 2.4 GHz Wi-Fi or 5.7 GHz U-NII band systems that are intended to work over short ranges, especially indoors through walls and floors.

SIM makes use of four basic propagation primitives: line-of-sight rays, wall transmission loss, corner diffraction, and attenuation due to partial Fresnel zone obstruction. The first calculation the model carries out is a determination of whether the path between the transmitter and receiver is line-of-sight or obstructed. If the path is LOS, the path loss at distance d is given by

$$L_0(d) = L(d_0) + n_1 10 \log_{10}\left(\frac{d}{d_0}\right) \text{dB for } d \leq d_{FR} \tag{3.69}$$

$$L_0 = L(d_0) + n_1 10 \log_{10}\left(\frac{d_{FR}}{d_0}\right) + n_2 10 \log_{10}\left(\frac{d}{d_{FR}}\right) \text{dB for } d > d_{FR} \tag{3.70}$$

where $d_0 = 1$ meter and

$$L(d_0) = 20 \log_{10}\left(\frac{4\pi d_0}{\lambda}\right) \tag{3.71}$$

and d_{FR} is the distance at which the Fresnel intersects the floor and ceiling. The path loss coefficients (or exponents) n_1 and n_2 are chosen depending on the building type. For typical floor to ceiling heights of 3 m, the typical values used are $n_1 = 2$ and $n_2 = 3.5$.

For paths that are obstructed by one or more walls, there are two mechanisms whereby power can get to the receiver. With the first, the EM wave can pass through all the intervening walls and arrive at the receiver attenuated by the losses incurred from passing through those intervening walls. Second, the EM wave can get to the receiver from the transmitter by diffracting around the corner or edge of a wall. In this model, the power at the receiver from both mechanisms is calculated and added together to yield the power at the receiver.

For the obstructed path, the path loss L_{obs} equals the line-of-sight path loss plus the attenuation, A_k, due to each of k walls. The attenuation is based on the attenuation for the wall material type. This information can be found in references such as those cited in Section 2.9.1. In general, such wall penetration attenuation values are measured with an illumination angle normal to the wall surface. This is the incidence direction that will in general result in maximum transmission through the wall. For oblique incidence, the reflection will increase and the transmission will decrease. Thus, the transmission factor through each wall in this model is modified by the angle of incidence, ϕ, of the path through the wall. For each path–wall intersection, the specular reflection coefficient $R(k, \phi)$ is calculated and the transmission coefficient then approximated as one minus the reflection coefficient.

$$L_{obs} = \sum_{k=1}^{K} A_k[1 - R(k, \phi)] + L_0 \text{ dB} \tag{3.72}$$

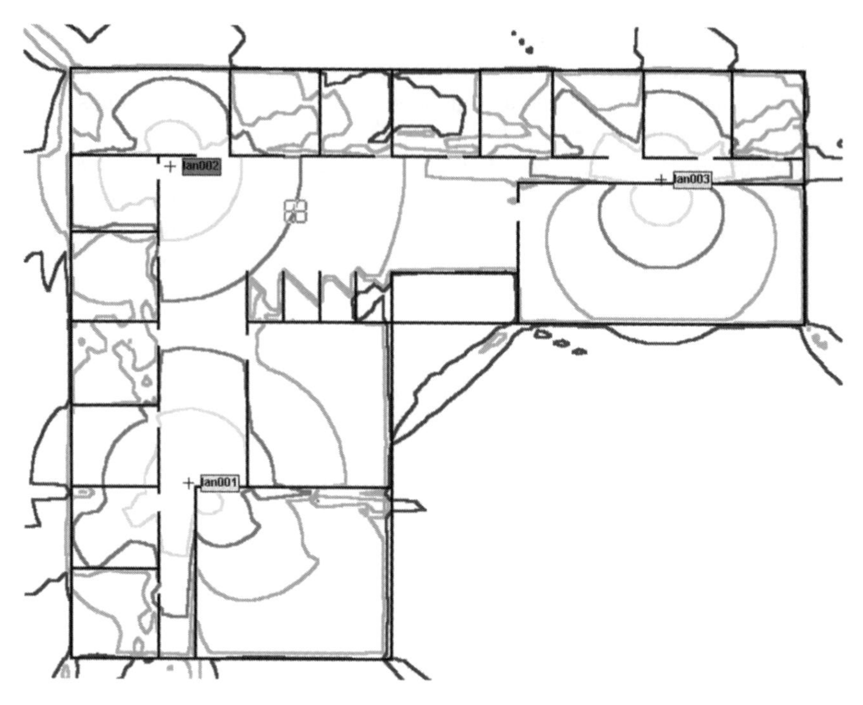

Figure 3.25 Signal strength contours for three access points for an indoor 802.11(b) 2.4 GHz wireless LAN system using simplified indoor model.

For diffracted paths, the UTD diffraction coefficient is calculated only for those corners that are *intervisible* to the transmitter and receiver using the wedge diffraction equations in Section 2.7. The power arriving at the receiver from the obstructed path and the diffracted path are added together and from that sum the total path loss is derived.

An example of signal level contours calculated using the SIM for an indoor floor plan with three access point transmitters is shown in Figure 3.25. The impact of the walls and corners on the signal strength is evident in the figure by the lobes of high signal levels down the corridors and through doorways in contrast to the pulled-in contours for signals passing through one or more walls.

3.6 CONCLUSIONS

The models presented in this chapter range from easily used empirical models to comprehensive 3D ray-tracing models that require some complexity to implement. The type of fixed broadband system that is being designed, and the goals of that design, will dictate the type of model that is best suited for the task. However, in general the physical models such as the dominant ray fixed broadband model in Section 3.5.6 offer the best all around choice. For NLOS systems in built-up or urban areas, the predicted loss of dominant ray fixed broadband model is refined with loss factors accounting for building penetration and foliage as discussed in Sections 2.9.1 and 2.9.2, respectively.

For simpler propagation paths that are known (or designed) to be LOS, the Free space + RMD model with free-space path loss, two-ray reflection, and partial Fresnel zone obstruction loss analysis is also appropriate. Models of this type have been used for decades in designing traditional point-to-point microwave links. As shown in [41], free-space path loss is also operative for short-range LOS paths in MMDS systems in suburban areas.

While propagation models generally provide predictions of mean signal strength, for fixed broadband wireless systems that are designed to provide high availability service in excess or 99.9% of the time, accurate models of multipath and rain fading are required to be used in conjunction with the basic mean signal level calculation. Such fading models are discussed in Chapter 4.

3.7 REFERENCES

[1] P.A. Bello, "Characterization of randomly time-variant linear channels," *IEEE Transactions on Communication Systems*, CS-11, pp. 360–393, December, 1963.
[2] G.L. Turin et al., "A statistical model for urban multipath propagation", *IEEE Transactions on Vehicular Technology.*, vol. VT-21, pp. 1–9, February, 1972.
[3] R.H. Clarke. "A statistical theory of mobile radio reception," *Bell System Technical Journal*, vol. 47, pp. 957–1000, 1968.
[4] H. Suzuki. "A statistical model for urban radio propagation," *IEEE Transactions on Communications*, vol. COM-25, no. 7, pp. 673–680, July, 1977.
[5] H. Hashemi, "Simulation of the urban radio propagation channel," *IEEE Transactions on Vehicular Technology*, vol. VT-28, no. 7, pp. 213–225, August, 1979.
[6] A.A.M. Saleh and R.A. Valenzuela, "A statistical model for indoor multipath propagation," *IEEE Transactions on Selected Areas in Communications*, vol. SAC-5, no. 2, pp. 128–137, February, 1987.
[7] A.M.D. Turkmani, D.A. Demery, and J.D. Parsons, "Measurements and modeling of wideband mobile radio channels at 900 MHz," *IEE Proc. I*, vol. 138, pp. 447–457, October, 1991.
[8] A. Ross and E. Zehavi, "Propagation channel model for personal communication systems," *Proceedings of 44th IEEE Vehicular Technology Conference*, Stockholm, pp. 188–189, June, 1994.
[9] FCC Rules, Part 73.313. CFR Title 47. United States Government Printing Office.
[10] VHF and UHF propagation curves for the frequency range 30 MHz and 1000 MHz. CCIR, Recommendation 370-5, vol. 5, XVI Plenary Assembly. 1986.
[11] M. Hata, "Empirical formula for propagation loss in land mobile radio services," *IEEE Transactions on Vehicular Technology*, vol. VT-29, no. 3, pp. 317–325, September, 1981.
[12] Y. Okumura, E. Ohmori, T. Kawano, and K. Fukuda, "Field strength and its variability in VHF and UHF land-mobile radio-service," *Rev. Elec. Communications Lab.*, vol. 16, no. 9–10, September-October, 1968.
[13] J.J. Egli, "Radio propagation above 40 Mc. over irregular terrain," *Proceedings of the IRE*, vol. 45, no. 10, pp. 1383–1391, October, 1957.
[14] R.E. Edwards and J. Durkin, "Computer prediction of field strength in the planning of radio systems," *Proc. IEEE*, vol. 116, no. 9, pp. 1143–1500, September, 1969.
[15] A. Bloomquist and L. Ladell. "Prediction and calculation of transmission loss in different types of terrain" NATO AGARD conf., Publication CP-144. Res. Inst. Nat. Defense Dept. 3, S-10450, Stockholm, Sweden, pp. 32/1–32/17, 1974.

[16] K. Allsebrook and J.D. Parsons. "Mobile radio propagation in British cities at frequencies in the VHF and UHF bands," *IEEE Transactions on Vehicular Technology*, vol. VT-26, no. 4, pp. 313–323, November, 1979.
[17] G.Y. Delisle, J.P. Leferve, M. Lecours, and J.Y. Chouinard, "Propagation loss prediction - a comparative study with application to the mobile radio channel," *IEEE Transactions on Vehicular Technology*, vol. VT-34, no. 2, pp. 86–96, March, 1985.
[18] J.D. Parsons. *The Mobile Radio Propagation Channel*. London: Pentech Press. 1992. Chapters 3 and 4.
[19] IEEE 802.16 Working Group. "Channel models for fixed wireless applications," Document 802.16.3c-01/29r4. July, 2001.
[20] J. Walfisch and H.L. Bertoni, "A theoretical model of UHF propagation in urban environments," *IEEE Transactions on Antennas and Propagation*, vol. AP-36, no. 12, pp. 1788–1796, December, 1988.
[21] F. Ikegami, S. Yoshida, T. Takuchi, and M. Umehira, "Propagation factors controlling mean field strength on urban streets," *IEEE Transactions on Antennas and Propagation*, vol. AP-32, no. 8, pp. 822–829, August, 1984.
[22] K. Bullington, "Radio propagation at frequencies above 30 Mc," *Proceedings of the IRE*, vol. 35, no. 10, pp. 1122–1336, 1947.
[23] A.G. Longley and P.L. Rice, "Prediction of tropospheric radio transmission over irregular terrain – a computer method," ESSA Technical Report ERL79-ITS67, 1968.
[24] Master Propagation System (MPS11) User's Manual. US Dept. of Commerce. NTIS Access no. PB83-178624.
[25] EDX Engineering, Inc. *SignalPro™ Reference Manual.* Appendix A, 1999.
[26] Telecommunications Industry Association (TIA/EIA), "Wireless Communications – Performance in Noise, Interference-Limited Situations – Recommended Methods for Technology-Independent Modeling, Simulation and Verification." TSB-88-1, June 1999, and Addendum 1, January, 2002.
[27] J. Epstein and D.W. Peterson. "An experimental study of wave propagation at 850 Mc.," *Proceedings of the IRE*, vol. 41, no. 5, pp. 595–611, May, 1953.
[28] J. Deygout, "Multiple knife-edge diffraction of microwaves," *IEEE Transactions on Antennas and Propagation*, vol. AP-14, no. 4, pp. 480–489, April, 1966.
[29] L.E. Vogler, "The attenuation of electromagnetic waves by multiple knife-edge diffraction," U.S. Dept. of Commerce, NTIA Rep. 81–86, 1981.
[30] J.W. McKown and R.L. Hamilton. "Ray-tracing as a design tool for radio networks," *IEEE Network Magazine*, pp. 27–30, November, 1991.
[31] H.R. Anderson, "A ray-tracing propagation model for digital broadcast systems in urban areas," *IEEE Transactions on Broadcasting*, vol. 39, no. 3, pp. 309–317, September, 1993.
[32] F. Ikegami, S. Yoshida, and M. Takahama, "Analysis of multipath propagation structure in urban area by use of propagation time measurements," *IEEE International Symposium on Antennas and Propagation*, Japan, 1978.
[33] F. Ikegami and S. Yoshida, "Prediction of multipath propagation structure in urban mobile radio environments," *IEEE Transactions on Antennas and Propagation*, vol. AP-28, no. 4, pp. 531–537, July, 1980.
[34] D.M.J. Devasirvatham, "Radio propagation studies in a small city for universal portable communications," *Proceedings of 38th IEEE Vehicular Technology Conference*, Philadelphia, pp. 100–104, June, 1988.
[35] H.R. Anderson. *Development and Applications of Site-Specific Microcell Communications Channel Modelling Using Ray-Tracing*. Ph.D. Thesis, University of Bristol, U.K., 1994, Chapter 4.
[36] IEEE Computer Society Working Group 802.16, "IEEE Recommended Practice for Local and Metropolitan Area Networks – Part 16: Air Interface for Fixed Broadband Wireless Access Systems." IEEE Standard 802.16–2001. *Institute of Electrical and Electronic Engineers*, April, 2002.

[37] IEEE Computer Society Working Group 802.16, "IEEE Recommended Practice for Local and Metropolitan Area Networks – Coexistence of Fixed Broadband Wireless Access Systems," IEEE Standard 802.16.2–2001, *Institute of Electrical and Electronic Engineers*, September, 2001.
[38] V. Erceg, et al. "An empirically based path loss model for wireless channels in urban environments," *IEEE Journal of Selective Areas in Communications*, vol. 17, no. 3, pp. 1205–1211, July, 1999.
[39] IEEE 802.16 Working Group. "Channels models for fixed wireless applications," Document 802.16.3c-01/29r4. July, 2001.
[40] L.J. Greenstein, V. Erceg, Y.S. Yeh, and M.V. Clark, "A new path-gain/delay-spread propagation model for digital cellular channels," *IEEE Transactions on Vehicular Technology*, vol. 46, no. 2, pp. 477–485, May, 1997.
[41] J.W. Porter and J.A. Thweatt, "Microwave propagation conditions in the MMDS frequency band," Proceedings of the 2000 *IEEE International Conference on Communications (ICC'2000)*, pp. 1578–1582, May, 2000.
[42] EURO-COST-231, Revision 2, "Urban transmission loss models for mobile radio in the 900 and 1800 MHz bands," September, 1991.
[43] H.R. Anderson, "Estimating 28 GHz LMDS Channel Dispersion in Urban Areas Using a Ray-Tracing Propagation Model," *Digest of the 1999 IEEE MTT-S International Topical Symposium on Technology for Wireless Applications*, Vancouver, pp. 111–116, February, 1999.
[44] Golden Software, Inc. *Surfer® 7 User's Guide*. Chapter 4.
[45] Federal Communications Commission. "Methods for prediction interference from response stations transmitters and to response station hubs and for supplying data on response stations systems." Report and Order in MM Docket No. 97–217. Appendix D. 1998.
[46] J.D. Parsons. *The Mobile Radio Propagation Channel*. 2nd Edition. New York: John Wiley & Sons. 2000. pp. 21–22.
[47] R.J. Luebbers, "Propagation prediction for hilly terrain using GTD wedge diffraction," *IEEE Transactions on Antennas and Propagation*, vol. AP-32, no. 9, pp. 951–955, September, 1984.
[48] P.L. Rice, A.G. Longley, K.A. Norton, and A.P. Barsis, "Transmission loss predictions for tropospheric communication circuits", National Bureau of Standards, Technical Note 101, January 1967, NTIS access number AD 687–820 and AD 687–821.
[49] G.A. Hufford. "Memorandum to users of ITS irregular terrain model", January, 1985.
[50] H.R. Anderson, "New 2D Physical EM Propagation Model Selected," *IEEE Vehicular Technology Society News*, vol. 44, no. 3, pp. 15–22, August, 1997.
[51] H.L. Bertoni. *Radio Propagation for Modern Wireless Systems*. Upper Saddle River, N.J.: Prentice-Hall PTR. 2002. Chapter 6.
[52] L.P. Felsen and N. Marcuvitz. *Radiation and Scattering of Waves*. Piscataway, N.J.: IEEE Press. 1994.
[53] A. Summerfeld. *Optics*. New York: Academic Press, 1954.
[54] J.B. Keller, "Geometric Theory of Diffraction," *Journal of the Optical Society of America*, vol. 52, pp. 116–130, 1962.
[55] R.G. Kouyoumjian and P.H. Pathak, "A uniform theory of geometric diffraction for an edge in a perfectly conducting surface," *Proceedings of the IEEE*, vol. 62, pp. 1448–1461, November, 1974.
[56] K.S. Schaubach, N.J. Davis, and T.S. Rappaport, "A ray-tracing method of predicting path loss and delay spread in microcell environments," *Proceedings of the 1992 Vehicular Technology Society Conference*, Denver, pp. 932–935, May, 1992.
[57] V. Erceg, S. Ghassemzadeh, M. Taylor, D. Li, and D.L. Schilling. "Urban/suburban out-of-sight propagation modeling," *IEEE Communications Magazine*, pages 56–61, June, 1992.
[58] J.C.S. Cheung, M.A. Beach, and S. Chard. "Propagation measurements to support third generation mobile radio network planning," *Proceedings of the 43rd Vehicular Technology Conference*, Secaucus, New Jersey, pp. 61–64, May, 1993.

4

Fading models

4.1 INTRODUCTION

'Fading' is a broad term that is applied to a wide range of variations observed in the signal amplitude, phase, and frequency characteristics. To some extent, what is classified as fading depends on the information that the propagation model itself can reliably furnish. For example, the term 'shadow fading' has been used to describe the decrease in signal strength that is observed when a mobile terminal is behind a building, resulting in an obstructed path between the terminal and the transmitter. If no database of buildings is available so that their locations are unknown and random, then the impact of buildings on the signal level (the shadow fading) must be described in some statistical way. However, if a building database is available and a physical model such as ray-tracing is used, the shadows are highly predictable. Consequently, there is no such thing as 'shadow fading' when such a model and database are used. From this perspective, a fading model is really a patch on the basic propagation model that describes those observed signal characteristics that the model is unable to adequately predict. The elements included in a fading model are therefore a function of the capabilities of the underlying propagation model.

Fading models are also used to describe the impact of physical mechanisms that defy our current technology in terms of predictability. For example, predicting atmospheric refractivity along a given link path would require a model that could take into account the temperature, the pressure, and the humidity at all points along the path and also take into account how the gases in the atmosphere respond – clearly an intractable problem with current calculation technology and data resources. Similarly, predicting when rain will cause a fade on a particular microwave link is as difficult as predicting the weather. In non-line-of-sight (NLOS) systems, signals are affected by moving cars, trucks, and buses. Their movements and their effect on system propagation are as unpredictable as traffic flow and must be treated with statistical models of some sort. All of these phenomena and many others that affect radio propagation are accounted for in the system planning process using fading models.

Fixed Broadband Wireless System Design Harry R. Anderson
 ISBN: 0-470-84438-8

The role of fading models in predicting system performance is significant. For a given design reliability criterion that may only allow an outage for a few minutes a year, the specific occurrence details of very low probability outage events, such as rain, is particularly important. Setting transmit power levels high enough to achieve an adequate *fade margin* impacts system cost as well as the potential for frequency reuse within a system and possible interference to neighboring systems.

Many fading models have been developed over the years; the most common employ Rayleigh, Rician, and lognormal probability distributions to describe signal amplitude variations. These models are widely used in designing mobile and cellular radio systems. For high-availability fixed link systems, however, the detailed shape of the low-probability 'tail' of the distribution function is particularly important so that simple, approximate statistical models like those used in mobile communications are generally not adequate for this purpose. Instead, special models have been developed to better predict the occurrence of very low probability events. This chapter will describe these specialized multipath and rain fading models in detail. Chapter 10 discusses how these models are used to predict link availability.

For NLOS systems at lower frequencies, customized empirical models of the 'tail' of the fading distribution are not available, so there is no choice but to make use of the traditional Rayleigh, Rician, and lognormal fading models familiar to cellular and mobile radio network design. The ability of these models to accurately characterize low probability fading events that are of most interest for high-availability link design is still an open question. These fading models are described in Section 4.6 of this chapter.

4.1.1 Link performance with fading

Since the amplitudes of the signal and the interference will vary (fade) with time following various statistical distributions, the signal-to-noise ratio (SNR) and signal-to-interference ratio (SIR) values also vary with time. The percentage of time the SNR and the SIR values are both above the desired thresholds is the *link availability*. The inverse of link availability is link outage. The threshold point ρ_{TH} where the performance is acceptable is usually the point where the raw bit error rate (BER) is too high for the error-correcting mechanisms to successfully remove essentially all the errors. The threshold will therefore vary as a function of link equipment design.

The link outage probability can be written as

$$\Pr(\text{outage}) = \Pr(\rho(t) < \rho_{\mathrm{TH}}) \tag{4.1}$$

where

$$\rho(t) = \frac{S(t)}{I(t) + N(t)} \tag{4.2}$$

The variables S, I, and N are random variables representing the desired signal, the interference, and the noise, respectively. If the probability distribution functions of these random variables are known, then the probability distribution of $\rho(t)$ can be found and the probability of an outage determined.

The amplitude of the desired signal $S(t)$ will experience variations or fading that depends on the propagation environment, the link geometry, the antennas used, and the bandwidth of the signal. For fixed broadband systems, the fading can be considered in two general, nonexclusive categories:

- *Line-of-Sight (LOS) Links*. For LOS links, the use of high gain directional antennas provides a degree of multipath rejection, which reduces the short-term fading depending on the link path length. In the case of long (>5 km) microwave paths subject to atmospheric fading as discussed in Section 4.2, the amplitude fading is described by empirical models that have been formulated using a vast collection of link performance data. These distributions are sometimes approximated using Rayleigh distributions. By contrast, in the case of short (<5 km) microwave links in urban areas where rooftop antennas provide significant elevation above the reflection surface, the large angular difference between the LOS direct signal and the reflections allows the antenna gain pattern to successfully suppress the reflections and thus reduce the fading and the time dispersion. The resulting signal envelope can be quite stable for long periods of time.
- *NLOS Links*. The signal amplitudes for links in obstructed non-LOS locations will exhibit fading that is similar to fading in mobile or cellular channels. This fading is usually described by a composite density consisting of Rice-distributed or Rayleigh-distributed envelope amplitude fading ('fast fading') due to multipath signals and lognormal fading of the mean values. The fast-fading distribution is due to shadowing by propagation environment features that are not accounted for by the propagation model. The necessity for statistical descriptions of shadow fading depends on the capabilities of the propagation model.

The interference term $I(t)$ in (4.2) can consist of one or more signals, each of which is subject to the same sort of fading mechanism as the desired signal $S(t)$. For LOS links, interference can exhibit the same sort of steady amplitude as the desired signal when high gain antennas are used on short, elevated links. Otherwise, for NLOS links the interference amplitude is described statistically in the same way as the desired signal amplitude. For the typical case of multiple interferers with different mean values and fading characteristics, the statistical description of the composite interference amplitude of the interference is imperative.

4.2 ATMOSPHERIC FADING MODELS

Fading due to atmospheric effects occurs for two primary reasons:

- changing propagation path curvature or alignment
- changing multipath conditions.

Both are due to changing atmospheric refractivity conditions. Fades due to changing path curvature or alignment is sometimes called a *power fade* since the signal power at the receiver is reduced because of the maximum effective radiated power from the transmitter terminal being redirected along a path that does not align with the maximum gain azimuth and elevation angles of the receiving antenna. The effect is similar to what occurs when the antenna-pointing directions are misaligned. When a link is initially installed, it is important to align antennas for maximum received power when the most commonly occurring atmospheric conditions for the link are present.

Multipath is a general term used to describe the process by which transmitted signals reach a receiver via multiple transmission paths. As discussed in Chapters 2 and 3, multipath occurs in broadband wireless systems in a variety of ways. It can result from reflection, diffraction, and scattering from fixed objects such as terrain and structures or from moving objects such as cars and buses. For microwave link systems, it can also result from reflections within atmospheric layers such as inversion layers, ducting, and various thermal layering effects. If the number, the amplitude, the phase, and the arrival times of the multipath signals arriving at the receiver were fixed (and predictable), the vector addition of these signals at the receiver would result in a net received signal that was invariant. If this level could be calculated, system design could be done using this level and there would be no need to account for signal level variability. System design would be simple and system performance perfectly predictable.

Of course, the multipath signals are not fixed in number, amplitude, phase, and arrival times, and the ability to predict these parameters is a continuing challenge for even the more advanced site-specific propagation models such as ray-tracing. For the standard propagation models used in fixed broadband wireless design, these basic multipath parameters must be treated using statistical or probabilistic descriptions. In other words, the question to be answered is 'what is the probability that the multipath signals will add vectorially with the directly received signal to produce a reduction in the signal of 20 dB (or 30 dB or 40 dB)?' If the probability density function (pdf) for the amplitude of the multipath fading is known, then this question can be answered with great accuracy. If this pdf is only approximately known, the answer to this question will be correspondingly approximate. The desired models for multipath fading are designed to present the pdf as accurately as possible.

Several multipath fading mechanisms and the models that describe them are discussed in the following sections.

4.2.1 Microwave multipath fading mechanisms

There are two primary mechanisms that create multipath in microwave systems:

- Atmospheric multipath in which the sharp changes or discontinuities in the refractivity gradient create reflection surfaces in the atmosphere.
- Multipath from ground reflections that vary as the refractivity gradient varies. These variations cause the strength and geometry of ground-reflected signals to change as the gradient changes.

As discussed in Section 2.10, most of the time the refractivity gradient of the atmosphere is a linear function of height. In this case the propagation direction of the wave follows a curved path that is approximately represented by an adjusted earth curvature or k factor. However, if the gradient is not uniform, this simple picture does not apply. Atmospheric layers with sharp transitions or discontinuities in the refractivity gradient

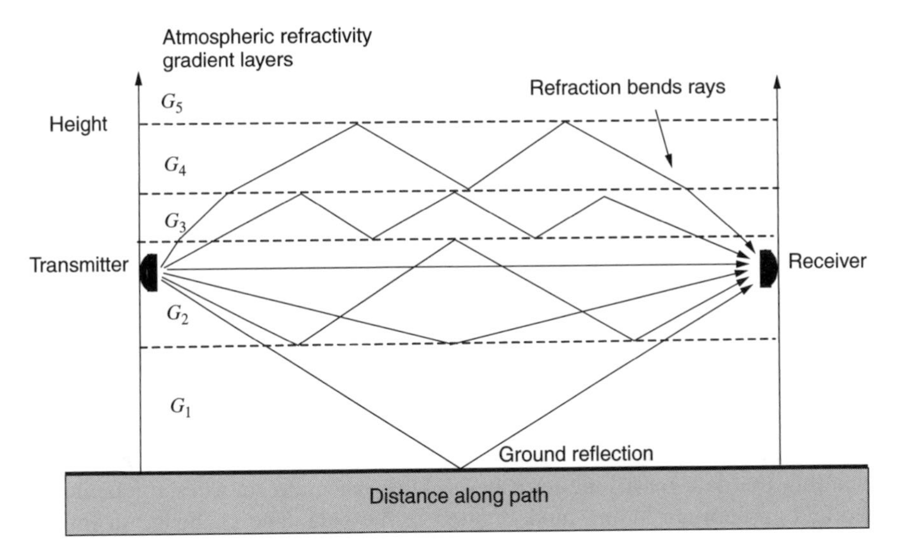

Figure 4.1 Schematic drawing of multipath due to refractivity gradient discontinuities and ground reflections.

due to abrupt temperature or humidity differences can create reflection layers that result in multiple signal paths and associated multipath fading. The source of the reflections is different from the ground or the wall reflections discussed in Chapter 2 and 3, but nonetheless, they may be treated statistically in much the same way. Figure 4.1 illustrates atmospheric multipath that results in rays reflecting from the inversion layers as well as from the ground. As the height, the location, and the refractivity gradient of the layers change with changing temperature and humidity, the location of the ray reflections change along with path length and corresponding phase delay (see Section 3.5.1.2 of Chapter 3). The vector sum of the signals arriving at the receiver will therefore change as these multipath components change.

The refraction mechanism that causes electromagnetic (EM) waves to change direction is illustrated in Figure 4.2 for a simple two-dimensional case. This is basically the same wave transmission mechanism discussed in Section 2.9. When the two media involved are atmospheric layers, Snell's Law can be used to predict the change in the angle of the ray path as it passes from one medium to the other.

$$\eta_1 \sin\theta_1 = \eta_2 \sin\theta_2 \tag{4.3}$$

where $\eta_{1,2}$ are the refractive indexes of the two media and $\theta_{1,2}$ are wave propagation directions through the media as shown in Figure 4.2. For time-dependent atmospheric fading to occur, the refractivity gradients (height, gradient value, etc.) must also change with time. The statistical models of fading described in the following sections attempt to describe the net impact on the received signal level from the fading caused by these mechanisms.

Much of the specialized fading models have been developed using measured data from actual operating links. Most microwave link systems use automatic gain control

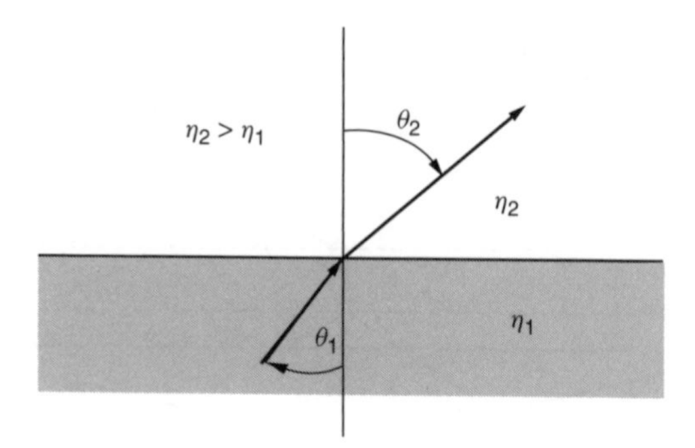

Figure 4.2 Wave refraction through different atmospheric refractivity layers.

(AGC) of some sort to try to keep the received signal level constant. By monitoring and recording the data on a long-term basis, fairly accurate statistics for fading can be gathered and appropriate fading models fitted to the data. The challenge in formulating such fading models is the same as the challenge in formulating the empirical propagation models discussed in Chapter 3. The fading statistics for each link are different, as are the statistics of technically identical links in different locations and climates. Analyzing this kind of data and distilling it into a simple model form is the task of the empirical model developer.

4.2.2 Vigants–Barnett model

The Vigants–Barnett method is not an administratively sanctioned multipath fading model as such, but is based on the published work of two microwave system researchers at AT&T Bell Labs [1,2]. This work used both analytical and experimental data to create semiempirical equations for fade depth probability for the received signal. This model was specifically created to fit the fading probability distribution for fade depths greater than 15 dB.

The probability of a fade of depth A using this method is given by

$$P_F = 6.0 \times 10^{-7} C f d^3 10^{\frac{-A}{10}} \tag{4.4}$$

where

P_F = probability of a fade as a fraction of time
d = path length in kilometers
f = frequency in gigahertz
C = propagation conditions factor
A = fade depth in decibel. This equation is only valid for fade depths of 15 dB or more.

Note that in the original Barnett paper [1] the factor $10^{-A/10}$ is shown as L^2. The two factors are equivalent since the normalized envelope value is defined in Barnett as $20 \log L$.

The propagation conditions factor, or C, is selected on the basis of the type of environment in which the link is to operate. The maps in Figures A.1 and A.2 (Appendix A) provide an indication of the appropriate C factor for the area where the link will be deployed.

Equation (4.2) has been developed for fade depths greater than 15 dB. For short-range paths using elevated directional antennas, such as those for a 28-GHz LMDS (Local Multipoint Distribution Service) system, such deep fading may not be relevant. For fade depths less than 15 dB, the fade probability can be drawn from the lognormal distribution. The standard deviation of the lognormal distribution can be adjusted so that at the 15-dB transition point the fade probabilities are equal, thus eliminating any step discontinuities at this point. The fade probability relative to the mean value for the lognormal distribution is given by

$$P_F = 0.5\left[1.0 - \text{erf}\left(\frac{A}{\sqrt{2}\sigma}\right)\right] \tag{4.5}$$

where erf(·) is the error function, A is the fade depth (in this case 15 dB), and σ is the standard deviation value being calculated. The first step is to find P_F for the Vigants–Barnett model using (4.2). Equation (4.5) can then be used to iteratively calculate P_F using different values of σ until the value of P_F from (4.5) matches the value from (4.4). The value of σ at this point will provide a fading distribution that can be used to find P_F for fade depths less than 15 dB. To find the outage probability, use this value of σ in (4.5) with the desired fade depth $A < 15$ dB.

Figure 4.3 shows a plot of the Vigants–Barnett fading model for a 6-GHz, 50-km link using a value of $C = 2.0$. The lognormal distribution fit to match at the 15-dB fade point.

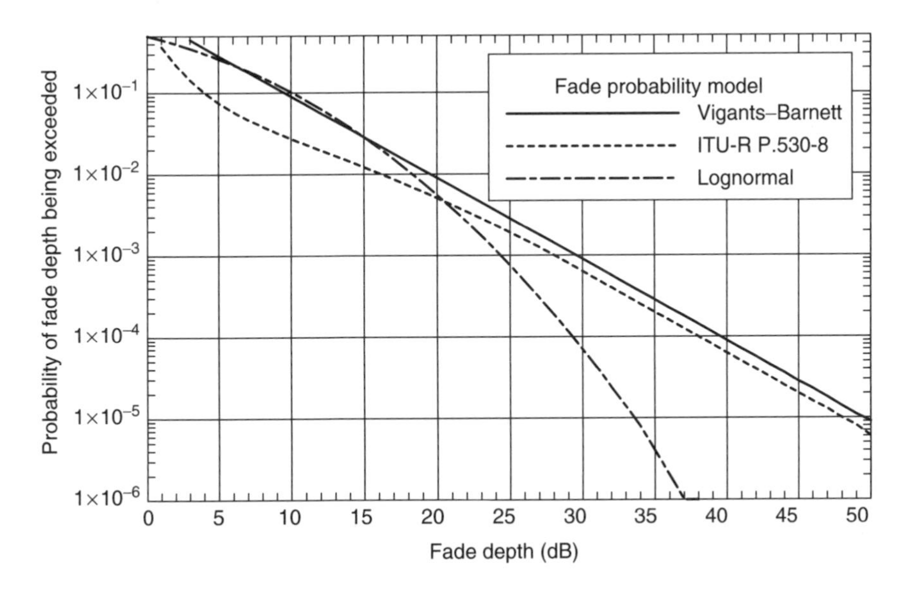

Figure 4.3 Fade probability versus fade depth for various multipath fade models.

4.2.3 ITU-R P.530-8 model

The multipath fade outage probability formula for the average worst month found in ITU-R Rec. 530-8 [3], (19) is

$$P_F = Kd^{3.6} f^{0.89} (1 + |\varepsilon_p|)^{-1.4} \times 10^{\frac{-A}{10}} \tag{4.6}$$

where

P_F = probability of a fade in percent. The resulting value is divided by 100 to yield fade probability as a fraction of time.
K = geo-climatic factor for worst fading month, as calculated below
d = path length in kilometers
f = frequency in gigahertz
ε_p = path inclination in milli-radians
A = fade depth in decibel. This equation is only valid for fade depths of 15 dB or more.

The path inclination angle ε_p in milli-radians is calculated using the transmitting and the receiving antenna heights above mean sea level as follows:

$$|\varepsilon_p| = \frac{|h_r - h_e|}{d} \tag{4.7}$$

The geo-climatic factor K is defined in terms of p_L, the percent of time the relative refractivity gradient is less than −100 N/km. This factor p_L can be found from the maps in Figures A.3 through A.6 (Appendix A) showing this value for four months representing each season. The month with the highest value of p_L should be used.

$$K = 5.0 \times 10^{-7} \times 10^{-0.1(C_0 - C_{Lat} - C_{Lon})} p_L^{1.5} \tag{4.8}$$

If the type of terrain is unknown, the following estimates of the coefficient C_0 can be used in (4.8):

$C_0 = 1.7$ for the lower altitude antenna in the range of 0 to 400 m above mean sea level.

$C_0 = 4.2$ for the lower altitude antenna in the range of 400 to 700 m above mean sea level.

$C_0 = 8.0$ for the lower altitude antenna more than 700 m above mean sea level.

The coefficient C_{Lat} in (4.8) is based on latitude ϕ and is given by

$C_{Lat} = 0$ dB for $\phi \le 53^\circ$ north or south

$C_{Lat} = -53 + \phi$ dB for ϕ greater than 53° north or south and less than 60° north or south

$C_{Lat} = 7$ dB for $\phi \ge 60^\circ$ north or south.

The coefficient C_{Lon} in (4.8) is based on longitude and is given by

$$C_{Lon} = 3.0\,\text{dB for longitudes of Europe and Africa}$$

$$C_{Lon} = -3.0\,\text{dB for longitudes of North and South America}$$

$$C_{Lon} = 0.0\,\text{dB for all other longitudes.}$$

The equation for K in (4.8) is basically for inland paths over terrain and not water. For paths over various kinds of terrain, the coefficient C_0 can be adjusted as set forth in Table 4.1. For paths near or over small or large bodies of water, there are additional correction factors to K, or alternate means of calculating K to improve fade outage predictions in such circumstances. Details of these specialized correction factors are found in [3]. Using this as a guide, the environment type and coefficients can be chosen to match the intended link environment.

Table 4.1 C_0 factors for various types of altitude and terrain for use with ITU-R 530-8 fade outage model from [3]

Altitude of lower antenna and type of link terrain	C_0 (dB)
Low altitude antenna (0–400 m) – plains:	
Overland or partially overland links, with lower-antenna altitude less than 400 m above mean sea level, located in largely plains areas	0
Low altitude antenna (0–400 m) – hills:	
Overland or partially overland links, with lower-antenna altitude less than 400 m above mean sea level, located in largely hilly areas	3.5
Medium altitude antenna (400–700 m) – plains:	
Overland or partially overland links, with lower-antenna altitude in the range 400–700 m above mean sea level, located in largely plains areas	2.5
Medium altitude antenna (400–700 m) – hills:	
Overland or partially overland links, with lower-antenna altitude in the range 400–700 m above mean sea level, located in largely hilly areas	6
High altitude antenna (>700 m) – plains:	
Overland or partially overland links, with lower-antenna altitude more than 700 m above mean sea level, located in largely plains areas	5.5
High altitude antenna (>700 m) – hills:	
Overland or partially overland links, with lower-antenna altitude more than 700 m above mean sea level, located in largely hilly areas	8
High altitude antenna (>700 m) – mountains:	
Overland or partially overland links, with lower-antenna altitude more than 700 m above mean sea level, located in largely mountainous areas	10.5

As noted in [3], (4.6) has been derived from fading data for paths in the 7 to 95 km range, at frequencies from 2 to 37 GHz, and path inclinations from 0 to 24 milli-radians. It may also remain reasonably valid for paths up to 237 km and for frequencies as low as 500 MHz. An estimate of the lower frequency bound is given as

$$f_{\min} = \frac{15}{d}\ \text{GHz} \tag{4.9}$$

where d is the path length. Like the Vigants–Barnett method, (4.6) is intended to be used when the fade depth is greater than 15 dB. For short-range paths such as those for LMDS systems in which the path length may only be on the order of 5 km or less, system reliability calculations at fade depths less than 15 dB may be relevant. The ITU-R 530-8 in [3] provides a supplemental method for calculating the probability of fade depths in this range. This method is similar to the one described for use with Vigants–Barnett because the shallow-fading distribution is merged with the deep-fading distribution described by (4.6) to create a composite distribution that is also valid at low fade depths.

The first step in this process requires calculating P_F using (4.6) and a fade depth of 0 dB ($A = 0$). This value of P_F is designated p_0. The next step is to calculate the value of the fade depth, A_t, at which the transition occurs between the deep-fading distribution and the shallow-fading distribution as predicted by the empirical interpolation procedure:

$$A_t = 25 + 1.2 \log p_0 \tag{4.10}$$

The procedure now depends on whether A is greater than or less than A_t. If the fade depth, A, of interest is equal to or greater than A_t, then the percentage of time that A is exceeded in the average worst month is calculated as

$$P_F = p_0 \times 10^{-A/10}\ \% \tag{4.11}$$

If the fade depth, A, of interest is less than A_t then the percentage of time, p_t, that A_t is exceeded in the average worst month is calculated as

$$p_t = p_0 \times 10^{-A_t/10} \% \tag{4.12}$$

Next, the quantity q'_a is calculated from the transition fade depth A_t and transition percentage time p_t:

$$q'_a = -20 \log\left[-\ln\left(\frac{100 - p_t}{100}\right)\right] \Big/ A_t \tag{4.13}$$

Next, the quantity q_t is calculated from q'_a and the transition fade depth A_t:

$$q_t = (q'_a - 2)/\lfloor (1 + 0.3 \times 10^{-A_t/20}) 10^{-0.016A_t} \rfloor - 4.3(10^{-A_t/20} + A_t/800) \tag{4.14}$$

Next, q_a is calculated from the fade depth of interest A:

$$q_a = 2 + [1 + 0.3 \times 10^{-A/20}][10^{-0.016A}][q_t + 4.3(10^{-A/20} + A/800)] \tag{4.15}$$

Table 4.2 Parameters for the ITU-R fade probability calculations shown in Figure 4.3

Parameter	Value
C_0	1.7 dB
C_{Lat}	0.0 dB
C_{Lon}	−3.0 dB
p_L	15.0%
ε_p	0.0
f	6 GHz
d	50 km

Finally, the fade probability percentage, P_F, that the fade depth A is exceeded in the average worst month is calculated as

$$P_F = 100[1 - \exp(-10^{-q_a A/20})] \tag{4.16}$$

This value can be divided by 100 to get fade probability as a fraction of time. According to [3], provided that $p_0 < 2000$, the above procedure produces a monotonic variation of P_F versus A, which can be used to find A for a given value of P_F using simple iteration procedure. Figure 4.3 shows the fade probability using the ITU-R model. Table 4.2 lists the parameters used to calculate the curve shown in Figure 4.3. As Figure 4.3 shows, the ITU-R and Vigants–Barnett methods give fade probabilities that are fairly close for typical path atmospheric conditions.

4.2.4 Dispersive (frequency-selective) fading

The generic channel model in Chapter 3 demonstrated that multipath fading is frequency-selective; that is, the depth of a fade for any given multipath condition at a given point in time will be different at different frequencies. If the bandwidth of the signal is large enough, these differences can result in some parts of the occupied transmission spectrum being faded more than others. This condition is illustrated in Figure 4.4 for a simple two-ray fading case. For the simple two-ray case, the frequency span between spectrum response minima Δf is found from the reflection time delay τ_d in seconds:

$$\Delta f = \frac{1}{\tau_d}\ \text{Hz} \tag{4.17}$$

The time domain representation of the two-ray situation is shown in Figure 4.5 using a conventional raised cosine pulse as the transmitted signal. The delayed reflection pulse will result in *inter-symbol interference* (ISI) when its voltage interferes with the correct detection of a subsequently transmitted pulse. This mechanism is discussed in some detail in Chapter 7.

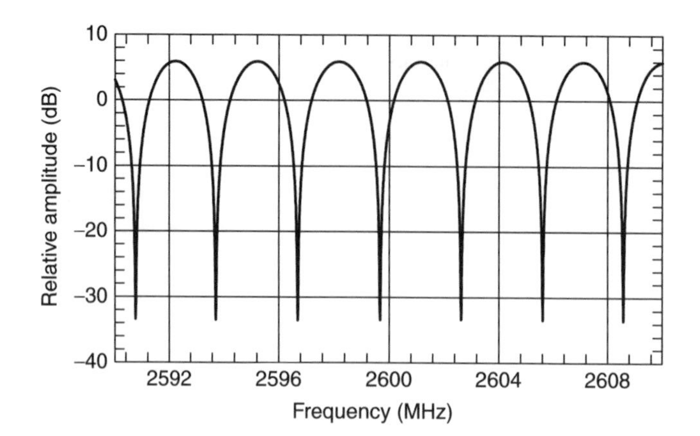

Figure 4.4 Frequency-selective fading in a simple two-ray propagation model situation. Reflection ray $\Delta r \cong 100\,\text{m}$.

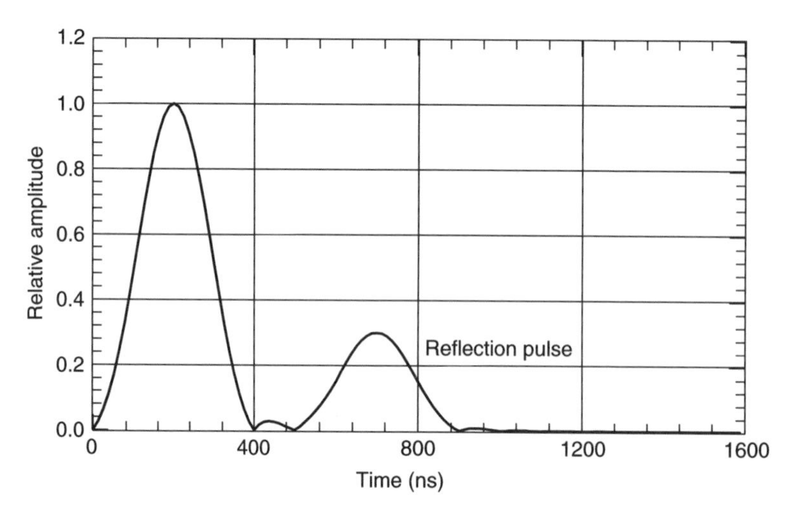

Figure 4.5 Time signature with one reflection ray with an amplitude equal to 30% of the directly received ray.

4.2.4.1 Coherence bandwidth

For most realistic propagation environments with many rays of different amplitudes and time delays, the *coherence bandwidth*, B_c, is another metric for describing the degree of time dispersion or frequency-selective fading in a channel. The coherence bandwidth can be found by evaluating the correlation coefficient of the time domain channel response as a function of frequency. As the frequency separation increases (as in Figure 4.4), the (narrowband) time domain response of the channel at one frequency versus another will be increasingly dissimilar or uncorrelated.

In Section 3.3.1.3 of Chapter 3, the time-variant transfer function, $T(f, t)$, of the channel was defined. The *time–frequency correlation function* of the channel is defined as

$$R_{TT}(f, f+\Delta f; t, t+\Delta t) = E[T(f, t)T^*(f+\Delta f, t+\Delta t)] \tag{4.18}$$

The *correlation coefficient function* (a function of time or frequency) is actually the covariance function normalized so that its values fall in the range of -1 to 1. The covariance function is given by [4]

$$\begin{aligned}\text{cov}[T(f, t), T(f+\Delta f, t+\Delta t)] &= E[T(f, t), T(f+\Delta f, t+\Delta t)] \\ &\quad - E[T(f, t)]E[T(f+\Delta f, t+\Delta t)]\end{aligned} \tag{4.19}$$

substituting (4.18)

$$\begin{aligned}\text{cov}[T(f, t), T(f+\Delta f, t+\Delta t)] &= R_{TT}(f, f+\Delta f; t, t+\Delta t) \\ &\quad - E[T(f, t)]E[T(f+\Delta f, t+\Delta t)]\end{aligned} \tag{4.20}$$

For wide sense stationary uncorrelated scattering (WSSUS) channels in which $T(f, t)$ is not dependent on the absolute value of time or frequency, this R_{TT} becomes

$$R_{TT}(f, f+\Delta f; t, t+\Delta t) = R_{TT}(\Delta f, \Delta t) \tag{4.21}$$

To create the correlation coefficient function, $\rho(\Delta f, \Delta t)$, the covariance function is normalized by dividing it by the square root of the product of the variances [4]. The variance of $T(f, t)$ is given by

$$\sigma^2_{T(f,t)} = E(\{T(f, t) - E[T(f, t)]\})^2 \tag{4.22}$$

and

$$\sigma^2_{T(f+\Delta f,t+\Delta t)} = E(\{T(f+\Delta f, t+\Delta t - E[T(f+\Delta f, t+\Delta t)]\}^2) \tag{4.23}$$

so

$$\rho(\Delta f, \Delta t) = \frac{R_{TT}(\Delta f, \Delta t) - E[T(f, t)]E[T(f+\Delta f, t+\Delta t)]}{\sqrt{\sigma^2_{T(f,t)}\sigma^2_{T(f+\Delta f,t+\Delta t)}}} \tag{4.24}$$

If (4.24) is used to consider the correlation coefficient as a function of frequency difference Δf only and not time (i.e. $\Delta t = 0$), then (4.24) becomes $\rho(\Delta f, \Delta t) = \rho(\Delta f)$.

From a practical perspective, the correlation coefficient as a function of Δf can be calculated using N samples taken of the time-dependent received signal fading envelope $y(t)$ at frequencies f and $f + \Delta f$ with the correlation at Δf given as

$$\rho(\Delta f) = \frac{\sum_n [y(f, t_n) - \overline{y}(f)][y_2(f+\Delta f, t_n) - \overline{y}(f+\Delta f)]}{N\sqrt{\sigma^2_{y(f)}\sigma^2_{y(f+\Delta f)}}} \tag{4.25}$$

where $\overline{y}(\cdot)$ is the average of the time samples at frequencies f and $f + \Delta f$ and σ_y^2 is the corresponding variances of these sample sets. If $\rho(\Delta f)$ is calculated for increasing values of Δf, the frequency correlation function is found.

The coherence bandwidth is usually defined as the value of Δf where $\rho(\Delta f)$ falls below a value of 0.5 (some use a value of 0.9 here). The coherence bandwidth is inversely proportional to the real microwave systems (RMS) delay spread (see Section 3.3.1).

$$B_c \propto \frac{1}{\sigma_\tau} \tag{4.26}$$

The value of σ_τ itself averages out much of the fine detail of the power delay profile that can have an important impact on the coherence bandwidth. If some well-behaved function is assumed for the shape of the power delay profile, then estimates of the coherence bandwidth as a function of σ_τ can be found. A classically used approximation for the delay spread as a function of τ is the exponential function [5]

$$P(\tau) = \frac{1}{2\pi\sigma_\tau} e^{-\tau/\sigma_\tau} \tag{4.27}$$

As shown in [5], this results in a correlation coefficient function of

$$\rho(\Delta f, \Delta t) = \frac{J_o^2(2\pi f_m \Delta t)}{1 + (2\pi\Delta f)^2\sigma_\tau^2} \tag{4.28}$$

where J_0 is a Bessel function of the first kind and zero order, the variable f_m is the maximum Doppler shift frequency, where $f_m = v/\lambda$, v is the velocity of the receive terminal, and λ is the wavelength. For a fixed receive terminal, $v = 0$, so $f_m = 0$ and $J_0^2(2\pi f_m \Delta t) = 1$. Thus, (4.28) reduces to

$$\rho(\Delta f, \Delta t) = \frac{1}{1 + (2\pi\Delta f)^2\sigma_t^2} \tag{4.29}$$

Figure 4.6 shows the correlation coefficient as a function of frequency spacing Δf for several values of delay spread. The coherence bandwidth in the 20 ns case is about 8 MHz, while the coherence bandwidth for an RMS delay spread of 150 ns is about 1.1 MHz. From Section 3.4.3, the average measured RMS delay spread for NLOS channel using a directional antenna was about 140 ns.

The curves shown in Figure 4.6 were created with the assumption that the power delay profile decreases exponentially with delay. In [6], a ray-tracing propagation model was used to investigate coherence bandwidth using more realistic site-specific power delay profiles. These studies showed that coherence bandwidth can vary significantly from one location to another depending on the prevailing propagation mechanisms at that location.

4.2.4.2 Dispersive fade margin

Dispersive fade margin (DFM) is actually a characteristic of the receiving equipment rather than the propagation channel, but its meaning is directly tied to the preceding

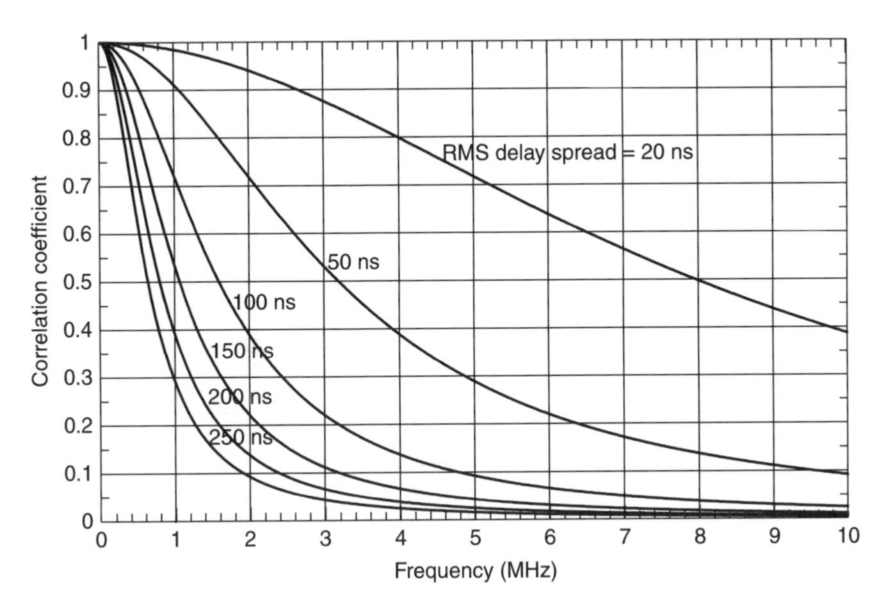

Figure 4.6 Correlation coefficient as a function of frequency separation Δf and delay spread σ_τ.

discussion of frequency-selective fading, so it is appropriate to present it here. DFM is a measure of the ability of the receiver to produce adequate error rates in the presence of frequency-selective fading.

Rummler [7,8] devised a simple, pragmatic model to characterize dispersive fading and the ability of the receiving equipment to successfully perform under dispersive fading conditions. Rummler proposed a two-ray propagation model with a direct ray and a single reflected ray in which the reflected ray is delayed by 6.3 ns. The 6.3 ns is not a magical number; it was simply the average value of observed reflection delay times for operating microwave links where such data had been collected. As such, Rummler's approach is particularly tuned to situations of relatively long microwave links operation in the 4-, 6-, and 11-GHz microwave bands. This should be kept in mind when considering dispersive fading in short-range microwave systems such as LMDS, or NLOS systems in the 2.6 or 5.7 GHz bands.

Using the Rummler two-ray model, the receiving equipment can be tested to a signal consisting of a directly received carrier and a reflected carrier delay by 6.3 ns. The carrier frequency can then be moved across the passband of interest and the amplitude of the reflection signal adjusted until the detected BER increases to a predefined level, usually 10^{-3}. The depth of the notch created by the reflected carrier is then recorded as a function of frequency to create the dispersion performance curve. A typical dispersion performance curve is shown in Figure 4.7. Because of its shape, these signature curves are often called W-curves. They provide a useful basis for comparing the performance of one receiver to another.

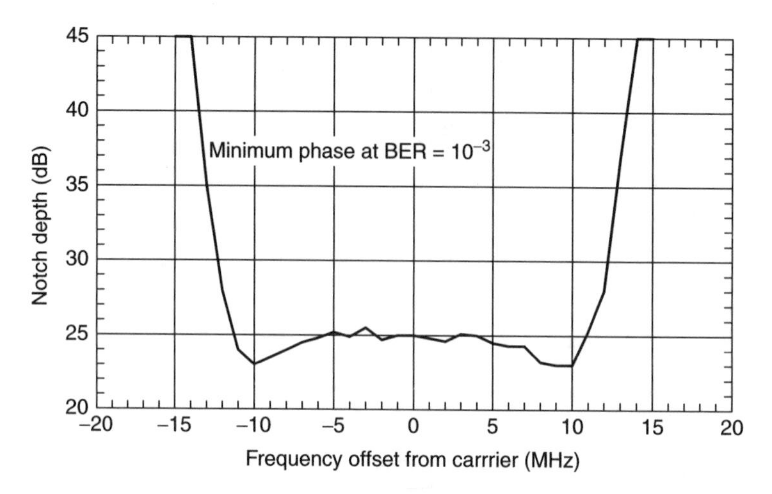

Figure 4.7 Representative measured dispersion performance signature.

In considering the directly received signal and the reflected signal, two conditions can exist: the reflection may arrive at the receiver before the directly received signal or after the directly received signal. If the reflection arrives after the directly received signal, it is known as the *minimum-phase* condition; if the reflection arrives before the directly received signal, it is known as the *non-minimum-phase* condition. Given the geometry of the propagation path, especially when the directly received signal is refracted and has a longer path to the receiver, both conditions are possible. Since both conditions can occur with RMS, it is important to assess the receiver performance under both conditions. A plot of the non-minimum-phase condition is often designated as the W-curve, while a plot of the minimum-phase condition is called the M-curve. When the two curves are plotted on the same graph, the M-curve can be drawn with the decibel scale reversed from that shown in Figure 4.7, so it takes on the general shape of an M-curve.

For more general NLOS situations, there is no direct path but merely a maximum amplitude path that may arrive at the receiver via reflection, diffraction, scattering, or a combination of all of these mechanisms. It could easily be delayed much longer than a signal arriving via a secondary path that is weaker in amplitude.

Using the W-curve and the M-curve, Rummler defined the DFM as

$$\text{DFM} = 17.6 - \log\left(\frac{S_w}{158.4}\right) \text{ dB} \tag{4.30}$$

where

$$S_w = \int_{f_{\min}}^{f_{\max}} \left(\exp\left(\frac{-B_n(f)}{3.8}\right) + \exp\left(\frac{B_m(f)}{3.8}\right)\right) df \tag{4.31}$$

The terms $B_n(f)$ and $B_m(f)$ are the dispersive performance curves for the non-minimum-phase and minimum-phase conditions, respectively, and the minimum and maximum frequencies for the measurement are selected to cover the operating bandwidth of the receiver. Equation (4.31) is basically a specialized average of the areas under the W-curve and the M-curve, which show the radio performance under both minimum-phase and non-minimum-phase conditions. The greater the area, the better the performance of the receiver.

For digital radios to perform well under dispersive fading conditions, they must be equipped with an *equalizer* of some sort. There are many types of time domain and frequency domain equalizers that will be discussed further in Chapter 7. Equalizers use a pilot training symbol to probe the channel response. When the training symbol is sent, the receiver knows ahead of time what the received symbol should be. Anything that differs from the expected symbol, such as a reflection symbol as illustrated in Figure 4.5, must result from dispersion in the channel. In simplest terms, an equalizer is a device with multiple taps and the ability to control the relative signal gain of those taps such that a reflection occurring at a particular time can be 'subtracted out' and the originally transmitted symbol restored. Once trained, the equalizer can perform the same operation on real transmitted data instead of the training symbol and thus counteract the effect of the time dispersion in the channel. Of course, if the channel response is time-varying, the equalizer must also be periodically retrained.

The design of optimum algorithms for training and adjusting equalizers is an area of significant research since it has a dramatic impact on the error rate performance of the link. Receivers equipped with equalizers typically have DFM values between 50 and 60 dB, while those without an equalizer have DFM values typically between 30 and 35 dB.

It should also be noted at this point that multipath signals are not necessarily a negative element in link performance. Multipath signals represent additional signal energy that can potentially be exploited to improve rather than degrade overall system performance. A class of receivers called *Rake receivers* are designed with this concept. Rake receivers are used in currently deployed code division multiple access (CDMA) cellular phone systems. In the context of W-CDMA technology for NLOS fixed broadband wireless systems, Rake receivers will be discussed in more detail in Chapter 7.

4.3 RAIN FADING MODELS

Unlike fading due to atmospheric conditions, signal fades due to rain are *flat fades* in that they are not frequency-selective. For any practical bandwidth of interest, the fade depth will be essentially the same at all frequencies within the band. This makes modeling of rain fades more straightforward than modeling atmospheric fading; however, it also means that desired signal power at all frequencies is weaker, so the use of equalizers and space diversity are not effective at combating rain fades. Instead, the conventional means of combating rain fades is with *automatic power control* (APC), which increases the transmit power as necessary to maintain the desired received BER. While this can be effective for the link being subjected to rain (up to the maximum of the power control range), for high-density systems this process can lead to increased interference to some other link that must also respond by raising its transmit power. This link coupling effect is an important aspect of analyzing high-density fixed broadband wireless systems that will be discussed in some detail in Chapter 11.

The electromagnetic process that results in a rain fade and depolarization was outlined in Section 2.12. To take this process into account in wireless system design, models are needed that can predict the rain fade impact. There are two models that are primarily used:

- Crane rain fade model
- ITU-R rain fade and depolarization model.

Each of these approaches will be discussed below.

The degree of attenuation due to rain is a function of the *rain intensity* or *rain rate*. Rain rates are given in terms of the probability that the average millimeters of rain that fall over an hour will be exceeded. Climates that have high total rainfall over a year could have fewer rain fade outages on a wireless link than climates with a lower total rainfall, but the rainfall comes during short, intense storms or cloudbursts. Rain fade models that rely on rain intensity data for the system will operate. Worldwide maps showing zones with particular rain rate profiles have been developed over the years, but are necessarily inexact when considering rain intensity for a highly localized area where a fixed broadband system operates. For reference, rain rate zone maps that are used with the Crane and ITU-R models are included in Appendix A.

4.3.1 Crane rain fade model

The Crane method for predicting the outage due to high intensity rainfall is based on a publication that appeared some years ago [9]. It was more recently updated with new rain zone definitions and rain rate values [10]. The rain attenuation is calculated by the Crane method as follows:

$$A_R = kR_p^\alpha\left(\frac{e^{\mu\alpha d}-1}{\mu\alpha} - \frac{b^\alpha e^{c\alpha d}}{c\alpha} + \frac{b^\alpha e^{c\alpha D}}{c\alpha}\right) \text{ dB for } d \le D \le 22.5\,\text{km} \tag{4.32}$$

$$A_R = kR_p^\alpha\left(\frac{e^{\mu\alpha D}-1}{\mu\alpha}\right) \text{ dB for } D < d \tag{4.33}$$

where

$$\mu = \frac{\ln(be^{cd})}{d} \tag{4.34}$$

$$b = 2.3R_p^{-0.17} \tag{4.35}$$

$$c = 0.026 - 0.03\ln(R_p) \tag{4.36}$$

$$d = 3.8 - 0.6\ln(R_p) \tag{4.37}$$

R_p = rain rate in millimeters per hour

D = path length in kilometers

Table 4.3 Regression coefficients for calculating k, α

Frequency (GHz)	k_H	k_V	α_H	α_V
1	0.0000387	0.0000352	0.912	0.88
2	0.00154	0.000138	0.963	0.923
4	0.00065	0.000591	1.121	1.075
6	0.00175	0.00155	1.308	1.265
7	0.00301	0.00265	1.332	1.312
8	0.00454	0.00395	1.327	1.31
10	0.0101	0.00887	1.276	1.264
12	0.0188	0.0168	1.217	1.2
15	0.0367	0.0335	1.154	1.128
20	0.0751	0.0691	1.099	1.065
25	0.124	0.113	1.061	1.03
30	0.187	0.167	1.021	1
35	0.263	0.233	0.979	0.963
40	0.35	0.31	0.939	0.929
45	0.442	0.393	0.903	0.897
50	0.536	0.479	0.873	0.868
60	0.707	0.642	0.826	0.824
70	0.851	0.784	0.793	0.793
80	0.975	0.906	0.769	0.769
90	1.06	0.999	0.753	0.754
100	1.12	1.06	0.743	0.744

k, α = regression coefficients from Table 4.3 for the frequency of interest. Note that these values are taken from ITU-R Rec. 838 and are different from the coefficients Crane used in his original paper.

The rain rate values in millimeters per hour are directly related to the geographical area where the link will operate. The map in Figure A.7 in Appendix A shows Crane rain regions throughout the world with letter designations A through H. The rain rates for each region as a percentage of time each year are given in Table A.1.

To find the rain outage percentage with the Crane rain model, (4.32) or (4.33) is iteratively calculated with increasing rain rates for the appropriate zone until the value of rain attenuation A_R equals the value that will result in a link outage (A_R equals the *rain fade margin*). The percentage of time corresponding to that rain rate is the outage percentage. Most often, linear interpolation is used between values of A_R that bracket the rain fade margin with a corresponding interpolation of the outage percentages. Alternately, the worst-case (highest) outage percentage can be used instead of the interpolated value.

This rain fade probability calculation procedure is for path lengths of $D = 22.5$ km or shorter. For path lengths greater than 22.5 km, the calculation is done for $D = 22.5$ km and the final adjusted outage probability is calculated as

$$P_R = P_{22.5}\left(\frac{D}{22.5}\right) \quad \text{where } D \text{ is the actual path length in km} \tag{4.38}$$

where P_R is the final rain outage percentage and $P_{22.5}$ is the rain outage percentage for a path length of 22.5 km. The annual outage probability may be multiplied by the number of seconds in a year to yield the total number of seconds per year that the link is unavailable because of rain outages.

4.3.2 ITU-R P.530-8 model

Rain attenuation by the ITU-R method is done in a similar way to the Crane analysis. However, the basic equation and the rainfall region map are different. The basic equation for rain attenuation is taken from ITU-R P. 530-8 [11] and Rec. 838 [12] as follows:

$$\gamma_R = kR^{\alpha} \text{ dB/km} \tag{4.39}$$

where k and α are the coefficients found from values in Table 4.3 for the relevant frequency and R is the rain rate in millimeter per hour for 0.01% of the time for a given rain region of the world. The maps in ITU-R P.837-2 [13] show 0.01% time rain rates for different parts of the world. The world map of ITU-R rain regions in Figure A.8 can also be used with the rain rate table in Table A.2 to find the 0.01% time rain rate. The values in Table 4.3 are used to find k and α using the following equations:

$$k = \frac{[k_H + k_V + (k_H - k_V)\cos^2\Theta\cos 2\tau]}{2} \tag{4.40}$$

$$\alpha = \frac{[k_H\alpha_H + k_V\alpha_V + (k_H\alpha_H - k_V\alpha_V)\cos^2\Theta\cos 2\tau]}{2k} \tag{4.41}$$

where Θ is the path elevation angle and τ is the polarization tilt angle relative to the horizon. For linear horizontal or vertical polarization used here, the polarization tilt angle τ is either 0 or 90°, respectively. Likewise, for terrestrial radio paths, the path elevation angle is typically very small, especially for path distances where rain attenuation might be important.

To find the total attenuation on the path due to rain, the attenuation factor γ_R must be multiplied by the effective path length, d_{eff}, which is defined as

$$d_{eff} = \frac{d}{\left(1 + \dfrac{d}{d_o}\right)} \tag{4.42}$$

where d is the actual path length in kilometer and for $R_{0.01} \leq 100$ mm/h (the rain rate 0.01% of the time),

$$d_o = 35e^{-0.015R_{0.01}} \tag{4.43}$$

The total path attenuation for 0.01% of the time is then given by

$$A_{0.01} = \gamma_R d_{eff}\ dB \tag{4.44}$$

The loss for 0.01% of the time can be converted to some other percentages, p, ranging from 0.01 to 1% using the following formula from [11], (4.44), for links operating at latitudes above 30° (north or south):

$$A_p = A_{0.01} 0.12 p^{-(0.546+0.043 \log_{10} p)} \text{ dB} \tag{4.45}$$

For latitudes below 30° (north or south), the corresponding formula is

$$A_p = A_{0.01} 0.07 p^{-(0.855+0.139 \log_{10} p)} \text{ dB} \tag{4.46}$$

As with the Crane method, (4.45) or (4.46) is solved iteratively to find a rain-caused attenuation that is equal to or exceeds the rain fade margin. When this value is found, the corresponding probability is used as the rain outage probability, P_R, for calculating link availability. This approach is only valid from 0.001 to 1% on the basis of the restrictions stated in ITU-R P.530-8.

Figures 4.8 and 4.9 show some comparisons of the Crane and ITU-R methods over range of frequencies and rain rates for a short 5-km path. For the regression coefficients in Table 4.3, the attenuation for vertical polarization can be substantially less than that for horizontal polarizations. Although these models show this distinction, according to Crane [14] such differences are not observed with operating systems.

4.3.3 Short-range rain fading

The Crane and ITU-R methods for computing rain attenuation as a function of path length, frequency, and rain rate can be generally applied for fixed broadband systems operating above 5 GHz. However, much of the data used to develop these models was taken for longer paths at which a particular point rain rate is used to integrate the loss over the

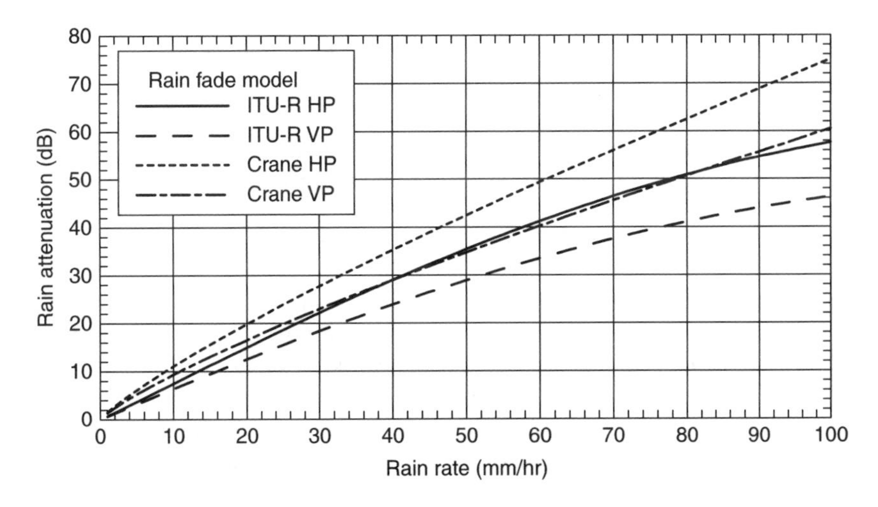

Figure 4.8 Rain attenuation versus rain rate for a 5-km path at 28 GHz.

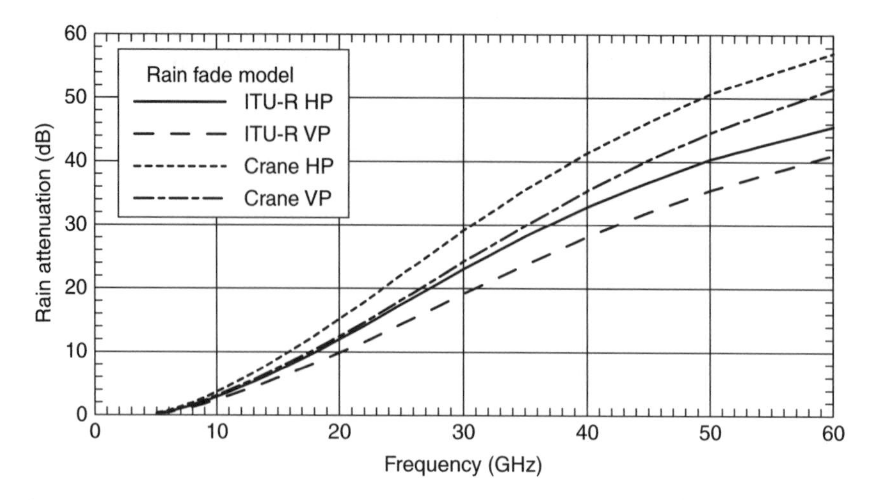

Figure 4.9 Rain attenuation versus frequency for a 5-km path with a rain rate of 28 mm/h.

entire path. Rain cells are generally amorphous shapes in which the rain intensity can vary considerably along a path. Deriving the total path attenuation from measurements or an approximate estimate requires an integration or expectation of what total loss will be along the path. Both the Crane and ITU-R models make use of this approach, although the ITU-R model is more dependent on an ensemble of attenuation measurements along a wide variety of paths, whereas the Crane model is formed from the statistics of the variation of the rain rates along the path.

For paths shorter than 5 km that will probably be employed for fixed broadband systems above 5 GHz, various researchers have published measurement results for operating paths that provide more specific information for such systems. In [15], measurements for a few short links were done at 38 GHz that show an upper bound on this additional attenuation of 2.7 and 5.2 dB above the Crane attenuation for an LOS link of 605 m and a partially obstructed link of 262 m, respectively. This additional attenuation could potentially be due to wet antennas or radomes as described in Section 4.3.4. The work in [15] also suggested a Rician distribution for short-term received power distribution during rainstorms, with the k factor of the distribution proportional to rain rate in millimeter per hour:

$$k = 16.88 - 0.04R \tag{4.47}$$

In [16] rain attenuation results are reported from a two-year study in England using systems operating at 42 GHz. The study on which the rain result in [16] is based also included measurements and comparisons of point rain rates and rain rates averaged over circular areas of 2.5 and 5 km in radius. Table 4.4 shows the results of these measurements. The average rain rates differ from the point rates significantly for high rain rates, but for low rain rates, the difference is much smaller. This suggests that high rain rates occur in very localized areas for a very short length of time. It also suggests that path diversity, that is, using an alternate link routing during such short-term events, is potentially a successful

Table 4.4 Point and area average rain rates for England (from [16])

Percentage of time	Point rain rate (mm/h)	Area-averaged rain rate (mm/h)	
		2.5-km radius area	5-km radius area
0.001	65.6	36.0	33.0
0.003	46.2	29.0	23.4
0.01	29.9	19.4	17.1
0.03	18.1	16.3	12.6
0.1	9.8	9.5	8.5
0.3	5.0	4.9	4.8
1	2.0	2.1	2.1

strategy for mitigating outage from these rain fades. Path diversity will be discussed in Chapter 10.

4.3.4 Other precipitation losses

In addition to the scattering that occurs from raindrops as they fall along the propagation path, there are other precipitation-related mechanisms that introduce attenuation into the link budget of a communication system. Chief among them is the water layer that collects and flows off of antenna *radomes* or reflectors.

Radomes are antenna aperture covers made of nonconductive material designed to protect the antenna components from the weather. Under clear sky conditions, radomes normally introduce only a fraction of decibel of additional signal loss. When the radome is wet, however, this loss can be considerably higher. Several studies have been done to this effect [17,18,19]. The amount of loss is dependent on the way the water is present on the radome surface. If the water is in the form of a laminar flow across the surface, the loss is highest as demonstrated theoretically by Gibble [20]. If the water 'beads up' into droplets rather than forming into a sheet, the attenuation for a given rain rate is much smaller. Activity has therefore been directly toward radome coating materials that are hydrophobic, that is, that are resistant to supporting a continuous water layer or flow. Figure 4.10 shows a graph of additional attenuation from wet radome versus rain rate.

Snow and ice on radomes are also a consideration, but as reported in [17], such losses are considerably less than those due to a rain surface layer. At 20 GHz, 1/2 inch of dry snow produced a loss of 0.1 dB; 1/4 inch of ice produced a loss of 1 dB.

4.3.5 Cross-polarization discrimination fading model

As discussed in Section 2.12, rain not only attenuates the desired signal but also scatters the signal such that EM waves with random polarizations are created. If a vertically polarized wave is transmitted, during a rainstorm horizontally polarized fields are created. For high-density broadband wireless systems that use polarization discrimination to increase system capacity, the decrease in cross-polarization discrimination can be significant.

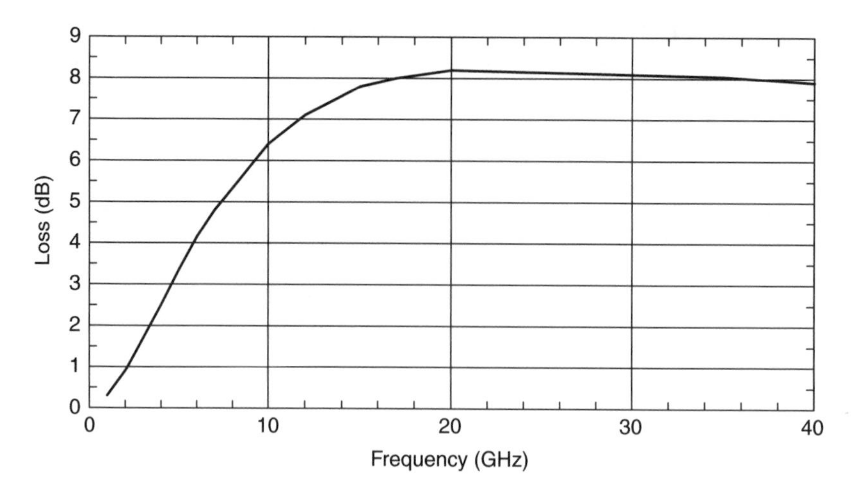

Figure 4.10 Wet radome loss as a function of frequency for a laminar water film (after [18]).

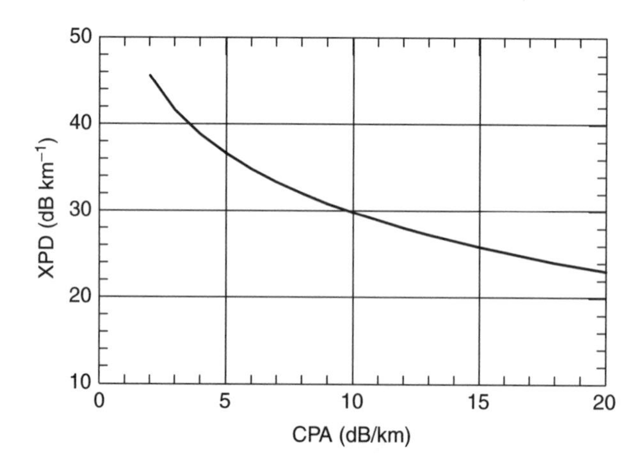

Figure 4.11 Decrease in cross-polarization discrimination versus copolarized attenuation from ITU-R formulas. Frequency = 28 GHz.

The commonly used model for fading in cross-polarization discrimination is found in [11], as follows:

$$\text{XPD} = U - V(f)\log(CPA) \tag{4.48}$$

The XPD is the cross-polarization discrimination, the CPA is the copolarized attenuation (standard rain fade), f is in GHz, and the factors U and $V(f)$ can be approximated as

$$U = U_0 + 30\log f \tag{4.49}$$

$$V(f) = 12.7 f^{0.19} \quad \text{for } 8 \le f \le 20\,\text{GHz} \tag{4.50}$$

$$V(f) = 22.6 \qquad \text{for } 20 < f \leq 35\,\text{GHz} \tag{4.51}$$

The value used for U_0 in (4.49) is typically from 9 to 15 dB. A graph showing XPD as a function of CPA is found in Figure 4.11 for a system at 28 GHz. No information is currently published on the decrease in cross-polarization discrimination due to wet radomes.

4.4 CORRELATED RAIN FADING MODEL

As noted above, rain fades result from the radio transmission path traversing an area where rain is falling. These areas are usually referred to as rain cells. Rain cells can be widespread or highly localized, with rain intensity varying within the cell. The shape of the cell is irregular (amorphous) and the cell is often moving in the direction of the prevailing wind during a storm.

Radio paths that transit the same rain cell will experience attenuation that is proportional to the path length, rain rate, and frequency. High-density fixed wireless systems will generally be *interference-limited*, that is, the performance of any particular link is controlled by the interference it receives rather than by the thermal noise in the receiving equipment itself. The ratio of the desired carrier signal strength, S, to the interfering signal strength, I, which may be the composite of many interfering sources, is denoted as the S/I ratio. If the interfering signal path to the *victim* receiver crosses the same rain cell as the desired signal path, then it will be attenuated owing to rain at the same time as the desired signal. This is correlated rain fading and the result is that during a rain fade, the desired carrier-to-noise (S/N) ratio may be decreasing as the desired signal fades, but the S/I ratio may not be changing, or may be changing to a small degree depending on the differences between the desired signal path and the interfering signal path.

Figure 4.12 illustrates the situation of a desired signal path and an interfering path arriving at a victim receiver across a rain cell. The extent to which the rain fade on the two paths are correlated will be a function of the angular difference α between the desired and the interfering paths. When $\alpha = 0$, the desired and the interfering signals are arriving along the same path and the rain fades are necessarily correlated. Under such conditions, the S/I ratio will not degrade as the desired signal is attenuated by the rain, depending on where the interfering transmitter lies along the desired signal path.

As the angular separation increases, the fading correlation on the desired and the interfering paths is reduced. The extent to which it is reduced will depend on the shape, the size, and the intensity of the rain within the cell. It will also depend on the lengths of the paths involved. When both paths are short (a few kilometers), it is more likely that they are both experiencing fades from the same rain cell. Unfortunately, there is little useful statistical information on the size, the shape, the intensity, and the movement of rain cells that could be applied to this problem.

A simple, first-order model for the correlation factor ρ_R as a function of angular separation can be written as

$$\rho_R \propto |\cos\alpha| \tag{4.52}$$

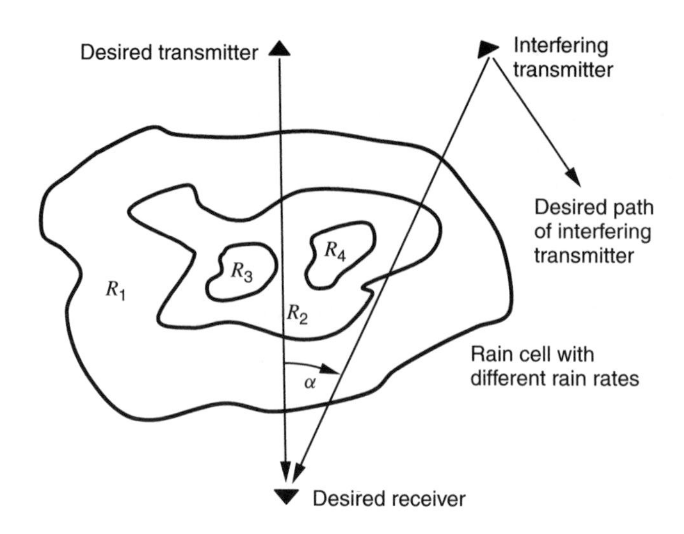

Figure 4.12 Desired and interference paths crossing a rain cell.

There are no published studies available on correlated rain fading as a function of path arrival difference. However, from Figure 4.12 it is clear that off-boresight antenna pattern gain discrimination by the desired receiving antenna and the interfering transmitting antenna will reduce the significance of any interfering source where α is very large, so the degree of correlated fading for larger values of α may be irrelevant given the reduced interfering signal level. This particular fading issue falls in the larger context of calculating interference in fixed wireless systems, especially those with high link densities where it is more likely to have pairs of desired interfering paths with small values of α. Calculating interference in such high-density systems will be treated in Chapter 11.

4.5 FREE SPACE OPTICS FOG FADING MODELS

The main atmospheric phenomenon that limits the range and the availability of a free space optics (FSO) link is fog as explained in Section 2.13. The fading statistics for attenuation due to fog are given by the probability that a particular fog density will occur. Table 4.5 from [21] shows international visibility definitions and the amount of path attenuation that can be expected because of fog of that density. Table 4.5 also shows FSO losses associated with rain for various rain visibility categories.

Unfortunately, there are no comprehensive databases of fog density for various parts of the world similar to those for atmospheric refractivity and rain intensity. If a database of the probability of various fog densities were available, it could then be used along with an FSO link budget to predict the probability of a link outage as was done with rain outages.

Currently one of the primary sources for visibility data are weather stations at airports, but this data is generally averaged over long periods of time, and fog conditions at

Table 4.5 International visibility codes for weather conditions and precipitation

Weather condition	Precipitation description	Rain visibility	Rain rate (mm/h)	Rain loss (dB/km)	Fog visibility	Fog loss (dB/km)
Dense fog		—		—	0, 50 m	271.65
Thick fog		—		—	200 m	59.57
Moderate fog		—		—	500 m	20.99
Light fog	Cloudburst	1000 m	100	9.26	770 m	12.65
Thin fog	Heavy rain	2.0 km	25	3.96	1.9 km	4.22
Haze	Medium rain	4.0 km	12.5	1.62	2.8 km	2.58
Light haze	Light rain	10 km	2.5	0.44	5.9 km	0.96
Clear	Drizzle	20 km	0.25	0.22	18.1 km	0.24
Very clear		50 km		0.06	23.0 km	0.19

the airport may not be indicative of the fog conditions in which the FSO links will be deployed. In general, the coarse approach of assessing visibility by human observation is still widely used. In general, an FSO link will successfully operate at a distance that is twice as far as a human can see.

4.6 FADING MODELS FOR NLOS LINKS

The tapped delay line wideband channel model discussed in Section 3.3.1.2 of Chapter 3 can be easily simplified to create a model that approximates the narrowband fading observed in NLOS links. Such links are typical of mobile and cellular networks and can also be applied to fixed NLOS links. The fading or signal level uncertainty that occurs comes from two sources:

- Fading due to unknown local changes in the propagation environment such as people moving around the room, passing vehicles, trees moving in the wind, and so on. The movement causes changes in the amplitude and the phase of signals that reflect and scatter from these objects. The variations manifest themselves as time variations in the signal envelope level even though they result from changes in the locations of elements in the propagation environment. This component of fading model is often called multipath fading or 'fast fading'.
- The mean value of the multipath fading distribution also varies because of mean path loss differences from location to location that cannot be predicted by propagation models. For example, inside a house, a remote terminal may be moved from room to room, with each location having a different building penetration loss. Outside, the remote terminal may be moved among different locations, some of which are line of sight to the hub transmitter and some of which are shadowed by tall buildings, trees, and so on. If the propagation model and propagation environment model are not refined enough to account for these static mean path loss variation, for the purpose of calculating link outage, they must be accounted for in some other way. The way

they are usually accounted for is to add a 'shadow fading' component to the NLOS link fading model.

Taken together, multipath fading and shadow fading form a composite fading model. The components of the composite model are described below.

4.6.1 NLOS multipath fading models

Multipath fading is usually modeled using one of several well-known probability distributions to describe the variations of the envelope voltage of the signal. The models assume that the signal is sufficiently narrowband such that the fading is not frequency-selective. Even in channels in which the bandwidth of the signal and the nature of the propagation environment result in frequency-selective fading, flat fading models such as these can still be used to describe the fading that occurs in time segments (windows or bins) or narrow frequency segments of the overall signal bandwidth. As discussed in Chapter 7, some multicarrier modulation schemes such as Orthogonal Frequency Division Multiplexing (OFDM) are designed so that the bandwidth of each of the multiple carriers is confined to that bandwidth that can be characterized as flat fading.

A flat fading model can be derived from the tapped delay line model in Section 3.3.1.2 of Chapter 3 by simply removing all the delay taps in Figure 3.8 except the first one. This delay now represents the single net propagation delay for the signal to travel from the transmitter to the receiver. The single multiplicative function $r(t)$ becomes a description of the narrowband fading envelope. For a channel in which something (terminal locations, propagation environment, refractivity, etc.) is changing in an unknown way, $r(t)$ is a random process that can have multiple realizations. Over the years, several statistical probability density functions (pdfs) have been used to represent $r(t)$. The most common are

- Rayleigh
- Rician
- Nakagami.

Each of these distributions is discussed in the following sections. From Figure 3.8, it is clear that variations in $r(t)$ will directly affect the detected amplitude of the output signal $y(t)$. However, this variation does not have a frequency dependence; that is, the amplitude of $y(t)$ will vary in the same way with $r(t)$ regardless of frequency and hence a flat fading model is achieved.

4.6.1.1 Rayleigh distribution

In radio systems engineering, the distribution of the flat fading envelope has in the past usually been assumed to be a Rayleigh distribution [22]. The Rayleigh distribution describes the pdf of the magnitude of the resultant vector sum of two independent, zero-mean Gaussian random variables in quadrature, shown in Figure 4.13 as $n_x(t)$ and $n_y(t)$. By the Central Limit Theorem, the distribution of the sum of a large number of

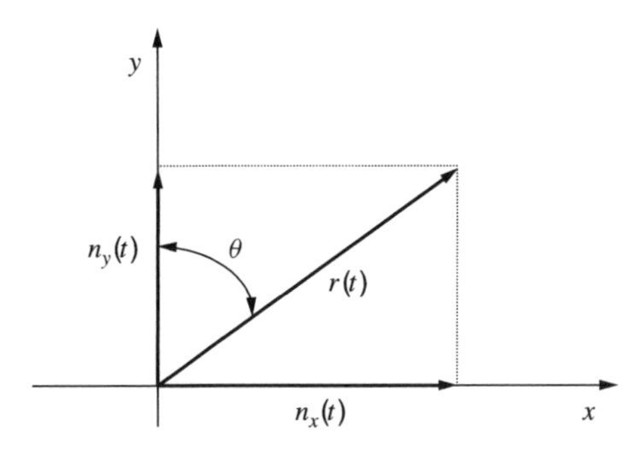

Figure 4.13 Rayleigh-distributed vector r.

random variables will yield a random variable with a Gaussian distribution regardless of the distributions of the individual random variables in the sum. Even with as few as six randomly phased sinusoids, the envelope of the sum is essentially Rayleigh [23]. From this perspective, when a large number of multipath components in (3.5) are considered, the Rayleigh distribution can be a justifiable choice.

Using this geometry, the pdf of the resultant vector can be derived into the following basic form:

$$p_r(r) = \frac{r}{\sigma^2} \exp\left(\frac{-r^2}{2\sigma^2}\right) \tag{4.53}$$

where σ^2 is the variance of the Gaussian distributions (both assumed to be the same). It can be shown [24] that the phase of the angle θ is uniformly distributed from 0 to 2π. A plot of the Rayleigh distribution is shown in Figure 4.14. The median of the distribution is found by finding the value for r for which 50% of the area of the distribution is above and below this value of r:

$$0.5 = \int_r^{\infty} p_r(r)\, dr = \exp\left(\frac{-r^2}{2\sigma^2}\right) \tag{4.54}$$

The median is then found as $r = 1.17\sqrt{\sigma^2}$. In a similar way, the mean $\bar{r}$ and the variance b^2 for the Rayleigh distribution can be found by integrating $p(r)$ to yield

$$\bar{r} = \sqrt{\frac{\pi}{2}}\sigma = 1.25\sigma \tag{4.55}$$

$$b^2 = \sigma^2\left(2 - \frac{\pi}{2}\right) \tag{4.56}$$

The ratio of the mean to the variance of the Rayleigh distribution is

$$\frac{\bar{r}}{b^2} = \frac{1.25}{\sigma(2 - \pi/2)} = \frac{2.91}{\sigma} \tag{4.57}$$

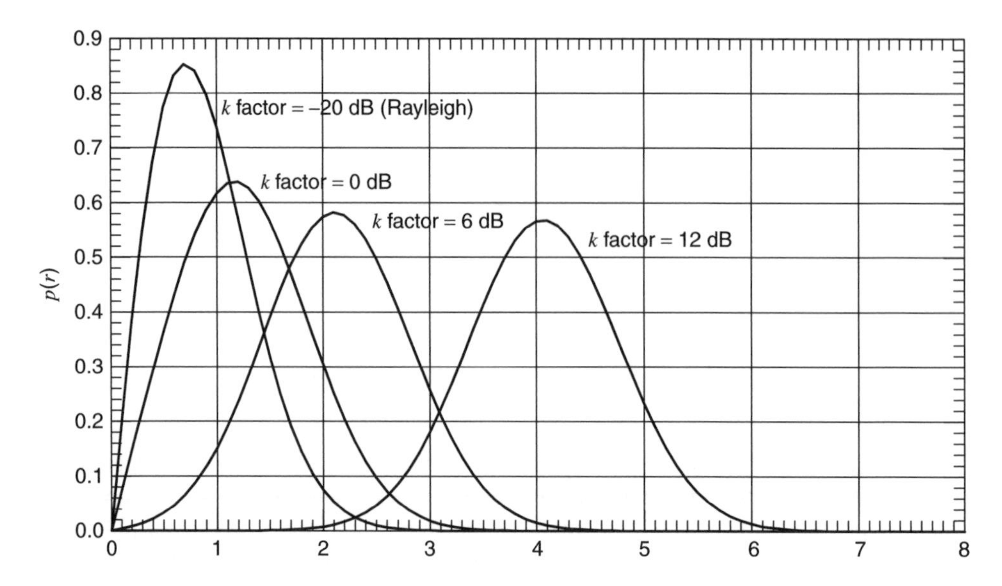

Figure 4.14 Rayleigh and Rician distribution pdfs for several values of k factor.

For analyzing the SNR or SIR, it is often most convenient to describe the distribution of the instantaneous power in the fading signal rather than its envelope voltage. The pdf of the instantaneous power can be found by using the probability function conversion. For the function

$$Y = cX^2 \tag{4.58}$$

where X has any distribution $p_X(x)$ (Rayleigh in this case), the distribution for Y (power) is given by [25]

$$p_Y(y) = \frac{p_X(\sqrt{y/c}) + p_X(-\sqrt{y/c})}{2\sqrt{cy}} \quad \text{for } y \geq 0 \tag{4.59}$$

The resulting pdf for the instantaneous power $s(t) = |r(t)|^2$ is then

$$p_s(s) = \frac{1}{2\sigma^2} \exp\left(-\frac{s}{2\sigma^2}\right) \tag{4.60}$$

A Rayleigh distribution is the pdf that describes the envelope of two Gaussian-distributed variants in quadrature. However, there are circumstances when a single ray in (3.5) will be much stronger than the others. This occurs close to the transmitter in LOS conditions when the ray received directly from the transmitter is stronger than the others. It can also occur in some NLOS receive sites if some characteristic of the propagation environment causes a single ray to be much higher in amplitude than the others. A receive location at the peak of the diffraction field of a corner diffraction source (see Figure 2.13(a) or 2.14(a)) could result in a single strong ray compared to the others being received. The distribution of the amplitudes of such a signal is best modeled with the Rician distribution.

4.6.1.2 Rician distribution

When a single strong constant amplitude component is included in the sum along with a number of weaker components, the distribution of the sum can be described by the Rician distribution [26]. The Rician distribution was developed specifically to describe the pdf of the envelope of the vector sum of a single constant amplitude sine wave and Gaussian noise. The Rician distribution is given by

$$p_r(r) = \frac{r}{\sigma^2} \exp\left(\frac{-(r^2 + A_c^2)}{2\sigma^2}\right) I_0\left(\frac{rA_c}{\sigma^2}\right) \tag{4.61}$$

where A_c is the amplitude of the constant amplitude sine wave component, σ^2 is the variance of the Gaussian noise, and I_0 is the modified Bessel function of the first kind and zero order. When A_c is small compared to σ^2, the pdf $p(r)$ is essentially Rayleigh. Several examples of the Rician pdf with different k factors are plotted in Figure 4.14. When A_c is zero, (4.61) is identically the same as the Rayleigh distribution in (4.53). The Rayleigh distribution therefore can be viewed as simply a special case of the Rician distribution. When A_c is large compared to σ^2, the pdf is essentially the same as a Gaussian distribution with a mean value of A_c. The ratio of the constant amplitude component to the variance is called the Rician k factor of the distribution and is calculated as follows:

$$k = \frac{A_c^2}{2\sigma^2} \tag{4.62}$$

As with the Rayleigh distribution, it is useful to have the pdf of the instantaneous power $s(t) = |r(t)|^2$, which is found in the same way as described above using (4.59):

$$p_s(s) = \frac{1}{2\sigma^2} \exp\left(-\frac{s + A_c^2}{2\sigma^2}\right) I_0\left(\frac{A_c\sqrt{s}}{\sigma^2}\right) \tag{4.63}$$

The average power is given by the sum of the powers in the constant amplitude component and the variable component, $E[s] = A_c^2 + 2\sigma^2$. The cumulative distribution function (CDF) for the Rician distribution is given by

$$F(r < b) = 1 - Q\left(\frac{A_c}{\sigma}, \frac{b}{\sigma}\right) \tag{4.64}$$

where $Q(a, b)$ is the Marcum Q function [27]. Figure 4.15 shows the probability that a Rician-distributed received signal envelope will fade to various amplitudes for several values of Rician k factor. The fade depth values in decibels are relative to the total average power, $P_{\text{ave}} = A_c^2/2 + \sigma^2$. For high k factors, deep fades are much less probable indicating that the link could be designed with a lower fade margin if permitted by the rain fade outage probability. By comparing the curves shown in Figure 4.15 with those

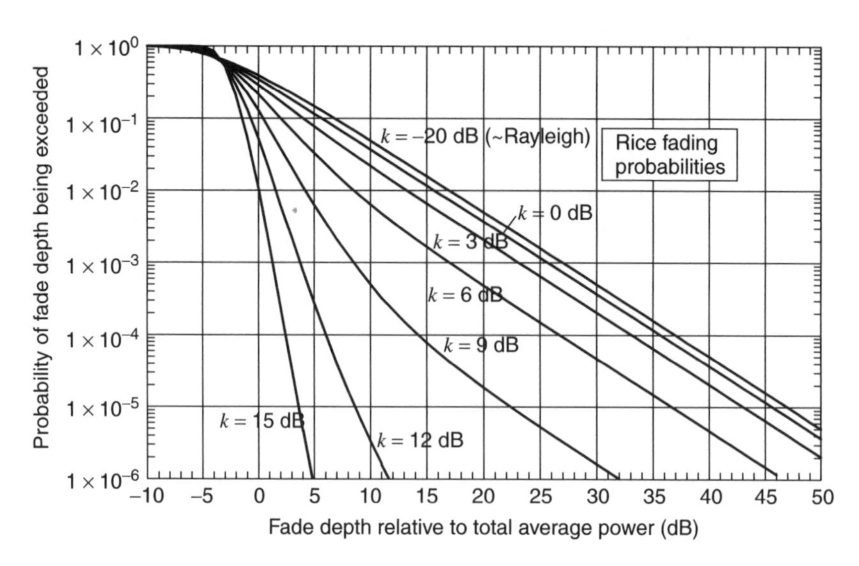

Figure 4.15 Fade probability distribution for Rician-distributed fading model with various k factors.

shown in Figure 4.3, it is apparent that the Vigants–Barnett and ITU-R fading models are comparable to Rayleigh fading.

Figure 4.16 shows a 3-D graphic of Rician-distributed signal level variations over an area of about 10 wavelengths predicted with the ray-tracing propagation model described in Section 3.5.7. This graph is representative of rapid fading from which the parameters of a fading distribution can be estimated. Predicting the exact locations of the signal envelope maxima and minima would require a propagation environment model that is sufficiently accurate to allow the ray path lengths and the associated phase delays to be correctly calculated. With a wavelength of just 11.5 cm at 2.6 GHz, the current database technology does not permit this level of precision except for very simple model environments in controlled circumstances (an anechoic chamber, for example).

4.6.1.3 Nakagami distribution

The third distribution considered here for modeling the multipath-induced voltage envelope variations represented by the random variable r is the Nakagami or Nakagami-m distribution [26]. Unlike the Rayleigh and Rician distributions, which are derived from real physical quantities in nature (Gaussian noise and the sine wave), the Nakagami distribution is a mathematical construct with no physical foundation. The Nakagami distribution is given by

$$p_r(r) = \frac{2}{\Gamma(m)}\left(\frac{m}{\Omega}\right)^m r^{2m-1}\exp\left(\frac{-m}{\Omega^2}r^2\right) r \geq 0, m \geq 0.5 \tag{4.65}$$

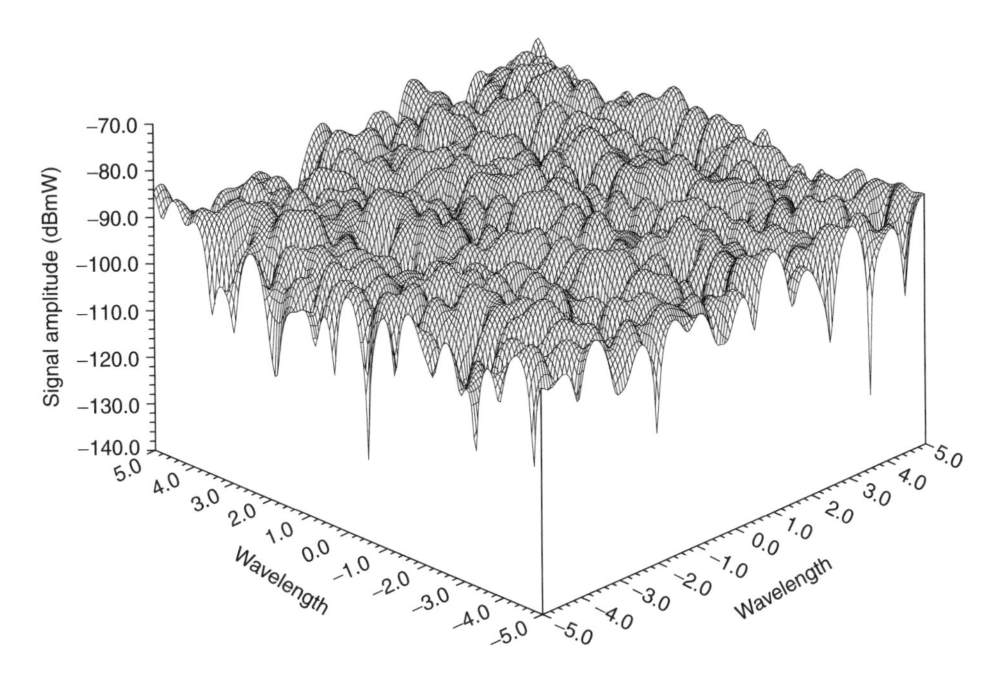

Figure 4.16 3-D representation of Rician fading of the signal envelope amplitude over a $\pm 5\lambda$ area predicted using the ray-tracing propagation model.

where m is the parameter that controls the basic shape of the distribution and Ω is the mean square value. The function $\Gamma(m)$is the Gamma function defined by

$$\Gamma(x) = \int_0^{\infty} t^{n-1} e^{-t} dt \ x > 0 \tag{4.66}$$

Tabulations of the gamma function can be found in mathematical tables. Also, recursive relationships and polynomial approximations are available. For the special case of $x = 0.5, \Gamma(x) = \sqrt{\pi}$. Using this value in (4.64) and taking m as 0.5, (4.64) becomes the Gaussian distribution for $x > 0$. For $m = 1.0$, it becomes the Rayleigh distribution with $\Omega = \sigma^2$.

Compared to the Rayleigh and Rician distributions, the Nakagami distribution is more flexible in creating a wide range of pdf shapes and is to some extent more mathematically tractable than the Rician distribution because the modified Bessel function is absent. However, the physical rationalization of the Rician distribution is satisfying when it is tied to a physical approach to modeling communication channels. The artificial mathematical construction of the Nakagami distribution is a useful curve-fit to experimental data. It can be the most convenient distribution to use in many cases.

4.6.2 NLOS shadow fading models

Shadow fading is the variation in the mean value of the multipath fading distribution. Confusion sometimes arises because the mean value can be the mean value of the multipath fading envelope voltage $E[r(t)]$ or the mean value of the envelope power $E[r(t)^2]$. The former approach is used here. The shadowing distribution is usually modeled as a lognormal distribution that, for communication engineering purposes, describes the variation of the decibel value of the mean signal as a normal or Gaussian distribution. If the mean value of the envelope $w = E[r(t)]$, its distribution is then

$$p_w(w) = \frac{1}{\sqrt{2\pi\sigma_w^2}} \exp\left(-\frac{(w-\mu_w)^2}{2\sigma_w^2}\right) \quad \text{for } w \text{ in dB} \tag{4.67}$$

The standard deviation σ_w for lognormal shadowing can vary from 5 to 12 dB depending on the environment. A value of 8 dB is typically used. The mean value μ_w can be taken as the expected median signal level found using the path loss predicted by the propagation model, for example, an empirical model such as the IEEE 802.16 SUI models in Section 3.4.1. The shadow fading pdf provides a way to account for path loss factors that a path loss model itself cannot explicitly predict. Using (4.67), curves of the probability of a signal level below the mean value for a lognormal distribution with several values of standard deviation are plotted in Figure 4.17.

Many measurement studies are available in the literature giving the standard deviation of the shadow fading for cellular (800–900 MHz) and personal communication service (PCS) (1800–1900 MHz) frequency bands. Similar studies for multipoint, multi-channel distribution service (MMDS) frequencies at 2.5 to 2.7 GHz and U-NII frequencies at 5 to

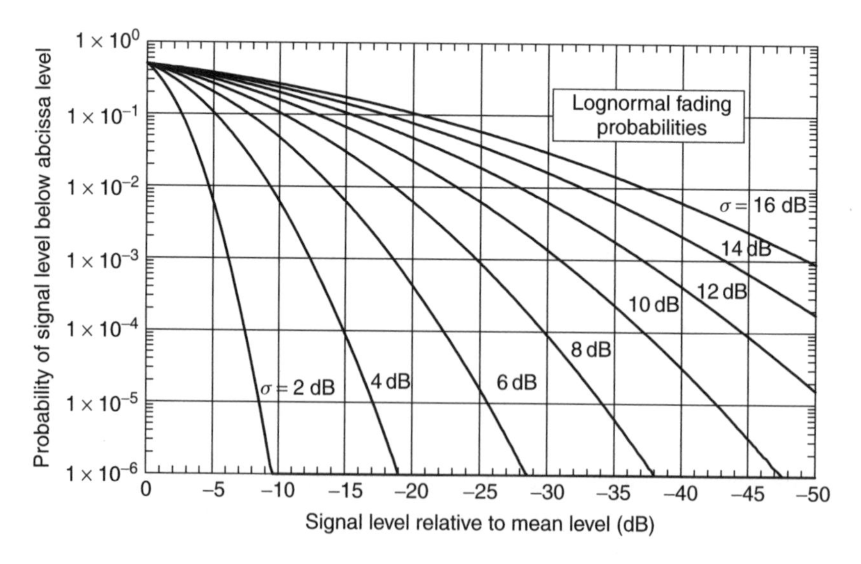

Figure 4.17 Lognormal fading signal level probabilities.

6 GHz are scarce, mainly because the shadow fading is normally associated with a mobile moving along the streets between buildings rather than a remote terminal located inside a building or on the outside wall or roof of a building.

The lognormal distribution can also be used to describe the path loss associated with particular variable factors in a physical propagation model that affects the signal strength. The building clutter, foliage, and building penetration loss terms in the NLOS model in Section 3.5.6 are examples of path loss factors that can be described individually by some mean loss value and an associated standard deviation of that loss.

The variations in signal levels that might be encountered throughout a neighborhood or throughout a building or house as described by a lognormal distribution must be taken into account when planning an NLOS link that achieves a given high location availability percentage. Chapter 10 contains information on how lognormal fading is incorporated into link budget calculations for predicting link availability.

4.6.3 Composite fading–shadowing distributions

Considering the multipath fading signal envelope variations described by a Rayleigh or a Rician distribution and the variations of the mean of that distribution described by the lognormal shadowing distribution, the complete description of the fading becomes a composite distribution that incorporates both these distributions. The distribution of the composite $p(z)$ can be found as a conditional distribution in which the Rayleigh or Rician distribution is conditioned on the mean value that is determined by the lognormal distribution. This can be written as

$$p_z(z) = \int_0^{\infty} p_r(z|w)p_w(w)\,dw \tag{4.68}$$

In the case when the multipath fading is represented by a Rayleigh distribution, the mean value is given by

$$\bar{r} = \sqrt{\frac{\pi}{2}}\sigma = w \text{ or } \sigma = \frac{2}{\pi}w^2 \tag{4.69}$$

Substituting the value for σ into the (4.53) gives

$$p_r(z|w) = \frac{\pi z}{2\,w^2}\exp\left(-\frac{\pi z^2}{4\,w^2}\right) \tag{4.70}$$

Substituting (4.67) and (4.70) into (4.68) results in the compound distribution. For a composite Rayleigh-lognormal distribution, the composite distribution is then given by [28]

$$p_z(z) = \int_0^{\infty} \frac{\pi z}{2\,w^2}\exp\left(\frac{\pi z^2}{4\,w^2}\right) \times \frac{2M}{w\sigma_z\sqrt{2\pi}}\exp\left(-\frac{(10\log w^2 - \mu_w)^2}{2\sigma_w^2}\right)dw \tag{4.71}$$

where $M = \ln 10/10$ and the variables have the same definitions as used in the preceding sections. This distribution is often called the *Susuki distribution*. A similar approach can be used to find the pdf for Rician-lognormal compound distribution. The mean value of the Rician distribution is given by

$$E[r(t)] = \sigma\sqrt{\frac{\pi}{2}}\left[(1+k)I_0\left(\frac{k}{2}\right) + kI_1\left(\frac{k}{2}\right)\right]e^{-k/2} \tag{4.72}$$

The same process described above can be used with this mean to find the compound Rician-lognormal distribution.

For NLOS links, a compound distribution can be used to find the probability that the desired signal will fall below a signal level performance threshold that is established by the maximum tolerable error rates. As discussed in Chapter 7, these thresholds are usually separately established for fading against thermal noise and interference.

In practice, compound distributions such as (4.71) are difficult to apply to real link design problems since knowledge of both the multipath and the shadow distribution parameters are required. Normally for NLOS links as contemplated for fixed broadband systems, multipath fading can be effectively combated with space diversity antennas. Many inexpensive 2.4-GHz cordless telephones now use space diversity antennas. If multipath fading is greatly reduced through diversity reception, what remains is a signal level described by lognormal fading distribution in (4.66). This greatly simplifies the NLOS link outage analysis.

4.7 CONCLUSION

Models of fading phenomena are essential for predicting the reliability of wireless communication systems. For fixed broadband systems that intend to be competitive with wireline systems such as twisted pair digital subscriber line (DSL) lines or broadband cable, reliability percentages above 99.9% are needed to satisfy the service users – in some cases the expectation will be 99.99 or 99.999% as with point-to-point microwave and FSO systems. Such reliability expectations far exceed those of typical mobile or cellular system designs. The details of the probability distribution shape, especially in the low probability 'tails' of fading distributions, are critical to assessing this link reliability.

The atmospheric and rain fading models described in this chapter have been developed over a number of years to provide accurate fade depth descriptions at these low probability levels. Where accurate fading descriptions are not available, the system designer must necessarily over-design to ensure the link's reliability, potentially resulting in higher equipment costs and increased interference. The trade-offs involved in making such design decisions are discussed in Chapter 10.

The NLOS fading models are applicable to links in networks in which there are a variety of objects that affect the instantaneous signal level at the receiver but which cannot be explicitly predicted by available propagation models using typical propagation environment. In Chapters 10 and 11 these models are used to describe fading in MMDS band, U-NII band, and similar bands that are targeted for NLOS technologies. Even though these models are drawn from conventional mobile radio modeling techniques, they are currently the only viable approach to predicting fading in NLOS fixed broadband wireless networks.

4.8 REFERENCES

[1] W.T. Barnett, "Multipath propagation at 4, 6 and 11 GHz," *Bell System Technical Journal*, vol. 51, no. 2, pp. 311–361, February, 1972.

[2] A. Vigants, "Space-diversity engineering," *Bell System Technical Journal*, vol. 54, no. 1, pp. 103–142, January, 1975.

[3] International Telecommunications Union (ITU-R), "Propagation data and prediction methods required for the design of terrestrial line-of-sight systems," Recommendation ITU-R P.530-8, 1999.

[4] P. Beckmann. *Elements of Applied Probability Theory*. New York: Harcourt, Brace & World, Inc. 1968. pp. 128–131.

[5] W.C. Jakes. *Microwave Mobile Communications*. Piscataway, New Jersey: IEEE Press, 1994. (re-printed), pp. 50–52.

[6] H.R. Anderson, et al., "Direct calculation of coherence bandwidth in urban microcells using a ray-tracing propagation model," *Proceedings of the Fifth Annual Symposium on Personal, Indoor and Mobile Communications*, The Hague, September, 1994.

[7] C.W. Lundgren and W.D. Rummler, "Digital radio outage due to selective fading- observation vs. prediction from laboratory simulations," *Bell System Technical Journal*, pp. 1073–1100, May-June 1979.

[8] W.D. Rummler, "Characterizing the effects of multipath dispersion on digital radios," *IEEE Globecom Proceedings*, pp. 1727–1732, 1988.

[9] R.K. Crane, "Prediction of attenuation by rain," *IEEE Transactions on Communications*, vol. COM-28, no. 9, pp. 1717–1732, September, 1980.

[10] R.K. Crane. *Electromagnetic Wave Propagation through Rain*. New York: John Wiley & Sons. 1996.

[11] International Telecommunications Union (ITU-R), "Propagation data and prediction methods required for the design of terrestrial line-of-sight systems," Recommendation ITU-R P.530-8, 1999.

[12] International Telecommunications Union (ITU-R), "Specific attenuation model for rain for use in prediction methods," Recommendation ITU-R P.838-3, 1999.

[13] International Telecommunications Union (ITU-R), "Characteristics of precipitation for propagation modeling," Recommendation ITU-R P.837-2, 1999.

[14] R.K. Crane. *Electromagnetic Wave Propagation through Rain*. New York: John Wiley & Sons. 1996. pp. 144.

[15] H. Xu, T.S. Rappaport, R.J. Boyle, and J.H. Strickland, "38 GHz wideband point-to-multipoint radio wave propagation study for a campus environment," *Proceedings of the 49th IEEE Vehicular Technology Society Conference*, Houston, CD-ROM, May, 1999.

[16] International Telecommunications Union (ITU-R), "Propagation data and prediction methods for the design of terrestrial broadband millimetric radio access systems operating in the frequency range of about 20–50 GHz," ITU-R P.1410-1, 2001.

[17] I. Anderson, "Measurements of 20-GHz transmission through a radome in rain," *IEEE Transactions on Antennas and Propagation*, vol. AP-23, no. 5, pp. 619–622, September, 1979.

[18] J.A. Effenberger, R.B. Strickland, and E.B. Joy, "The effects of rain on a radome's performance," *Microwave Journal*, pp. 261–274, May, 1988.

[19] M.M.Z. Kharadly and R. Ross, "Effect of wet antenna attenuation on propagation data statistics," *IEEE Transactions on Antennas and Propagation*, vol. 49, no. 8, pp. 1183–1191, August, 2001.

[20] D. Gibble, "Effects of rain on transmission performance of a satellite communications system," *IEEE International Convention*, New York, March, 1964.
[21] H. Willebrand and B.S. Ghuman. *Free-Space Optics: Enabling Optical Connectivity in Today's Networks*. Indianapolis: SAMS Publishing, 2002.
[22] W.C. Jakes. *Microwave Mobile Communications*. IEEE Press: Piscataway, New Jersey. 1994 (re-published). Chapter 1.
[23] M. Schwartz, W.R. Bennett, and S. Stein. *Communication Systems and Techniques*. New York: McGraw-Hill, 1966. page 329.
[24] M. Schwartz, W.R. Bennett, and S. Stein. *Communication Systems and Techniques*. New York: McGraw-Hill, 1966. page 24.
[25] P.Z. Peebles. *Probability, Random Variables, and Random Signal Principles*. New York: McGraw-Hill. 1980. page 72.
[26] S.O. Rice, "Statistical properties of a sine wave plus random noise," *Bell System Technical. Journal*, vol. 27, no. 1, pp. 109–127, January 1948.
[27] M. Schwartz, W.R. Bennett, and S. Stein. *Communication Systems and Techniques*. New York: McGraw-Hill, 1966. Appendix A.
[28] G.L. Stuber. *Principals of Mobile Communications*. Boston: Kluwer Academic Publishers. 1996.

5

Propagation environment models

5.1 INTRODUCTION

Except for a vacuum such as outer space, wireless system wave propagation is affected by the elements and characteristics of the real environment in which the networks are deployed. All empirical and physical propagation models to a greater or lesser extent rely on information about the propagation environment to operate. It is obvious that the models that take into account more information about the propagation environment have a better chance of accurate predictions than those that take into account less information. The challenge of developing propagation models is to successfully make use of this greater detail to provide more accurate predictions.

The four main categories of propagation environment information that are used for designing fixed broadband wireless systems are

- terrain elevations or topography
- buildings and other structures
- land use/land cover or morphology (also referred to as *clutter*)
- atmospheric and meteorological conditions.

Each of these elements will be discussed in the following sections along with the current techniques for developing databases containing such information. While having perfect, high-resolution data describing terrain and buildings would be ideal, as a practical matter such databases may be difficult or impossible to obtain, or prohibitively expensive for wireless system planning purposes. Inevitably, some trade-offs are necessary. For example, spending a great deal more money for a higher resolution building database may only yield marginal (and perhaps statistically insignificant) improvements in signal predictions and overall system design given the underlying capabilities of the propagation model.

Fixed Broadband Wireless System Design Harry R. Anderson
 ISBN: 0-470-84438-8

5.2 TOPOGRAPHY

It has been recognized for nearly a century that planning wireless communications systems that use space waves propagating through the atmosphere requires knowledge of the topography. Hills, mountains, and other features can block and severely attenuate radio signals. Mountains can also reflect and scatter transmitted signals creating multiple paths for them to arrive at the receiver (multipath). The geometry of such additional reflection paths can result in very long signal delays (time dispersion) on the order of several tens of microseconds.

5.2.1 Topographic maps

Topographic maps are the fundamental source of information of terrain elevations. Figure 5.1 shows an example of a topographic map for an area near Eugene, Oregon. The map contains many types of information, including lines of constant elevation, or contours, that undulate across the map. The contour lines are usually drawn at regular elevation spacings so that relative line density is an indicator of how steep or how shallow the terrain slope is.

Before terrain elevation databases were available for use on computers, determining the elevations along a path between a transmitter and receiver was a matter of drawing a line across one or more topographic maps connecting the transmitter and receiver locations. A tabulation was then created of the distances to each of the contour lines that were crossed, along with the elevation of that contour line. The result was a profile of terrain elevations along the path as shown in the examples in Figures 2.22 and 2.23. By placing the transmitting and receiving antennas at their correct elevations at each end of the path and adjusting the intervening elevations along the path for earth curvature k factor, as discussed in Section 2.10, a straight line can be drawn between the transmitter and

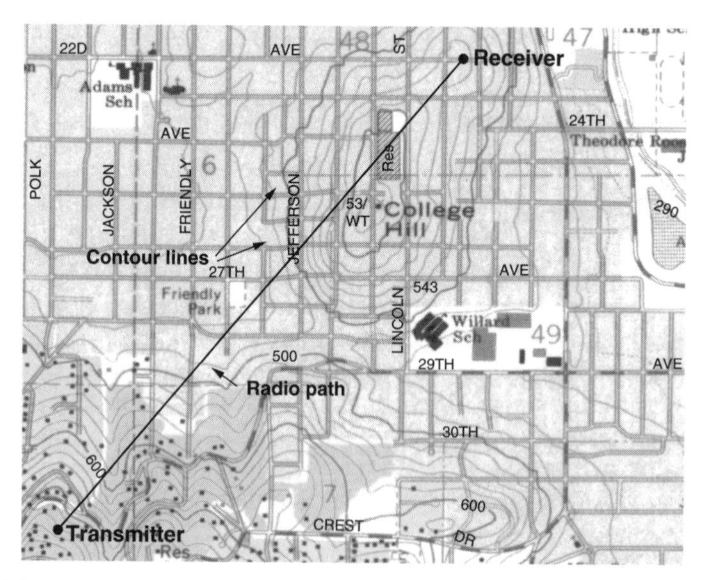

Figure 5.1 Sample topographic map showing elevation contour lines.

the receiver to determine whether any terrain points along the profile obstruct the path. In this way, 0.6 first Fresnel zone clearance can be assessed. Of course, making this assessment for multiple refractivity gradients (multiple k factors) would require redrawing the elevation profile with a different curvature.

For long paths, there is an additional issue. The radio path between the transmitter and the receiver is actually a great circle path along the surface of the Earth. Depending on the map projection used for the topographic maps, a line drawn on the maps may not necessarily represent a great circle route. For short paths this was not of much consequence, but for longer paths across many map sheets, it was necessary to introduce corrections to create a piecewise linear curved line across the maps that would approximately represent the great circle path.

This very laborious, mistake-prone manual process was necessarily used for many years before the advent of computer databases that contained terrain elevation data. The original databases were simple digitized representations of the contour lines on the map, the digitizing itself a time-consuming process that was usually undertaken by government agencies. The raw description of the contours consisted of an elevation and a string of latitude and longitude coordinates where the elevation was found along the contour. The raw contour data could be used directly with computer programs that could automatically calculate the contour crossing intersections in much the same way as the manual method.

5.2.2 Terrain DEMs

While the raw contour approach could be used successfully, the preferred method that has emerged for storing and using the terrain elevation data is a fixed grid digital elevation model (DEM). As shown in Figure 5.2, a DEM consists of a matrix of elevation points with a fixed spacing (the resolution) in either meters or seconds of latitude and longitude.

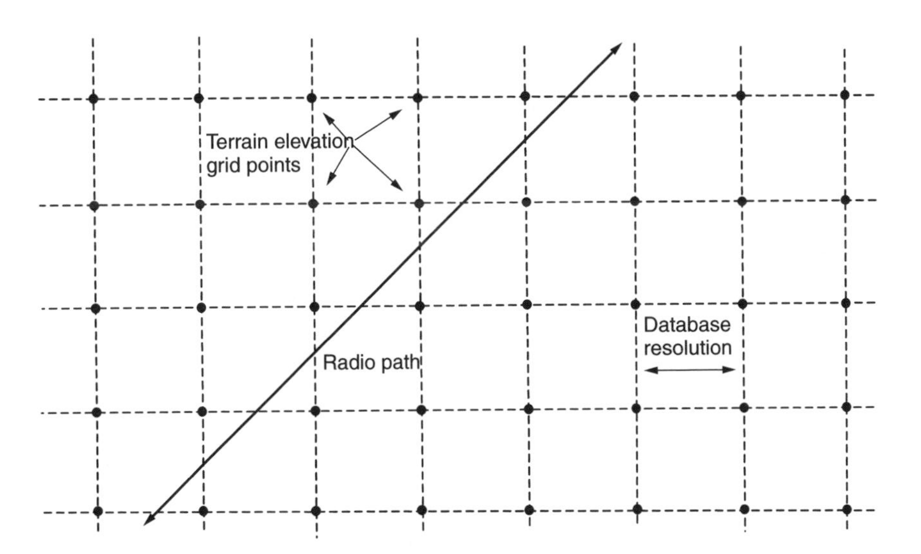

Figure 5.2 DEM, a regular grid of terrain elevation points.

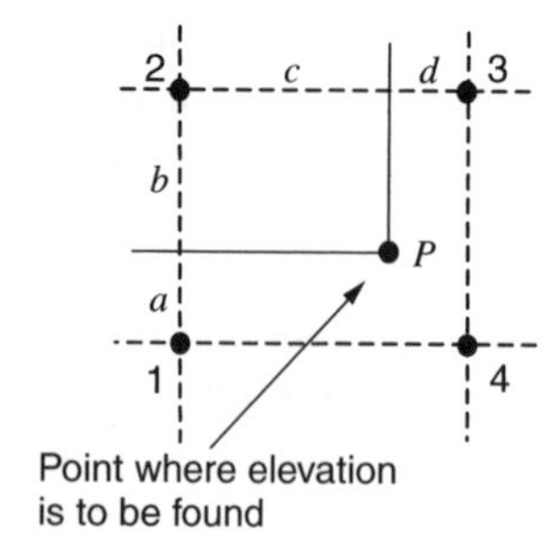

Figure 5.3 Bilinear interpolation.

To create a terrain profile, a radio path line is 'drawn' through the grid. In this case, 'drawn' means a great circle path constructed with a computer algorithm that follows latitude and longitude points through the elevation point matrix. The elevations can then be found at fixed distance points along the line by interpolating between nearby grid points. Many approaches have been used for carrying out this interpolation, including some elaborate methods that apply different weighting factors to several nearby points based on their distance from the point on the radio path where the elevation is being determined. The mostly commonly used method employed by the Federal Communications Commission (FCC) for multipoint, multi-channel distribution service (MMDS) and other studies done in the United States is the so-called 'bilinear' interpolation method in which the four points in the elevation matrix surrounding the desired point are used with simple linear interpolation as shown in Figure 5.3. The elevation z_P at the desired point surrounded by grid points $z(1)$, $z(2)$, $z(3)$, and $z(4)$ is given by

$$q1 = z(1) + [a/(a + b)] \cdot [z(2) - z(1)] \tag{5.1}$$

$$q2 = z(4) + [a/(a + b)] \cdot [z(3) - z(4)] \tag{5.2}$$

$$z_P = q1 + [c/(c + d)] \cdot (q2 - q1) \tag{5.3}$$

The validity of the radio path clearance analysis over terrain is directly related to the quality of DEM. As an elevation model, it is just that, a model of the real terrain, *not* the real terrain. If the resolution of DEM is low (matrix points are widely spaced), the peak of a mountain, ridge, or hill could be missed or inaccurately represented. The analysis of path clearance using DEM could show the clearance to be adequate when in fact it is not.

Figures 5.4 and 5.5 show two terrain profiles along an identical radio path, one derived from a database with approximately 100-m point spacing and the second from a database with 30-m point spacing. The flattened peaks, filled-in valleys, and other errors in Figure 5.4 when compared to Figure 5.5 are clear, especially between 20 and 30 km along the profile. Even the elevations at the transmitter and receiver sites are 30 to 40 m lower when using the coarser database. Errors of this type could potentially lead to an erroneous conclusion about path clearance.

Depending on the quality of the database and the criticality of the wireless system being deployed, a field check of the path clearance using a theodolite or other surveying

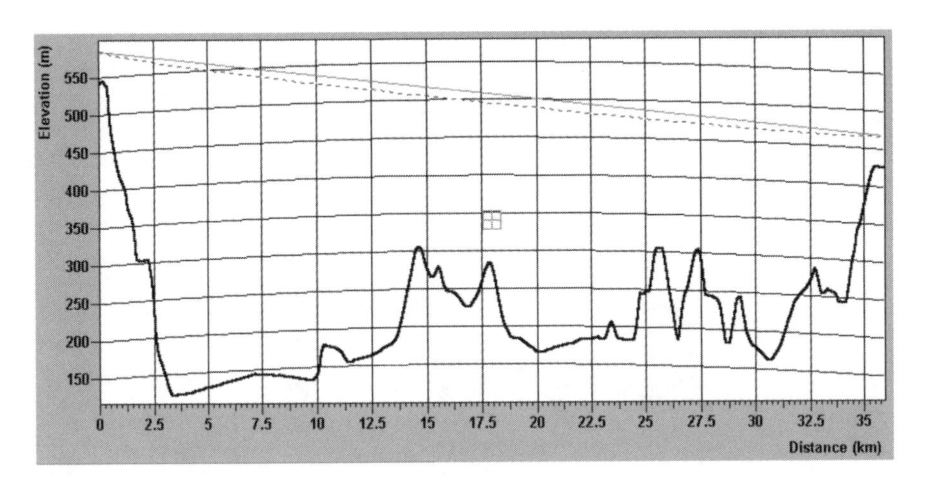

Figure 5.4 Terrain profile created using DEM with 3 arc second (~100 meter) grid spacing.

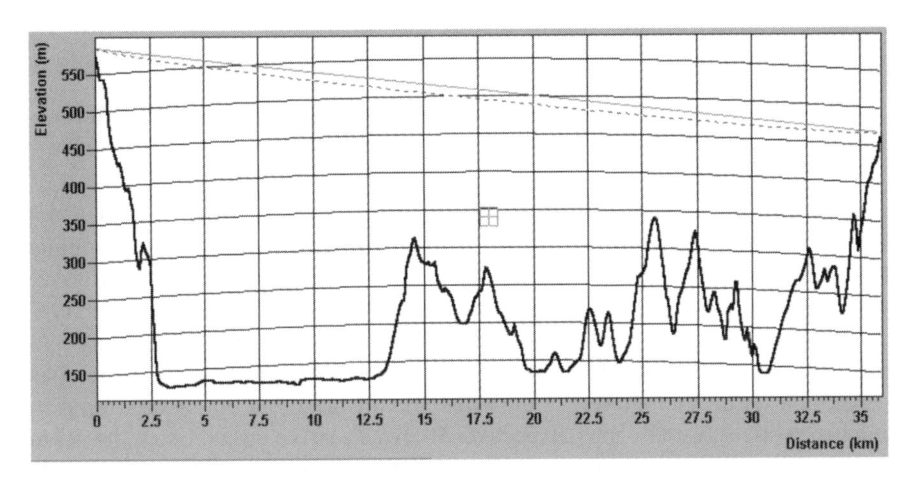

Figure 5.5 Terrain profile created using DEM with 30 meter grid spacing.

tools might be necessary. In such a field check, a visit to the intended transmitter and receiver sites is made and a visual sighting taken along the path. A visual confirmation of path clearance at optical light frequencies cannot exactly account for refractivity at radio frequencies, but in some cases it is the only means to assure link clearance. As databases improve, the necessity for a field confirmation is reduced.

5.2.3 DEM data from satellite and aerial imagery

Topographic maps have been created for the entire world with varying resolutions and quality. However, for various political and security reasons, the data may not be accessible

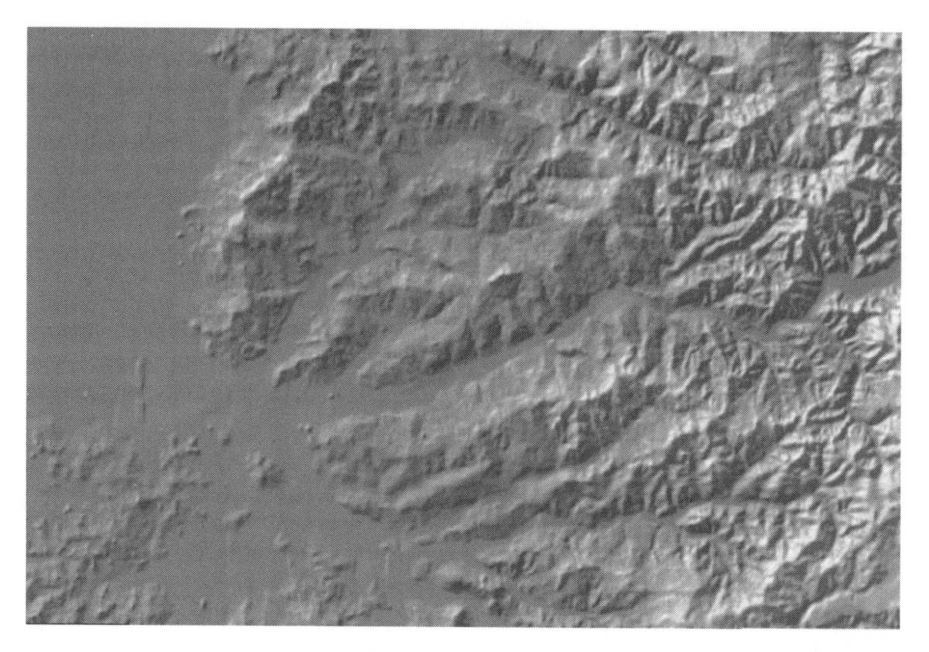

Figure 5.6 Terrain relief map using 30-m DEM. Illumination is from the northwest.

for normal commercial wireless system planning purposes. In such cases, DEM for a given area can be developed using satellite imagery. Several government and commercial satellites circle the Earth in low Earth orbits ranging from 150 to 500 km above the surface. These satellites are continuously taking photographs and using other types of sensors to collect a vast array of information about the surface they are passing over.

An elevation model can be developed from two photographs that are a 'stereo pair'. A stereo pair is a set of photographs of the same area of the Earth but taken from different viewing perspectives, usually the perspectives from successive orbits. Using the geometric relationship of the viewing locations to the same feature on the photograph, the relative height or elevation of that feature toward the viewing point can be determined. Applying this approach in a rigorous way yields a three-dimensional DEM of the area covered by the stereo pair. A detailed description of this process can be found in [1].

The optical resolution of the photos taken by the first generation of commercial imagery satellites was on the order of 10 m, adequate for most wireless system planning needs. The new generation of commercial imagery satellites that were first launched in 2000 can achieve photo resolutions on the order of 1 to 2 m.

While satellites have the advantage of being able to photograph any portion of the Earth without violating any nation's airspace restrictions, the cost of DEMs developed in this way can be quite high. The same process of using a stereo pair of photos to create a DEM also works for aerial photography. Given the much lower height above ground, and with the absence of cloud obscuring some areas, aerial-based photography and remote sensing is currently the best means to highly accurately create DEMs. A relatively recent aircraft-based remote sensing technique called light detection and ranging (LIDAR) can also be used to directly create terrain DEMs without using stereo photogrammetric methods. Thus far LIDAR's primary use has been to develop databases of structures and other features in urban areas.

5.3 BUILDINGS AND OTHER STRUCTURES

Buildings are the primary factor affecting short-range wireless communication links in urban areas. The design of systems including individual point-to-point links, consecutive point networks (CPN), and local multipoint distribution service (LMDS) point-to-multipoint (PMP) networks in urban areas all require very detailed information about the location and heights of building features in order to assess path clearance and calculate interference.

As discussed in the preceding section, comprehensive terrain database DEM coverage at reasonable resolutions has been developed for a variety of environmental and other uses by national governments. Consequently, these databases are widely available and relatively inexpensive. Development of building databases, however, has not proceeded in the same way. Most have been privately created for particular purposes and thus are not readily available. Creating new databases from original source material like aerial photographs is expensive. Depending on the system design circumstances, an important initial decision is whether buying or creating a building data is a worthwhile expense. In making this decision, there are a number of technical factors that must be taken into account. Foremost is the number of links being designed. If only a few links are being planned between known locations for intrabusiness communications among buildings or for cellular backhauls, a site visit and visual inspection of the path is most efficient. However, for PMP networks in which links from hubs to several as-yet undefined remote terminal locations will be installed on an ongoing basis, the use of a building database with an appropriate wireless planning tool is a much more efficient way to establish the viability of individual links as well as plan the optimum locations of hubs to achieve maximum rooftop visibility. It will also allow interference between various links to be accurately calculated, something not possible with a simple visual line-of-sight (LOS) determination of a path. Accurately taking interference into account is important for achieving maximum spectrum utilization in high-density, interference-limited wireless systems.

Building databases can be produced in a variety of ways but ultimately exist in one of the two basic forms:

- Vector databases in which individual walls, roofs, and other surfaces are represented in the dataset by their latitude, longitude, and height coordinates. The horizontal coordinates may also be in x, y meters. The individual walls taken together form 3-D polygon shapes for representing buildings.
- 'Canopy' databases in which the buildings, foliage, highway overpasses, and so on are represented by a very fine resolution regular grid of elevation points in much the same way as terrain DEMs represent varying terrain features.

Each of these approaches is discussed in the following sections.

5.3.1 Vector building databases

The vectors that represent the location of walls are usually defined by the x, y coordinates of each vector endpoint and the height of the base and the top of the wall. The main advantage of such databases is that they can represent wall positions with a high degree of resolution – wall coordinates can be specified to a few centimeters if desired. However, databases with this resolution require excellent source photography and considerable processing to develop. Taking advantage of this high resolution may only be useful when every aspect of the radio link analysis is specified to the same resolution. For designing indoor wireless systems, this resolution may be important. For outdoor, interurban fixed wireless systems, building database resolutions on the order of 1 m is as precise as a designer can reasonably use. For some purposes, a coarser database may be sufficient.

Vector databases can also be easily specialized to represent only those structures that are significant to the system design objective. For example, for most urban fixed system planning, a database that consists of just structures of more than two stories (about 6–8 m) in height is all that is needed for successful planning. Including single-story office buildings, residential structures and other low-rise structures that are not tall enough to be link path obstructions does not improve system planning; it only adds to the development cost, the storage size, and the complexity of the database in addition to the time it takes to process the database in the planning tool. By filtering out irrelevant structures, the database can be much more manageable in development time and cost. For sprawling metropolitan areas that may cover many square kilometers, a vector database that includes only those widely spaced clusters of high-rise office buildings can be an appropriate compromise.

Vector databases are also best suited to ray-tracing (see Chapter 3) and other propagation models that make use of, and take advantage of, precise information about the propagation environment. Canopy databases discussed in the next section are difficult to use with such models, and inherently limit their accuracy.

Fixed broadband wireless systems are often deployed in an incremental fashion so that service is first provided to those locations where the revenue potential is the greatest. When vectors forming polygons are used to represent individual buildings, the polygons can be assigned ID or reference numbers of some sort that can be tied to databases containing other information about individual buildings. Address information, tenant lists, current availability of optical fiber service, and many other characteristics can be geographically tied to the building. From a system planning perceptive, this type of information is highly valuable in determining which buildings are the best candidates for base station or hub locations and which buildings have the greatest revenue potential from customers who will access the services the wireless system intends to provide. See Chapter 9 for information about using building data for estimating potential traffic for the wireless system.

The disadvantage of vector databases is that they are not well suited for representing amorphous shapes such as trees and other foliage or odd-shaped structures such as highway overpasses. Depending on the system, these relatively low-lying environment features may or may not be important to the planning process.

The heights of building features can be referenced to local ground level (above ground level or AGL) or to mean sea level (above mean sea level or AMSL). In the former case, a high-resolution terrain DEM is also needed so that the absolute height of the buildings

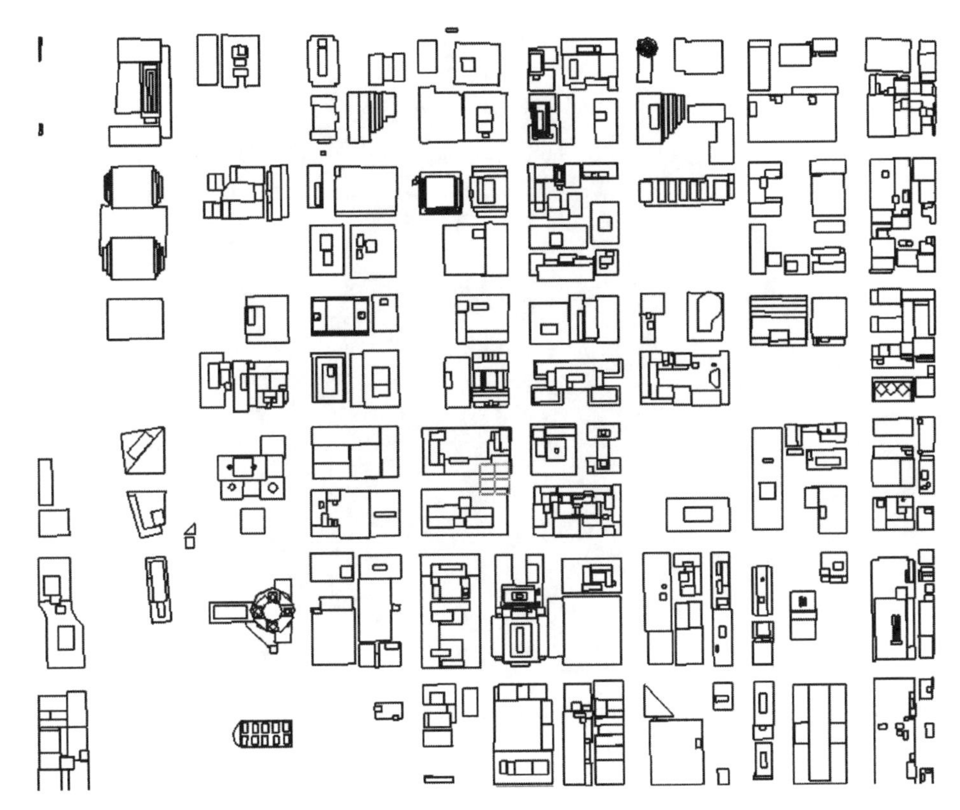

Figure 5.7 High-resolution building vector database for a section of downtown Chicago. (Data from [2].)

can be obtained. This DEM is usually a by-product of the vector database development process, so it does not add significantly to the cost. However, this approach does require wireless planning software tools that can properly combine the elevations from a uniformly spaced terrain DEM with the heights of the vectors representing the buildings.

Figure 5.7 shows a vector building database for a downtown section of Chicago. Figure 5.8 is a 3D view from the southwest of the same area. The high resolution allows relatively small cross-sectional features to be included such as the two broadcasting towers on top of the Sears building.

5.3.2 Canopy building databases

A canopy database is formed by essentially draping a grid or a canopy over all the features in the propagation environment. The height of whatever is underneath at a given x, y grid location is the elevation assigned to that grid location. This will include buildings, trees, height structures, open terrain, or anything else. A useful analogy to this conformal mapping process is 'shrink-wrapping' or 'vacuum-sealing' in which a membrane is forced to conform to all the detailed shapes of the object being wrapped.

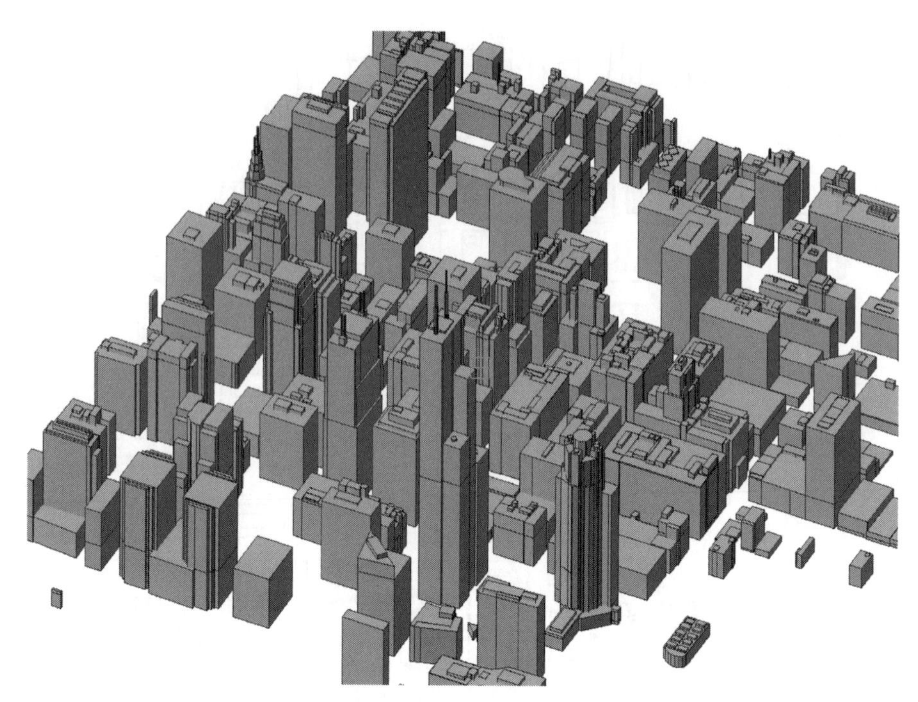

Figure 5.8 3D view of high-resolution building vector database for a section of downtown Chicago. (Data from [2].)

In order to be useful for link planning in urban areas, the resolution of the grid draped over the buildings, trees, and so on must be sufficiently fine to capture and accurately represent those features that could be link obstacles and thus be relevant to the wireless planning process. This dictates a grid point spacing on the order of 1 or 2 m. Even with a fine resolution of 1 or 2 m, perfectly vertical walls of buildings (most commonly encountered) are necessarily represented as slopes since the grid of fixed resolution cannot perfectly represent what amounts to a step-function transition from ground level to roof level. In this respect, creating a canopy elevation grid is a two-dimensional sampling process of the physical environment in which many of the familiar aspects of the signal sampling theory – such as the Nyquist criteria – apply. An analysis of errors resulting from coarse grid spacing in canopy databases is included in the following section.

The advantage of this approach when compared to the vector approach is that all features in the environment, including those that are difficult to describe with vectors, can be easily represented. The resulting canopy grid of points can also be formatted in the same way as terrain DEMs, so planning software tools can readily access them.

Canopy databases are not well suited for use with ray-tracing propagation models that rely on precise knowledge of the location and the slope of building walls (reflecting surfaces) and building corners (diffracting wedges). They also cannot be used for representing indoor environments in which floors, walls, ceilings, doorways, and so on are important to system studies.

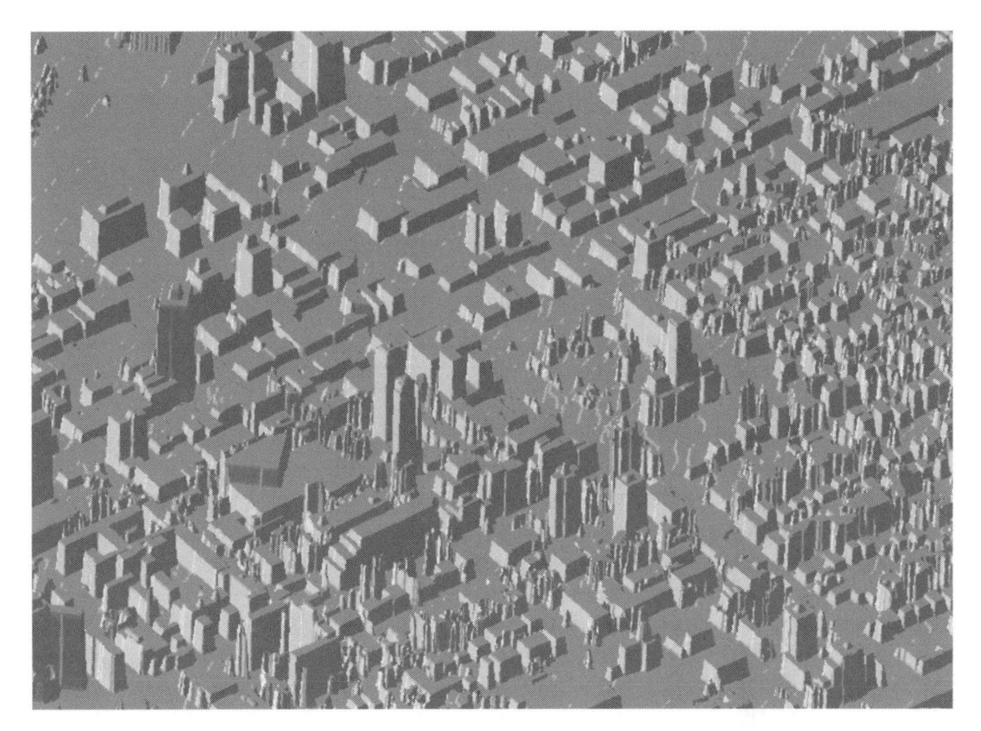

Figure 5.9 1-m building canopy database for Vancouver, B.C., Canada. Odd shapes with flat tops are foliage. (Data from [2].)

A 3D example of a 1-m canopy database for the Vancouver, B.C., Canada, area is shown in Figure 5.9. Buildings, trees, and linear features are clearly visible. This particular database was created using aerial photography and advanced photogrammetry methods.

A newer, more automated technique called LIDAR mentioned earlier makes use of multiple laser beams mounted on a pod at the bottom of an aircraft. The laser beams are launched from the aircraft at known angles. The time delay of the beams reflected back to the aircraft is precisely measured and from this, the exact location of whatever the beam is reflected from can be determined. A discussion of the use of LIDAR for creating canopy databases for wireless planning can be found in [3].

5.3.3 System analysis errors from using canopy databases

A recent analysis [4] has shown that as the canopy grid spacing gets coarser, the errors in system analysis grow, as might be expected. Figure 5.10(a) shows a link profile across a set of buildings derived from a vector database. The walls are perfectly vertical. Figure 5.10(b) through 5.10(d) shows the same building profile using canopy databases with increasing degrees of coarseness.

The profiles shown in Figure 5.10 are examples taken from each of five databases covering the same urban area where there are approximately 1,600 buildings. The databases

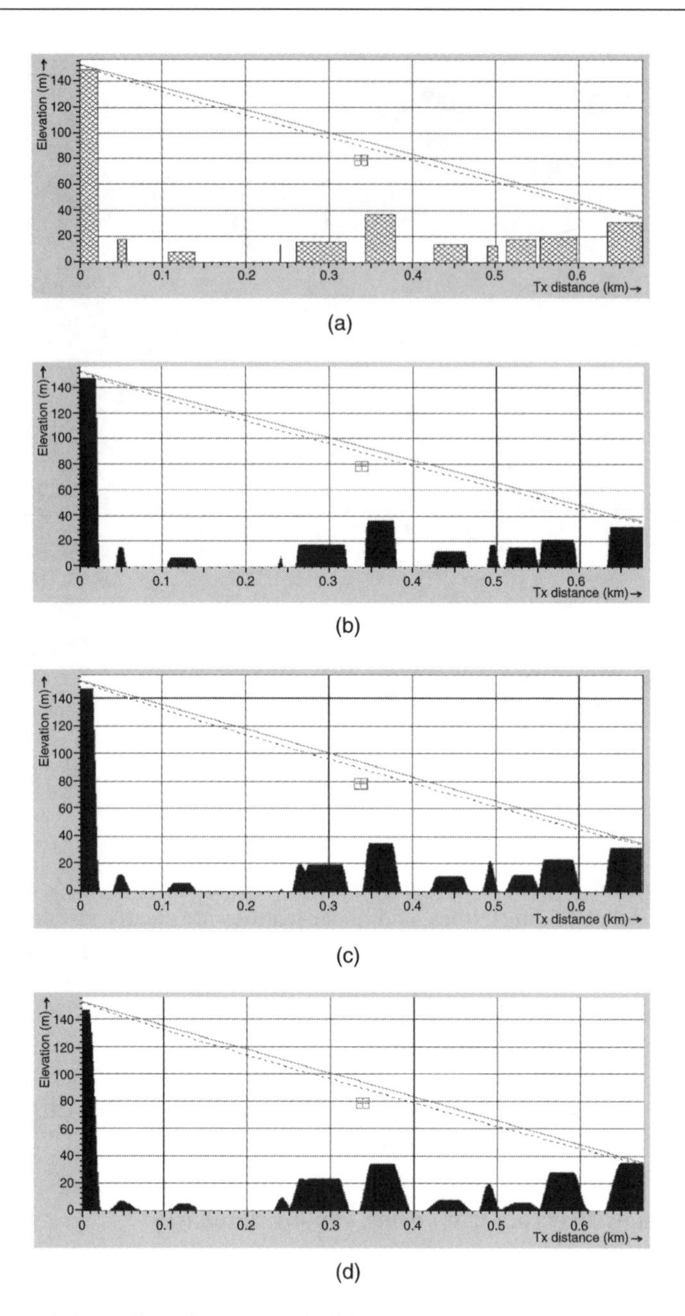

Figure 5.10 (a) Link profile using vector building database. (b) Link profile using a canopy database with 2-m resolution (2-m grid point spacing). (c) Link profile using a canopy database with 5-m resolution (5-m grid point spacing). (d) Link profile using a canopy database with 10-m resolution (10-m grid point spacing).

only differ in the type and coarseness of the canopy grid. A hypothetical LMDS system with five hubs was configured in this environment connecting all 1,600 buildings where a CPE (customer premises equipment) was located. The best server hub for each of the 1,600 buildings was assigned and the performance of these links assessed using each of the databases. Assuming that the system analysis done with the vector database is an error-free baseline, the extent of errors introduced by the canopy approximation relative to this baseline can be determined for each of the four canopy databases.

As presented in [4], Table 5.1 shows the results of this comparison of the simple count of how often a link path was mistakenly classified as LOS or non-line-of-sight (NLOS).

A further analysis in [4] examined the errors in the best server assignment and percent service availability of each link. The best server assignment is indicative of where a CPE has the best LOS path. Percent service availability is based on fade margin, which in turn is based largely on C/(I + N) ratios. The percent availability analysis thus takes into account interference paths as well as desired link paths. Table 5.2 shows the results of this analysis.

Tables 5.1 and 5.2 both show that system-design errors increase with increasingly coarse canopy resolution as expected. This can be attributed to the increasing slopes of the building walls that impact the path-clearance analysis and result in LOS paths being falsely classified as NLOS and an improper, nonoptimum server assignment is made. The database selection objective is to determine the point where the errors introduced by the canopy database are tolerable and represent a valid trade-off with the database cost. From this analysis it appears that canopy resolutions of 1 to 2 ms can provide performance comparable to the vector database.

Table 5.1 LOS/NLOS path assessment errors with canopy database

Database	LOS to NLOS (%)	NLOS to LOS (%)
vector	0	0
1-m canopy	2.15	1.68
2-m canopy	2.73	2.65
5-m canopy	3.42	5.12
10-m canopy	5.22	9.51

Table 5.2 Assignment errors with canopy database

Database	Number of CPE's changing servers	Average change in % service availability
vector	0	0.00000
1-m canopy	141	0.00337
2-m canopy	217	0.00251
5-m canopy	337	0.00485
10-m canopy	556	0.00137

5.4 MORPHOLOGY (LAND USE/LAND COVER OR CLUTTER)

Morphology or clutter databases contain information that generally classifies the character of the land cover or land use at a particular location on the Earth. Classifications such as 'urban', 'densely urban', 'residential', 'forest', and 'agricultural' are typically found in clutter databases but unfortunately there are no standards for what classifications should be included in the dataset or how these classifications are defined. When clutter data represent buildings, it actually is a very coarse substitute for having actual specific building information in either vector or canopy form as discussed in Section 5.3.

Clutter databases are almost always in a grid matrix form similar to the terrain DEM in which a clutter classification number is assigned to a point, which then represents an area of perhaps 100 m by 100 m. This grid resolution can vary widely, however, just as in the case of terrain DEMs.

The databases are usually applied to propagation/channel models by associating a given 'clutter loss' factor or attenuation in dB to each clutter category as a function of frequency. A 'clutter height' value is also sometimes included in the database to provide another parameter to differentiate a given category based on average building height. The clutter classification can also be used to set the slope of a point slope type propagation model, or even determine time dispersion or delay profile information.

However, owing to the lack of standardization, clutter databases are difficult to rigorously apply to propagation modeling. What would be regarded as 'urban' to some is 'densely urban' or 'commercial/industrial' to someone else. The clutter loss factors versus frequency are similarly not standardized. Useful application of clutter databases thus becomes a localized process in which the clutter categories and associated loss factors must be empirically determined for each database. Once determined, clutter databases can provide a degree of improved prediction over those obtained using terrain data alone. Clutter databases are also much less expensive to obtain than building databases, so clutter data essentially provides an accessible intermediate level of prediction improvement at an intermediate cost.

Because of the limitations of clutter databases, and because traditional fixed broadband wireless systems typically used physical propagation models and systems with terminal deployment 'above the clutter', clutter databases have not been widely used for fixed broadband system planning. However, the advent of NLOS PMP systems, discussed in Chapter 1, has changed this perspective. NLOS systems are intended to serve terminals that are located in and behind the clutter so an accurate clutter database could actually be quite useful for designing such systems. As discussed in Chapter 3, empirical channel models developed for 2 to 4 GHz systems have approximate environment classification associates with them. However, the nonstandard classification problems persist, so using a clutter database for NLOS system design currently remains a localized process that will require independent measurement data to properly calibrate it for a given frequency band and system type.

Figure 5.11 Sample of land use (clutter) data for Lisbon, Portugal. An additional map layer showing basic roads and airport runways has been added. (Data from [5].)

A sample of land use/land cover for Lisbon, Portugal is shown in Figure 5.11. The highly localized island of different land use types is indicative of real propagation environments. When accurate clutter loss and time dispersion values can be associated with the different land use/clutter categories, this data can be a very effective way of refining propagation prediction and channel modeling in any frequency band.

5.5 ATMOSPHERIC AND METEOROLOGY FACTORS

In Chapter 2, the effects of changing metrological conditions on predicting the performance of fixed broadband links, especially those at microwave frequencies, were discussed. Atmospheric refraction, rain, and fog all have significant consequences on link performance. In order to consider these propagation mechanisms in fading models that are used for system design, it is necessary to have databases that describe what conditions are likely to occur in the area where a wireless link or network is being deployed. To this end, several researchers over the years have endeavored to collect data and develop maps, charts, tables, and other presentation formats to make the data accessible to system design engineers.

5.5.1 Atmospheric refractivity

Atmospheric refractivity depends on a number of factors such as temperature, pressure, and humidity. Measurements of this information have been distilled into world maps that show mean values of N_S (see 2.51) during different seasons. These maps can be found in the ITU-R Recommendation P.453 to 8 [6]. Also included in this recommendation are mean values of the refractivity gradient ΔN for four different seasons and the percentage time when the refractivity gradient is less than $-100N$-unit/km, again for four different seasons. The latter type of data is a useful indicator of how often super-refractivity and ducting may occur. Finally, [6] includes data on the refractivity gradient that does not exceed 10% to 90% of the time. These values again can be used to estimate how often subrefractivity and superrefractivity conditions may occur so that fixed-link design can be adjusted as appropriate. Appendix A contains many of these refractivity maps.

The data in [6] is currently the comprehensive location for this type of refractivity data. Unfortunately, as of this writing, only some of the world maps in [6] have been translated into computer data files so they can be directly accessed by link-planning software. Those that have been translated are represented on a data grid that is 1.5 degrees latitude by 1.5 degrees longitude. This scale resolution is appropriate for a world scale database, and given the approximately thousand reporting stations used to develop the raw measurements, this resolution is probably all that can be justified. Nonetheless, with 1.5 degrees latitude equal to about 167 km, clearly much fine-grained detail about local refractivity conditions will be missed, especially in coastal areas and other areas with bays and significant water features. The data sets should therefore be used with caution and be supplemented with local knowledge of refractivity conditions where available.

For selected locations, raw refractivity gradient statistics graphs have been published in [7]. Although these data were collected sometime ago, climatic changes have not been significant enough to make this information obsolete.

5.5.2 Rain rates

Rain is usually the primary factor in limiting the range of fixed broadband links operating in the upper microwave range above 8 GHz. The situation with rain-rate databases is similar to that of refractivity databases; there are coarse world maps for rain rates that do not very accurately depict highly localized conditions in which radio systems may be deployed. Rain rates are given for the worst month in terms of the percentage time a given rain rate in millimeters per hour (mm/hr) occurred.

The worldwide maps normally used for estimating rain rates are divided into zones with letter designations. The two most widely used maps are found in [6,7]. These are also included in Appendix A.

The maps in the ITU-R Recommendation [8] are actually computer-generated for a 0.01% of the time. The 0.01% datasets can also be used to calculate link outage for other rain occurrence percentages using the method described in [8]. The worldwide datasets themselves are grids of rain-rate parameters that use a resolution of 1.5 degrees as with the refractivity databases discussed in Section 5.5.1.

The rain-rate databases model rain as stationary events; they do not describe the changing size, shape, and movement of rain cells. The statistics of rain rates during the worst month do not show how the change in rain rates occur during a storm, which is the factor that controls fading on microwave links. If such data were available, it could potentially be used to design systems that could adapt to the anticipated conditions and more probably maintain the system's reliability objectives.

5.5.3 Fog data

Propagation through fog is not an important issue for microwave frequencies, but it is the limiting factor for free-space optics (FSO) systems. Ideally, a high-resolution database would exist showing fog density (visibility range) that occurs for a given percentage of time in a typical year. No such database exists. The only current data sources that provide useful visibility information in this regard are available from [9,10].

5.6 MOBILE ELEMENTS OF THE PROPAGATION ENVIRONMENT

Except for rain and atmospheric effects, the elements of the propagation environment that have been discussed thus far are generally stationary. On a macroscopic timescale, terrain, building, and clutter characteristics may change as new buildings are built or demolished, terrain features change with road constructions, and land use changes with land use policy and development, but they do not change over the short period of time that is relevant to wireless system operation.

As noted in Chapters 2 and 3, electromagnetic (EM) wave reflections play a role in propagation model formulation and signal strength predictions. Roadways that at some times are free of cars, truck, and buses, but at other times are clogged with traffic, can affect signal propagation. Asphalt and concrete road surfaces become metallic surfaces at a height that is 2 or more meters higher that the road surface. A few studies have been done on the differences in propagation conditions with and without significant traffic, primarily with regard to signal levels available to pedestrians who are walking alongside the roadway in question.

At present there is no viable statistical analysis that could be employed to account for these transitory effects. Moreover, for fixed broadband systems, a mobile pedestrian user is not an intended operation scenario, so the transitory reflection issue is of very little significance. Nonetheless, it is conceivable that a given terminal may be successfully installed during traffic-free conditions but suffer from reduced performance when traffic is present. This possibility is an additional motivation to deploy systems that are automatically adaptable to these short-term, unpredictable changes in the characteristics of the propagation environment.

5.7 MAPPING FUNDAMENTALS

The information used to describe propagation environment models is specified in geographic terms; that is, it describes terrain, land use, buildings, refractivity, and so on at some position on the Earth's surface. Unfortunately, there are many ways to describe a position on the Earth's surface, each having been adapted over the years to provide some level of accuracy or convenience for a given task. The same physical position on the Earth's surface can have many types of position coordinates describing it. Even with latitude/longitude coordinates, the same physical position can have different latitude/longitude coordinates depending on which datum is used.

In general, describing the position on the Earth's surface is done with maps that attempt to translate the curved surface of the Earth into a flat representation on a piece of paper. This translation or *projection* process always creates distortions and errors, the extent of which depends on the extent of the map area and the methods used for the translation. The methods used for the translation are chosen to minimize particular kinds of errors. The Earth can generally be described as an ellipsoid with major and minor axis values, but over a localized area, the curvature can be more accurately described with different shapes, or different ellipsoid constants. In this way, more accurate mapping is achieved. For example, Switzerland, which covers a relatively small, mountainous area, uses different parameters for its national mapping than France, which covers a larger area and is less mountainous on an average. Another example is a country with a mostly an east–west extent (such as Turkey) using a different map projection than a country that has a mainly north–south extent (such as Sweden). Extending this concept worldwide, a multitude of approaches for localized and wide-area mapping have been created over the years.

To make the most accurate use of propagation environment databases for wireless system design, it is important to be able to very accurately synchronize the location of transmitter terminals, receiver terminals, and other system nodes with the features of the environment. As was discussed in Section 5.3.3, accuracies on the order of a few meters can have an important impact on the design and analysis of microwave link systems in urban areas where the question of whether a link is LOS or NLOS may pivot on such small distances. Therefore, accurately and consistently relating the position of propagation environment data to system node locations is essential to successful system design.

Generally, wireless engineers are trained in the geographic aspects of mapping that are relevant to the intelligent use of propagation environment model databases. This section outlines some of the basic factors and terminology used in mapping and how it applies to terrain, buildings, clutter, and other databases that are used for system design. There are four primary aspects in creating a flat map of a portion of the Earth's curved surface which are:

- spheroids, ellipsoids, and geoids
- geodetic systems and datums
- map projections
- coordinate systems

Each of these topics is discussed in the next section.

5.7.1 Spheroids, ellipsoids, and geoids

All mapping starts with an assumption about the shape of the Earth. The simplest and oldest approximation for the shape of the Earth is a *sphere* where the three radial axes are equal. For accurate local mapping, however, the inaccuracies or distortions in this assumption are significant.

A *spheroid* is similar to a sphere except one axis is not equal to the other two. A spheroid is a much closer approximation to the shape of the Earth, and, in fact, is the most widely used shape for mapping but is usually called an ellipsoid. From a mathematical perspective, the correct name for such shapes is not *spheroid* but *ellipsoid of revolution*.

An *ellipsoid* is like a spheroid except that all three axis lengths are different. As it happens, ellipsoids provide only a slightly better approximation to the Earth's shape than spheroids. The mathematics involved in using ellipsoids for map projections are also considerably more complex than the mathematics needed for map projections based on spheroids. For this reason, true ellipsoids are rarely used for mapping. Unfortunately, the terms ellipsoid and spheroid are used interchangeably in the mapping community even though they are clearly different. The term 'ellipsoid' will be used for this discussion since it is consistent with the standard usage in the mapping community. As shown below, all the ellipsoids used for mapping discussed here are specified using only two axis lengths (not three), so this alone makes it clear they are really spheroids and not ellipsoids.

A *geoid* is the final term that is sometimes used to describe the Earth. A geoid is simply an irregular three-dimensional shape that can include all the hills, valleys, oceans, and other irregularities of the Earth's surface. Most commonly, the geoid is defined as the surface around the Earth's center of mass where the gravitational force is a constant value, or an equipotential gravitation surface. It will therefore track the undulations of the Earth's surface to some extent. The most common geoid used in mapping is the one representing mean sea level, which is used as a reference for elevations. The elevations found in the DEMs described in Section 5.2.2 are referenced to mean sea level.

As mentioned above, the shape of the Earth can generally be described as an ellipsoid. The most commonly used ellipsoids that have been developed over the last 175 years are listed in Table 5.3.

The *flattening* factor is the ratio between the major and minor axis of the ellipsoid, and related to the *eccentricity* of the ellipsoid, as follows:

$$f = 1 - \sqrt{1 - e^2} \tag{5.4}$$

$$b = a(1 - f) \tag{5.5}$$

$$e = \sqrt{1 + b^2/a^2} \tag{5.6}$$

where a and b are major and minor axis lengths, respectively, f is the flattening, and e is the eccentricity.

5.7.2 Geodetic systems, datums, and datum transformations

A geodetic system is a framework for determining the coordinates of points with respect to the Earth. A geodetic system may be global or local, but each has associated with

Table 5.3 Reference ellipsoids (spheroids)

Ellipsoid name	Major axis length (m)	1/flattening(f)
Airy 1830	6,377,563.396	299.3249646
Modified airy	6,377,340.189	299.3249646
Australian national	6,378,160.0	298.25
Bessel 1841	6,377,397.155	299.1528128
Clarke 1866	6,378,206.4	294.9786982
Clarke 1880	6,378,249.145	293.465
Everest 1956 (India)	6,377,302.243	300.8017
Modified Fischer 1960	6,378,155.0	298.3
Helmert 1906	6,378,200.0	298.3
Hough 1960	6,378,270.0	270.0
Indonesian 1974	6,378,160.0	298.247
International 1924	6,378,388.0	297.0
Krassovsky 1940	6,378,245.0	298.3
GRS 80	6,378,137.0	298.257222101
South American 1969	6,378,160.0	298.25
WGS 72	6,378,135.0	298.26
WGS 84	6,378,137.0	298.257223563

it a reference ellipsoid and a geoid model. The geoid is used as a reference surface for elevations.

Local geodetic systems consist of a horizontal datum for determining the horizontal position and a vertical datum for determining elevations. The intent of a local geodetic system is to improve the accuracy of local mapping by taking into account particular attributes of the Earth's shape in that area as with the Switzerland example given earlier. Prior to satellite measurement techniques, local horizontal datums had an origin at some point in the Earth's surface. The North American Datum of 1927 (NAD27) used a position at Meade's Ranch, Kansas, as the surface datum origin. With satellite measurement, global horizontal datums were developed that have an origin at the center of the reference ellipsoid. The World Geodetic System datum of 1972 (WGS72) and of 1984 (WGS84) are examples of global datums.

It is possible to determine the relationship between a local horizontal datum and a global horizontal datum such as WGS84 by using measurements at ground control points. For basic transformations, two points are needed; for more advanced transformations, additional control points are needed. Table 5.4 shows examples of a few local datums, the associated reference ellipsoid, and the three parameters that shift the relative position of the ellipsoid center point in the x, y, and z directions relative to the Earth's center point. The Δx, Δy, and Δz shifts can be positive or negative and range from a few meters to several hundred meters. The shift values are used to carry out a transformation from the local datum and the global WGS84 datum.

Latitude, longitude, and height coordinates derived from one datum can be converted to latitude, longitude, and height coordinates in another datum using one of several transformations. A comprehensive transformation assumes the datums can have

Table 5.4 Examples of datums and datum shifts

Datum name	Area	Ellipsoid	Δx	Δy	Δz
NAD27	Continental U.S.	Clarke 1866	−8	160	176
OSGB 1936	England	Airy 1830	371	−112	434
European 1950	Average for Europe	International 1924	−87	−98	−121
South American 1969	Brazil	South American 1969	−60	−2	−41
Tokyo	Japan	Bessel 1841	−148	507	685
WGS72	Global	WGS72	0	0	0
WGS84	Global	WGS84	0	0	0

arbitrary orientations, center points, and scales, and requires seven parameters to completely describe the datum differences. However, if the assumption is made that the axes of the two datums are parallel (no rotation difference), then a simpler 3-parameter transformation can be used which assumes that only the center point needs to be shifted in the x, y, and z directions. If the appropriate Δx, Δy, and Δz shifts are known, the *Molodensky* 3-parameter transformation can be used. For a point with latitude ϕ, longitude λ, and height h in one datum with radius a and flattening f, the shifts in latitude $\Delta\phi$ (in radians), longitude $\Delta\lambda$ (in radians), and height Δh (in meters) to get the coordinates in the new datum with radius a' and flattening f' are given by

$$\Delta\phi = \frac{A + B + C}{(M + h)} \tag{5.7}$$

$$\Delta\lambda = [-\Delta x \sin\lambda + \Delta y \cos\lambda]/[(N + h)\cos\phi] \tag{5.8}$$

$$\Delta h = \Delta x \cos\phi \cos\lambda + \Delta y \cos\phi \sin\lambda + \Delta z \sin\phi - \Delta a(a/N) + [\Delta f(1 - f)N \sin^2\phi] \tag{5.9}$$

where

$$A = -\Delta x \sin\phi \cos\lambda - \Delta y \sin\phi \sin\lambda + \Delta z \cos\phi \tag{5.10}$$

$$B = \Delta a(N\varepsilon^2 \sin\phi \cos\phi)/a \tag{5.11}$$

$$C = \Delta f[(M/(1 - f)) + N(1 - f)] \sin\phi \cos\phi \tag{5.12}$$

$$\varepsilon^2 = 2f - f^2 \tag{5.13}$$

$$W = 1 - \varepsilon^2 \sin^2\phi \tag{5.14}$$

$$M = a(1 - \varepsilon^2)/W^{1.5} \tag{5.15}$$

$$N = a/W \tag{5.16}$$

$$\Delta a = a - a' \tag{5.17}$$

$$\Delta f = f - f' \tag{5.18}$$

The latitude, longitude, and heights in the new datum are given by

$$\phi' = \phi + \Delta\phi \tag{5.19}$$

$$\lambda' = \lambda + \Delta\lambda \tag{5.20}$$

$$h' = h + \Delta h \tag{5.21}$$

5.7.3 Map projections

A map projection is a means of transferring a representation of a section of the curved Earth's surface to a flat map. The map projection is independent of the geodetic system. The objective of a map projection is to produce a flat representation of the mapped area with a minimum of distortion. Since some distortion always results, choosing an appropriate map projection to make a given map involves a number of trade-offs. A projection can be selected that provides accurate angle or distance measurements between points on the map, or that accurately represents relative shape and size of areas (like countries).

The map projection can be chosen so that it is exactly accurate at one point, or on a line on the surface to be mapped. Such lines are called *standard* (reference) longitudes or latitudes. As a simple example, consider the cylindrical Mercator projection for a world map. This map is created by wrapping a cylinder of paper around a globe such that it touches the globe at the equator. The features of the Earth are then projected onto the paper. The map is very accurate at the equator where the paper touches the surface, but becomes increasingly distorted moving away from the equator toward the poles. The result is some significant distortions such as Canada and Greenland appearing to be vastly larger than they actually are. Several common map projections are listed in Table 5.5. As the projection names attest, some have been created with a specific objective in mind in terms of minimum distortion. The mathematic construction of these map projections can be quite involved. Reference [11] provides detailed information on the history and construction of many of these projections. Since a map projection transfers the curved Earth surface to a flat map, as discussed in the next section, it allows the use of linear coordinate systems that make determining position much more convenient.

Table 5.5 Map projections

Projection name	
Albers equal-area conic	Mercator
Azimuthal equidistant	Miller
Cylindrical equal area	Mollweide
Equidistance conic	Sinusoidal
Hotine oblique Mercator	Stereographic
Lambert Azimuthal equal area	Transverse Mercator
Lambert conformal conic	

5.7.4 Coordinate systems

The basic geodetic coordinate system is a spherical coordinate system that consists of two angles and a radius. In this case the two angles are called latitude and longitude, and the radius value is the height value above the geoid surface at that latitude/longitude location. The angles are given in degrees of arc, and the radius is given in meters or kilometers.

Because latitude and longitude are part of a spherical coordinate system, the latitude and longitude lines become curved lines on a flat map. Such curved lines on a map are difficult to use for determining coordinate position, so alternate linear coordinate systems are defined for most map projections. The coordinate systems are devised to be convenient for the projection type and the flat map being created. Such linear coordinate systems have a defined origin, which is usually the point or line where the map projection distortion is minimized. In some cases, a 'false' origin is used so that all the linear coordinate values are positive. A good example is the coordinate system used by the Ordnance Survey in England. The datum used in England is the Ordnance Survey, Great Britain (OSGB) 1936 datum as noted above. The map projection used for many of its maps is the Transverse Mercator projection. The linear coordinate system is in x, y meters with a true origin at 2-degrees west longitude, 51-degrees north longitude. The false coordinate system origin is shifted 400,000 m west and 100,000 m south from the true origin so that all x, y coordinates falling on the land surface of England are positive numbers.

5.8 CONCLUSIONS

The accuracy and effectiveness of propagation and channel modeling for wireless system design is directly dependent on the precision and accuracy of the models that describe the propagation environment in which the system will operate. The more accurate the terrain, building, land use, atmospheric, and other databases are, the more accurate the system design can be. Accurate databases are an important ingredient in achieving the service operator's commercial objectives with the lowest infrastructure cost.

Conversely, the choice of databases and appropriate accuracy depends on the propagation model that is being used. If an empirical model being used cannot take advantage of high-resolution terrain data, there is no point in obtaining high-resolution terrain data. This perhaps more directly points to the propagation model choice rather than the database choice, as discussed at the end of Chapter 3. If high-resolution terrain and building data are available, the logical action is to choose a physical propagation model that can take full advantage of this information in designing the system.

While high accuracy is usually desirable, it comes at a correspondingly high cost, leaving the design engineer with the task of making trade-off decisions about the required accuracy for the system to be designed. This is especially true for high-resolution building databases that are used for LMDS, CPN, and other point-to-point and PMP networks in urban areas. To some extent the problem can be mitigated with the network design. With PMP networks with many hubs, PMP networks with repeaters, or mesh networks, there are potentially multiple paths for connecting a particular customer building to the network so the need to know *a priori* precisely where buildings are located is greater. As specific system design issues are treated in the following chapters, these trade-off decisions will be discussed in the context of link and network performance calculations.

5.9 REFERENCES

[1] P.R. Wolf and B.A. Dewitt, *Elements of Photogrammetry with Applications in GIS*. Boston: McGraw-Hill. 2000.
[2] Vexcel Corporation, Boulder, Colorado. U.S.A. http://www.vexcel.com.
[3] J.L. Kirtner, "Using LIDAR data in wireless communication system design," *ASPRS 2000 Annual Conference Proceedings*, Washington, May, 2000.
[4] H.R. Anderson, "The impact of building database resolution on predicted LMDS system performance," *1999 IEEE Radio and Wireless Conference*, Denver, August, 1999, pp. 81–84.
[5] ISTAR Corporation. Sophia Antipolis, France. http://www.istar.com.
[6] International Telecommunications Union, "The radio refractivity index: its formula and refractivity data," Recommendation ITU-R P.453-8, 2001.
[7] C.A. Samson, "Refractivity Gradients in the Northern Hemisphere," U.S. Department of Commerce, Office of Telecommunications, April, 1975, Access number AS-A009503.
[8] International Telecommunications Union (ITU-R), "Characteristics of precipitation for propagation modeling," Recommendation ITU-R P.837-2, 1999.
[9] National Climate Data Center (NCDC), a division of the National Oceanographic and Atmospheric Administration (NOAA), Solar and Meteorological Surface Observational Network (SAMSON) dataset. CD-ROM. Data from 1961–1990.
[10] National Climate Data Center (NCDC), a division of the National Oceanographic and Atmospheric Administration (NOAA), International Surface Weather Observations (ISWO), CD-ROM. Data from 1982–1997.
[11] J.P. Synder. *Map Projections – A Working Manual*. Washington: United States Geological Survey. Professional Paper 1395. 1997 reprint.

6

Fixed wireless antenna systems

6.1 INTRODUCTION

Antennas are physical devices that convert radio frequency power into electromagnetic (EM) waves at the transmitting terminal and convert electromagnetic waves into radio frequency power at the receiving terminal. Every wireless communication system requires antennas of some sort at the transmitter and receiver. Within this broad context, antenna systems can take on many different forms to achieve specific objectives.

Generally, the antenna radiation pattern or antenna directivity is the primary attribute of an antenna used in wireless communication systems. Tailoring the radiation pattern to a set of desired characteristics is a means of sending the transmitted power in only those directions that are desired to achieve the intended service objectives. For point-to-point (PTP) systems, sending the transmitted power only towards the receiver terminal is the objective, so very highly directional antennas are used to radiate power over a very limited angular arc. For Local Multipoint Distribution Service (LMDS), multipoint, multi-channel distribution service (MMDS), non-line-of-sight (NLOS) system hubs, or any systems where the antenna is designed to communicate with geographically dispersed receiving terminals, broader beamwidth antennas that radiate power over a controlled arc of 45, 90 or more degrees are appropriate.

At the receiving terminal, the objective is to receive the transmitted signal from only the desired transmitting source and reject signals coming from other sources. For essentially any wireless system, a highly directional receive antenna is desirable, although for many cases such as cellular or mobile systems, it is physically difficult to have such receive antennas. For fixed PTP and Point-to-Multipoint (PMP) systems, however, such antennas are possible. This is one of the reasons that fixed wireless systems can potentially achieve higher capacities than mobile systems.

To be commercially viable, modern wireless communication systems are almost all two-way systems; that is, each node or terminal is capable of both transmitting and receiving

Fixed Broadband Wireless System Design Harry R. Anderson
 ISBN: 0-470-84438-8

signals. The characteristics of the transmitting and receiving antennas at a given terminal may, in fact, be different, although most systems are designed to be symmetrical in their transmitting and receiving characteristics.

While antennas are physical devices that control where EM radiation is sent and how it is received, in the larger network context they can provide *processing gain* that increases the overall system capacity. This is accomplished by increasing the signal-to-interference ratio (SIR) through pattern directivity. The processing gain is achieved without the expense of additional spectrum, as with spread spectrum systems such as code division multiple access (CDMA), and generally without a significant increase in noise. Increasing the processing gain without increasing the occupied bandwidth or noise power is one reason that antenna system design, and in particular, adaptive antennas, is one of the most active areas of current wireless research and development. However, such antennas are not completely without cost. Directional antennas of any type require more physical space than nondirectional antennas. Adaptive or smart antennas also require signal processing electronic circuitry that will increase the cost and complexity of the terminal.

This chapter will first review antenna theory fundamentals and basic concepts of gain and directivity. Traditional fixed antennas are discussed next. For line-of-sight (LOS) systems, where the desired propagation path direction is known, fixed antennas are now used almost exclusively, although this may change as adaptive antenna technology matures. High gain, narrow beam antennas such as parabolic dish and horn antennas are discussed, as well as lower gain broad beam antennas used at hubs.

Adaptive antennas are also treated in this chapter with a presentation of their basic theory of operation and some of the ways adaptive techniques are employed for system improvement. Multiple-input, multiple-output (MIMO) antenna systems are also discussed here along with their potential capability for increasing the transmitter–receiver link data transmission capacity.

Finally, this chapter will discuss waveguides, transmission lines, and radomes as they are considered part of the antenna system. Appropriate selection of the waveguide or transmission line can have an important impact on the overall link budget and the resulting system performance. Radomes are designed to provide physical protection to antennas, but when wet, can attenuate the signal at microwave frequencies. The losses associated with radomes (wet and dry) must be factored into the link budget and system performance calculations.

6.2 ANTENNA SYSTEM FUNDAMENTALS

An antenna is basically any device that can transform time-varying electrical power into a radiating EM wave. Antennas are devices designed to efficiently accomplish this function in a controlled way. There are many devices that inadvertently and inefficiently convert electrical power into radiating EM waves. Suppressing such inadvertent EM radiation in all kinds of devices from personal computers to microwave ovens is a significant area of electrical engineering activity. Such background radiation, in fact, can affect wireless communication system performance, but is beyond the scope of the discussion in this book. Only antennas with intended radiation characteristics will be addressed here.

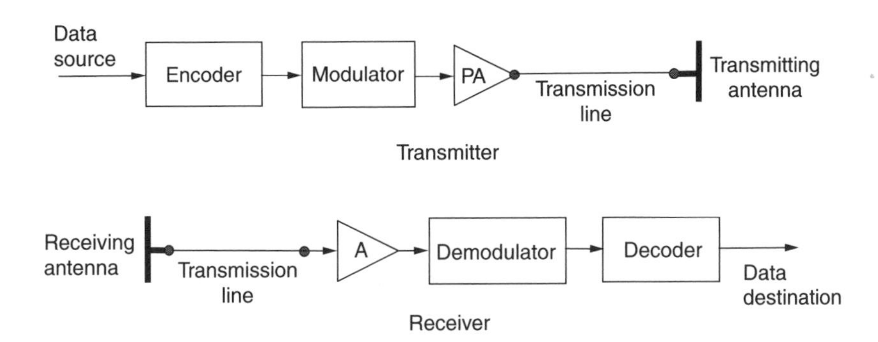

Figure 6.1 Basic components of a wireless communication transmitting and receiving terminal.

Further information on unintended electromagnetic interference (EMI) and suppression can be found in [1].

A basic block diagram of the wireless communication antenna system is illustrated in Figure 6.1. The transmitter terminal consists of a data source, encoder, modulator, power amplifier, transmission line or waveguide, and the antenna. The receiver terminal consists of an antenna, transmission line or waveguide, radio frequency (RF) amplifier, demodulator, decoder, and data destination. The antenna system includes the transmission line or waveguide and the antenna on each end of the communications link.

Traditional microwave link systems built over the last several decades, often have the antenna mounted high on a tower structure with the transmitting equipment located some distance away at the base of the tower. The purpose of the transmission line or waveguide is to convey the RF power from the transmitter power amplifier to the antenna. The transmission line or waveguide always introduces power loss in the system, which depends on its size and design. Generally the loss increases with increasing frequency and length of the line or waveguide. For this reason, many newer LMDS, consecutive point networks (CPN), and other microwave systems, especially those intended to be mounted on building rooftops, have the power amplifier integrated into single outdoor units (sometimes referred to as an ODU) that can directly be attached to the antenna, thus obviating the transmission line or waveguide and its associated losses. The diagram in Figure 6.1, can be regarded in general terms, with the understanding that the physical configuration of the equipment may permit the elimination of the waveguide or transmission line component.

The element of Figure 6.1 that is the subject of this section is the antenna. From Maxwell's equations presented in Chapter 2, it can be shown that electromagnetic radiation occurs when electric charges (electrons) are accelerated. The emission of photons from the acceleration is necessary for conservation of energy. Charges can be accelerated in a variety of ways. In antennas, the acceleration a charge experiences usually results from the charge changing direction. In a representative wire antenna shown in Figure 6.1, the charges are forced to change direction (accelerate) when they reach the end of the wire. For wireless systems, the charges are subjected to sinusoidally varying voltages that continually force a change in the direction of electron travel in response to the changing voltage potential. The EM field created by the power radiating from the antenna changes in the same way, thus resulting in a varying EM field at the variation frequency of the impressed voltage, and with the particular amplitude and phase (signal) characteristics of that voltage.

6.2.1 Radiation from an elemental dipole antenna

A basic theoretical antenna that is useful for illustrating several antenna concepts is the small infinitesimally thin dipole, also called a Hertzian dipole. This is a dipole whose length and wire diameter are very small compared to the wavelength of the sinusoidal power source. Figure 6.2 shows this short antenna element aligned along the z-axis and the spherical coordinate system normally used to describe the fields around antennas. The angle θ is the elevation angle while ϕ is the azimuth angle.

When analyzing the fields from this dipole, it is assumed that the current density along the wire is uniform at all points, that is, equal to a constant value I_0. This current is an *impressed* or source current density (J_i in 2.2), which results from the power coming from the transmitter. In real antennas, the current density distribution can be quite complicated. For a simple dipole, the current must be zero at the end of the wire. The elemental dipole with a constant current density, although theoretical, can be used to analyze more complex wire antennas using a superposition of multiple elemental dipoles, each with a different constant current density that conforms to the overall antenna current distribution.

By solving Maxwell's equations in the time-harmonic form for radiation problems, it can be shown [2] that the fields around this dipole are given by

$$E_r = \frac{I_0 \Delta z}{4\pi} e^{-jkr} \left(\frac{2\eta}{r^2} + \frac{2}{j\omega\varepsilon r^3} \right) \cos\theta \tag{6.1}$$

$$E_\theta = \frac{I_0 \Delta z}{4\pi} e^{-jkr} \left(\frac{j\omega\mu}{r} + \frac{\eta}{r^2} + \frac{1}{j\omega\varepsilon r^3} \right) \sin\theta \tag{6.2}$$

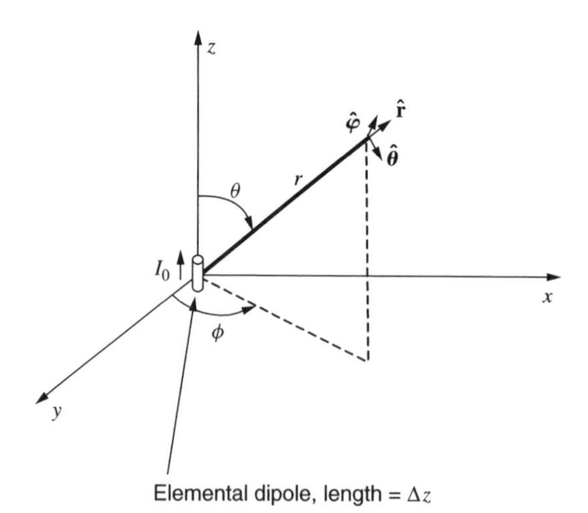

Figure 6.2 Antenna spherical coordinate system with elemental dipole antenna.

$$H_\phi = \frac{I_0 \Delta z}{4\pi} e^{-jkr} \left(\frac{j\beta}{r} + \frac{1}{r^2} \right) \sin\theta \tag{6.3}$$

$$H_r = 0; H_\theta = 0; E_\phi = 0 \tag{6.4}$$

where ε, μ, and ω are, as already mentioned, the pemittivity and permeability of the medium (air for a wireless antenna), and radian frequency of the transmitted signal, respectively, and

$$k = 2\pi/\lambda \text{ (the wave number)} \tag{6.5}$$

$$\eta = \sqrt{\frac{\mu}{\varepsilon}} = Z_0 = \sqrt{\frac{\mu_0}{\varepsilon_0}} = \frac{E_\theta}{H_\phi} \cong 377 \text{ ohms} \tag{6.6}$$

Equations (6.1) through (6.3) show several interesting things about the fields around antennas. In addition to the $1/r$ terms that were expected from the free-space propagation discussion in Chapter 3, there are additional terms that vary as $1/r^2$ and $1/r^3$. At large distances from the antenna, these terms become diminishingly small, so that the radial component of the radial electric field term in (6.1) is zero, and (6.2 and 6.3) simplify to

$$E_\theta \cong \frac{I_0 \Delta z}{4\pi} e^{-jkr} \left(\frac{j\omega\mu}{r} \right) \sin\theta \quad r \gg \lambda \tag{6.7}$$

$$H_\phi \cong \frac{I_0 \Delta z}{4\pi} e^{-jkr} \left(\frac{j\beta}{r} \right) \sin\theta \quad r \gg \lambda \tag{6.8}$$

The region at distances r where the approximations in (6.7 and 6.8) are sufficiently valid is called the *far field*; the region at distances closer to the antenna is called the *near field.* A commonly used threshold for the near-field/far-field demarcation distance r_{FF} is where

$$r_{FF} = \frac{2D^2}{\lambda} \tag{6.9}$$

where D is the diameter of the smallest sphere that entirely encompasses all the elements of the antenna.

The time-average power in the radiated field at a given distance is given by $S_{ave} = (\mathbf{E} \times \mathbf{H}^*)/2$. Using the far-field equations in (6.7 and 6.8), the total power flowing out of a sphere encompassing the antenna P_t (the total radiated power) is found by integrating the fields over the surface of the sphere.

$$P_t = \frac{1}{2} \iint \mathbf{E} \times \mathbf{H}^* \, ds \tag{6.10}$$

This results in a total radiated power for the elemental dipole of

$$P_t = \frac{\omega\mu k}{12\pi} (I_0 \Delta z)^2 \tag{6.11}$$

This is real power that is dissipated since it is radiated and not recovered. For the near field, where the additional $1/r^2$ and $1/r^3$ terms in (6.1) through (6.3) are relevant, there is additional power in the electric and magnetic fields. Performing the same power integration in (6.10), using the fields given by (6.1) through (6.3), shows that the power resulting from these terms is imaginary or reactive power. This indicates that power is actually being stored in the electric and magnetic fields in the near-field region.

6.2.2 Directivity and gain

Antenna *directivity* is of primary interest in wireless system design. It is used not only to determine the amount of power radiated in the desired direction of the receiver but also the ability of the antenna to reject undesired signals coming from other directions. The concept of directivity is illustrated in Figure 6.3. The spherical shape in Figure 6.3(a) is a (hypothetical) *isotropic* antenna that radiates with equal intensity in all directions from a given point. The shaded sphere is indicative of this radiation.

Figure 6.3(b) shows an antenna with directivity greater than that of an isotropic antenna. Considering the geometry of a sphere, a solid angle is defined as $d\Omega = d\theta \cdot d\varphi \sin\theta$ and has units of square radians or steradians. The total solid angle represented by the sphere can be found by integrating $d\Omega$ over all angles θ and ϕ

$$\iint d\Omega = \int_0^{2\pi} d\phi \int_0^{\pi} \sin\theta \, d\theta = 4\pi \tag{6.12}$$

If $P(\theta, \phi)$ is the radiated power per unit solid angle in the direction θ, ϕ, then the directivity D is defined as the ratio of the power per unit solid angle in direction θ, ϕ to the total average power per unit solid angle, or

$$D(\theta, \phi) = \frac{P(\theta, \phi)}{P_{\text{ave}}} \tag{6.13}$$

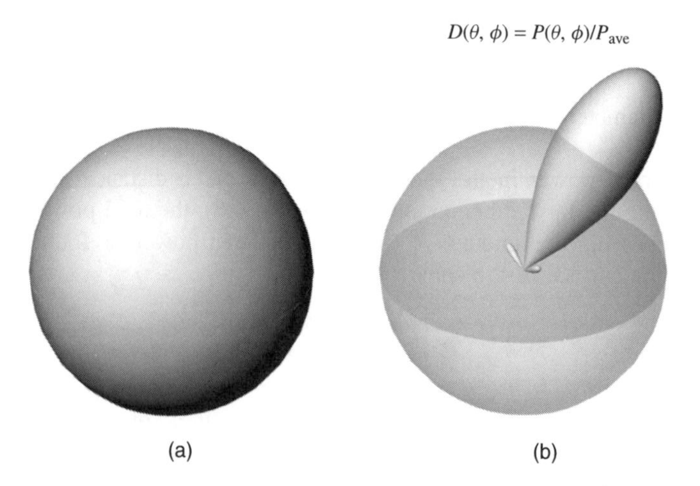

Figure 6.3 (a) Isotropic antenna and (b) directional antenna.

where $P_{ave} = P_t/4\pi$. The maximum *directivity* D_m of the antenna can be represented as the ratio of the maximum radiation power P_m per unit solid angle to the average radiated power with:

$$D_m = \frac{P_m}{P_{ave}} \tag{6.14}$$

An isotropic antenna has a directivity of 1.

The *gain* $G(\theta, \phi)$ of the antenna is closely related to directivity, but it also takes into account any power losses in the antenna. If the antenna is lossless, then $G(\theta, \phi) = D(\theta, \phi)$. More generally, the gain in any direction is the ratio of the radiated power density in that direction to the input power normalized on a per unit solid angle basis

$$G(\theta, \phi) = \frac{P(\theta, \phi)}{(P_{in}/4\pi)} \tag{6.15}$$

where P_{in} is the input power to the antenna. Gain is the most commonly used description of an antenna for normal wireless system design since it directly gives the amount of power radiated in a given direction using only the RF power at the input terminals of the antenna. The input power to the antenna is found from the transmitter power less any loss in the transmission line or waveguide. As such, this power is easily calculated or measured. Antenna gain is normally given in decibel terms

$$G(\theta, \phi) = 10 \log\left(\frac{P(\theta, \phi)}{P_{ave}}\right) \text{ dB} \tag{6.16}$$

Antennas are routinely compared on the basis of their maximum gain since this provides not only an indication of how directive they are but also whether the received signal level using the maximum gain will be sufficiently high to achieve acceptable link performance. For fixed broadband systems, the antenna gain in dB is usually referenced to an isotropic radiator, or dBi. Occasionally, however, the antenna gain is referenced to a half-wave dipole, or dBd. The gain in dBi is approximately 2.15 dB higher than the gain in dBd.

6.2.3 Antenna radiation patterns

Besides gain, the antenna characteristic that is most significant for wireless system design is the antenna radiation pattern. The antenna radiation pattern is the normalized relative magnitude of the electric field at some constant far field distance r as a function of the elevation angle θ and azimuth angle ϕ. The electric field is directly related to the radiated power density S by

$$S = \frac{|E|^2}{Z_0} \text{ W/m}^2 \tag{6.17}$$

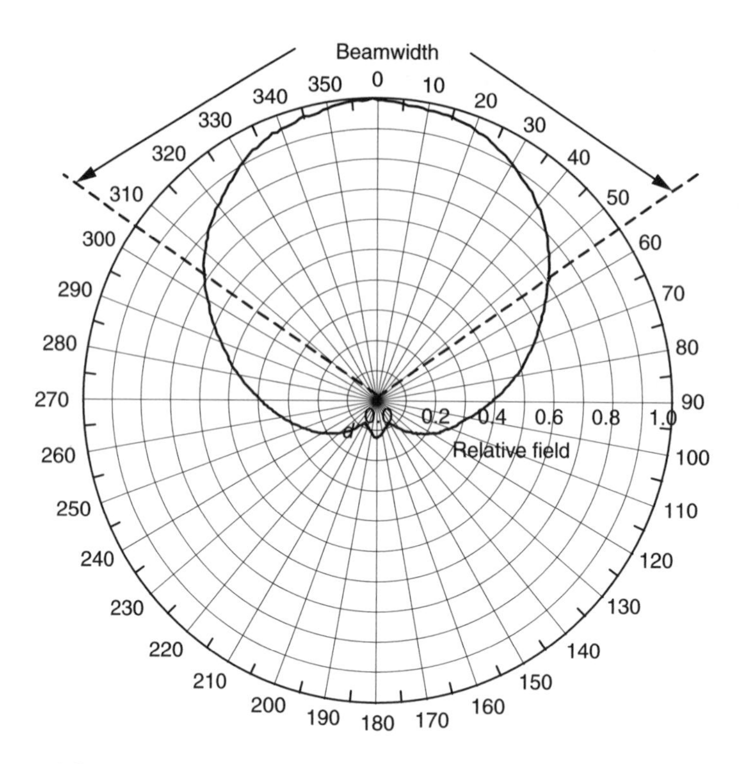

Figure 6.4 Sample horizontal plane antenna radiation pattern.

An example of the radiation pattern as a function of ϕ drawn on a polar graph is shown in Figure 6.4. This particular pattern is typical of those antennas that might be selected for use as a sector hub antenna in fixed broadband wireless systems. Since the z direction is usually oriented along the vertical axis, this ϕ-dependent pattern is also called the horizontal plane pattern, while the θ-dependent pattern is called the vertical plane pattern.

The horizontal plane pattern for the elemental dipole discussed earlier is circular, or constant with ϕ, since the field equations in (6.7 and 6.8) have no ϕ dependence. The vertical plane for both E and H vary with $\sin\theta$. Figure 6.5 shows a sample vertical plane pattern as a function of θ drawn on a polar graph for an antenna with 8 radiating elements spaced at a distance of $\lambda/2$ vertically. The portion of the pattern where the maximum gain occurs is often referred to as the *main lobe* of the pattern.

Antennas with multiple elements are called antenna arrays. As discussed in standard antenna array design reference books such as in [3], increasing the number of elements would cause the main lobe of the pattern to narrow and the maximum gain to increase. Conversely, decreasing the number of elements would cause the main lobe of the pattern to become wider and reduce the maximum gain.

The polar plots shown in Figures 6.4 and 6.5 are very useful for wireless system design. They provide a rapid and convenient way to judge the approximate orientation of the antenna to meet certain service objectives, which can later be confirmed with detailed

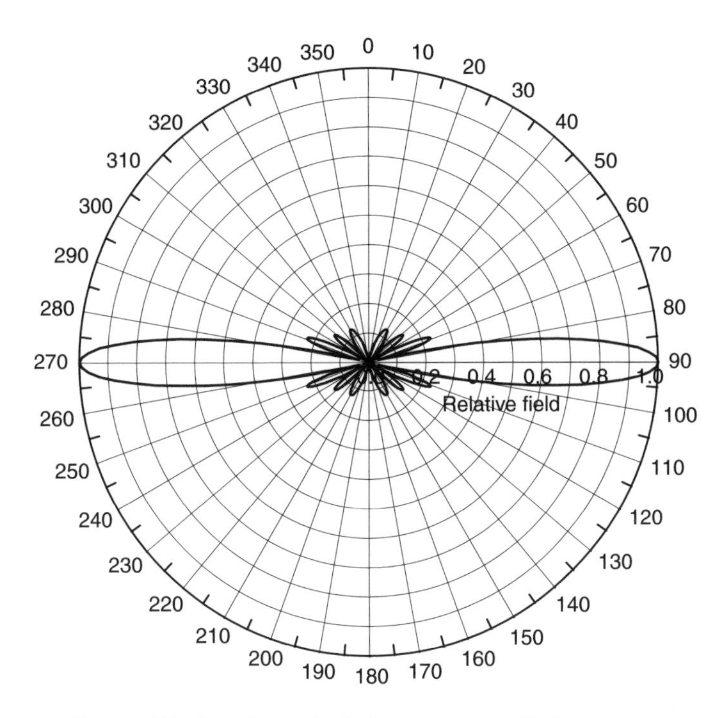

Figure 6.5 Sample vertical plane antenna radiation pattern.

propagation and system performance analysis. Many wireless system planning software tools provide planning map icons at transmitter sites that are miniature versions of the plot shown in Figure 6.4 as an aid to quickly assessing the direction of maximum radiated power in relation to the intended service area.

For narrow beam antennas, the radiation pattern is often drawn using a rectangular graph instead of a polar graph. This allows different angle scales to be used on the horizontal axis to show better the pattern shape details when the pattern values are rapidly changing. Figure 6.12 shows an example of such an antenna pattern graph. Note that the cross-polarized pattern is also shown. Cross-polarized patterns will be discussed in a later section.

The antenna half-power *beamwidth* is often used to describe the degree of directionality of an antenna. It is defined as the angle arc in degrees between the points on the antenna pattern in which the radiation falls to half of the maximum power value. For a normalized relative field pattern like that shown in Figure 6.4, this occurs at the point in which the relative field falls to a value of 0.707. The half-power beamwidth of the antenna in Figure 6.4 is approximately 110 degrees.

6.2.4 Polarization

Unlike mobile systems, where it must be assumed that the mobile handset and its antenna have an arbitrary orientation with fixed systems, some degree of knowledge about the

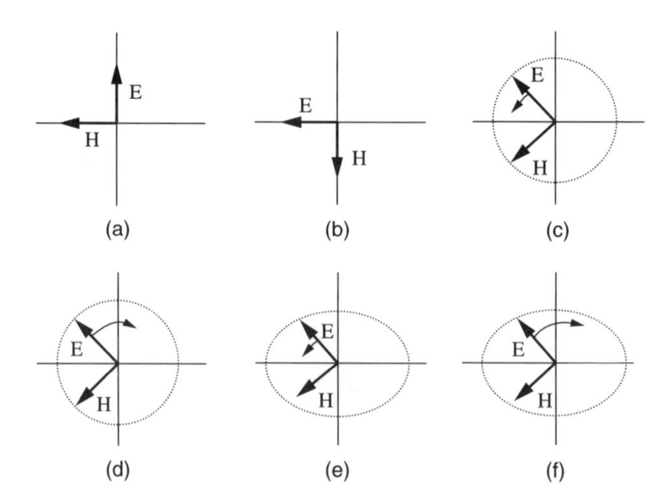

Figure 6.6 (a) Vertical polarization, (b) horizontal polarization, (c) right-hand circular polarization, (d) left-hand circular polarization, (e) right-hand elliptical polarization, and (f) left-hand elliptical polarization.

antenna orientation can be exploited to increase system capacity. For fixed LOS systems where the antenna system installations are specifically engineered rather than *ad hoc*, polarization discrimination can be effectively used to essentially double the capacity of the system for a given bandwidth and modulation/multiple access scheme.

As presented in Chapter 2, EM waves can have linear, circular or elliptical polarizations. Linear polarization can be either horizontal or vertical. The convention is that the orientation of the electric field vector establishes the polarization of the wave. Figure 6.6(a) illustrated the various polarizations from the perspective of the receiving antenna (i.e. the direction of radiation or Poynting vector is coming out of the page towards the reader). Circular polarization can be thought of as a special case of the more general elliptical polarization.

The orientation of the electric field vector is related to the construction and orientation of the antenna. In the short dipole example previously discussed, **E** was oriented along the z-axis parallel to the direction of current flow. The dipole could be physically mounted so that **E** took on any arbitrary orientation from purely vertical to purely horizontal. For orientations at angles in between, the electric field vector is resolved into two orthogonal components – one horizontal and one vertical.

If two dipoles are used and mounted at right angles to each other – one vertical and one horizontal – then vertically and horizontally polarized waves are created. If the phase of the current to one of the dipoles is altered so that it is 90 degrees ahead or behind the phase of the power feeding the other dipole, then the composite electric field vector from the two dipoles will rotate in a circular fashion as shown in Figure 6.6(c). Depending on whether the 90-degree phase shift leads or lags, determines whether right-hand or left-hand circular polarization is created. If the power to one dipole is further altered so that it has more power or less power than the other dipole, then elliptical polarization is created, again with either left-hand or right-hand elliptical polarization.

These simple configurations of dipoles illustrate some basic approaches to creating linear and elliptical polarized waves. Practical antennas for fixed broadband using parabolic reflectors, horns, panels, and other constructions to create antennas that launch EM waves of different polarizations will be discussed in a following section.

Antennas are normally designed to create the intended polarization state in the main lobe or maximum gain direction of the antenna. Even with this design objective, some power in the cross-polarized plane will still be radiated. The ratio of the copolarized field to the cross-polarized field is called the cross-polarization discrimination, or XPD, of the antenna. By design, the XPD is usually maximized in the main lobe, but at other azimuth and elevation angles XPD usually decreases. At angles in which there are pattern nulls or *side lobes*, XPD can be considerably reduced, in fact, it may be 0 dB. The rectangular pattern graph in Figure 6.12 is an apt example of the reduction in XPD at angles away from the main lobe. For fixed wireless systems designs that rely on XPD for interference rejection and capacity increase, it is essential to have the antenna patterns for both the copolarized and cross-polarized radiation, like those shown in Figures 6.4 and 6.5, for all the antennas used in the system.

Section 2.6 demonstrated the differences in reflection characteristics for horizontally and vertically polarized waves. The 180-degree phase reversal that occurs with parallel polarization under particular incident angle circumstances will reverse the rotation direction of a reflected circularly polarized wave. Consequently, it will be attenuated by a circularly polarized receiving antenna with the opposite rotational sense. This mechanism, along with other factors, lead the US television broadcasting industry to deploy transmitting stations using circular polarization in the belief that undesirable multipath reflections, or 'ghosts', could be reduced.

6.2.5 Antenna efficiency and bandwidth

As mentioned above, the gain of the antenna takes into account the fact that some power sent to the antenna is not radiated but lost through heating in the antenna structure itself, or is stored as reactive power in the near field of the antenna. An efficient antenna radiates most of the power delivered to it since this power is useful for achieving communication while power dissipated elsewhere in the antenna is wasted. The efficiency of the antenna is sometimes given the symbol e so that

$$G(\theta, \phi) = eD(\theta, \phi) \tag{6.18}$$

Figure 6.7 shows the equivalent circuit for the signal power source and the antenna where R_s and X_s are the resistance and reactance of the source, respectively, and R_R, R_0, and X_0 are the radiation resistance, ohmic (heating) resistance, and reactance of the antenna, respectively. The efficiency e is then given by

$$e = \frac{P_t}{P_{\text{in}}} = \frac{R_R}{R_R + R_0} \tag{6.19}$$

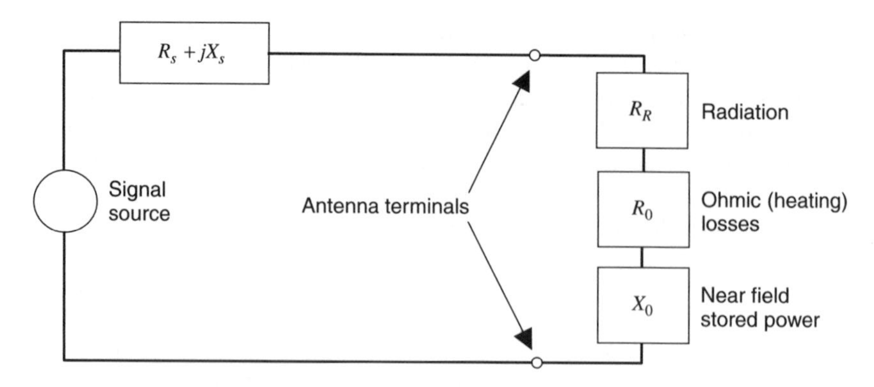

Figure 6.7 Equivalent circuit of transmitter power source and antenna.

If the source impedance and the antenna impedance are complex conjugates, maximum power transfer to the antenna is achieved and the antenna is said to be *matched* to the transmitter source. Depending on the frequency, the power is delivered to the antenna by a transmission line so that the match is made between the characteristic impedance of the transmission line and the antenna. Transmission lines have standard characteristic impedances discussed in Section 6.8 so that antennas are designed with the same standardized input impedances, typically 50 ohms. The radiation resistance for the short dipole discussed above is in [3]

$$R_R = 20\pi^2 \left(\frac{\Delta z}{\lambda}\right)^2 \text{ ohms} \tag{6.20}$$

If the reactance X_0 of the antenna is zero at a given frequency, the antenna is said to *resonant* at that frequency. Maintaining uniform impedance across the band of frequencies in which the antenna must operate is one important objective of antenna design. This band of frequencies is referred to as the antenna *bandwidth*. The antenna bandwidth is often defined in terms of the voltage standing wave ratio, or VSWR, which is the ratio of the reflected standing wave voltage to the incident standing wave voltage, or

$$\text{VSWR} = \frac{V_r}{V_i} = \frac{Z_a - Z_s}{Z_a + Z_s} \tag{6.21}$$

where $Z_a = (R_R + R_0) + jX_0$ and $Z_s = R_s + jX_s$. The antenna bandwidth is usually defined as the point where the VSWR is less than 2.0, however this depends very much on the purpose of the antenna. Even within this definition, substantial attenuation at some transmitted frequencies can occur, thus linearly distorting the power in the transmitted signal spectrum and diminishing the receiver's ability to successfully receive the signal.

The bandwidth of the antenna not only refers to how it accepts power from the source but also how the radiation pattern changes with frequency. This gives rise to the term *pattern bandwidth,* which is usually regarded as the frequency difference for which the radiation in a given direction is within 3 dB of the radiation at the nominal design center frequency of the antenna. This becomes increasingly difficult to achieve as the antenna gain increases. Antenna gain and pattern bandwidth are competing objectives that represent a trade-off for an antenna design engineer.

Wire dipole-type antennas have bandwidths on the order of a few percent of the carrier frequency, depending on its radius. It can be shown [4] that bandwidth increases with increased wire radius, a concept that extends to the design of panel antennas that use flat metal plates in one form or another instead of wires as the dipole elements, to improve the antenna bandwidth. Panel antennas are discussed in Section 6.4.2.

Practically, modern antennas for Ultra High Frequency (UHF) and microwave frequencies can be designed with low ohmic losses, so this rarely is a consideration in antenna selection. Bandwidth remains an important issue, especially for antennas intended to function very effectively over a bandwidth that is a large percentage (>5%) of the channel center frequency.

6.2.6 Electrical beamtilt, mechanical beamtilt, and null fill

The objective of any transmitting antenna is to focus as much of the radiated power as possible towards the desired receiving locations or service area and suppress it elsewhere. A horizontal plane radiation pattern such as that shown in Figure 6.4, is an example of the radiation in the horizontal plane being tailored so it is directed where desired.

The same concept applies for the vertical plane pattern such as that shown in Figure 6.5. Shaping the vertical plane pattern can be an important factor in antenna design when the intended receiver locations or service areas are *vertically* dispersed as well as spread out in distance. For example, an LMDS hub sector antenna mounted at the top of a tall building may be intended to serve remote terminal locations on the tops of buildings that are much lower than, or comparable in height to, the building where the hub is located. The large vertical span in the service area would dictate an antenna that could radiate the appropriate power over a large vertical arc represented by these buildings. A similar vertical angle spread occurs when the receive terminals are located at various distances ranging from very close to the base of the hub antenna structure to larger distances that are nearer to the horizon.

Modifying the vertical radiation of an antenna to meet these service objectives can be achieved by several means:

- Lower the antenna vertical gain so the vertical plane pattern is broader.
- Use *electrical beamtilt* such that the main lobe of the vertical plane pattern directs most of the power below the horizontal plane. Electrical tilt can be accomplished by adjusting the phase relationship among the currents delivered to the antenna elements. Electrical tilt usually decreases the maximum main lobe gain of the antenna.
- Add *null fill* to the antenna, which essentially removes the deep nulls in the vertical plane pattern (like those shown in Figure 6.5) so the pattern has a more uniform monotonic decrease in power as the elevation angles relative to the horizontal plane increase. Like electrical beamtilt, using null fill in the antenna design usually decreases the maximum main lobe gain of the antenna somewhat.
- With the use of *mechanical beamtilt*, the vertical axis of the antenna is physically mounted at a titled angle so that the main lobe radiation is directed below the

horizontal plane in the azimuth direction of the tilt. While this is a simple approach, the drawback is that the main lobe radiation in the azimuth direction opposite the tilt azimuth is elevated above the horizontal plane. For antennas with radiation greatly suppressed in this direction (like the pattern in Figure 6.4) this may not be significant. Consequently, mechanical tilt is a commonly used technique for adjusting service areas, and reducing interference, for fixed broadband, cellular, and other wireless hub sector antenna systems.

These techniques will be discussed in Chapter 11 that deals with specific design situations where these antennas attributes are relevant.

6.2.7 Reciprocity

It is convenient to discuss antenna principles in terms of transmitting antennas because it is easy to visualize power flowing from an antenna just as light (also EM radiation) flows from a light bulb. When the receiving antenna is immersed in the time-harmonic electric and magnetic field created by the transmitting antenna, power will be generated at the terminals of the receiving antenna. This is the function of the receive antenna design – it converts the field created by the transmitting antenna into power at its terminals. The power can then be transported to the receiver via a transmission line or waveguide, as shown in Figure 6.1. For a given transmitter–receiver pair, it is possible to swap the position of the transmitting and receiving antennas by connecting the transmitter to the receive antenna and the receiver to the transmitting antenna, and the same received signal as before will be present at the input to the receiver. This duality is called *reciprocity*.

As a consequence of reciprocity, all the characteristics of the transmitting antenna discussed earlier in this chapter including directivity and gain, antenna patterns, polarization, efficiency, impedance, and bandwidth apply equally to the receiving antenna in a wireless communication system. The directivity of the receiving antenna is explicitly exploited to maximize the power received from the desired direction (the transmitter) and suppress interference from other sources.

6.3 FIXED NARROW BEAM ANTENNAS

Narrow beam antennas are the most commonly used for fixed broadband wireless systems now in operation. Although the meaning of the term ‘narrow beam’ is relative, the basic objective of a narrow beam antenna is to send radiated power to one receiving location, or conversely to receive power from one transmitting location. In doing so, the antenna gain is maximized in the direction where it will do the most good for link performance, and limit the interference caused to other neighboring systems or receivers within the same system. The ideal narrow band antenna would have a beamwidth just large enough to fully illuminate the receive antenna aperture and send power nowhere else. Such an antenna would create the proverbial infinitesimally thin ‘invisible wire’ that connects the transmit and receive terminals. The lasers used for the free-space optics systems are the closest approximation to this ideal system with present-day technology. As will be discussed theoretically in Chapter 12, a wireless system with such idealized invisible wire links could connect an unlimited number of terminals, and on a system level, achieve infinite system communication capacity. Narrow beam antennas also reduce illumination of the propagation environment so that time dispersion effects due to multipath signals arriving at the receiver are reduced.

There have been many narrow beam antennas developed over the years. The few types selected for discussion below are those most commonly used today in fixed broadband services operating in the 2 to 40 GHz range.

6.3.1 Horn antennas

Typical horn antenna configurations are illustrated in Figure 6.8. From a rudimentary perspective, a horn antenna is a piece of waveguide that has been flared out and cut off so that the waves in the waveguide are launched into space as propagating EM waves. Horn antennas are popular for frequencies above 2 GHz because they can provide high gain, wide bandwidth, and low VSWR.

Horn antennas are part of the class of antennas known as *aperture* antennas. The far-field radiation is found by considering the tangential fields across the aperture. For real horn antennas with various field distributions across the aperture, this analysis can be complicated. For a rectangular aperture like those shown in Figure 6.8, the analysis can be simplified if the assumption is that the amplitude and phase of the electric field across the physical aperture is uniform. With the aperture dimensions shown in Figure 6.9, the aperture electric field $\mathbf{E}_a$ is

$$\mathbf{E}_a = E_0\hat{\mathbf{y}} \text{ for } |x| \leq \frac{L_x}{2}, |y| = \frac{L_y}{2} \tag{6.22}$$

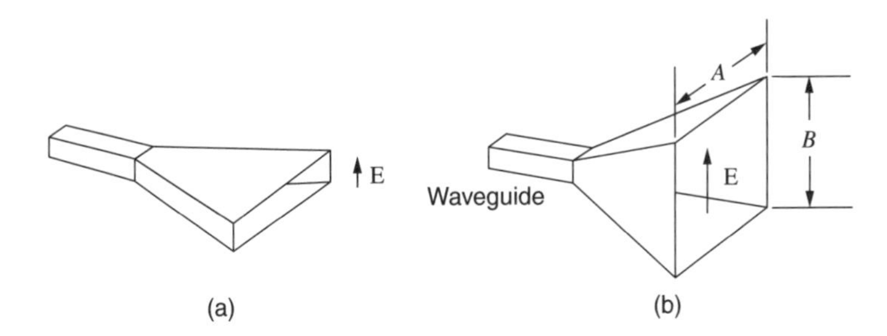

Figure 6.8 Examples of horn antenna configurations.

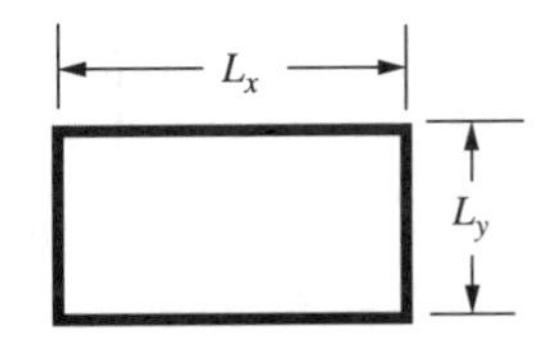

Figure 6.9 Uniform amplitude aperture.

The θ-plane and ϕ-plane radiation fields can then be found [5] as

$$E_\theta = jk\frac{e^{-jkr}}{2\pi r}E_0 L_x L_y \sin\phi \left[\frac{\sin(ukL_x/2)}{(ukL_x/2)}\right]\left[\frac{\sin(vkL_y/2)}{(vkL_y/2)}\right] \tag{6.23}$$

$$E_\phi = jk\frac{e^{-jkr}}{2\pi r}E_0 L_x L_y \cos\phi\cos\theta \left[\frac{\sin(ukL_x/2)}{(ukL_x/2)}\right]\left[\frac{\sin(vkL_y/2)}{(vkL_y/2)}\right] \tag{6.24}$$

where the variables $u = \sin\theta\cos\phi$ and $v = \sin\theta\sin\phi$. If these fields are considered in just the vertical $y - z$ plane ($\phi = 90°$), then (6.23) reduces to

$$E_\theta = jk\frac{e^{-jkr}}{2\pi r}E_0 L_x L_y \sin\phi \left[\frac{\sin(\sin\theta k L_y/2)}{(\sin\theta k L_y/2)}\right] \tag{6.25}$$

Similarly, in the $x - z$ plane ($\theta = 0°$) (6.24) becomes

$$E_\phi = jk\frac{e^{-j\beta r}}{2\pi r}E_0 L_x L_y \cos\theta \left[\frac{\sin(\sin\theta k L_x/2)}{(\sin\theta k L_x/2)}\right] \tag{6.26}$$

Equations (6.25 and 6.26) are in the familiar form of the $\sin x/x$ function, as plotted in Figure 6.10. The $\sin x/x$ pattern occurs frequently in describing the pattern of antenna arrays of multiple elements, so it is often called *the array factor* (AF).

The $\sin x/x$ factor leads to the interpretation of the far field pattern of the aperture on the θ, ϕ plane as the Fourier transform of the uniform electric field amplitude across the aperture as given by (6.22). In signal analysis, the Fourier transform of a single rectangular pulse in time is proportional to a $\sin x/x$ function in frequency with the width of lobes inversely proportional to the time width of the rectangular pulse. In the same way, the widened aperture in Figure 6.9 (in either plane) narrows the width of the pattern lobes.

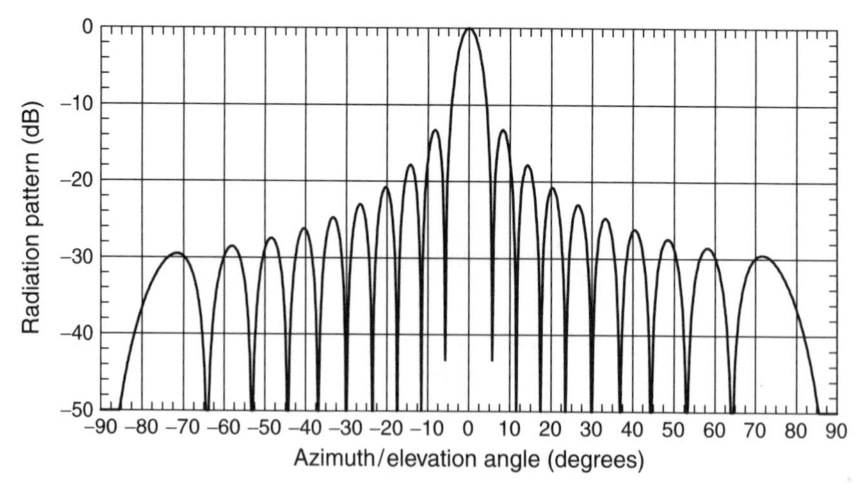

Figure 6.10 Radiation pattern for uniform rectangular aperture, $L_x = L_y = 10\lambda$.

As with signal pulse analysis, the spectrum side lobes are usually undesirable since they can cause interference to systems operating in adjacent areas. These side lobes are there to represent the high frequency power necessary to correctly represent the abrupt step-function transitions at the beginning and end of the pulse. To reduce the occupied spectrum, a pulse without these abrupt transitions is used. Typical smooth pulse shapes such as the raised cosine pulse have much lower spectral side lobes (see Chapter 7).

In an analogous way, the side lobes for a horn antenna can be reduced by modifying the field distribution across the aperture to remove the abrupt transition from no field outside the aperture to a constant value inside the aperture. This is a primary goal in horn antenna design and is usually accomplished by increasing the length of the horn compared to the aperture width.

With the assumption of a uniform field across the aperture, the gain of a horn antenna can be calculated with the simple approximation

$$G_m = e_{\text{ap}} \frac{4\pi}{\lambda^2} L_x L_y \tag{6.27}$$

where L_x and L_y are the dimensions of the rectangular aperture in meters and e_{ap} is the efficiency of the antenna. The efficiency of horn antennas ranges from 40 to 80%, with 50% efficiency being typical for horn antennas with optimized side lobe characteristics [6].

The half-power beamwidth can be found for the H-field plane (perpendicular to the E-field orientation) from the following formula

$$BW = 78 \frac{\lambda}{L_x} \text{ degrees} \tag{6.28}$$

For the E-field plane, the half-power beamwidth is

$$BW = 54 \frac{\lambda}{L_y} \text{ degrees} \tag{6.29}$$

In addition to being used as stand-alone antennas, horn antennas are often used as the focal point source for reflector antennas, as described in the next section.

6.3.2 Parabolic and other reflector antennas

Parabolic reflector antennas are the most widely used fixed broadband wireless narrow beam antennas. They usually are configured with the feed point or driven element situated at the focal point of the parabolic-shaped reflector, as shown in Figure 6.11. The feed-point antenna is often a horn or two-element colinear antenna, which essentially provides a uniform illumination of the reflector surface.

If the diameter of the parabolic reflector is large compared to a wavelength, that is, $a \gg \lambda$, ray-tracing techniques can be used to evaluate the reflection properties of the parabola. The slope of the parabolic surface is such that the reflection images are located behind the reflector so the source waves are reflected parallel to the axis of the parabola, as shown in Figure 6.11. The result is that all signal rays are reflected

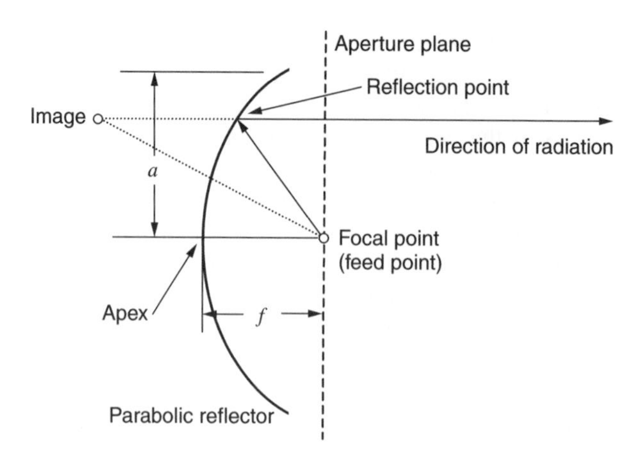

Figure 6.11 Geometry for a parabolic reflector antenna.

in the same direction, creating a high gain antenna with a narrow beam of radiation. The gain of the antenna increases with increasing parabolic radius a. Very large radius parabolic antennas have been used for space communications with distant spacecraft where high gain is required. However, to achieve the maximum gain from the antennas, the parabolic surface must be a perfect parabola within a small fraction of a wavelength. As the frequency increases, the surface must be perfect within a fraction of a centimeter. For large parabolic reflectors several tens of meters across, this can become a significant mechanical engineering challenge, especially when differential heating and metal expansion during the day are considered.

The radiation pattern of a parabolic dish antenna can be found with aperture techniques or by considering the current distribution on the surface of the reflector. Using this method, an approximation to the radiation characteristics can be formed using ray-tracing and replacing the reflector with an infinite number of image sources behind the reflector. Since the total ray trajectory distance through the reflection is the same for all the reflections on the surface, the reflections are in phase at the aperture plane. The result is a uniform (in-phase) plane wave at the aperture plane with a cross section that is circular with the same radius as the reflector. Using Huygen's principle (see Section 2.7.2), the fields on the aperture can be considered sources for a secondary wavefront that makes up the far-field radiation pattern.

Using this approach, the radiation pattern for the particular parabolic dish aperture is [7]

$$E(\phi) = \frac{\lambda}{\pi a} \frac{J_1[(2\pi a/\lambda)\sin\phi]}{\sin\phi} \tag{6.30}$$

where J_1 is the first-order Bessel function of the first kind and ϕ is the azimuth angle to the field point. The angular arc ϕ_0 between the first nulls in the pattern is given by the point in which the Bessel function has its first zeros, which occurs when the argument is 3.83. The result is

$$\phi_0 = 2\sin^{-1}\left(\frac{1.22\lambda}{2a}\right) \tag{6.31}$$

Equation (6.31) makes it clear that the beamwidth narrows as the radius of the parabolic dish increases. The maximum directivity of the parabolic antenna may be calculated with a simple formula

$$D_m = \frac{4\pi}{\lambda^2} \times \text{area} \tag{6.32}$$

by including the antenna efficiency, the maximum gain is found as

$$G_m = e_{\text{par}}\frac{4\pi}{\lambda^2} \times \text{area} \tag{6.33}$$

where area is the parabolic dish aperture area. The efficiency of a parabolic reflector antenna is typically 55 to 60%. The formula in (6.33) is identical to the one for the rectangular horn aperture in (6.27), leading to the observation that (6.33) is a reasonable approximation for the gain of any aperture antenna, regardless of its particular shape.

The equations given earlier for the parabolic reflector antenna are useful for understanding the characteristics of the antenna and how they affect its operation in a wireless system. However, these are approximations that ignore several practical details about the antenna, including edge diffraction at the edges of the reflector, nonuniform illumination of the surface by the feed point antenna, and the fact that blockage from the presence of the feed point antenna will, to some extent, upset the uniform array of secondary sources on the aperture plane.

Normally for system design, these equations are not needed since accurately measured radiation patterns for antennas are available from antenna vendors. The pattern shape data is often an 'envelope' of the actual achieved pattern results so that it represents the worst case that could be expected from a standard production antenna. Figure 6.12 shows a typical envelope pattern graph for a parabolic reflector antenna available from one of the larger manufacturers of such antennas. The pattern response in the form of data files usually can be imported directly into software planning tools that are used for calculating signal levels and interference in a network design. From Figure 6.12, it is important to note that the XPD decreases substantially at angles away from the main lobe of radiation.

Besides parabolic dish antennas, there are several other types of reflector antennas that are used for various purposes. The corner reflector, Cassegrain dual reflector, and horn-reflector antenna are shown in Figure 6.13. Like the example cited earlier, these reflector antennas can be analyzed to a first approximation as aperture antennas using ray-tracing techniques.

The horn-reflector antenna is one of antennas types used for the original AT&T cross-country microwave system completed in 1951. Because of its high gain, low noise, and low side lobe characteristics, it is widely used for backbone microwave link systems in the US and other countries.

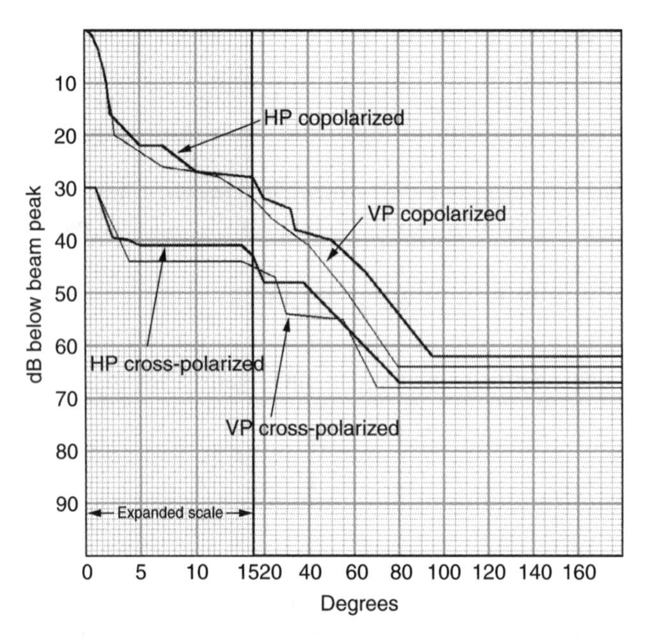

Figure 6.12 Radiation pattern envelope for a 0.3-m (1 foot) diameter parabolic antenna at 28.5 GHz. Horizontal (HP) and vertical (VP) copolarized and cross-polarized patterns shown. (Data from [8]).

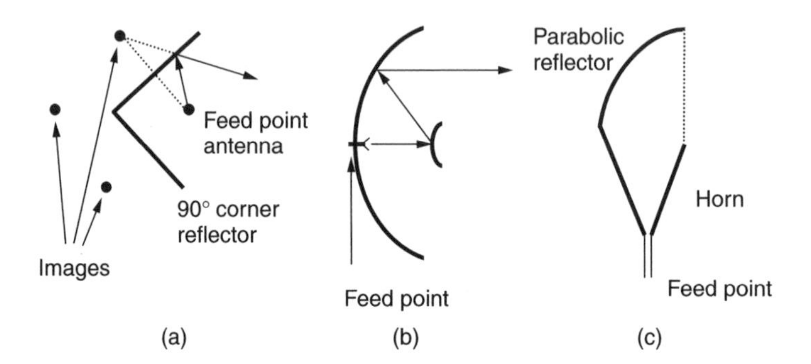

Figure 6.13 Examples of narrow beam reflector antennas: (a) corner reflector, (b) Cassegrain dual reflector, and (c) horn-reflector.

6.4 FIXED BROAD BEAM ANTENNAS

Broad beam antennas in a fixed wireless system are generally used at cell site or hub locations where the idea is to communicate with several specific remote terminal customer premise equipment (CPE) locations or to provide service to an area where the specific locations of the CPE's are not yet known. From a technical perspective, like 'narrow beam', the term 'broad beam' is somewhat nebulous. Broad beam antennas can have beamwidths that vary from 15 degrees up to omnidirectional antennas with 360-degree beamwidths. When used as broad beam antennas at hub sectors, the number of sectors and the angular coverage arc of each sector will determine the required antenna beamwidth. A 4-sector hub intended for service in all directions will require broad beam antennas at each sector with a nominal beamwidth of 90 degrees, similar to that shown in Figure 6.4. A hub with 12 sectors will need antennas with 30-degree beamwidths.

The number and beamwidth of the sector antennas will also have an important impact on the system capacity. This will be further explored in Chapters 11 and 12 where the trade-offs between hub and CPE antenna beamwidth and the impact of hub sectorization on network capacity will be discussed.

The antenna radiator types used to achieve broader beamwidth sector antennas depend on the particular frequency band. The following discussion, therefore, is divided into radiator types that are most suitable for frequencies above and below 10 GHz. Hub sector antennas also must be capable of handling reasonable power levels and bandwidths, a fact that largely excludes some currently popular antenna types such as microstrip (patch) antennas because of power limitations. A full discussion of the advantages and disadvantages of patch antennas and the associated engineering details can be found in [9].

6.4.1 Horn antennas for hub sectors above 10 GHz

For the higher microwave frequencies above 10 GHz, the hub sector antenna types are adapted from the narrow beam antennas described in the previous section. For example, the rectangular horn antennas described in Section 6.3.1 can achieve beamwidths of 30 degrees if the physical aperture is small enough. For the LMDS band at 28 GHz, from (6.27) a horn aperture width of 2.60λ would be required. With a wavelength of about 1 cm, this aperture width is about 2.6 cm, a narrow but certainly feasible aperture dimension.

There are currently several commercially available hub antennas for LMDS that use horn radiators, usually with other elements to further optimize first side lobe suppression of greater than 35 dB below the main lobe and back lobe suppression of greater than 50 dB. The concept can generally be applied to other frequency bands like 24 GHz in the US where PMP service is authorized.

6.4.2 Hub sector antennas for MMDS and U-NII bands

At lower frequencies there are more options for the fixed radiator types that can be effectively used for sector antennas. Most of the broad beam antennas used in these bands are arrays of dipoles or slots, either in a linear or planar configuration. Such antennas can be made with very thin profiles so that wind load is reduced and a wider variety of mounting and beamtilt options are possible. To achieve the necessary bandwidth, the antenna elements are often wide, flat plates (see Section 6.2.5). This gives the antenna array the shape of a panel, so such antennas are often generically described as panel antennas.

An infinitesimally thin wire dipole like that described in Section 6.2 has a relatively narrow bandwidth of perhaps a few percent. For an MMDS system operating in the 2.5 to 2.7 GHz range, a bandwidth of almost 10% is needed. To achieve broader bandwidths, dipole elements that have low length-to-diameter ratios are used; in fact, flat plate dipole elements are a commonly used option. Depending on the power requirements, the dipole elements can be etched onto a nonconductive substrate, which lowers the cost and permits automated design techniques to be used to achieve a wide variety of pattern shapes.

Slots can also be used as radiator elements. Slot radiators in a semi-infinite conducting sheet are efficient elements that radiate with essentially the same pattern as dipoles, except the E field is transverse to the longitude direction of the slot. Antenna arrays made from slotted transmission lines or waveguides have become a popular approach to broad beam antennas for television broadcasting because of their high-power handling capabilities (which may be more than one hundred kilowatts) and because of their mechanical ruggedness.

These simple radiators have limited directivity that makes them unsuitable for hub sector antennas by themselves. To achieve well-controlled, flexible radiation patterns that can meet the requirements of the system design, it is necessary to form these elements into arrays of elements. Depending on the horizontal and vertical radiation patterns desired, two types of arrays are commonly used – linear and planar. Each of these is discussed below.

6.4.2.1 Linear antenna arrays

Figure 6.14 shows a linear antenna array. The array consists of a line of basic antenna elements, typically dipoles or slots. The amplitude and phase of the power a_n, φ_n delivered to each element, and the spacing between the elements, can be adjusted to achieve a wide

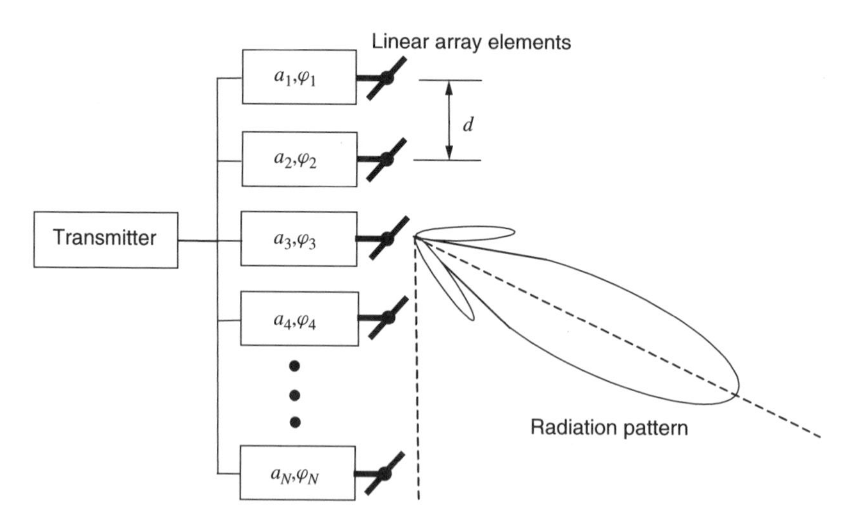

Figure 6.14 Linear antenna array. Changing values of amplitude and phase (a, φ) of the power at the elements changes the shape and direction of the pattern.

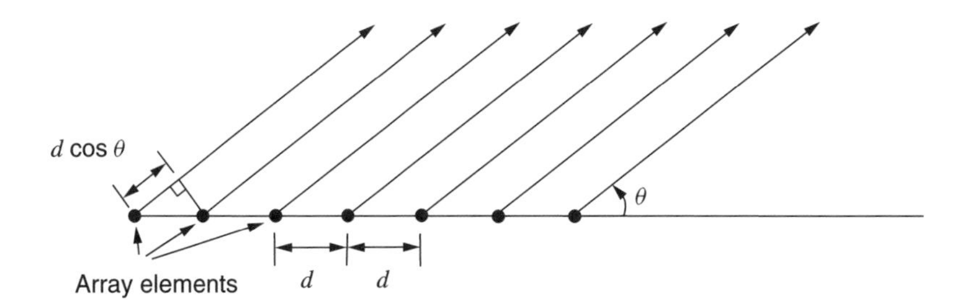

Figure 6.15 Geometry for linear array of isotropic radiators.

range of antenna pattern shapes. The bandwidth of the linear array is effectively the bandwidth of the individual elements. For a horizontally polarized antenna the dipole elements are oriented horizontally, making the direction of current flow horizontal and the polarization horizontal as with the elemental dipole in Section 6.2.1. For a vertically polarized antenna, the dipole elements are vertically oriented.

Referring to the geometry in Figure 6.15, the far field pattern for an array of isotropic radiators like that shown in Figure 6.3(a) is written as

$$E(\theta) = \sum_{n=1}^{N} a_n \exp[j(n-1)kd\cos\theta + \varphi_n] \tag{6.34}$$

with the first element taken as a phase reference and $k = 2\pi/\lambda$ as before. For arrays of dipoles or other real radiators instead of isotropic elements, (6.34) is actually the array factor, AF

$$\mathrm{AF} = \sum_{n=1}^{N} a_n \exp[j(n-1)kd\cos\theta + \varphi_n] \tag{6.35}$$

To make the quadrature vector summation more apparent, (6.35) can also be written as

$$\mathrm{AF} = \sum_{n=1}^{N} a_n[\cos((n-1)kd\cos\theta + \varphi_n) + j\sin((n-1)kd\cos\theta + \varphi_n)] \tag{6.36}$$

For the special case when all the amplitudes a_n are uniform (a_0), $d = \lambda/2$, and the phase shift to each element is uniform (φ_0), then (6.35) can be written as

$$\mathrm{AF} = a_0 \sum_{n=1}^{N} e^{-jn\psi} \tag{6.37}$$

where $\psi = kd\cos\theta + \varphi_0$. By multiplying both sides of (6.37) by $e^{j\psi}$, and through some simple manipulations [10]

$$\mathrm{AF} = a_0 \frac{e^{jN\psi} - 1}{e^{j\psi} - 1} = a_0 \frac{e^{jN\psi/2}}{e^{j\psi/2}} \frac{e^{jN\psi} - e^{-jN\psi/2}}{e^{j\psi/2} - e^{-j\psi/2}} \tag{6.38}$$

$$\mathrm{AF} = a_0 e^{jN\psi/2} \frac{\sin(N\psi/2)}{\sin(\psi/2)} \tag{6.39}$$

Since the radiation pattern is the objective here, the phase factor $e^{jN\psi/2}$ is not important. For a value of θ in which the fields from all the elements are in phase, the amplitude of the radiation pattern is Na_0. AF can therefore be normalized by dividing by this amount, leaving the normalized AF for the uniform amplitude linear array as

$$\mathrm{AF} = \frac{1}{N} \frac{\sin(N\psi/2)}{\sin(\psi/2)} \tag{6.40}$$

This has a form similar to the $\sin x/x$ function that was also the AF for the uniform rectangular aperture as given in (6.25 and 6.26). The radiation pattern graphic in Figure 6.10 is therefore indicative of the far field pattern for the linear array with uniform power distribution on all the elements.

As noted above, (6.40) is the AF that is also the far field pattern if the array elements are isotropic radiators. For nonisotropic radiators such as dipoles or slots, the AF is multiplied by the element radiation pattern. For a dipole, the element radiation pattern perpendicular to the longitudinal axis of the dipole is circular (omnidirectional). A vertical stack of vertically oriented dipoles will therefore be omnidirectional in the horizontal plane – all the directivity will come in the vertical plane because of the array factor and the element pattern of the dipole parallel to the dipole axis. For a half-wave dipole, the element pattern is shown in Figure 6.16. Other configurations of linear dipole arrays (horizontal and vertical) can be analyzed in the same way to find the resulting vertical and horizontal plane patterns.

For sector antennas that are intended to have half-power beamwidths of 90 degrees or less, techniques such as using reflector-backed or screen-backed dipole arrays can be used to shape the pattern in the plane perpendicular to the line of elements.

6.4.2.2 Planar antenna arrays

For hub antennas with particularly narrow beamwidths, additional linear arrays can be added alongside the first to achieve increased directivity in that plane. This creates a planar array as shown in Figure 6.17. The radiator elements shown in this schematic are assumed to all be situated in the same plane rather than arbitrarily located, hence the array name 'planar'.

The amplitudes and phases in Figure 6.17 can be adjusted to steer the main beam of the radiation in two dimensions – both the horizontal and vertical planes. This provides significant flexibility in adjusting the maximum gain pointing direction but clearly with an increase in array complexity.

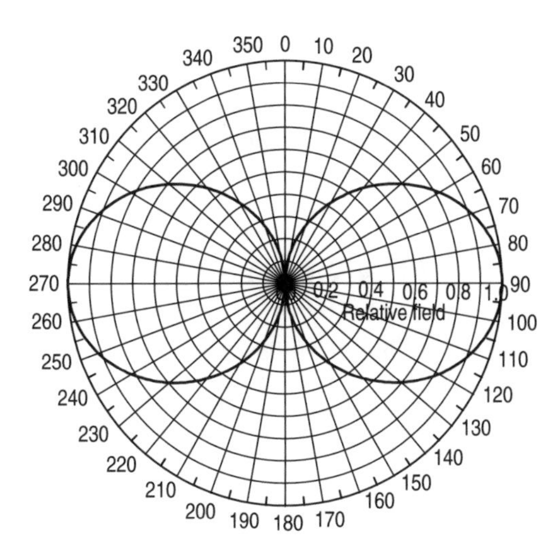

Figure 6.16 Horizontal plane radiation pattern for a horizontally oriented dipole element.

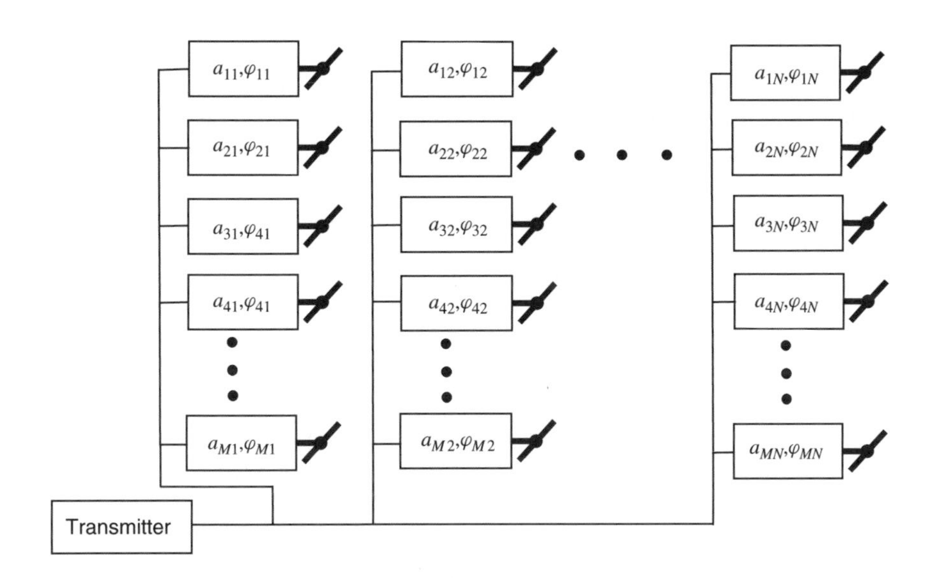

Figure 6.17 Planar antenna array.

For the linear and planar array antennas discussed in this section, it was assumed that the components needed to adjust the amplitudes and phases of the power to each array element are static or fixed, so the resulting radiation pattern is also fixed. For the system design this fixed radiation pattern is used along with the transmitter power and propagation models to predict the signal levels throughout the coverage area, or for the uplink, to predict where signal reception from remote terminals will be adequate. If the amplitude and phase of the signals from the networks in the linear or planar array could be automatically and electronically adjusted rather than being fixed, the radiation pattern characteristics could be similarly adjusted and optimized for best link margins or best desired signal-to-interference (S/I) ratios. Such adaptive arrays are discussed in Section 6.6.

6.5 DIVERSITY ANTENNA SYSTEMS

The antenna arrays discussed in the preceding sections were described in terms of their operation as hub sector transmitting/receiving arrays. However, multiple antennas can also be used for *space diversity* reception. Antenna space diversity is one of several diversity methods used in wireless communication systems. The others include frequency, time (multipath), angle, polarization, and field diversity. With space diversity, the signals received from two or more antennas are combined in some constructive fashion to reduce the effects of the signal amplitude fading that occurs due to multipath propagation (see Chapter 4). Fading is one of the primary factors limiting link availability.

This potential benefit can be illustrated by considering two uncorrelated, narrowband Rayleigh-fading envelope waveforms in Figure 6.18 – one shown as solid line, the other as a dashed line. These waveforms can represent the signals at the terminals of two antennas or two antenna elements within the same antenna array. If the fading is uncorrelated, it is intuitive that the waveforms could be combined so that when one is faded the other is not. The combined signal envelope will then have a much smaller probability of an unacceptable deep fade than one signal alone. For a fixed microwave link system that is subject to atmospheric multipath fading as described in Chapter 4, two receive antennas could be used, spaced some distance apart, to produce the two signals needed for the diversity system. The concept can be extended to more than two antennas or, generally

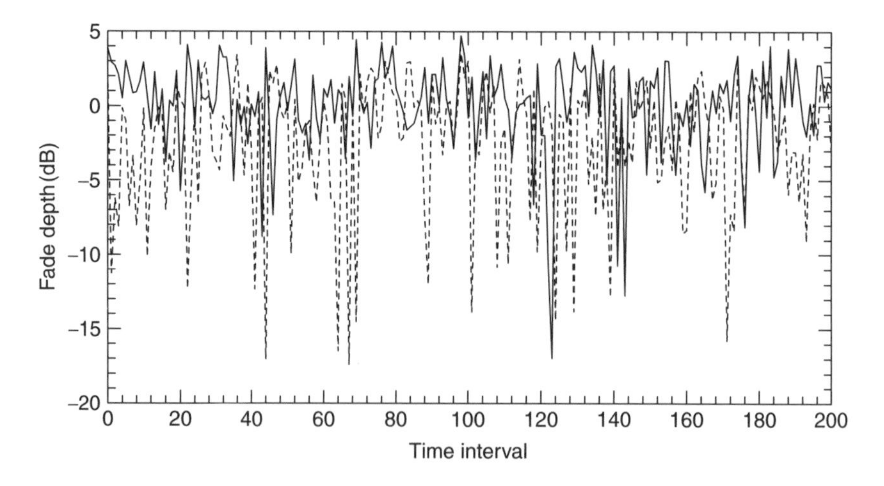

Figure 6.18 Two uncorrelated Rayleigh-fading signal envelopes.

speaking, two or more uncorrelated representations of the signal from whatever source. These representations are generally called *diversity branches.*

Space diversity is a technique that has been used for decades to improve the reliability of all types of communications systems. Three basic linear diversity-combining techniques are used for constructively combining the signals

- switched or selection combining
- maximal ratio combining (MRC)
- equal gain combining.

There are many books that provide detailed discussions of the operation and relative merits of these different combining methods (for example, see [11]). For brevity, only the results are presented here. The impact of using diversity can be seen from the probability distribution $p(\gamma)$ of the signal-to-noise ratio γ at the output of the combining network. For switched diversity in which the combiner switches to the diversity branch with the strongest signal, the result is

$$p(\gamma) = \frac{L}{\Gamma} \exp\left(\frac{-\gamma}{\Gamma}\right)\left[1 - \exp\left(\frac{-\gamma}{\Gamma}\right)\right]^{L-1} \tag{6.41}$$

where Γ is the average signal-to-noise ratio for each diversity branch and L is the number of diversity branches.

Rather than a simple on–off selection of the best antenna of diversity branch signal, MRC uses the sum of the signals from all the branches, but weights them so that the branches with the better signal-to-noise ratio are weighted most heavily. For MRC the resulting $p(\gamma)$ distribution is in [12]

$$p(\gamma) = \frac{\gamma^{l-1} e^{-\gamma/\Gamma}}{\Gamma^L (l-1)!} \tag{6.42}$$

And finally for equal gain combining the simplified expression found in [13] is given by

$$p(\gamma) = \frac{L}{a\Gamma} \exp\left(-\frac{\gamma}{a\Gamma}\right)\left[1 - \exp\left(-\frac{\gamma}{a\Gamma}\right)\right]^{L-1} \tag{6.43}$$

where $a = \sqrt{L/1.25}$ for $L \geq 2$ and 1 otherwise. Equation (6.43) is an efficient formula that avoids the numerical convolution normally required for assessing the results for equal gain combining.

As already mentioned, the diversity techniques produce the most improvement when the branches are uncorrelated, although worthwhile improvement can still be achieved with correlation coefficients as high as 0.5 to 0.7. For the present discussion, in which the diversity branches are the outputs of two or more antenna elements separated in space, the correlation between the signals from these elements as a function of spacing is important. For a link system design where space diversity is being considered, the problem is to find the required antenna separated distance that provides the needed diversity improvement or gain.

From [12], the correlation coefficient ρ_D between the signals from two antenna elements in a linear array separated by distance D is given by

$$\rho_D = \int_0^{2\pi} \exp(jkD\cos(\alpha - \phi))p(\alpha)\,d\alpha \qquad \int_0^{2\pi} p(\alpha)\,d\alpha = 1 \tag{6.44}$$

where α is the incident angle of the plane wave signal and ϕ is the angle of orientation of the linear array (α and ϕ are relative to 0 degrees azimuth). If it is assumed that the probability distribution of the arrival angles for the signals at the array is uniform; for example, $p(\alpha) = 1/2\pi$, then (6.44) yields

$$\rho_D = J_0(kD) \tag{6.45}$$

where J_0 is the zero order Bessel function of the first kind and D is in wavelengths. Equation (6.43) as drawn in Figure 6.19 shows that the signals from the antenna elements should have zero correlation with an element spacing of about 0.4λ using these assumptions.

Of course, the problem with (6.44) is that the distribution of arrival angles in not uniform over 2π, in fact, it is never close to uniform for outdoor NLOS situations as discussed in [14] and demonstrated by measurements reported in [13]. The $p(\alpha)$ distribution is actually highly nonuniform with a few peaks or preferred directions that change with the position of the remote terminal and hub sector relative to the propagation environment. The antenna element spacing needed to achieve uncorrelated diversity branches also varies. Usually the spacing is selected to be as wide as convenient, given the mounting circumstances for the antennas on a tower or on the remote terminal itself. A desktop

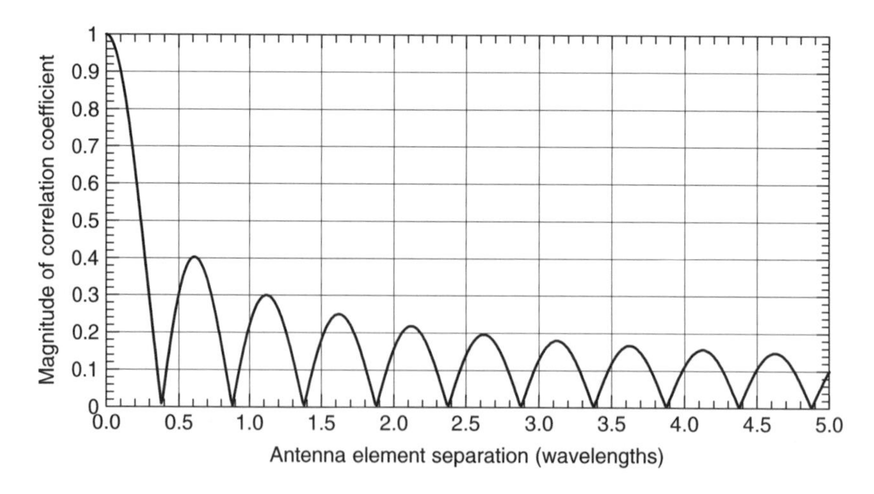

Figure 6.19 Correlation coefficient for antenna element signals with $p(\alpha) = 1/2\pi$.

terminal unit for an MMDS system at 2.6 GHz can readily achieve the 0.4λ spacing of 4.6 cm. Such diversity antennas are now commonly employed in the base units for cordless phones operating in the 2.4-GHz band.

6.5.1 Empirical microwave link diversity improvement

For the practical problem of predicting diversity gain improvement in PTP microwave links as a function of antenna spacing, some empirical formulas have been devised using measurements from links where the two diversity antennas were generally parabolic reflector antennas. For the diversity case with switched combining, the diversity improvement is given as in [15]

$$I_{SD} = 1.2 \times 10^{-3}\left(\frac{f}{d}\right)s^2v^2 10^{A/10} \tag{6.46}$$

where d is the path length in km, f is the frequency in GHz, s is the vertical antenna separation in meters, and v is the difference in the amplitudes of the main and diversity branches which, in most cases, will be the differences in the main and diversity antenna gains (note v in 6.46 is not in dB). In (6.46) A is the nominal thermal fade margin (see Chapter 10). Usually the diversity selection is done by comparing baseband signals, which means that two complete receiver RF/IF components are required. However, this also provides redundancy in case of equipment failure – a feature usually included with high-reliability microwave circuits that would also resort to diversity antennas to achieve the required high link availability objectives.

For the case when optimum combining is done at the intermediate frequency (IF) frequency, the diversity improvement is given by a formula similar to (6.46)

$$I_{SD} = 1.2 \times 10^{-3}\left(\frac{f}{d}\right)s^2\frac{(16v^2)}{(1+v)^4}10^{A/10} \tag{6.47}$$

where the variables have the same meaning as before. Chapter 10 will demonstrate how (6.46 and 6.47) are applied to link design and link budget calculations. The ITU-R has also developed methods for calculating diversity improvement for both narrowband analog and wideband digital systems [16].

6.6 ADAPTIVE ANTENNAS

Adaptive arrays are antenna systems that automatically adjust themselves to achieve some predetermined performance criterion such as maximizing the S/I ratio or the link margin. There are three general categories of adaptive antennas:

- *Switched Beam*: With this approach several antennas at the hub sector are available for signal transmission and reception to a given remote terminal. The patterns for the individual antennas are fixed, so this scheme is easy to implement and analogous to switched diversity that has been used for many years with only two antennas. Because the pattern shapes and pointing angles of the available antennas are fixed, the degree of signal gain and interference suppression that can be achieved is limited. The optimum beam for communication with a remote terminal is chosen by comparing the signals from each and choosing the one with the best signal.

- *Beam Steering*: This approach seeks to direct the maximum gain direction of the hub antenna toward the remote terminal to improve the link margin and S/(I + N) ratio. It is not specifically designed to suppress interference, although this can be a consequence of its operation. Beam steering is more complex than switched beam but is less complex than optimum combining. Beam steering seeks to maximize the signal to/from the remote terminal. For mobile terminals, this requires that the expected power range be periodically controlled. For fixed remote terminals where the mean path loss is much less variable than for mobile systems, control can be reduced or eliminated.
- *Combining for Optimum* $S/(I+N)$ *Ratio*: This approach is basically an optimum linear spatial filter where the antenna is adaptively adjusted so that the final output closely matches a reference signal. The filtering process therefore attempts to suppress any artifact that is not part of the desired incoming signal, including noise and interference. As a feedback system, it requires a periodically updated reference signal to compare with the incoming signal. It also requires a high rate of filter parameter updates (weighting coefficients) to account for fading. For cellular systems, the update rate is of the order of the fading rate of the channel. Again, because the channel is more stable in fixed links, the processing power required to update the weighting coefficients will be less demanding than in the mobile system case.

Adaptive antennas have been employed in military applications for several decades. Beam steering and tracking antennas for ship-borne and aircraft-borne radar and tracking systems are common. The rapid growth in engineering activity centered on adaptive antennas for cellular and other commercial wireless systems has come about in the last 10 years largely because of the advances in fast integrated circuit technology, especially digital signal processor (DSP) chips, which have made the rapid calculations feasible. A good overall description of the types of adaptive antennas and their role in mobile communications can be found in [17].

For mobile systems, one of the larger problems to be addressed is tracking the mobile as it moves and its propagation circumstances change. In free space the angle of arrival (AOA) of the signal from a mobile is a single direction from the array. In a real propagation environment, the mobile is moving from LOS to NLOS areas, or into areas where all the signals arriving at the hub are from scattering sources (reflections and diffractions from buildings, terrain, and so on). In such environments, the AOA from the mobile is not a single unambiguous direction but a range of angles with a particular angular spread around a mean AOA, with a standard deviation of AOA and associated statistics that change as the mobile moves. A similar situation exists for interfering signals from other mobiles so that the problem of optimally filtering the desired signal from interference is a challenge. This challenge has been at the core of adaptive antenna research for mobile systems for the last few years.

Applied to fixed broadband systems, the use of adaptive antennas is different from those for mobile systems in two important respects:

- With cellular systems, it is very difficult with current technology to put an adaptive antenna on a handset owing to the limited space and required processing power. This restricts the adaptive array to the base station sector where it can be used for both receiving and transmitting. For fixed systems it is assumed the remote terminal is not a mobile handset but a fixed (and portable or nomadic) device that is not moving while in operation. As such, it can be designed as a desktop device of some scale that permits the antenna separation necessary for an adaptive antenna. The adaptive antennas can therefore be applied to both the hub and remote ends of the link.
- For fixed broadband systems it is assumed the remote terminal is not moving during operation, so the problems with angular dispersion should be greatly reduced. However, the problems are not entirely removed since various features of the propagation environment can still move, including people in the same room and in the vicinity of the antenna. Outside, reflecting sources mentioned before, such as moving trucks, busses and cars can also introduce time-varying ambiguities in the AOA in both the uplink and downlink.

The discussion included here will focus on the application of adaptive antennas to fixed broadband systems. Of the three types of adaptive antennas listed above, optimum combining offers the greatest potential for system capacity improvement and is therefore discussed in some detail.

6.6.1 Optimum combining

The theory and technology for adaptive antennas have become fairly advanced. It is the intent of this section to present some of the basic mathematical formulas and nomenclature for adaptive antennas to better understand their capabilities and performance. In a later chapter, the performance of practical NLOS systems in real propagation environments with and without adaptive antenna technology will be discussed.

Figure 6.20 shows the basic schematic for an adaptive antenna. Signals from each of the antenna elements are first down converted to baseband and then converted to digital form through an analog-to-digital converter (ADC). The down-converter and ADC are not shown in Figure 6.20. The noise associated with the front end of the down-converter and other sources is added to the signal – the sum is a complex baseband signal $x_m(k)$. The $x_m(k)$ signal is then multiplied by a complex weighting function $w_m(k)$ and finally summed with similar signals from the other antenna elements. This sum signal is compared with a previously derived (and periodically updated) reference signal. The error signal $\varepsilon(k)$ is used to adjust the weight values $w_m(k)$ to reduce $\varepsilon(k)$ to zero.

The configuration in Figure 6.20 is designed to produce optimal S/(I + N) ratio for a single received signal $s_1(k)$. The remaining arriving signals are interference. The configuration can be replicated for each signal $s_j(k)$ that is to be recovered – each will function in an identical way with the desired signal optimized and others considered as interferers.

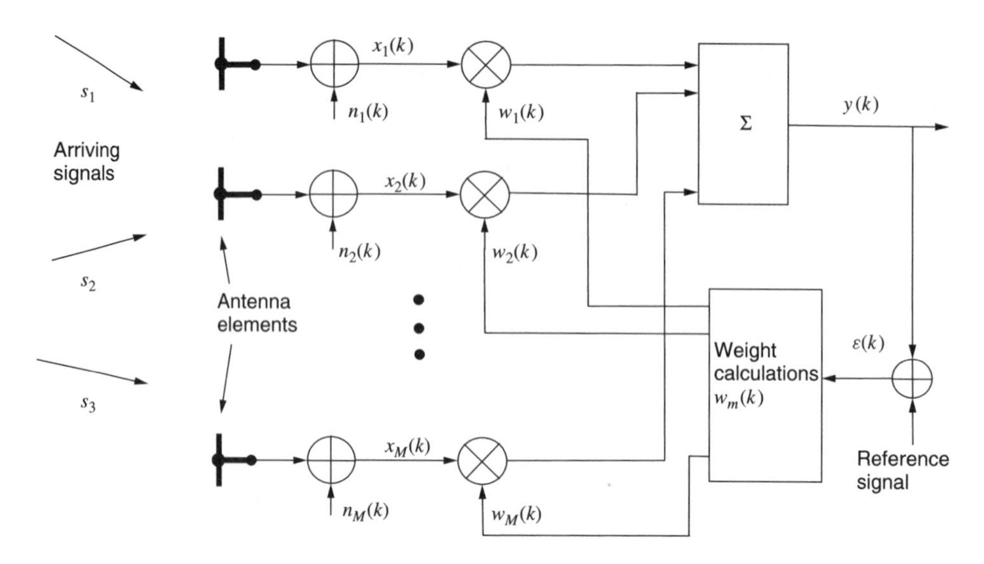

Figure 6.20 Adaptive antenna with optimum combining.

An adaptive antenna can also be constructed as an analog antenna with the amplitude and phase adjustments (weighting functions) done at RF or IF frequencies. However, current technology takes the baseband approach so that digital versions of the signals can be used for the calculation and processing functions in the antenna.

Figure 6.20 is specific to an optimum combining array, but a beam-steering array is essentially identical to the system in Figure 6.20. For a beam-steering adaptive array, the goal is to maximize the signal from the desired signal taken as $s_1(k)$. As such, it does not require the reference signal shown in Figure 6.20, but as mentioned, it does require periodic calibration to know where the maximum signal range is. It may also inadvertently enhance the signal from an interferer and thus not improve SIR as much as the optimum combining case.

The signals at each element of the antenna are the sums of the signal from the desired terminal and the signals from all the interfering terminals $s_j(k)$. The independent variable k can indicate a time interval step, or transmitted symbol or code block. Since the described adaptive array uses the baseband digital approach, ultimately the processing is quantized into time steps whose interval will depend on the rate at which the antenna performance must be updated. As noted above, the rate at which the antenna must be updated will be different between cellular mobile applications and fixed systems.

The arriving signals are multiplied by the antenna element gain in the direction of the arriving signal and by the channel response on the propagation path between that terminal location and the adaptive sector hub antenna. The propagation channel response includes the mean path loss as well as fading and time-delayed multipath components due to scattering. These sometimes complex effects can be modeled using the techniques described in Chapter 3. The signals on the M antenna elements are then

$$c_{m,j}(k) = G_{m,j}(k)p_{m,j}(k)a_{m,j}(k) \tag{6.48}$$

$$\mathbf{c}(k) = \begin{bmatrix} c_{11}(k) & c_{12}(k) & \dots & c_{1J}(k) \\ c_{21}(k) & c_{22}(k) & \dots & c_{2J}(k) \\ \vdots & \vdots & \ddots & \vdots \\ c_{M1}(k) & c_{M2}(k) & \dots & c_{MJ}(k) \end{bmatrix} \tag{6.49}$$

where the matrix $\mathbf{c}(k)$ is the response of all M elements to the signal source s_j, $G_{m,j}(k)$ is the gain of the mth element to signal s_j, and $p_{m,j}(k)$ is the channel response (including path loss and fading) for the signal s_j to the mth antenna element. The term $a_{m,j}(k)$ is the phase shift associated with the signal s_j on the mth element. It is often called the *steering vector*, and for a linear array with uniform element spacings d is given by

$$\mathbf{a}_j(k) = \begin{bmatrix} 1 \\ \exp(-j\beta d \sin\theta_j) \\ \exp(-j\beta 2d \sin\theta_j) \\ \vdots \\ \exp(-j\beta(M-1)d \sin\theta_j) \end{bmatrix} \tag{6.50}$$

Note that in (6.50) the wave number $\beta = 2\pi/\lambda$ has been used instead of k as used earlier in this chapter to distinguish it from the k used as the independent time or data block variable. The signals at the individual antenna elements from a given source may in fact be different (uncorrelated) even though they have the same AOA. The degree to which they are uncorrelated plays an important role in MIMO antenna system operation discussed in the next section. For adaptive antennas the element spacing is usually set to about $\lambda/2$ for a linear array to avoid extraneous grating lobes in the resulting radiation pattern, however, other configurations of elements such as circular or square, rather than linear, and with nonuniform spacings can also be used.

The signals in an adaptive antenna are normally written in matrix form, so that for M antenna elements there are $M \times 1$ column matrices

$$\mathbf{s}(k) = \begin{bmatrix} s_1(k) \\ s_2(k) \\ \vdots \\ s_J(k) \end{bmatrix}, \ \mathbf{x}(k) = \begin{bmatrix} x_1(k) \\ x_2(k) \\ \vdots \\ x_M(k) \end{bmatrix}, \ \mathbf{n}(k) = \begin{bmatrix} n_1(k) \\ n_2(k) \\ \vdots \\ n_M(k) \end{bmatrix}, \text{ and } \mathbf{w}(k) = \begin{bmatrix} w_1(k) \\ w_2(k) \\ \vdots \\ w_M(k) \end{bmatrix} \tag{6.51}$$

The signal inputs $x_m(k)$ to the weighting function multiplication are written as

$$\mathbf{x}(k) = \mathbf{c}(k)\mathbf{s}(k) + \mathbf{n}(k) \tag{6.52}$$

where the elements of $\mathbf{n}(k)$ are independent Gaussian noise contributed by the receiver and $\mathbf{c}(k)$ is the $M \times J$ channel response matrix whose elements are given by (6.48). The output signal $y(k)$ can then be written as

$$y(k) = \mathbf{w}^H(k)\mathbf{x}(k) \tag{6.53}$$

where $y(k)$ is the array output signal. The transpose matrices are required here so that the matrix multiplication in (6.52 and 6.53) can be carried out.

The objective is to find a weight matrix $\mathbf{w}$ such that $y(k)$ matches as closely as possible the desired signal $s_1(k)$. This is done by minimizing the error $\varepsilon(k)$ between the combiner output and the reference signal given by

$$\varepsilon^2(k) = [d^*(k) - y(k)]^2 \tag{6.54}$$

The optimum solution is achieved when the error is zero. Finding the expected values $E[\cdot]$ of both sides of (6.54) and taking the derivative relative to the weighting matrix $\mathbf{w}(k)$ (the dependent variable in this case) yields the optimum weight matrix given by [18] (the time or data block dependence has been suppressed)

$$\mathbf{w}_{opt} = \mathbf{R}_{xx}^{-1}\mathbf{r}_{xd} \tag{6.55}$$

where

$$\mathbf{R}_{xx} = E[\mathbf{x}\mathbf{x}^H] \quad \hat{\mathbf{R}}_{xx} = \frac{1}{K}\sum_{i=1}^{K}\mathbf{x}^*(i)\mathbf{x}^{\mathrm{T}}(i) \tag{6.56}$$

and

$$\mathbf{r}_{xd} = E[d^*(k)\mathbf{x}(k)] \quad \hat{\mathbf{r}}_{xd} = \frac{1}{K}\sum_{i=1}^{K}d^*(i)\mathbf{x}(i) \tag{6.57}$$

The superscript ^{H} indicates the *Hermitian* or transpose matrix with the complex conjugates. The superscript $^{-1}$ indicates the inverse matrix. The matrix $\mathbf{R}_{xx}$ is the covariance or correlation matrix that may or may not be singular so its inverse may not exist. It is normally assumed that $\mathbf{R}_{xx}$ is nonsingular. The solution in (6.55) is a general Wiener solution for an optimum filter.

The reference signal $d^*(k)$ can be created using a training sequence, which may be an implicit part of the wireless system signal protocol for other purposes such as equalizer training. The reference signal can also be viewed as the channel propagation and element response characteristics for the desired signal. If signal 1 is the desired signal, the reference is given by

$$\mathbf{r}_{xd} = \mathbf{c}_1(k) \tag{6.58}$$

Of course, if the remote terminal locations are not known, the channel propagation response and element gain can only be deduced by determining the array response to a training signal. For a fixed broadband system, where it is possible to approximately catalog the locations of the remote terminals, it may be possible to have *a priori* information to assist in this determination. For example, if the locations of users accessing a hub sector are known through some initial logon or registration process, an initial arrival angle and steering vector may be calculated on the basis of these locations, with adjustments to that initial value made to account for localized scattering on NLOS links that may change or broaden the perceived AOA.

If no reference is available, it is also possible to use 'blind' adaptive beam forming in which an implicit reference signal is derived in a variety of ways. Some techniques for blind beam forming that produce an implicit reference are described in [18].

As set forth in (6.55), the optimum weighting matrix can be found by direct matrix inversion (DMI) of the covariance matrix $\mathbf{R}_{xx}$. Estimates of the covariance matrix and reference vector $\hat{\mathbf{R}}_{xx}$ and $\hat{\mathbf{r}}_{xd}$ can be found by averaging over a sample period K as given in (6.56 and 6.57), respectively. The DMI approach has the fastest convergence but is computationally complex. Matrix inversion is a commonly encountered problem in mathematics and engineering. To make the process efficient, several matrix inversion methods have been devised that usually rely on some particular known characteristics of the matrix to achieve their efficiency. Most standard texts on computational mathematics contain descriptions of these methods.

A more widely used approach to finding the weight matrix values is the least-mean-square (LMS) algorithm, which essentially uses a steepest descent optimization approach to find the elements of $\mathbf{w}$. The updated values of the weight matrix are calculated on the basis of the preceding values and some incremental adjustment based on the magnitude of the remaining error [19]

$$\mathbf{w}(n+1) = \mathbf{w}(n) + \mu x^*(k)\varepsilon(k) \tag{6.59}$$

where μ is a constant that sets the adjustment step sizes. Those weight value changes that reduce the error to the greatest degree are kept, and the process is then repeated. The LMS algorithm is considered to converge more slowly than the DMI approach, but is computationally more efficient since matrix inversion is not required. The slow convergence issue is also less important when the variations in the channel are slow, as can be expected with a fixed wireless channel as compared to a mobile channel.

From a fixed broadband system planning perspective, the net effect of an adaptive antenna is to improve the S/(I + N) ratio and potentially the fade margin calculated from the link budget. The degree of improvement will depend on how effective the array is at enhancing the desired signal and suppressing interference when $\mathbf{w}_{opt}$ is adjusted to its optimum value. An adaptive antenna array with M elements can suppress $M - 1$ interferers. Fading, scattering, and noise all inhibit the link gains that can be achieved with the adaptive process. The realized S/(I + N) ratio ultimately achieved has a statistical distribution that depends on these time-varying channel parameters. Considering the S/(I + N) ratio in this way for system design purposes is the approach that will be taken in later chapters where the impact of adaptive antennas on individual link and overall network performance is considered.

6.7 MIMO ANTENNA SYSTEMS

MIMO antenna systems are similar to adaptive antennas in that they consist of multiple-element antennas that are connected to an optimizing array processor. The term 'input'

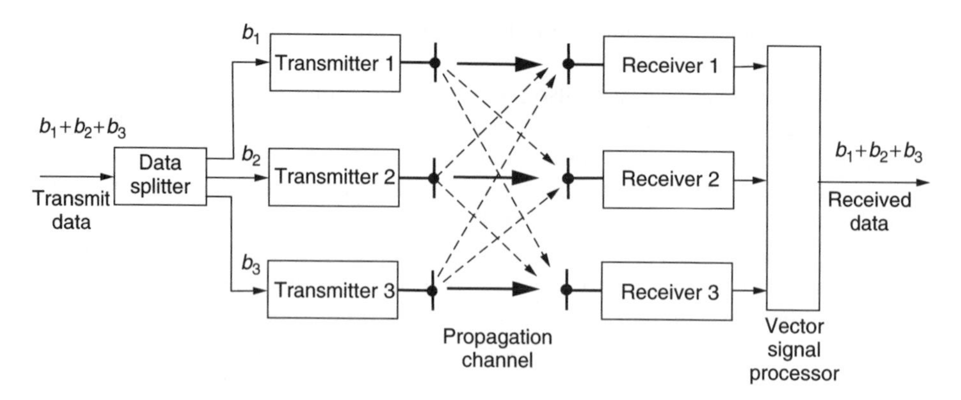

Figure 6.21 Schematic diagram of a MIMO antenna system.

in this case means the transmitting antennas or input to the wireless channel, while the term 'output' refers to the receiving antenna or output of the channel. For a single link the usual element configuration is a MIMO configuration, but SIMO (single-input, multiple-output) and MISO (multiple-input, single-output) arrangements are also possible. As will be explained, the attraction of MIMO antennas systems is the large gains in link transmission capacity that can be achieved within a given channel bandwidth.

Figure 6.21 shows a diagram for a MIMO system where 3 antennas each are used on the transmitting and receiving ends of the link. The basic idea of a MIMO system is that the channel response between the various channels (nine in the system in Figure 6.21) created by the different antenna combinations will be sufficiently uncorrelated that the vector signal processor will be able to establish distinct, essentially non-interfering channels between each one of the transmitters and one of the receivers. Ideally, for Figure 6.21 with 3×3 antennas this process can create three distinct wireless channels where only one existed with a single transmitter-receiver antenna combination, with a corresponding threefold increase in link capacity. The information bit stream to be transmitted is divided into three separate streams b_1, b_2, and b_3. The bit streams are often coded using *space-time codes* (STC) designed to add transmit-diversity gain and improve performance in fading channels (see Section 7.6.7).

If there are N transmitters and M receivers, the output of the system can be written as

$$\mathbf{y}(k) = \mathbf{H}(k)\mathbf{s}(k) + \mathbf{n}(k) \tag{6.60}$$

where

$$\mathbf{s}(k) = \begin{bmatrix} s_1(k) \\ s_2(k) \\ \vdots \\ s_N(k) \end{bmatrix}, \ \mathbf{y}(k) = \begin{bmatrix} y_1(k) \\ y_2(k) \\ \vdots \\ y_M(k) \end{bmatrix} \tag{6.61}$$

$$\mathbf{H}(k) = \begin{bmatrix} h_{11}(k) & h_{12}(k) & \dots & h_{1N}(k) \\ h_{21}(k) & h_{22}(k) & \dots & h_{2N}(k) \\ \vdots & \vdots & \vdots & \vdots \\ h_{M1}(k) & h_{M2}(k) & \dots & h_{MN}(k) \end{bmatrix} \tag{6.62}$$

The vector $\mathbf{s}(k)$ represents the transmitted signals, and the noise vector $\mathbf{n}(k)$ is the same as given in (6.51). The matrix $\mathbf{H}(k)$ represents the overall channel response from each transmitter to each receiver antenna. For the system in Figure 6.21 with three antennas (6.60) expands to

$$\begin{bmatrix} y_1 \\ y_2 \\ y_3 \end{bmatrix} = \begin{bmatrix} h_{11} & h_{12} & h_{13} \\ h_{21} & h_{22} & h_{23} \\ h_{31} & h_{32} & h_{33} \end{bmatrix} \cdot \begin{bmatrix} s_1 \\ s_2 \\ s_3 \end{bmatrix} \tag{6.63}$$

where h_{mn} is the response for the channel between transmitter n to receiver m. In elementary terms, the objective for the vector signal processor in Figure 6.21 is to reduce the off-diagonal values in $\mathbf{H}(k)$ to zero, or

$$\begin{bmatrix} y_1 \\ y_2 \\ y_3 \end{bmatrix} = \begin{bmatrix} h_{11} & 0 & 0 \\ 0 & h_{22} & 0 \\ 0 & 0 & h_{33} \end{bmatrix} \cdot \begin{bmatrix} s_1 \\ s_2 \\ s_3 \end{bmatrix} \tag{6.64}$$

The result will be three separate channels and three times the data capacity. From Forchini and Gans [20], the theoretical capacity C of a MIMO system is given by

$$C = \ln \det[\mathbf{I}_M + (\rho/N) \cdot \mathbf{H}\mathbf{H}^H] \text{ bps/Hz} \tag{6.65}$$

where ρ is the signal-to-noise ratio, N and M are the number of transmit and receive antenna elements as before, $\mathbf{I}_M$ is the identity matrix, and "bps" is bits per second. For the special ideal case like (6.64) above where $\mathbf{H} = \mathbf{I}_M$ so $M = N$, equation (6.65) becomes

$$C = N \ln[1 + (\rho/N)] \text{ bps/Hz} \tag{6.66}$$

The capacity in (6.66) varies directly with the number of antenna elements as compared to Shannon's classic capacity for a single channel [21]

$$C = \ln(1 + \rho) \text{ bps/Hz} \tag{6.67}$$

which gives 2.39 bps/Hz for $\rho = 10$ dB. Clearly the capacity available from (6.66) exceeds (6.67) for $N > 1$.

For cases where the covariance matrix $\mathbf{R}_{HH} = \mathbf{H}\mathbf{H}^H$, the capacity will be less than that given by (6.66). The challenge is to devise a vector signal processor that can get as close to the ideal as possible given the degree of correlation between the channels. In free space (such as outer space), the channels will be perfectly correlated since there is no scattering mechanism to make them different. In such circumstances, a MIMO system would not achieve any increase in capacity. Similarly, for uncluttered terrestrial propagation environments found in rural areas, the capacity gain that might be achieved with a MIMO system would be very modest and probably not warrant the increased system complexity. In a PTP LOS link where the channels would be highly correlated most of the time, the capacity gain would be similarly modest. The system circumstance in which MIMO offers the greatest potential is in NLOS systems where a high degree of scattering leads to highly uncorrelated channels, even when the antennas have fractional wavelength spacing.

Given the wide range of channel correlations that may be encountered in real systems, the capacity really becomes a statistical variable that depends on $\mathbf{R}_{HH}$. The covariance matrix in turn will depend on the nature of the propagation channel. For the planning process, the objective is to create a system that meets the service objectives of the operator. If maps or databases were available showing MIMO channel covariance as a function of location, this could then be incorporated in the planning process to assess the enhanced capacity available from MIMO as a function of service location. Such maps do not exist, but the results from initial field trials of MIMO systems in various LOS and NLOS propagation circumstances have been reported. In [22], a 4-antenna symmetrical MIMO system operating in a typical outdoor 1.9 GHz cellular system produced channel capacities of 4 times the single antenna system capacity. In [23], a measurement campaign of a 2.5 GHz outdoor 2×2 MIMO system in a suburban area gathered both channel-fading statistics (Rician k factor), XPD, and net channel-capacity gain. The net capacity depended on the fading statistics but in general fell in the range of 6 to 8 bps/Hz for 90% of the test locations. This is 3 times greater than the equivalent non-MIMO system using the same bandwidth with the same modulation efficiency. As additional measurement data is gathered, a relationship between land use/land cover or clutter (see Section 5.4) and channel correlation could potentially be established that could be directly exploited in the system planning process.

As noted above, adaptive antennas achieve network capacity gain by suppressing interference while MIMO systems achieve capacity gain by exploiting the decorrelated signals in the antenna elements rather than suppressing interference. When interference is considered, a recent study [24] shows that the capacity gain achieved by an adaptive antenna is similar to that of a MIMO system. For a capacity improvement perspective in network design, the nature of the *planned* interference environment may indicate that straightforward interference suppression with adaptive antennas is the more efficient approach to enhancing overall network capacity.

6.8 WAVEGUIDES AND TRANSMISSION LINES

For traditional PTP microwave links and other systems that use antennas mounted on towers, the connection from the transmitter to the antenna is potentially a significant source of signal attenuation – both for transmitting and for receiving. For this reason, and to reduce cost, systems are designed to minimize the height of the antenna on the tower and the resulting distance from the transmitter/receiver to the antenna. With the length minimized, the system design engineer has the task of choosing an appropriate waveguide or transmission line for making this connection with the lowest signal attenuation.

The operating frequency, the length of the connection, and the physical connection circumstances will determine whether a waveguide or coaxial transmission line is used. Transmission lines are almost never used above 3 GHz except for short equipment interconnections. Conversely, waveguides are rarely used below 2 GHz because of the size requirements and the wind load that might result. However, a waveguide has been used for MMDS operations at 2.5 to 2.7 GHz in which the antennas are mounted a few hundred feet from the transmitter. Waveguide size requirements are discussed in Section 6.8.1. For MMDS installations with shorter connections, a transmission line is used more often. Even though a coaxial transmission line has more attenuation for a given length than a waveguide, it usually does not need to be pressurized and is therefore much easier to install and maintain.

In recent years, the issue of waveguide and transmission line connections from the transmitter/receiver to the antennas has been reduced because the antenna is directly attached to an outdoor unit (ODU), an all-weather box that contains the transmitter power amplifier and perhaps lower level circuitry, and the receiver preamplifier and RF/IF modules. For such units, there is no need for a waveguide or transmission line, and the associated attenuation is eliminated from the link budget calculations. This has partly come about because of advances in solid-state devices that develop reasonable power levels at microwave frequencies without the power supply requirements of traveling wave tubes (TWT's), klystrons or some of the other early microwave power devices. As solid-state technology advances, this trend will continue.

6.8.1 Waveguides

Once the path has been designed with minimum tower heights to achieve adequate path clearance, the choice of waveguide is a straightforward process based on the system operating frequency. A chart with typical waveguide type numbers, their frequency ranges, and the associated attenuation is shown in Figure 6.22.

Waveguides are designed to propagate EM waves with very little attenuation. The only substantial attenuation is due to the small penetration of the wave into the metal walls of the waveguide (the 'skin' effect). Waveguides can support an infinite number of propagation modes that are designated TE_{mn} (transverse electric) or TM_{mn} (transverse magnetic) modes in which the subscripts indicate the dimension of the waveguide in integer half-wavelengths for a given waveguide dimension. On the basis of the mode number, the lower cutoff frequency for the rectangular waveguide (WR) is given by [25]

$$f_{\min} = \frac{1}{2\sqrt{\mu\varepsilon}}\sqrt{\left(\frac{m}{a}\right)^2 + \left(\frac{n}{b}\right)^2} \tag{6.68}$$

where $a > b$ and

a = inside width in meters
b = inside height in meters
m = number of half-wavelength variations of fields across the a dimension (wavelength in meters)

n = number of half-wavelength variations of fields across the b dimension (wavelength in meters)
μ = permeability of the material in the waveguide. Usually this is air, but sometimes other gases are used. For air, $\mu \cong 4\pi \times 10^{-7}$ H/m.
ε = permittivity of the material in the waveguide. Usually this is air, but sometimes other gases are used. For air, $\varepsilon \cong 8.854 \times 10^{-12}$ F/m.

The lowest loss mode is the TE_{01} mode; in general, waveguides are selected to operate in that mode. For this mode, the lower cutoff frequency is given simply by

$$f_{\min} = \frac{1}{2a\sqrt{\mu\varepsilon}} \text{ Hz} \tag{6.69}$$

For air, $1/\sqrt{\mu\varepsilon} \cong c \cong 3 \times 10^8$m/s. For a circular waveguide (WC) with radius r, the lower cutoff frequency for the TE_{01} mode is in [22]

$$f_{\min} = \frac{0.609}{r\sqrt{\mu\varepsilon}} \text{ Hz} \tag{6.70}$$

The type number of a waveguide indicates whether it is rectangular (WR), circular (WC), or elliptical (EW).These type numbers are used in Figure 6.22.

6.8.2 Transmission lines

Transmission lines can be built in many ways, including open parallel wires and coaxial geometries to suit a variety of purposes. High-power transmission lines that deliver power at 60 Hz across the country are the most prevalent example of open wire transmission lines. For high-frequency communication purposes, transmission lines are coaxial with a copper inner conductor and a solid or braided outer conductor (shield), also usually made of copper. The dielectric material between the inner conductor and outer shield can be either air or low-loss solid dielectric foam material. For outdoor use, air dielectric lines require pressurization to prevent moisture accumulating in the dielectric space, so foam lines are preferred for ease of maintenance even though their attenuation is somewhat higher than air dielectric lines. Most coax lines have a characteristic impedance of 50 ohms, with attenuation that increases with increasing frequency and decreasing line diameter.

As mentioned above, transmission lines are rarely used for transmitter/receiver feeder connections to antennas at frequencies above 3 GHz. At higher frequencies, the smaller diameters have much higher losses than waveguides and the larger diameters have upper cutoff frequencies because they start to act like waveguides. However, coax lines can be used for short distances to connect equipment modules or when the antenna and transmitter/receiver are close to each other. Table 6.1 shows a selection of transmission lines for some common fixed broadband frequencies. A ‘ – ’ in the table indicates that the line size cannot be used in the listed frequency band.

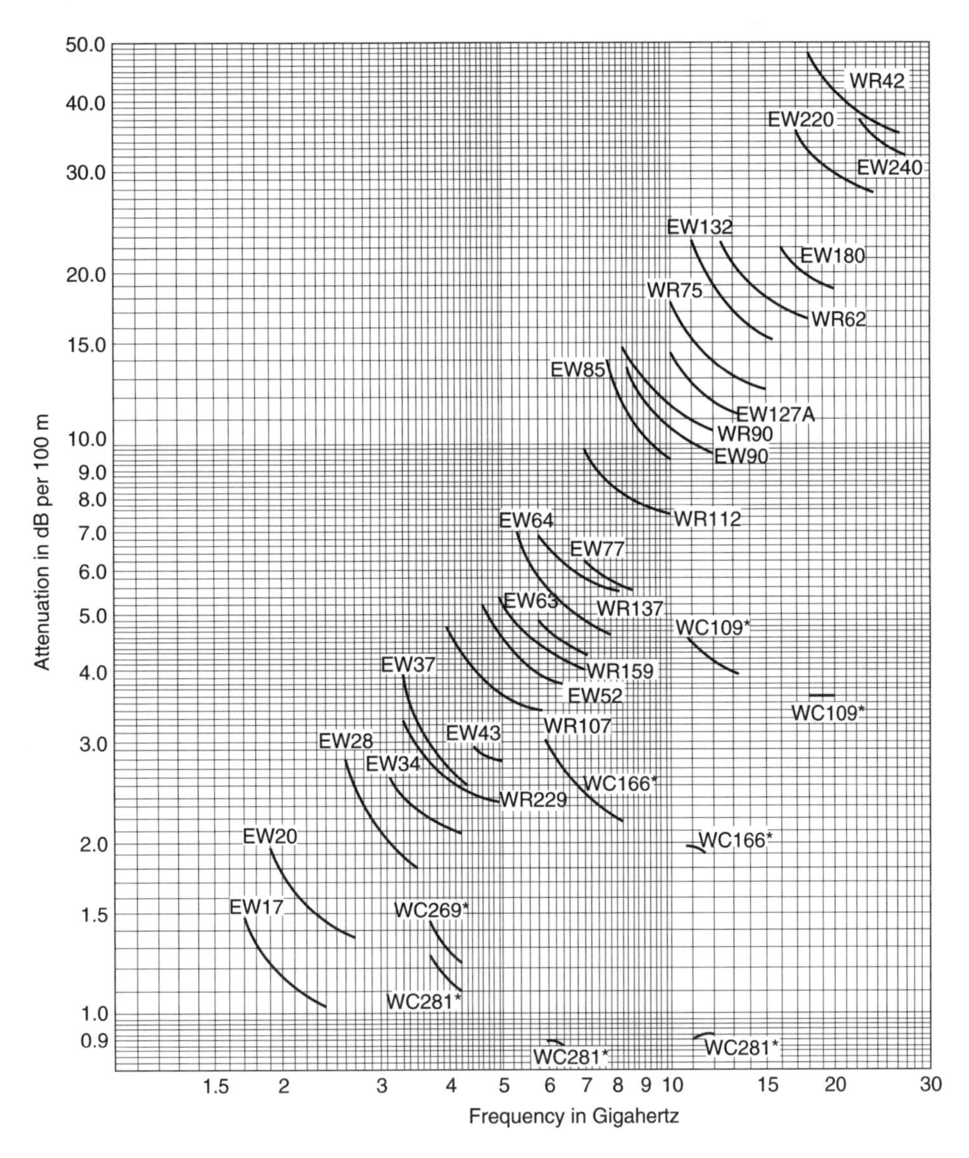

Figure 6.22 Waveguide frequency bands and attenuation.

6.9 RADOMES

Radomes are covers on antennas that serve two purposes: (1) they protect the antenna elements from environment factors such as inclement weather, birds, etc. and (2) they reduce the wind load of the antenna by creating a relatively streamlined profile for the wind to flow around. The latter factor is particularly important for parabolic reflector

Table 6.1 Transmission line loss at different frequencies

Line type	Attenuation for a line length of 10 m (dB)		
	2.6 GHz	3.5 GHz	5.8 GHz
1/2 inch air dielectric	1.49	1.78	2.51
7/8 inch air dielectric	0.72	0.87	1.21
1 5/8 inch air dielectric	0.41	—	—
2 1/4 inch air dielectric	—	—	—
3/8 inch foam dielectric	1.96	2.34	3.16
1/2 inch foam dielectric	1.32	1.57	2.13
7/8 inch foam dielectric	0.76	0.91	—
1 1/4 inch foam dielectric	0.56	0.67	—
1 5/8 inch foam dielectric	0.48	—	—

antennas that by their intrinsic shape have substantial wind resistance. This situation is exacerbated with high-performance parabolic dish antennas that include a shroud around the outer edge of the reflector, in effect, creating a very effective wind scoop. Radomes cover the aperture of the antenna so the scoop effect is eliminated. Because of wind load factors, radomes should always be used on any solid parabolic reflector antenna. Figure 6.23 is a photo of a parabolic dish reflector antenna with a shroud and radome.

Figure 6.23 Photo of a parabolic reflector 'dish' antenna with shroud and radome.

A compromise approach in dealing with the wind load problem is to construct the parabolic reflector from a grid of metal wires or tubes. If the spacing of the grid openings is small enough, the antenna performs essentially as well as a solid metal reflector. Most grid reflectors have the tubes running one direction making the antenna only capable of polarizations in the same orientation as the grid tubes. For high microwave frequencies, the grid reflector approach is not practical, but antennas of reasonable gain are also smaller resulting in less wind load for the solid reflectors so the wind load is lower in any event.

Radomes are generally made of fiberglass or some other material that is lightweight and essentially transparent to EM radiation. Depending on frequency it is prudent to include a fraction of a dB of loss in the link budget to account for any incidental attenuation by the radome. This situation changes when the radome is wet, as discussed in Section 4.3.4. A wet radome, where a thin water layer forms a sheet or laminar flow across the radome, can result in additional attenuation of as much as 8 dB. To prevent this, the radome surface is coated with a hydrophobic material that causes the moisture to bead up and flow off in rivulets. When the moisture beads up as discussed in the references for Section 4.3.4, the wet surface loss is considerably reduced to 2 dB or less. For the link budget used in Chapter 10, conservative calculations will include additional radome loss during rain events.

6.10 ENGINEERED AND *AD HOC* ANTENNA INSTALLATIONS

The antennas described in this chapter are generally used in 'engineered' antenna installations. For the purposes of this book, an engineered antenna installation is one in which some deliberate effort has been made to position and orient the antenna to achieve a particular signal level or signal quality. For fixed wireless systems, usually this deliberate effort is intended to yield an LOS path, or at least the lowest loss path, for the link. As a practical matter, an engineered antenna installation usually requires the efforts of a technician on site, which implies a substantial initial expense associated with the installation of each remote terminal. PTP microwave and LMDS are examples of LOS systems in which engineered antenna installations are necessary for both the transmitter and receiver.

In the direct broadcast satellite industry, some success has been achieved by having the system end user install the receiving antenna themselves with some simple instructions on how to orient it for adequate performance. While this approach may be workable in some cases, it is still a large potential inconvenience that may dissuade customers from signing up for the service.

An *ad hoc* antenna installation as used in this book is one in which no special effort is made to locate or orient the antenna to achieve a given level of performance. *Ad hoc* formally means 'for a particular purpose' although in common usage it also means temporary or arbitrary. *Ad hoc* installations are normally associated with NLOS links in which the system end user places the antenna and transmitter/receiver modem wherever convenient based on how it will be used. A cellular handset when in use is a good example of an *ad hoc* NLOS antenna installation. For fixed broadband, an *ad hoc* installation could be on a desktop next to the computer, or it could be a separate location of a wireless or wire line LAN network hub that distributes signals throughout an office or a residence. For a system design, it is not advisable to rely on any assumptions about how good or bad an *ad hoc* installation will be, and therefore it is necessary to use a statistical distribution for the quality of the installations, the performance of the antennas system, and the resulting quality of the transmit and receive signals. Lognormal distributions of mean signal level in NLOS links are often used for this purpose as discussed in Section 4.6. The cellular system design must make a similar assumption since the location of a mobile handset could be literally anywhere. One factor that has driven the development of adaptive antennas, MIMO systems, and adaptive link modulation, has been the need for wireless systems that adjust themselves to optimize performance in whatever installation circumstance may exist for a given end user.

Whether an antenna placement is an engineered LOS or an *ad hoc* NLOS installation has a profound effect on how the link and network design is done. For that reason, the system design process including propagation channel models, fading, radio technology, and many other factors will be discussed separately in later chapters for systems with LOS and NLOS remote terminal antenna installations.

6.11 CONCLUSIONS

Antennas are the physical devices that give wireless systems interface access to the radio spectrum. The choice of antenna for a given system design is based on its efficient frequency of operation, its bandwidth, and its directivity characteristics. The choice of antenna will have a fundamental impact on the capacity of a fixed broadband wireless system. The increased system processing gain and capacity achieved with directional antennas comes at no expense to bandwidth or system noise when compared to advanced modulation, spread spectrum, multiple access, error correction, and other signal processing techniques.

In the last few years a new level in antenna system design has been advanced with the introduction of adaptive antennas. With the use of digital signal processing, these antenna systems become spatial filters that can respond to the changing characteristics of the propagation environment and thus achieve S/(I + N) ratios that cannot be realized with static, fixed antennas in similar changing circumstances. Also using adaptively driven antenna elements, MIMO systems exploit the decorrelated channel response between separate transmitting and receiving antenna elements in a scattering environment to create dissimilar transmission channels that can support separate data streams, thus achieving higher net link capacity. Though there are limited commercial deployments as of this writing, these systems offer the potential to more fully exploit the network processing gain and capacity improvements that directional antennas generally offer.

6.12 REFERENCES

[1] C.R. Paul. *Introduction to Electromagnetic Compatibility*. New York: John Wiley-Interscience. 1992.
[2] W.L. Stutzman and G.L. Thiele. *Antenna Theory and Design*. New York: John Wiley & Sons. 1981. Chapter 1.
[3] A. Ishimaru. *Electromagnetic Waves, Propagation, Radiation, and Scattering*. Englewood Cliffs, New Jersey: Prentice-Hall. 1991. pp. 127.
[4] W.L. Stutzman and G.L. Thiele. *Antenna Theory and Design*. New York: John Wiley & Sons, 1981. Chapter 5.
[5] W.L. Stutzman and G.L. Thiele. *Antenna Theory and Design*. New York: John Wiley & Sons, 1981. Chapter 8.
[6] W.L. Stutzman and G.L. Thiele. *Antenna Theory and Design*. New York: John Wiley & Sons, 1981. page 396.
[7] J.D. Kraus. *Antennas*. New York: McGraw-Hill, 1950. pp. 344–345.
[8] RPE antenna data sheet, LMDS Antenna Model BCP-030-275. Andrew Corporation, Orland Park, Illinois, U.S.A.
[9] C.A. Balanis. *Antenna Theory: Analysis and Design*. 2nd Edition. New York: John Wiley & Sons. 1997. Chapter 14.
[10] W.L. Stutzman and G.L. Thiele. *Antenna Theory and Design*. 2nd Edition. New York: John Wiley & Sons, 1998, page 94.
[11] M. Schwartz, W.R. Bennett and S. Stein. *Communication Systems and Techniques*. New York: McGraw-Hill, 1966. pp 468 ff.
[12] W.C. Jakes. *Microwave Mobile Communications*. IEEE Press: Piscataway, New Jersey. 1994 (republished). page 319.
[13] H.R. Anderson. *Development and Applications of Site-Specific Microcell Communications Channel Modelling Using Ray-Tracing*. Ph.D. Thesis, University of Bristol, U.K., 1994, Chapter 3.
[14] T.-S. Chu and L.J. Greenstein, "A semiempirical representation of antenna diversity gain at cellular and PCS base stations," *IEEE Transactions on Communications*, vol. 45, no. 6, pp. 644–646, June 1997.
[15] A. Vigants, "Space-Diversity Engineering," *Bell System Technical Journal*, vol. 54, no. 1, January, 1975, pp. 103–142.
[16] International Telecommunications Union (ITU-R), "Propagation data and prediction methods required for the design of terrestrial line-of-sight systems," Recommendation ITU-R P.530-8, 1999.
[17] G.V. Tsoulos, "Smart antennas for mobile communications systems: benefits and challenges," *Electronics and Communication Engineering Journal*, 1999, **11**, pp. 84–94.
[18] J. Litva and T.K.-Y. Lo. *Digital Beamforming in Wireless Communications*. Norwood, MA: Artech House. 1996.
[19] J.H. Winters, "Signal acquisition and tracking with adaptive arrays in digital mobile radio system IS-543 with flat fading," *IEEE Transactions on Vehicular Technology*, vol. 42, no. 4, pp. 377–384, November 1993.
[20] G.J. Foschini and M.J. Gans, "On limits of wireless communications in a fading environment when using multiple antennas, *Wireless Personal Communications*, vol. 6, no. 3, p. 311, March 1998.
[21] C.E. Shannon, "Mathematical theory of communications," *Bell Systems Technical Journal*, vol. 27, pp. 379–423, July 1948, pp. 623–656, October, 1948.
[22] C.C. Martin, J.H. Winters and N.R. Sollenberger, "Multiple-input multiple-output (MIMO) radio channel measurements," *Proceedings of the IEEE Vehicular Technology Conference*, Fall, 2000.

[23] V. Erceg, P. Soma, D.S. Baum and A.J. Paulraj, "Capacity obtained from multiple-input multiple-output channel measurements in fixed wireless environments at 2.5 GHz," *IEEE International Conference on Communications*, New York, CD-ROM, May, 2002.

[24] S. Catreux, P.F. Driessen and L.J. Greenstein, "Simulation results for an interference-limited multiple-input multiple-output cellular system," *IEEE Communications Letters*, vol. 4, no. 11, pp. 334–336, November, 2000.

[25] S. Ramo, J.R. Whinnery and T. Van Duzer. *Fields and Waves in Communication Electronics*. New York: John Wiley & Sons. 1965. Chapter 8.

7

Modulation, equalizers, and coding

7.1 INTRODUCTION

The preceding chapters have dealt with the mechanisms that result in transmitted electromagnetic signals arriving at a receiver and the modeling techniques that can be used to represent these mechanisms in the system design process. The quality of the wireless communication circuit depends on the physical propagation mechanisms and on the nature of the transmitted signal itself. Since communication engineers have complete control over how the transmitted signal is created and how it is detected, many of the advances in communications technology in recent years have been in the way signals are coded, modulated, demodulated, decoded, and otherwise structured to maximize efficient, multiple access use of the spectrum. Many of these developments have been specifically designed to provide more robust performance in the presence of the propagation channel variations and distortions described in earlier chapters.

The original point-to-point microwave systems that were used for long-distance, multihop telephone circuits were analog systems in which a large number (up to 1,800) of single-sideband (SSB) 3-kHz voice channels were multiplexed together into a single analog signal that was used to frequency-modulate (FM) a microwave radio carrier. Such analog systems dominated the microwave industry for a few decades but progressively were replaced by systems using digital multiplexing and modulation in the 1970s and 1980s. The ability to correct transmission errors using coding techniques and the ability to regenerate a noise-free signal at each microwave repeater made the digital radios a logical improvement.

This change was paralleled in the cellular radio industry. The first-generation cellular systems built in the early 1980s used analog voice signals to frequency-modulate a radio frequency (RF) carrier operating in channels 25 or 30 kHz wide. These were later replaced with second-generation digital systems using various modulation and multiple access

Fixed Broadband Wireless System Design Harry R. Anderson
 ISBN: 0-470-84438-8

methods. Even the radio and TV broadcast industries, which have used analog transmission for several decades, are now slowly transitioning to digital transmission systems. Although analog still plays a role when used by older incumbent fixed systems, new fixed broadband system deployments will undoubtedly be digital systems; so only digital modulation will be treated in this chapter.

Because modulation, equalizers, and coding are features of the system that can be completely controlled, the effort that has been devoted to this area in communication system design has been enormous, with an associated enormous volume of literature on the subject. There are several books available at both the beginner and the advanced level that deal with modulation, equalizers and coding. Several of these books are used as references here and listed at the end of the chapter. The information in this chapter is intended to extract and focus on those basic elements of modulation, equalizers, and coding that are germane to fixed broadband wireless system design. Specifically, this includes the performance of various technologies in the presence of noise and interference, and the ability of equalizers and coding to improve performance by mitigating the effects of noise, interference, and channel response impairments. For the system design process, the signal-to-interference + noise ratio (S/(I + N) ratio) is a fundamental design objective. The S/(I + N) ratio or signal-to-interference + noise ratio (SINR) objective is also a direct indicator of the quality and robustness of the signal received by the end user.

Modern digital communication systems including those for fixed broadband usually have adaptable modulation that responds to the conditions of the channel (also called *link adaptation*). Under nominally good channel conditions, the transmission will use the highest data rate available. As the channel deteriorates because of a rain fade or some other reason, the system will automatically revert to a more robust modulation and coding scheme with associated lower net data throughput to maintain the communication link. Adaptability creates a degree of complication for the system designer, which will be addressed in later chapters.

7.2 DIGITAL MODULATION – AMPLITUDE, FREQUENCY, AND PHASE

A transmitted signal has three fundamental characteristics – frequency, amplitude, and phase. These characteristics can be changed individually or in combination in response to the information that is to be transmitted. It is assumed that the information itself is already in digital form. For an information signal that is basically analog like speech, a digital-to-analog converter (DAC) is used along with some means of compressing the signal so that the fewest number of bits necessary are used to achieve communication with the desired fidelity. Regardless of the means used to create the information signal, it ultimately becomes a time sequence of binary digits ('bits') that can have only one of two states: 0 or 1.

Digital modulation is the process of modifying the frequency, phase, or amplitude of a signal in response to whether a 0 or a 1, or a string (block) of 0s and 1s, is to be sent. From the terminology of the earliest digital systems (telegraph and the radio Morse code), this process is still called 'keying', so that fundamental modulation types can be grouped as:

- *ASK*: Amplitude Shift Keying in which the amplitude of the signal is changed among discrete amplitude levels to represent a pattern of bits. This includes the simplest form, on–off keying (OOK), in which the signal is turned on and off (from maximum amplitude to zero amplitude) in response to a bit pattern. OOK is used by free space optic systems.
- *FSK*: Frequency Shift Keying in which the frequency of the signal is changed among discrete frequencies to represent a pattern of bits.
- *PSK*: Phase Shift Keying in which the phase of the signal is changed among discrete phase states to represent a pattern of bits.

A wide variety of modulation methods are derived from these basic forms. As the number of available discrete amplitude levels, frequencies, or phase states increases, a greater number of more information bits can be conveyed with one signal state. For example, with a simple ASK system with two possible amplitudes, only one bit can be conveyed at a time – a 1 with the high amplitude and a 0 with the low amplitude. The voltage/frequency/phase that is transmitted to represent a bit pattern is called a *channel symbol* or simply *symbol*. The binary amplitude symbol in this case can represent one bit. If the symbol could have four amplitude levels instead of two, then each symbol could represent two bits (00, 01, 10, and 11). Similarly, as the number of possible states increases, the number of bits that can be conveyed increases such that

$$n = \log_2(M) \text{ or } M = 2^n \tag{7.1}$$

where M is the number of signal states a symbol can have and n is the number of information bits it can convey. This leads to the obvious observation that to increase the data rate or capacity of a link, symbols with a large number of signal states should be used. In a world with no noise, no interference, and no channel impairments, this would be the clear thing to do. The problem, of course, is that there is noise, interference, and channel impairments in a real system, which makes it increasingly difficult to correctly detect which signal state has been transmitted as the number of signal states increases (for a given transmitter power). Detection errors potentially lead to a higher bit error rate (BER) at the receiver and to a decrease in the ultimate signal quality delivered to the end user. The impact of noise and interference, and the associated trade-offs with modulation type, will be discussed in Section 7.4.

7.3 FIXED BROADBAND WIRELESS MODULATION METHODS

Given the broad flexibility that comes from being able to change the amplitude, frequency, or phase of the signal, there have been a large number of modulation schemes developed over the years. Of these, a few have emerged as the most attractive for fixed broadband wireless use as well as general wireless communication such as cellular and other mobile radio applications. The discussion of the modulation types is confined to these types.

7.3.1 BPSK, QPSK, $\pi/4$-DQPSK

Two of the simplest and the most robust modulation methods are binary phase shift keying (BPSK) and quadrature phase shift keying (QPSK). As the names imply, BPSK switches between two phase states to convey the bit pattern, while QPSK switches between four phase states. The associated signal states or *signal constellations* for these two types are shown in Figures 7.1(a and b), respectively. BPSK can take on phase values of 0 and 180 degrees. QPSK can have phase values of 45, 135, 225, and 315 degrees. Since each phase state for QPSK conveys two bits of information, the data used with QPSK can be coded so that adjacent phases differ by just one bit as shown in Figure 7.1(b). Since most detection errors result in an adjacent phase being detected (not two phase states away), this encoding approach means that a symbol detection error will result in one bit error rather than two. This type of coding is called *Gray encoding* [1].

From Figure 7.1(b) it is clear that certain bit pattern transitions will cause the signal vector to go through zero (00 to 11, for example), which in turn will cause the carrier amplitude to go to zero. This transition through zero carrier amplitude can cause adjacent channel spectral components that result in adjacent channel interference, depending on the linearity of the power amplifier (PA) over this signal power transition range. To eliminate the phase transition through zero, a form of QPSK called $\pi/4$-DQPSK is used in which the 2-bit pattern is conveyed by phase transitions among eight-phase states (separated by an angle of $\pi/4$) rather than an absolute phase as depicted in Figure 7.1(b). Because the transmitted bits are detected by the phase difference from one symbol to the next, it is *differential* QPSK or DQPSK. The source bit sequence translates to the phase transitions shown in Table 7.1.

In addition to BPSK and QPSK, eight-phase modulation, or 8PSK, has grown in interest recently because of its choice of an enhanced cellular system with higher data rates called EDGE. With 8PSK, the phase states are separated by 45 degrees rather than 90 degrees as with QPSK. With 8 states, 3 bits of information are conveyed with each channel symbol.

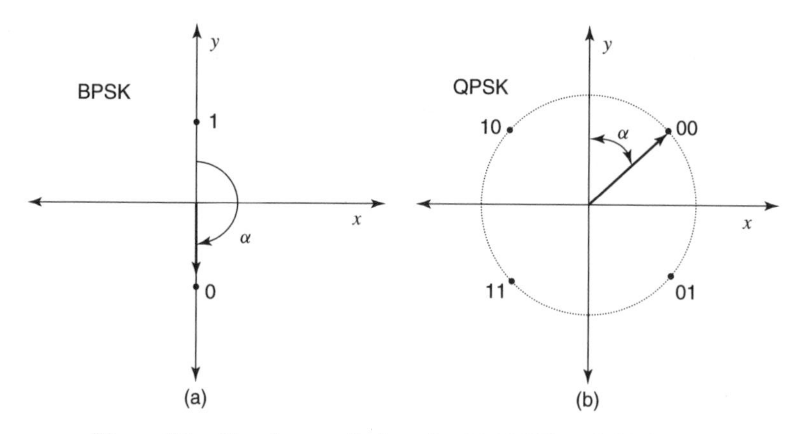

Figure 7.1 Signal constellations for (a) BPSK and (b) QPSK.

Table 7.1 $\pi/4$-DQPSK phase transitions

Bit pattern	Phase transition
00	$-3\pi/4$
01	$-\pi/4$
10	$+\pi/4$
11	$+3\pi/4$

The power spectra of BPSK and QPSK depend primarily on the symbol pulse shape that is used in the transmitter. There are a wide variety of choices here but the one most commonly used is the *raised cosine pulse*. The time domain representation of this pulse is given by

$$x(t) = \frac{\sin \pi t/T}{\pi t/T} \frac{\cos(\pi \alpha t/T)}{1 - 4\alpha^2 t^2/T} \tag{7.2}$$

where T is the symbol time length, α is a pulse shape or *rolloff* parameter ($0 \leq \alpha \leq 1.0$), and t is the time. The spectrum of this pulse is given by [2]

$$X(f) = \left\{ \begin{array}{ll} T & \text{for } 0 \leq |f| \leq (1-\alpha)/2T \\ \dfrac{T}{2}\left[1 + \cos\left(\dfrac{\pi T}{\alpha}\right)\left(|f| - \dfrac{1-\alpha}{2T}\right)\right] & \text{for } \dfrac{1-\alpha}{2T} \leq |f| \leq \dfrac{1+\alpha}{2T} \\ 0 & \text{for } |f| > (1+\alpha)/2T \end{array} \right\} \tag{7.3}$$

The time and frequency domains for the raised cosine pulse are drawn in Figures 7.2(a and b), respectively, for a few different values of α . Figure 7.2(a) shows that the complete time domain function is noncasual (starts before zero time). This is normally addressed by injecting a delay of 3 or 4 symbol times to permit a reasonably accurate representation of the leading tail of the pulse to be transmitted and the expected spectrum shape to be closely realized.

When this pulse shape is used with BPSK and QPSK, the resulting power spectral density (PSD) for the transmitted signal is given by [3]

$$S(f) = \frac{A^2}{T}|X(f)| \tag{7.4}$$

This is basically the same shape as Figure 7.2(b) except that it is scaled for the pulse width and amplitude. Note that the PSD for QPSK is the same as that for $\pi/4$-DQPSK.

The analysis of the error rates for BPSK and QPSK as a function of noise and interference follows in Section 7.4.

7.3.2 16QAM, 64QAM, and 256QAM

The number of bits per transmitted symbol with BPSK and QPSK is rather modest for a given bandwidth. To increase the data rate, symbols that convey more information

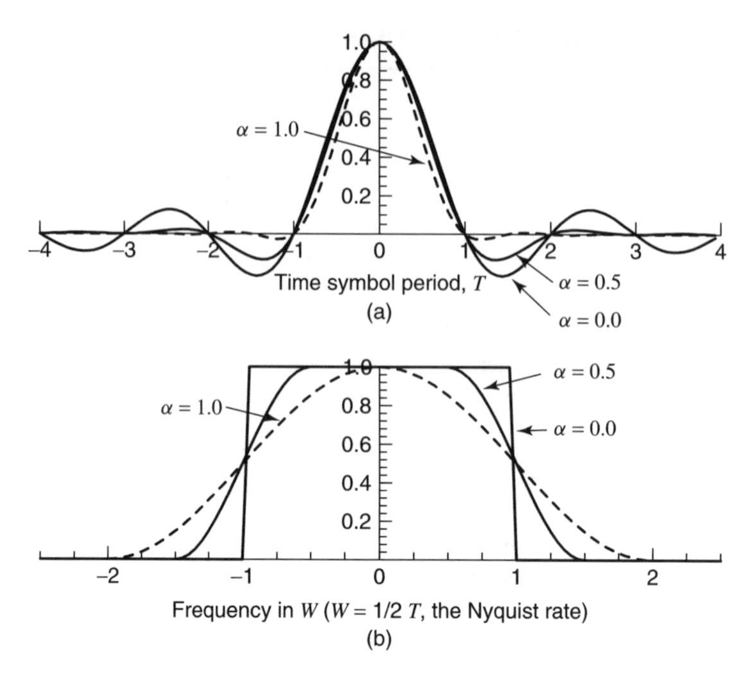

Figure 7.2 Raised cosine spectrum: (a) time domain and (b) frequency domain.

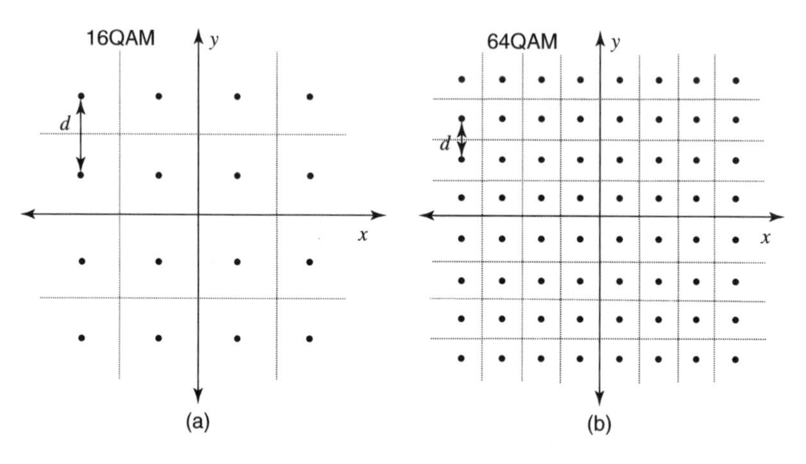

Figure 7.3 Signal constellations for (a) 16QAM and (b) 64QAM.

bits with more signal constellation states are required. The family of modulation types known as *quadrature amplitude modulation* (QAM) is the usual choice for achieving higher symbol efficiencies. Figure 7.3 illustrates the signal constellations for 16QAM and 64QAM. 256QAM is the same as 64QAM except with a 16×16 square array of constellation states.

From the constellations it is apparent that the QAM family of modulations are closely related and in fact can be generated using the same basic in-phase/quadrature (I/Q) modulator circuitry. Comparing Figures 7.1(b) and 7.3(a), it is also evident that QPSK is a member of this family that could be called 4QAM and is generated using the same I/Q modulator. With these four symmetric constellations – 4QAM, 16QAM, 64QAM, and 256QAM, the symbol transmission efficiencies ranging from 2 bits/symbol to 8 bits/symbol are possible. Other constellations such as 32QAM and 128QAM are also possible and sometimes used, but the four symmetric constellations described here are the most common because of the ease with which they are generated and because of their ability to be readily transitioned to higher or lower efficiency QAM constellations.

The PSD for QAM modulation is the PSD of the transmitted symbol provided the transmitted symbols are uncorrelated; that is, the autocorrelation function of the input data sequence is 0 for $\tau > T/2$ and $\tau < -T/2$ where τ is the time offset variable. If the raised cosine pulse described in the last section is used for the symbol shape, the spectrum shown in Figure 7.2(b) applies to QAM.

The analysis of the error rates for the QAM modulation as a function of the noise and the interference levels follows in Section 7.4.

7.3.3 Orthogonal frequency division multiplexing (OFDM)

Orthogonal Frequency Division Multiplexing (OFDM) is a multicarrier modulation method that uses a number of carriers (called subcarriers), each operating with a relatively low data rate. Taken together, the composite data of all the subcarriers is comparable to the data rate possible using the same basic modulation at a higher data rate with a single carrier in the same channel bandwidth. The main advantage of OFDM is that the symbol duration can be much longer so that the susceptibility to errors from intersymbol interference (ISI) due to multipath time dispersion is greatly reduced. From a frequency domain perspective, a frequency-selective fade will cause a problematic fading depth on only a few subcarriers. Errors will potentially occur on the few bits associated with those subcarriers but not the others, so the net error rate for all subcarriers taken together can still be made acceptably low if coding is also employed.

If a block of N_s symbols is considered, in OFDM each is used to modulate a subcarrier, so there are also N_s subcarriers. If each symbol has duration T_s, the time domain signal representation for the OFDM during T_s can be written as

$$s(t) = \sum_{N_1}^{N_2} d_{i+N_s/2} \exp\left[j2\pi \frac{n}{T_s}(t - t_s)\right] \text{ for } t_s \leq t \leq t_s + T_s \tag{7.5}$$

where $N_1 = N_s/2$, $N_2 = N_s/2 - 1$, and t_s is a symbol start time. Since the amplitudes and phases of the subcarriers in the transmitted spectrum $S(f)$ represent the N_s data symbols, the time domain in (7.5) is actually the inverse Fourier transform (IFFT) of the data block. As discussed in [4], the transform can be performed by an inverse fast Fourier transform (IFFT) or inverse discrete Fourier transform (IDFT), the choice being driven simply by which of these is the most computationally efficient.

It can be observed from (7.5) that this is a sum of N_s quadrature cosine and sine functions with frequencies $2\pi n/T_s$. These frequencies are integer multiples of each other so that they are inherently orthogonal, or

$$\int_{t_s}^{t_s+T_s} \exp(j2\pi nt/T_s)\exp(j2\pi mt/T_s)\,dt = 0 \quad n \neq m \tag{7.6}$$

$$\int_{t_s}^{t_s+T_s} \exp(j2\pi nt/T_s)\exp(j2\pi mt/T_s)\,dt = T_s \quad n = m \tag{7.7}$$

where n and m are integers. The receiver detector has the same subcarrier frequencies that are synchronized with the transmitted carriers. The demodulated OFDM signal using the orthogonal subcarriers recovers the original data symbols as implied by (7.7). The N_s detected data symbols can then be reassembled into the data block.

The number of subcarriers is determined by the total required data rate and the maximum delay spread that will be experienced in the channel. The individual subcarriers can be modulated using PSK or QAM as described in the preceding sections. The error performance of the separate subcarriers can be considered independently as long as the subcarriers remain orthogonal and there is no intercarrier interference.

At the receiver the amplitude and phase distortions introduced by the channel must still be equalized. In the case of differential QPSK, only the phase distortions in the channel are relevant to detection and they may be corrected using a differential equalizer that considerably simplifies the receiver [5]. The use of QAM or other amplitude-dependent constellations on a subcarrier requires amplitude equalization as well, so there is strong motivation to use BPSK, QPSK, or other phase-only constellations. The use of amplitude-dependent modulation like QAM also aggravates the peak-to-average power ratio problem discussed in Section 7.3.3.1.

For channels with deep fades at some frequencies, the subcarriers at these faded frequencies will experience very high error rates because of low signal-to-noise ratio (SNR) values. The error performance on these weak channels can dominate the overall error rate, thus setting a lower bound on the error performance of OFDM. Therefore, for reasonable error rate performance, it is essential for OFDM systems to have coding and interleaving. Coded OFDM is known as COFDM. Coding and interleaving are discussed in Section 7.6.

OFDM can also be sensitive to carrier frequency errors that degrade the orthogonality between subcarriers. This is known as intercarrier interference (ICI). This requirement can impose carrier accuracy demands on OFDM systems that are several orders of magnitude higher than those for single-carrier systems of similar capacity [6].

To ensure that time dispersion in the form of multipath does not introduce intersymbol interference (ISI – see Section 7.4.4), a guard interval is usually added into the system that extends the symbol duration. This guard interval is actually a cyclic extension of the subcarrier wave. The guard time will be determined by the delay spread of the channel that the system is expected to accommodate. Since the guard interval is not used to transmit information, it directly reduces the data capacity of the OFDM link.

The power spectrum of an OFDM system is the composite of the power spectra of the individual subcarriers. As the number of subcarriers increases, the spectrum approaches the rectangular band-limited spectrum of ideal Nyquist filtering. This is one of the advantages of OFDM in applications in which the spectrum is densely occupied and in which controlling adjacent channel interference is important. An example of an OFDM spectrum for 16 subcarriers is shown in Figure 7.4 for a rectangular pulse amplitude shape for the transmitted symbol ($\alpha = 0.0$) and with a raised cosine pulse shape with $\alpha = 0.1$. The out-of-band spectrum can be further attenuated using the raised cosine pulse, but this erodes the orthogonality of the subcarriers, which in turn degrades the received error rate [3]. Note that the raised cosine function is applied to the time domain pulse, and not the frequency domain as shown in Figure 7.2.

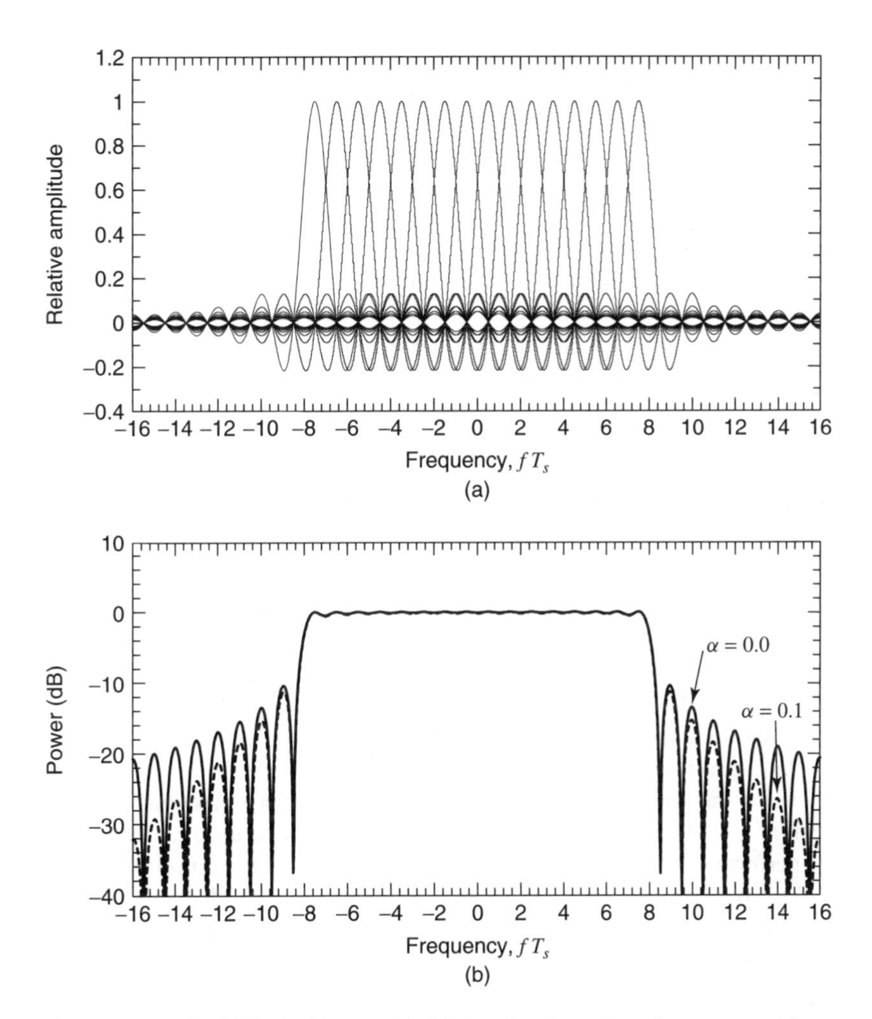

Figure 7.4 Spectrum for OFDM with $N = 16$. (a) Overlapping subcarrier spectra with rectangular pulse and (b) composite power spectrum with rectangular pulse and raised cosine pulse with $\alpha = 0.1$.

7.3.3.1 OFDM peak-to-average power ratio

Depending on the data signal being transmitted, it is possible that several of the subcarriers could be at their maximum values during a symbol time. If the sum of an unlimited number of (uncorrelated) subcarriers is considered, the probability density function (pdf) of the resulting signal envelope is a Rayleigh distribution with the probability of high peak values diminishing as the value increases. The peak power needed to represent this summation without distortion could be considerably higher than the average power of the waveform, greatly increasing the maximum power and linearity requirements on the transmitter power amplifier (PA).

A number of approaches for dealing with this problem have been explored over the past several years. The simplest is to clip the signal amplitude in a clipping 'window', which is what the PA ultimately does anyway, but this produces distortion and out-of-band emissions that are undesirable. As discussed in [7], several variations on the windowing approach show various levels of out-of-band emission. As shown in [8], it is possible to apply a code to the transmitted data block that essentially prevents subcarriers from adding simultaneously to limit the peak to a certain value above the average power level. A suitable code set turns out to be complementary Golay codes [9]. The coding method has a relatively small coding overhead (a 3/4 rate code can achieve a peak-to-average ratio (PAR) of 4 dB) and is currently the preferred approach.

These various methods have not eliminated the PAR problem but have reduced it to a manageable level. OFDM has thus become a viable and important option for fixed broadband wireless systems, especially for non-line-of-sight (NLOS) point-to-multipoint systems.

7.4 ERROR PERFORMANCE WITH NOISE AND INTERFERENCE

A fundamental measure of the capability of a digital modulation scheme is its BER in the presence of noise and interference. The BER is a function of the ratio of the desired signal level to the noise and interference levels, decreasing as the ratio decreases. In the system design process, accurate predictions of the S/N ratio (SNR) and the S/I ratio (signal-to-interference ratio (SIR)) are sought so that the link BER can be accurately predicted and the overall link quality assessed.

In considering the impact of interference and noise on signal detection errors, the underlying objective is to develop system design criteria that can be used to determine whether the outage performance of a particular link will meet the service objectives of the system operator. The error performance of different modulation types degrade differently with noise and interference. The higher-level modulation type where the constellation states are closer together will obviously have more detection errors at lower noise and interference levels than low-level modulations with more widely spaced constellation states.

For practical system design, the required SNR and SIR levels for a given maximum tolerable error rate threshold is usually given by the equipment manufacturer's specifications. These values can normally be employed directly in the link design. However, it is useful to understand the fundamental way in which errors in digital systems are caused by noise and interference so that trade-offs between high-level, high-efficiency modulation with higher error rates versus low-level, low-efficiency modulation with lower error rates can be recognized. The purpose of this section is to provide this information using a fundamental description of the error-creating processes that occur in digital signals.

7.4.1 Error performance with Gaussian noise only

The BER for BPSK and QPSK with noise can be found in many books on digital communications. Regardless of the modulation type, the same basic signal geometry illustrated in Figure 7.5 is applicable. The received signal vector $s(t)$ is perturbed by Gaussian noise vectors in quadrature creating a resultant vector $r(t)$ that is presented to the detector. Since n_x and n_y are jointly Gaussian random variables with $\sigma_x^2 = \sigma_y^2 = N_0/2$, then n_x and $s + n_y$ are also jointly Gaussian with $\sigma_r^2 = N_0/2$. The joint pdf is then

$$f(n_x, s+n_y) = \frac{1}{2\pi\sigma_r^2}\exp\left[-\frac{(n_y^2+n_x^2)}{2\sigma_r^2}\right] \tag{7.8}$$

Since the phase $\theta = \tan^{-1}(n_x/(s+n_y))$ and the amplitude $r = \sqrt{n_x^2 + (s+n_y)^2}$, these variables can be used in (7.8) to get

$$p_{r,\theta}(r,\theta) = \frac{1}{2\pi\sigma_r^2}\exp\left(-\frac{r^2+s^2-2sr\cos\theta}{2\sigma_r^2}\right) \tag{7.9}$$

Equation (7.9) can be integrated over r from 0 to ∞ to get $p_\theta(\theta)$, or over 0 to 2π to get $p_r(r)$. For PSK modulation, the phase pdf $p_\theta(\theta)$ is of interest. The result of this integration is given by [10]

$$p_\theta(\theta) = \int_0^\infty p_{r,\theta}(r,\theta)\,dr \tag{7.10}$$

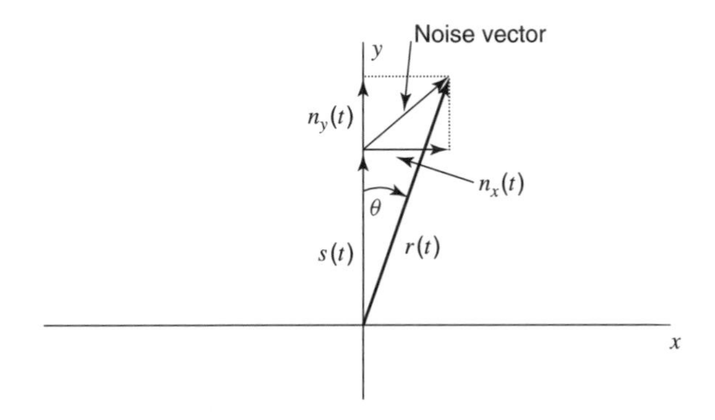

Figure 7.5 Signal vector with additive Gaussian noise.

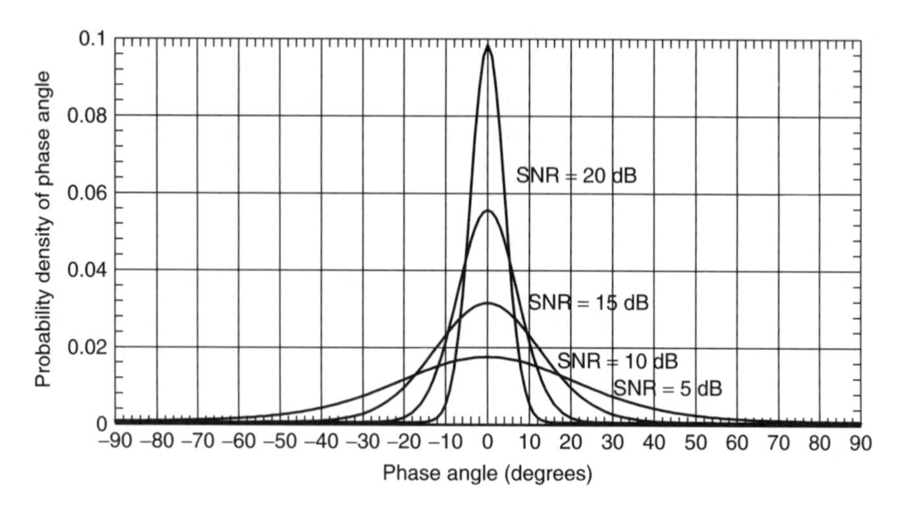

Figure 7.6 Probability of phase angle for PSK with Gaussian noise.

This is usually reduced to the form

$$p(\theta) = \frac{1}{2\pi} e^{-\gamma} \left(1 + \sqrt{4\pi\gamma} \cos\theta e^{\gamma \cos^2\theta} \frac{1}{\sqrt{2\pi}} \int_{-\infty}^{\sqrt{2\gamma}\cos\theta} e^{-r^2/2}\, dr \right) \tag{7.11}$$

If the symbol SNR is defined as $\gamma = S^2/N_0$, (7.12) can be evaluated numerically to find the probability densities. Figure 7.6 shows $p_\theta(\theta)$ for several values of γ in dB. The probability of an error due to noise is found as the area under the pdf curve for the phase angles outside the range of $\pm\pi/M$ around the transmitted phase, where M is the number of phase states (2 for BPSK, 4 for QPSK, 8 for 8PSK, etc.). The integral in (7.10) only reduces to a simple form for BPSK and QPSK – for other cases it can be readily evaluated numerically. The assessment of the phase error assumes that a perfect, noise-free reference is available for coherent detection. For differential DBPSK and DQPSK, the transmitted symbol is determined by the phase difference between successive symbols. This allows for a simpler receiver but slightly degrades error performance since the phase detection reference is also corrupted by noise.

The formulas for the symbol error rate for BPSK, QPSK, and *M-ary* QAM, and a few other modulation types as a function of the ratio of the signal energy per bit to the noise power per bit (E_b/N_0 or γ_b) are shown in Table 7.2, along with the source reference for each formula. The equations in Table 7.2 give symbol error rates; the bit error rates can be found assuming Gray encoding using

$$P_b \cong \frac{1}{k} P_M \text{ where } k = \log_2 M \tag{7.12}$$

Equation (7.12) is most accurate for high values of SNR because it assumes that a symbol error will result in an adjacent symbol being detected, which differs from the correct

Table 7.2 Symbol error probability for various modulation types based on γ_b

Modulation type	Symbol error rate equation	Source
Coherent BPSK	$P_M = P_b = \frac{1}{2}\text{erfc}(\sqrt{\gamma_b})$	Reference [1] eqn. (4.2.106)
Differential BPSK	$P_M = P_b = \frac{1}{2}\exp(-\gamma_b)$	Reference [1] eqn. (4.2.117)
Coherent QPSK	$P_M = \text{erfc}(\sqrt{\gamma_b})[1 - \frac{1}{4}\text{erfc}(\sqrt{\gamma_b})]$	Reference [1] eqn. (4.2.107)
Differential QPSK	$P_M = 2[Q(a,b) - \frac{1}{2}I_0(ab)\exp[-\frac{1}{2}(a^2+b^2)]]$ $a = \sqrt{2\gamma_b(1-1/\sqrt{2})} \quad b = \sqrt{2\gamma_b(1+1/\sqrt{2})}$	Reference [1] eqn. (4.2.118)
Coherent M-ary PSK	$P_M \cong \text{erfc}\left(\sqrt{k\gamma_b}\sin\frac{\pi}{M}\right) \gamma_b \gg 1$	Reference [1] eqn. (4.2.109) and eqn. (4.2.110)
Differential M-ary PSK	$P_M \cong \sqrt{\frac{1+\cos(\pi/M)}{2\cos(\pi/M)}}\text{erfc}\sqrt{k\gamma_b[1-\cos(\pi/M)]}$ for $M \geq 3$	Reference [11] eqn. (46)
Coherent M-ary QAM	$P_M = 2\left(1 - \frac{1}{\sqrt{M}}\right)\text{erfc}\left(\sqrt{\frac{3k}{2(M-1)}\gamma_b}\right) \times \left[1 - \frac{1}{2}\left(1 - \frac{1}{\sqrt{M}}\right)\text{erfc}\left(\sqrt{\frac{3k}{2(M-1)}\gamma_b}\right)\right]$	Reference [1] eqn. (4.2.144)
GMSK	$P_M = \frac{1}{2}\text{erfc}(\sqrt{\alpha\gamma_b}) \quad \alpha = 0.68$ for typical filter	Reference [12]

Note: erfc is the complementary error function; I_0 is the modified Bessel function of the first kind and order 0; $Q(a, b)$ is the Marcum Q function; M is the number of signal constellation states; $k = \log_2 M$; $\gamma_m = k\gamma_b$.

symbol by just one bit. With low SNR, the erroneously detected symbol may be farther away and two or more bits will be incorrect. The symbol error rates calculated from the formulas in Table 7.2 are plotted in Figure 7.7 for BPSK and QPSK and Figure 7.8 for QPSK, 16QAM, 64QAM, and 256QAM.

For a typical fixed wireless system, the amplitude of the signal will vary because of either multipath fading or rain fading events as described in Chapter 4. Since the noise power level from the receiving equipment is constant, the SNR will vary as the signal level fades. The relevant parameter, then, is the *instantaneous SNR* rather than the average SNR. The error rate then becomes a variable that follows the fading. When the signal level fades to a low level, the error rate rises during that fade, potentially resulting in an error burst that can destroy a block of data, or system synchronization, and timing. Considering that the 'bursty' nature of errors in these systems leads to the design concept of *fade margin* in wireless systems – the faded signal level that can occur before the error rate reaches an unacceptable threshold level. Fade margin is one of the most significant parameters in fixed broadband system design since it directly leads to a determination of the link availability. Fade margin will be employed in the link design discussion in later chapters.

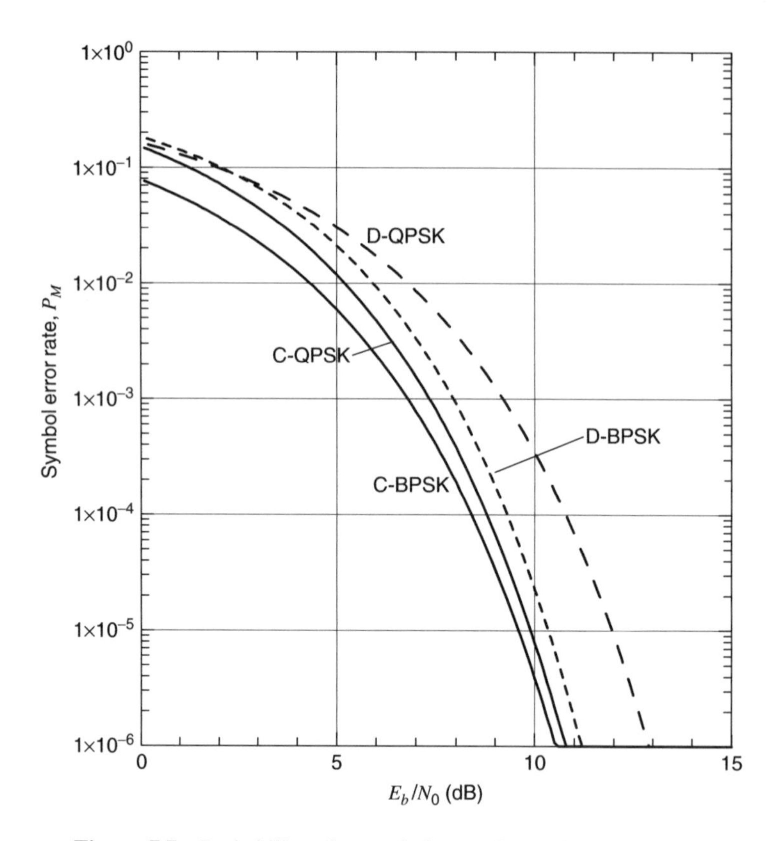

Figure 7.7 Probability of a symbol error for BPSK and QPSK.

The error rate performance shown in Figures 7.7 and 7.8 is the theoretical best-case performance for the indicated modulation type. For actual equipment, the error rate performance will usually be worse than the theoretical values due to a variety of nonideal aspects of the equipment design and operation. For example, the published specifications for a typical Local Multipoint Distribution Service (LMDS) system show an error performance for 16QAM of 10^{-6} for an SNR of 17 dB, 2 dB worse than the theoretical value. The theoretical curves therefore are best used as relative indicators of error rate performance and associated capacity. For link design, of course, the actual performance numbers of the equipment that will be deployed should be used. This is a parameter that can potentially differentiate one vendor from another.

7.4.2 Error performance with noise and constant amplitude interference

The vector approach used above can be extended to include a single sine wave interference vector as shown in Figure 7.9. The quadrature noise vectors are shown as n_x and n_y as before with the sine wave interference as a vector with amplitude b. The objective is to

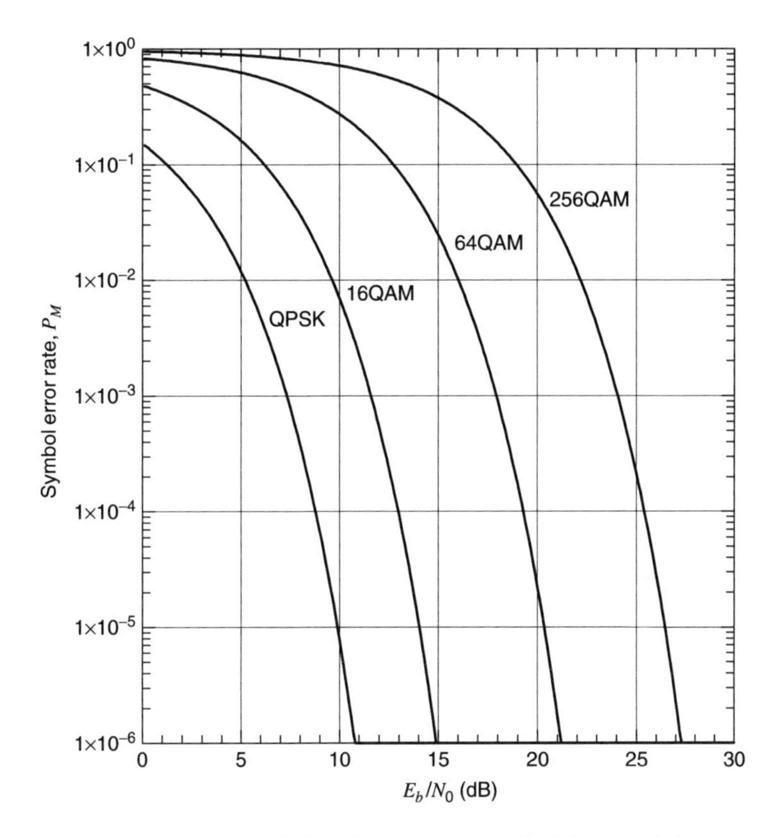

Figure 7.8 Probability of a symbol error for M-ary QAM.

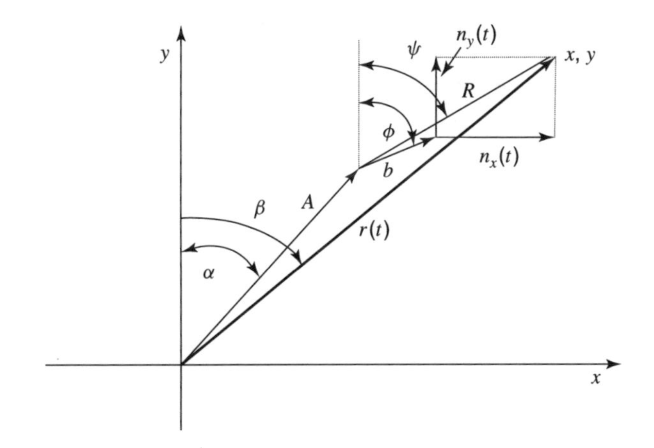

Figure 7.9 Signal, interference, and noise vectors for noise and sine wave interference case.

find the pdf of the phase and amplitude of this ensemble of vectors. Because an analysis of the pdf and error rate of digital modulation with a sine wave interferer and noise is rarely found in texts on digital communications, the details of the derivation of the pdf of x, y are included in Appendix B. The resulting pdf from Appendix B is

$$p(x, y \mid A, b) = \frac{1}{2\pi\sigma^2} \exp\left[-\frac{(x - A\sin\alpha)^2 + (y - A\cos\alpha)^2 + b^2}{2\sigma^2}\right] \times I_0\left(\frac{b\sqrt{(x - A\sin\alpha)^2 + (y - A\cos\alpha)^2}}{\sigma^2}\right) \quad (7.13)$$

where $I_0(\cdot)$ is the Bessel function of the first kind and of the order zero. To find the pdf of x or y alone, it is necessary to integrate (7.13) over one or the other random variables:

$$p(x) = \int_{-\infty}^{\infty} p(x, y \mid A, b)\, dy \quad p(y) = \int_{-\infty}^{\infty} p(x, y \mid A, b)\, dx \quad (7.14)$$

For mobile communications, the interference signal usually includes fading as does the desired signal, so a statistical description is normally used to describe the interference amplitude and the SNR. The aggregation of multiple random phase interferers, even if with constant amplitudes, also leads to a statistical description of the interference vector amplitude.

For fixed line-of-sight (LOS) systems, however, a single dominant interference signal can be received that exhibits very little fading due to high-gain antenna rejection of multipath components. The error rate analysis with a quasi-constant amplitude interferer is relevant to such systems.

The pdf in (7.13) does not easily yield a closed-form expression for the probability of error. However, it can readily be integrated by numerical methods to find the volume of the pdf in the error region for any particular modulation scheme, especially those in which the error boundaries between modulation states are rectilinear in x and y. As an example, this pdf will now be applied to finding error rates for 16QAM and QPSK when perturbed by noise and interference.

7.4.2.1 16QAM with noise and interference

As mentioned, a commonly used rectangular signal constellation in fixed wireless is 16QAM. The modulation constellation states are shown in Figure 7.3(a) along with the error boundaries (shown as dotted lines and the x-, y-axis). Coherent detection of *M-ary* QAM generally involves finding the Euclidean distance in the 2D signal space from the receive signal point to each of the M expected constellation states. The expected signal state that is closest to the received signal point is selected as the signal state that was transmitted. An error occurs when this distance is shorter to a signal state that is different from the transmitted signal state. The probability of error can be found by finding the volume of the pdf in (7.15), which crosses over the error boundaries in Figure 7.3(a) for a certain transmitted state.

With many modulation types such as 4QAM and *M-ary* PSK, the error boundaries relative to a signal state are the same for all of the M possible modulation states. Evaluating the probability of error P_e, or error rate, can then be done by considering the error for one signal state. However, for 16QAM it is clear from Figure 7.3(a) that the error boundaries are not the same for all signal states. There are actually three different types to be considered. First are the four corner points that only have error boundaries on two sides. Next are the eight 'edge' points with error boundaries on three sides. Finally, there are four interior points with error boundaries on four sides. The integration of the error region for each of these three types of signal states must be done with different integration limits that take into account the different error boundaries.

An approach to this integration that has simple limits is to integrate the volume under the 2D pdf inside the region where a given state has no error, and then to subtract that result from 1.0 to arrive at the error probability. However, with numerical integrations involving small numbers, this can be a difficult approach to make-work successfully. Instead, the straightforward approach of integrating the error region directly is used here, with the integration beginning in the low value 'tails' and progressing toward the higher values near the center of the pdf to avoid losing the contribution of these small values to the sum due to the round-off error.

The awkward integration limits can be resolved by choosing an integration region that easily encompasses all the relevant areas of the pdf in (7.13), both those where errors occur and those where no errors occur. In the case under study here the integration limits are set at $\pm 3d$. When the integration loops are sequenced through the values of x and y, each x, y pair is first evaluated to determine if it falls inside the 'no-error' limits. If it does, evaluating $p(x, y)$ and its contribution to the integration sum is skipped. This 'cut out' approach to integrating a complicated region works very effectively with the 16QAM case. It can also be used successfully when the error boundaries are not rectilinear as in the case of M-PSK modulation signal constellations and their derivative forms. The cut out region can even be defined by complicated polygons if necessary.

The numerical integration using the cut out approach can be written in the following form:

$$P_e = \sum_{-3d}^{3d} \sum_{-3d}^{3d} p(x, y) \Delta x \Delta y, \text{ subject to integration limits below} \tag{7.15}$$

The cut out integration limits the evaluation of (7.15) so that the results of the double summation are not included under the following conditions:

$$x > -d/2, y > -d/2 \text{ for the corner signal states}$$

$$-d/2 > x > d/2, y > -d/2 \text{ for the edge signal states}$$

$$-d/2 > x > d/2, -d/2 > y > d/2 \text{ for the interior signal states} \tag{7.16}$$

With these integration limits, the probability of a symbol error was calculated for 16QAM with 6 different levels of sine wave interference (signal-to-interference ratios or SIR) and a range of signal-to-noise ratios (SNR per bit). The results are shown in Figure 7.10.

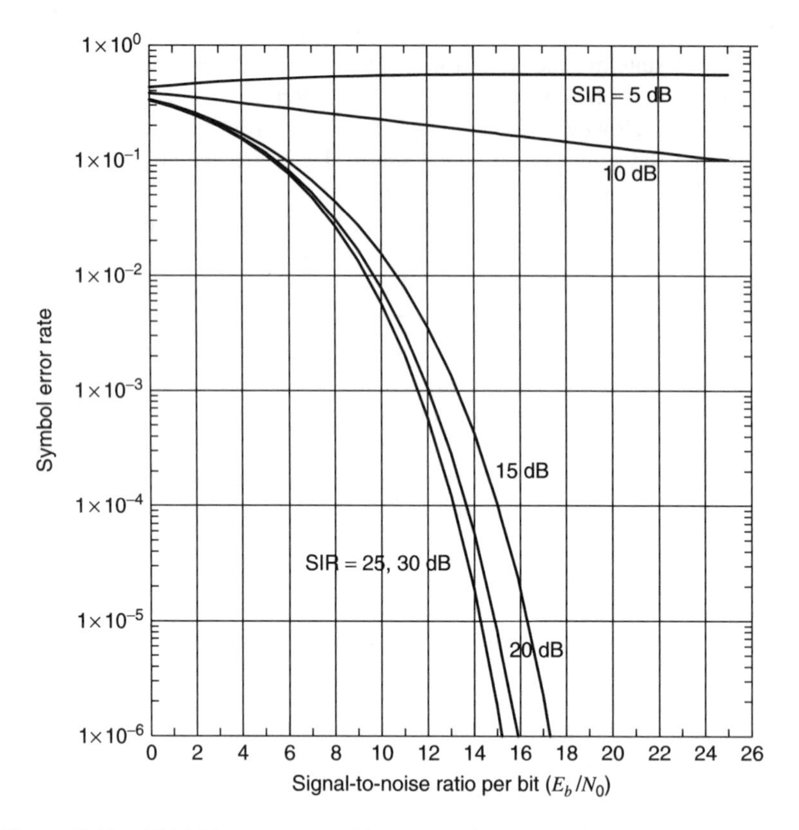

Figure 7.10 16QAM error rates with noise and constant amplitude interference.

The signal levels are the average signal levels for the 16 constellation states. The quantities SNR and SIR are defined as follows:

$$\mathrm{SNR} = 20\log\left(\frac{\overline{V}}{\sqrt{2}\sigma}\right)\ \mathrm{dB} \quad \mathrm{SIR} = 20\log\left(\frac{b}{\overline{V}}\right)\ \mathrm{dB} \tag{7.17}$$

where $\overline{V}$ is the average voltage of the signal modulation constellation. These values of SNR and SIR as defined in (7.17) will be used in the balance of this analysis.

As expected, with an SIR of 30 dB (essentially no interference) the error rate results are the same as the noise-only case and the curve in Figure 7.10 is in close agreement with [13]. As the interference level increases, the error rate also increases. At high levels of interference, a large portion of the pdf from (7.13) actually lies outside the no-error region even with no noise. At SIR = 5 dB, the error rate actually increases with *reduced* noise – an extraordinary condition in which the signal and interference pdf lies outside the interior signal state error boundaries such that only by adding noise does the signal occasionally fall within the 'no-error' region.

7.4.2.2 16QAM with 16QAM interference

The 16QAM error rate results for the constant amplitude interference can be applied to the problem of interference from another 16QAM signal in the same network. From Figure 7.3(a), it is clear that the interference vector b can take on one of the three discrete values, each with different probabilities. These are

- $b = d\sqrt{2}/2$ with probability 0.25
- $b = 3d\sqrt{2}/2$ with probability 0.25
- $b = d\sqrt{10}/2$ with probability 0.50

The error probability analysis described above can be repeated for each of these values of b, scaled by the overall S/I ratio. A similar approach can be applied to any single interferer with discrete amplitude values.

7.4.2.3 Coherent QPSK with noise and interference

The 2D pdf in (7.13) is given in terms of x and y, which is well suited to rectangular signal constellations like 16QAM. For angle constellations such as *M-ary* PSK, transforming (7.13) into polar coordinates is a more useful representation of the probability density. A typical example is coherent QPSK with a signal state constellation shown in Figure 7.1(b). With coherent detection it is assumed that some noise and interference-free phase reference is available against which the phase of the noise and interference-corrupted received signal can be evaluated.

The transformation from the rectangular coordinates of (7.13) to polar coordinates can be done by making the substitutions:

$$x = r\sin\beta \quad y = r\cos\beta \tag{7.18}$$

With these substitutions, (7.13) becomes

$$p(r,\beta \mid A,b) = \frac{1}{2\pi\sigma^2}\exp\left[-\frac{(r\sin\beta - A\sin\alpha)^2 + (r\cos\beta - A\cos\alpha)^2 + b^2}{2\sigma^2}\right] \times I_0\left(\frac{b\sqrt{(r\sin\beta - A\sin\alpha)^2 + (r\cos\beta - A\cos\alpha)^2}}{\sigma^2}\right) \tag{7.19}$$

Recognizing that $dx\,dy = r\,dr\,d\beta$, the pdf for the phase can be found by integrating over the range of r:

$$p(\beta) = \int_0^\infty p(r,\beta) r\,dr \tag{7.20}$$

Several examples of $p(\beta)$ are shown in Figures 7.11(a and b) for different values of SNR and SIR. The pdf in (7.13) can also be used to create visually instructive 3-D plots of the overall signal constellation pdf. An example of such a plot for coherent QPSK with the indicated values of SNR and SIR is shown in Figure 7.12. The 'rings' in the pdf

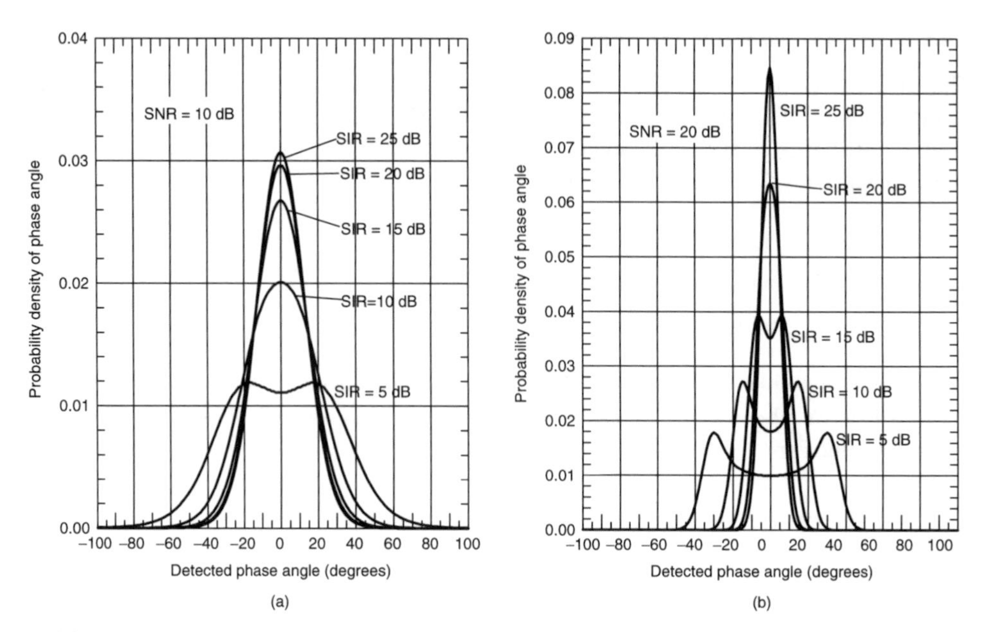

Figure 7.11 Phase pdf's for PSK with noise and single constant amplitude interference.

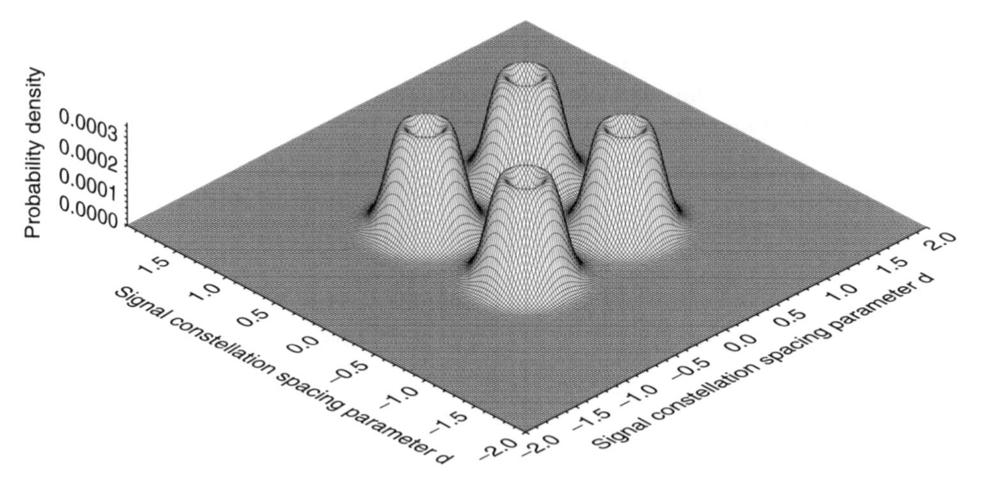

Figure 7.12 3D illustration of complete QPSK signal pdf with additive noise and single constant amplitude interference. SNR = 15 dB and SIR = 10 dB.

around each of the transmitted signal constellation points are clear. A similar 3D graph for 16QAM and an SIR value of 5 dB shows these ring pdf's crossing the error boundaries leading to the increasing error rates with lower noise as previously discussed.

Equation (7.20) can be evaluated numerically at a number of phase angles to provide a pdf description of the phase alone. The error rate P_e for an *M-ary* PSK system can then be found by integrating $p(\beta)$ over the error region as follows:

$$P_e = \int_{-\pi}^{-\pi/M} p(\beta)\, d\beta + \int_{\pi/M}^{\pi} p(\beta)\, d\beta \tag{7.21}$$

Considering the symmetry of $p(\beta)$, (7.21) can be simplified to

$$P_e = 2\int_{\pi/M}^{\pi} p(\beta)\, d\beta \tag{7.22}$$

Equation (7.22) can now be applied directly to *M-ary* PSK system. Using (7.22), the symbol error rate for coherent QPSK was computed for the same range of SIR and SNR values as that used for the 16QAM results in Figure 7.10. The results for QPSK are shown in Figure 7.13. As expected, the performance of QPSK with equivalent noise and interference is superior to that of 16QAM (ignoring the different capacity of the two modulation types). This is especially true for low values of SIR where QPSK has not yet reached the error region even with an SIR value of 5 dB.

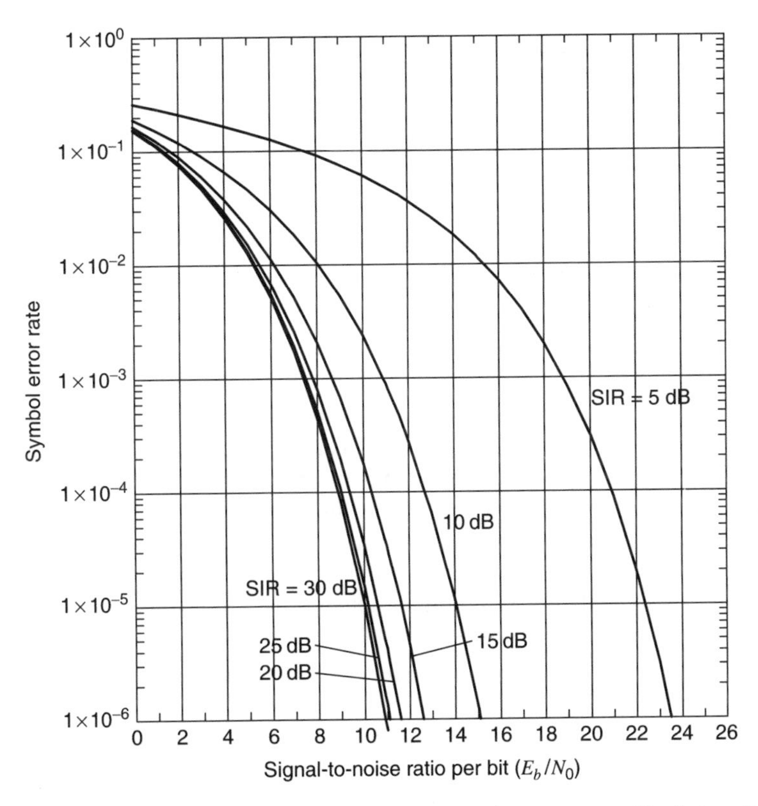

Figure 7.13 Coherent QPSK error rates with noise and constant amplitude interference.

The error rate results for QPSK were found using the pdf in (7.20). However, for this particular constellation they could have also been found using the joint pdf for x and y in (7.13). This approach would correspond to using only the corner cut out integration limits in (7.16).

7.4.2.4 Differential QPSK with noise and interference

The previous examples for predicting error rates in the presence of single-carrier interference were developed for coherent detection of 16QAM and QPSK, where a noise-free and an interference-free phase and amplitude reference was available. Many system architectures have been devised to recover a suitably uncorrupted phase and amplitude reference for use in wireless channels.

However, important classes of modulation schemes that offer somewhat simpler hardware implementation are differential techniques in which the information data is conveyed by the transition from one constellation state to another. The receiver then detects the difference between successive received symbols and interprets the differences in terms of the encoding scheme used at the transmitter. A commonly used example mentioned before is differential QPSK, or DQPSK, in which the change in phase, rather than the phase relative to a reference, is used to convey data.

The general technique described above for analyzing BER in the presence of noise and interference for coherent detection can be extended to the differential case by assuming that signal vectors for successive symbols are similarly perturbed by noise and interference as was shown in Figure 7.9. This approach was taken in [14] when it was assumed that the interference phase would undergo a fixed (but unknown) phase shift θ from symbol to symbol. An upper bound error rate result was then developed for DQPSK as a function of θ. The analysis can proceed in a manner similar to that given above with a resulting pdf for the phase *difference* between successive symbols. The error rate can then be determined by integrating the area under the pdf, which is outside the correct symbol detection bounds as before.

7.4.3 Error performance with flat-fading signal and interference

The analysis of the errors in the preceding section were done for discrete values of SNR and SIR as shown by the curves in Figures 7.10 and 7.13. As explained above, for short-range LOS links, this model may be applicable to the noise and interference circumstances of real links. For NLOS links, however, the desired signal will experience fading so that the SNR and SIR vary in the same way as the signal level variations assuming that the interference power is constant. Incorporating the pdf of the desired signal fading variations into the pdf for the overall signal amplitude given in (7.13) results in

$$p(x, y \mid b) = \int_0^\infty p(A)p(x, y \mid A, b)\, dA \tag{7.23}$$

Similarly, the fading distribution of the interference can be incorporated as

$$p(x, y) = \int_0^\infty p(b)p(x, y \mid b)\, db \tag{7.24}$$

Since there are no known closed-form solutions for (7.23) or (7.24), as a practical matter the error probability that results from the numerical integration of (7.13) for various values of A and b provides a tabular description that can be used as a lookup table in evaluating link error rates.

For example, Figure 7.10 shows the error rates for values of SNR and SIR. If the desired signal fades to a certain level with a given probability, the resulting SNR will occur with that probability so that the associated error rate will also occur with that probability. The probability of this error rate occurring can then be compared against the system performance criteria. The fading distribution for the interference can be handled in a similar numerical fashion. The degree of independence of the variations in SIR and SNR is important for fixed broadband systems, especially for correlated rain fading on desired and interfering paths as discussed in Section 4.4.

7.4.3.1 Noise approximation of interference

An accurate continuous or discrete (tabular) function for $p(A)$ or $p(b)$ is unlikely to be available for most systems, so the common approach is to make assumptions about the distribution of the desired and the interfering signal levels. This approach is discussed in the next section. The most common and straightforward approach to handling interference in fixed LOS broadband systems is to simply regard the interference power to be additional noise power. The instantaneous SNR $\rho(t)$ is

$$\rho(t) = \frac{s(t)}{N'(t)} \tag{7.25}$$

where the power of $N'(t)$ is the noise power plus the power of each of the interferers

$$N' = N + \sum_{j=1}^{J} I_j \tag{7.26}$$

The validity of this assumption improves as the number of interferers increases. Even with as few as six sine waves with random phases, the envelope begins to approach a Rayleigh distribution like noise [15]. When the interference amplitude is also varying, by the Central Limit Theorem, the probability distribution of the sum will approach a Gaussian distribution with a variance equal to the sum of the variances (power) of the individual interferers. This approximation also represents a conservative approach that results in higher error probabilities than that would occur if the interferers were treated with their real fading distributions. For real transmitters, the interfering powers are always necessarily limited in amplitude (a Gaussian distribution has no amplitude limits). For the occasion in which the interference is known to come from a single identifiable interferer, the explicit pdf approach detailed above will provide more accurate error probability predictions.

7.4.4 Error performance with frequency selective signal fading

The preceding sections discussed error rate performance in terms of noise and external independent interference sources. For a propagation environment that creates multiple

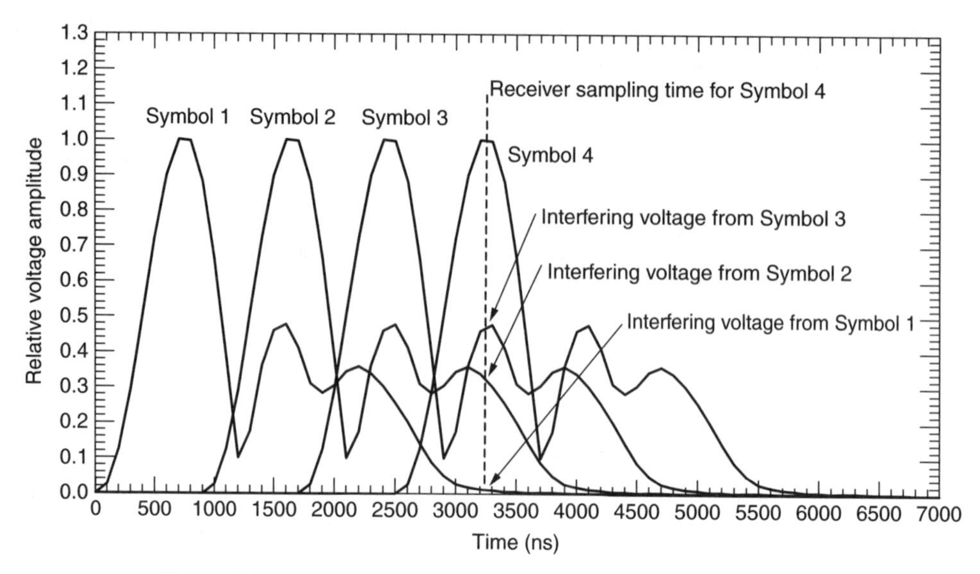

Figure 7.14 Intersymbol interference for a channel with multipath.

signal paths between the transmitter and the receiver, there is an additional interference mechanism due to components of the delayed multipath signals upsetting the proper detection of some subsequently transmitted symbol. This situation is illustrated in Figure 7.14. This is known as intersymbol interference or ISI.

The signal at the sampling instant can be modeled using the interference vector in Figure 7.9. The probability density for the composite signal vector developed in Section 7.4.2 can also be applied to find the error rate due to ISI. Of course, the multipath signals are typically not going to be static but vary with some distribution that can be postulated as Rayleigh or Rician (see Chapter 4).

The real microwave systems (RMS) delay spread (see Section 3.3.1) is often used as an indicator of the error rate for such time-dispersive channels. From the illustration in Figure 7.14, it is straightforward to envision two delay profiles each with the same value of RMS delay spread, but one with a low-amplitude multipath vector with a large delay, and the second with a high-amplitude multipath vector with a small delay. The high-amplitude, short-delay channel will have a higher error rate than the channel with the low-amplitude, long-delay multipath. RMS delay spread can therefore be a very a misleading indicator of error rate if no equalization or exploitation of the multipath is used in the receiver. For simple receivers without equalization, much more specific information about the nature of the multipath (amplitude, delay, variability) is needed before error rates can be found for such channels.

Unlike errors resulting from noise and external interference, the error rate resulting from the ISI cannot be reduced by increasing the transmitter power because this will simply increase the multipath signal amplitudes by the same proportional amount; for example, the SIR of the ISI stays the same. Instead, ISI can be effectively combated using equalizers of various designs as discussed in Section 7.5.

Multipath is not necessarily detrimental to system performance; multipath can also be exploited with two techniques:

- Using a Rake receiver, it is possible to reduce fading and improve error performance, essentially achieve time-diversity through combining of the signals on the branches of the Rake receiver. Rake receivers are discussed in Section 8.4.4.1 of Chapter 8.
- A multipath time-dispersive channel is indicative of an uncorrelated scattering channel that can be exploited to increase link capacity through the use of multiple-input, multiple-output (MIMO) technology, or increase link robustness with space-time codes (STCs). MIMO techniques are discussed in Section 6.7; STCs are discussed in Section 7.6.7.

A receiver must have sufficient bandwidth to take advantage of this additional decorrelated received signal energy provided via the multipath signals. For systems with limited bandwidth, time dispersion will manifest itself as narrowband frequency flat-fading for which alternate remedies such as diversity reception and pilot-assisted gain control are available to improve error performance.

7.5 EQUALIZERS

For systems in which multipath is expected and error rates must be maintained below an acceptable threshold level, there are equalizer solutions that essentially remove most or all the symbol distortion resulting from the multipath energy before the symbol detection is done. Two general types of equalizers are discussed here.

7.5.1.1 Time domain symbol equalizers

Time domain equalizers have been employed in digital point-to-point microwave link equipment for some time. In Section 4.2.4.2 of Chapter 4, the quantity 'dispersive fade margin' was defined along with the standard procedure for measuring it. The dispersive fade margin is directly related to the performance of the receiver equalizer and its ability to correct frequency-selective channel distortions. Microwave radios without equalizers typically have dispersive fade margins of 35 dB. Adding an effective equalizer can raise the dispersive fade margin to 55 or 60 dB.

A linear transversal equalizer has time-delayed signal taps as shown in Figure 7.15. The signals from the taps are multiplied by coefficients that are independently adjusted such that the multipath components for a received symbol are adjusted to a low level and the ISI is effectively removed. Within this general framework, there are several strategies for adjusting the taps. The earliest and simplest is the *zero-forcing* (ZF) equalizer [16] in which the signal additions due to the multipath are subtracted out by properly adjusting the tap coefficients c_j in Figure 7.15, forcing the ISI to zero. The equalizer uses a training symbol of known value to initially adjust the equalizer tap coefficients. If the channel is not varying, these tap settings can suffice. If the channel does vary as it normally does with wireless communications, the coefficients need to be periodically updated. Another

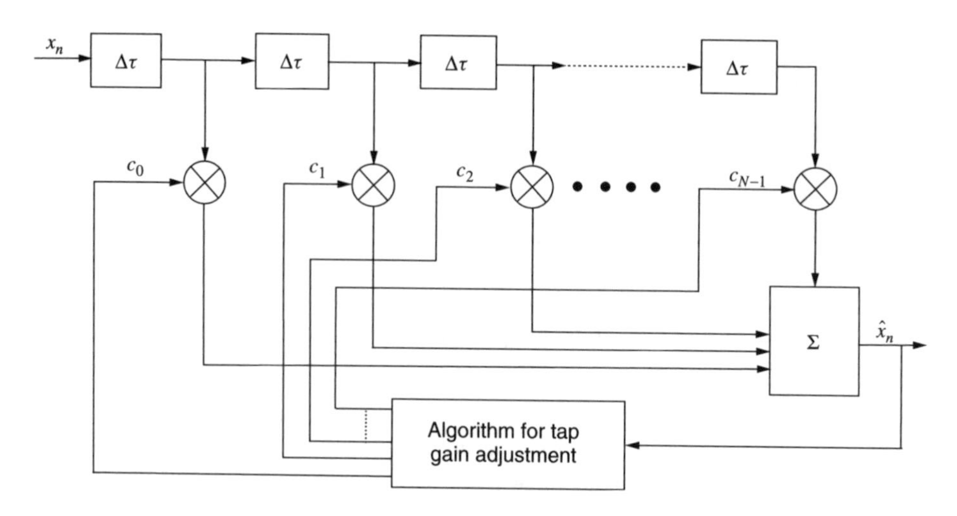

Figure 7.15 Linear transversal equalizer.

approach is the minimum mean-square error (MMSE) equalizer [17] in which the strategy is to minimize the mean-square error between the received signal, including residual ISI and noise, and the known training symbol.

The tap delay spacings in the original adaptive equalizers were set at the symbol transmission interval T. Since the multipath components can arrive at arbitrary times depending on the nature of the propagation environment, the time resolution of the equalizer adjustments was not ideal and resulted in high sensitivity to choosing the sampling time. This led to the development of *fractionally spaced equalizers* (FSE) in which the tap delay spacing was less than T. As described in [17], the FSE equalizer can perform as well as a symbol-spaced equalizer that is preceded by a matched filter. Since the matched filter is hard to realize in practice when the channel response is not known or varying, the FSE represents the more attractive solution.

Linear equalizers have the drawback of enhancing noise through the multiplicative process in the tap coefficients, especially when the channel has severe amplitude distortion leading to some high tap gains. This deficiency led to the development of the *decision feedback equalizer* (DFE), also called a nonlinear equalizer. A DFE uses previous detection decisions to eliminate the ISI caused by previously detected symbols on the symbol currently being detected. Figure 7.16 is a block diagram of a DFE consisting of two equalizers – a feedforward filter and a feedback filter. The feedforward filter is generally a fractionally spaced linear equalizer as described above. The feedback equalizer is a symbol-spaced equalizer whose input is the set of previously detected symbols. One drawback with DFE's is error propagation, which occurs because of the feedback of the detection decision from one symbol to the following symbol.

For all adaptive equalizers, the tap coefficients are periodically set to achieve some objective such as zero ISI or MMSE. Although the tap settings can be found by solving a set of linear equations, typically a search algorithm is used to find the optimum settings. There are several types of search algorithms, among them the *steepest descent method* and

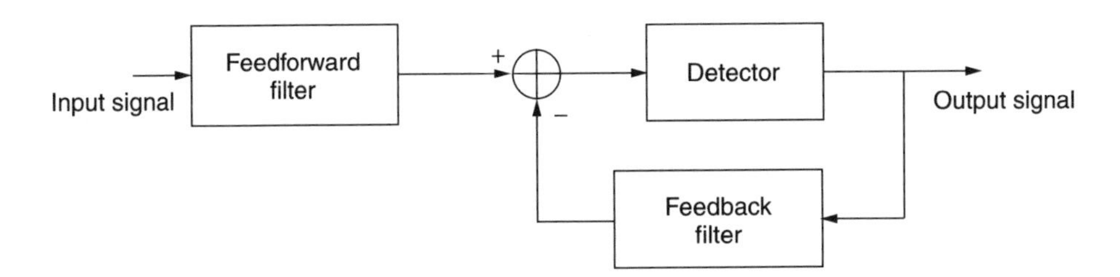

Figure 7.16 Nonlinear decision feedback equalizer DFE.

the *stochastic gradient algorithm,* also called the *least-mean-square* (*LMS*) *algorithm*. The intent of each is to rapidly find the coefficients with the least computation complexity. The LMS algorithm is the most popular approach because of its simplicity and stability [17].

These algorithms are iterative procedures in which an optimum solution can be approached arbitrarily closely depending on the number of iterations. Because each iteration takes one symbol time, and several hundred iterations may be required, the algorithm and gradient interval of the equalizer may be selected to be best suited for a particular type of wireless system. For example, with fixed broadband systems in which the channel response is reasonably assumed to be less variable than the channel response for a mobile radio system, the parameters of the adaptive algorithm for fixed broadband receivers can be adjusted to perform more efficiently in a more stable channel.

7.5.1.2 Frequency domain equalizers (FDE)

The objective of a frequency domain equalizer (FDE) is fundamentally the same as the time domain equalizer – to eliminate the ISI by correcting the channel response distortions. A frequency domain equalizer attempts to do this by equalizing the amplitude and phase of the channel response over the bandwidth of interest using a process that operates on a frequency domain representation of the signal. The concept of frequency domain equalization has actually been around for several decades; in fact, it is commonly found in consumer audio equipment at all levels in which the listener can manually adjust the amplitude or volume in various frequency bands to make up for frequency response deficiencies in the speaker system.

Ironically, adaptive FDE's were not considered a serious option for wireless communications until an OFDM system (Eureka-147) was deployed for digital audio broadcasting (DAB) in Europe in the 1980s. With the acceptance of the relative complexity of OFDM in a consumer system, especially the required fast Fourier transform (FFT's), the use of an FDE on a single-carrier system became not only plausible but potentially a more attractive alternative to the OFDM. For channels with long time delay ISI, FDE's are more effective and less complex than a DFE, to achieve the same net result.

A basic FDE is shown in Figure 7.17. The signal samples presented to the FFT can consist of one or more samples per received symbol. The M frequency domain samples from the FFT are multiplied by complex coefficients to correct the amplitude and phase of the frequency response. As with the time domain equalizer, a training symbol or sequence

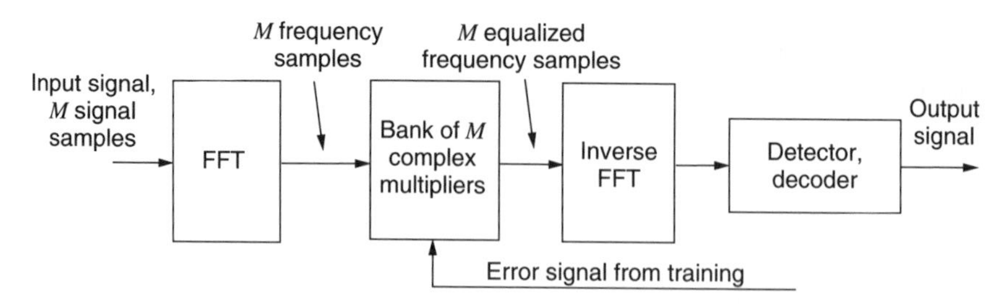

Figure 7.17 Frequency domain equalizer (FDE).

and similar steepest decent or LMS algorithms are used to find the optimum coefficients. Once equalized, the frequency domain samples are transformed back to the time domain using an inverse FFT where detection and decoding can be done.

As shown in [6], single-carrier QPSK with FDE (SC-FDE) can outperform uncoded OFDM in terms of error rate performance on both static and time-varying frequency-selective channels. This occurs because an FDE exploits the frequency diversity of the channel; OFDM requires coding to take similar advantage of frequency diversity. In [5], results were presented comparing SC-FDE to COFDM. With convolutional coding, interleaving, and weighted soft decisions, OFDM can produce error rate results that are comparable to SC-FDE in both static and time-varying frequency-selective channels. Further information on the application of FDE's to single-carrier modulation can be found in [18].

7.6 CODING TECHNIQUES AND OVERHEAD

The basic idea of coding is to add controlled redundancy to the transmitted signal that allows errors in data reception to be detected and corrected using forward error correction (FEC). Coding and FEC can detect and correct errors due to noise, fading, multipath, external interference, and other channel impairments. Codes can be designed that automatically correct a certain number of errors and detect an even larger number. For errors that are detected but not corrected, the receiver can request a retransmission of the data block that contained an error that could not be corrected. This is known as automatic retransmission request (ARQ) and is a necessary part of any wireless (or wired) data communications system. Of course, any retransmission detracts from the total link data throughput, so it is worthwhile to design a system to avoid such retransmissions. ARQ also causes a round-trip delay in the delivery of the data, which may make it unsuitable for real-time applications such as voice or video conversations. The usual emphasis, then, is to design coding schemes for use in wireless equipment that are sufficiently robust to accommodate both real-time and non real-time applications.

Any coding scheme necessarily adds 'overhead' to the signal. Overhead are data bits that the system needs for successful performance that do not carry information. Overhead also includes transmitted bits used for synchronization, equalizer training, system control and management, and so on. The bits that are exclusively used for information are sometimes referred to as the *payload*. The system overhead may vary from 10 to 50% or more, a fact that should be borne in mind when the total link and net link capacities are considered with respect to the data rate delivery expectations of the end user.

Current coding schemes fall into one of the three general categories: block codes, convolutional codes, and space-time codes. These coding methods are discussed in the following sections.

7.6.1 Block codes

A block encoder takes a sequence or block of k data symbols and encodes it into a sequence or block of n code symbols. The encoded block is a vector called a *code word*; the number of redundant symbols is $n - k$, and the code is referred to as an (n, k) code. The degree of redundancy introduced into the code word by this process is called the *code rate*, R_c, and is given by the ratio k/n where $n > k$. The inverse of the code rate also indicates the amount of overhead added to the transmission by the coding. A rate $3/4$ code requires 4 code word symbols for every 3 data symbols, so the coding overhead is approximately 33%.

The elements in the code word are drawn from an alphabet of q elements. If the alphabet has only two members, 0 and 1, the code is a binary code and the elements in the code word are bits. If the number of elements is greater than 2 ($q > 2$), the code is nonbinary. It is clear that if a code word has n elements, the number of possible code words is q^n. For a binary code, the number of possible code words is 2^n, the number of code words needed to represent the possible information bit sequences is 2^k. The encoder will map a set of k information bits into an n length code word selected from the 2^k code words.

In addition to codes being classified as binary or nonbinary, they can also be classified as linear or nonlinear. If two code words C_i and C_j are considered, and a_1, a_2 are two elements of the alphabet, a code is said to be linear only if $a_1C_i + a_2C_j$ is also a code word. This also indicates that a linear code must contain an all-zero code word.

An additional basic parameter that describes a code is the *code word weight*. The code word weight is the number of nonzero elements that it contains. Considering the weights of all the codes is called the weight distribution or weight enumeration.

The basic objective in adding the redundant $n - k$ symbols is to increase the *Hamming distance* between any two code words. The Hamming distance is calculated as the number of symbols that are different between two code words. If the binary code word C_i has the sequence 00110011 and binary code word C_j has the sequence 01010101, the Hamming distance between C_i and C_j is 4. For a linear code that necessarily has a zero code word, the weight of the code word is the Hamming distance to the zero code.

For a minimum Hamming distance of $d_{\min}$ between any two code words, the code can correct t errors where t is given by

$$t \leq \text{int}\left(\frac{d_{\min} - 1}{2}\right) \tag{7.27}$$

The function int(·) is the integer function that rounds the argument down to the nearest integer value (since it is not possible to correct a fraction of a symbol). The minimum

Hamming distance $d_{\min}$ depends on the number of redundant code symbols $n - k$. The upper bound of the Hamming distance is given by $d_{\min} \leq (n - k) + 1$. If 4 redundant data symbols are added, the largest $d_{\min}$ can be is 5, and the largest number of bits in a code word that can be corrected from (7.27) is 2.

A number of binary linear codes have been developed over the years that differ primarily in how the code words are constructed. A *Hamming code* is one in which

$$(n, k) = (2^m - 1, 2^m - 1 - m) \tag{7.28}$$

where $m = n - k$ is any positive integer ($m \geq 2$) and $d_{\min} = 3$. From (7.28), with this minimum distance a Hamming code is capable of correcting one bit error, a rather limited capability compared to other coding schemes.

A *Hadamard code* is one in which the code words are selected from the rows of an $N \times N$ Hadamard matrix. A Hadamard matrix is a matrix of 0s and 1s in which any row differs from any other row in exactly $n/2$ vector elements. This necessarily implies that N must be an even integer. As a linear code, one row of the matrix contains all zeros and the other rows contain $n/2$ ones and $n/2$ zeros.

7.6.1.1 Cyclic codes

Cyclic codes are a type of linear block code that have defined relationships between code words that makes encoding and decoding easier to implement. A cyclic code is one in which code word $\mathbf{C}_i = [c_{n-1}, c_{n-2}, c_{n-3}, \ldots, c_0]$ can be formed into code word $\mathbf{C}_j = [c_{n-2}, c_{n-3}, c_{n-4}, \ldots, c_0, c_{n-1}]$ by a cyclic shift of the code elements. All cyclic shifts of $\mathbf{C}$ are therefore valid code words. Generating cyclic codes follows a straightforward algebraic process using generator polynomials. A detailed description can be found in [19].

A *Golay code* is a special (23,12) binary linear code with $d_{\min} = 7$. By adding a parity bit, it becomes an extended (24,12) Golay code with $d_{\min} = 8$. A Golay code can be generated as a cyclic code using a generator polynomial given in [19].

The Bose–Chaudhuri–Hocquenghem (BCH) codes are a large group of cyclic codes that can use both binary and nonbinary alphabets. Binary BCH codes have the parameters

$$\begin{aligned} n &= 2^m - 1 \\ n - k &\leq mt \\ d_{\min} &= 2t + 1 \end{aligned} \tag{7.29}$$

These parameters allow a wide range of codes with different code rates and error correction capabilities to be created. As with other cyclic codes, there are specific generator polynomials to create the codes. The coefficients of the generator polynomial for various block lengths are tabulated in various texts on digital communications and coding such as the one in [19].

As mentioned, BCH can be binary or nonbinary. An important subset of nonbinary BCH codes is *Reed–Solomon* (RS) codes. For nonbinary codes, the alphabet consists of

q elements ($q > 2$). As before, the length of a nonbinary code is indicated by N; the number of information symbols is indicted by K. The minimum distance is given by $D_{\min}$. The nonbinary block code is thus specified as (N, K) with $N - K$ added code bits. If the code is a *systematic code*, the code symbols are appended on to the end of the information symbols in the form of parity check symbols.

RS codes are described by the following parameters:

$$\begin{aligned} N &= q - 1 = 2^K - 1 \\ K &= 1, 2, 3, \ldots, N - 1 \\ D_{\min} &= N - K + 1 \\ R_c &= K/N \end{aligned} \tag{7.30}$$

The number of symbol errors, t, that the RS codes can correct has an upper bound of

$$\begin{aligned} t &= (D_{\min} - 1)/2 \\ &= (N - K)/2 \end{aligned} \tag{7.31}$$

RS codes are attractive for wireless communications because of the good distance properties between code words; in fact, it is a *maximum-distance* code that produces the largest possible $D_{\min}$ for any given linear code (N, K) [20]. There are also efficient decoding algorithms that are readily implemented in hardware designs.

7.6.2 Concatenated codes

The block codes described above can be used in conjunction with, or concatenated with, each other, or with the convolutional codes described in the next section. The motivation for using concatenated codes is that the error-correcting performance achieved for a given complexity can be superior to that achieved using a single code alone, largely due to the wider range of different error patterns each code is most suited to correcting in high and low SNRs. When used with interleaving, this approach is also found to have good error-correcting properties for channels in which the errors occur in bursts, a property typical of most fading wireless channels.

The two concatenated codes are usually described as the *inner code* and the *outer code*. Typically a nonbinary code such as an RS code is used as the outer code and a binary block code or convolutional code is used as the inner code. The combined code is then denoted as an (Nn, Kk) code with code rate $R_c = Kk/Nn$ and minimum distance $d = D_{\min} d_{\min}$.

7.6.3 Interleaving

Interleaving is a systematic way of shuffling the order in which the symbols are transmitted over a wireless channel. Usually, the code words are arranged in an interleaving matrix and sent to the modulator in the opposite dimension they were entered. For example, if the code words were used to populate the rows of the matrix, the symbols would be sent to the modulator in a column-by-column order. For channels in which the errors are not uniformly distributed but instead come in bursts or clusters as might be expected with a sudden deep fade, interleaving results in the corrupted symbols being distributed among multiple code words. Since coding schemes are designed to correct a limited number of symbol errors within a code word, it is reasonable to spread the erroneous symbols over several code words instead of grouping them into a single code word where it is less likely that they can be corrected. Interleaving is a standard part of most modern digital communication systems. It requires no additional overhead bandwidth, but it does introduce a delay since all the code words that populate an interleaving matrix must be received before any can be extracted and sent on to the decoder.

7.6.4 Convolutional codes

The block codes described above are well suited to transmitted information that lends itself to being divided into blocks, such as data packets. However, convolutional codes have actually become more popular, especially for cellular radio applications. This is primarily due to the existence of very simple decoding algorithms (such as the Viterbi algorithm) that can achieve significant coding gain at near optimum theoretical performance. For fixed broadband wireless systems, both block and convolutional codes with be encountered in equipment design.

Convolutional codes can also be described in terms of (n, k) with a code rate $R_c = k/n$. Convolutional codes are distinguished from block codes in that the encoder output is a function not only of a present block of k information symbols but also of the previous information symbols. The dependence on previous information symbols results in the convolutional encoder being a finite state machine (FSM).

A convolutional encoder is shown in Figure 7.18. It consists of L banks of shift registers, each holding k information symbols. At each state of the encoder, the shift

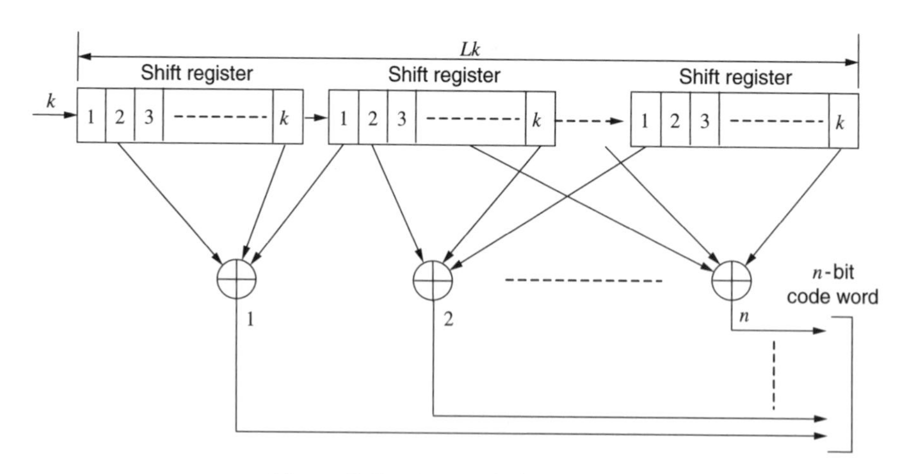

Figure 7.18 A convolution encoder.

registers contain k-bit sequences. The contents of the registers are added together to form the n-bit code word. Which registers are added together to form which code word elements depends on the code generation polynomial.

In addition to the information symbol sequence k and the code word length n, a convolutional code is also described by the constraint length L. The constraint length is simply the number of shift register groups that hold the sequence of information symbols, as shown in Figure 7.18.

Convolutional codes can be decoded in a variety of ways that represent states in a trellis decision tree. The most widely used of these in the Viterbi algorithm [21]. A more detailed discussion of convolutional codes, their encoding and decoding algorithms can be found in several texts such as those in [22,23].

7.6.5 Trellis-coded modulation (TCM)

Trellis-coded modulation (TCM) gained interest in the early 1980s with the published work of Ungerboeck [24–26]. TCM is an amalgamation of coding and modulation that provides a coding gain of 3 to 6 dB without the additional bandwidth necessary for the redundant coding symbols. This was an important innovation that was quickly adopted in many areas of wireless communications.

TCM uses a rate k/n convolutional encoder and directly maps those points onto a set of signal constellation points using a technique called *mapping by set partitioning*. This approach has these basic features:

- A modulation signal constellation (like 64QAM constellation in Figure 7.3b) is used that has more points than would be necessary for the information message alone at the same data rate. The additional signal constellation points essentially increase the modulation efficiency in bits/Hz without increasing the occupied bandwidth.
- The expanded signal constellation is partitioned into subsets in such a way that the Euclidean distance between subsets at each stage of the partitioning is maximized. The Euclidean distance is the magnitude of the multidimensional vector between the signal constellation points in the two subsets.
- The convolutional coding and signal mapping are arranged so that only selected sequences of signal points are used.

As shown earlier in this chapter, modulation with a larger number of signal constellations states will have a larger error rate for a given SNR (E_b/N_0). It seems counterintuitive, then, to use such a constellation. However, this *increase* in error rate is smaller than the *decrease* in error rate that is achieved by using coding among the constellation states. Fundamentally, this is how TCM achieves its advantage.

The performance of TCM has been analyzed in additive white Gaussian noise (AWGN) channels, and channels with flat Rician and Rayleigh fading. From this, rules have been formulated for designing trellis codes for fading channels as set forth in [27].

It should be clear from the above discussion that TCM requires modulation types other than simple binary modulation like BPSK. TCM is well suited to 16QAM, 64QAM, and 256QAM, which are already widely used in high-capacity digital microwave radio. The use of TCM for these radio systems provides coding gain without sacrificing information throughput to coding overhead.

7.6.6 Coding gain

The error rate improvement achieved through the use of coding is often called the *coding gain*. The coding gain in dB is the reduction of the SNR needed to achieve the same detected/decoded error rate as the uncoded raw error rate. For example, if a SNR of 10 dB is needed to achieve a raw error rate of 10^{-6}, and the same error rate can be achieved with coding and an SNR of 6 dB, the system is said to have a coding gain of 4 dB at an error rate of 10^{-6}. The coding gain will clearly be a function of the code rate and code design, especially the minimum Hamming distance of the code.

If the raw symbol error rate of the channel is denoted as P_M, the code word error rate can be found by viewing the symbols in the code word as independent Bernoulli trials. Independence of adjacent symbols is achieved through interleaving. If the code is guaranteed to correct t symbol errors, then the probability, P_{cw}, a code word will have an error is given by the standard probability for Bernoulli trials [28]

$$P_{cw} \leq \sum_{i=t+1}^{n} \binom{n}{i} P_M^i (1 - P_M)^{n-i} \tag{7.32}$$

where

$$\binom{n}{i} = \frac{n!}{i!(n-i)!} \tag{7.33}$$

Note that the inequality in (7.32) occurs because the code may in some cases be capable of correcting more than t errors but is not guaranteed to do so. The improvement in error rate from coding is clear from (7.32).

When a code word error is made, the symbol error probability is given by [29]

$$P_{MC} = \frac{1}{n} \sum_{i=t+1}^{n} i \binom{n}{i} P_M^i (1 - P_M)^{n-i} \tag{7.34}$$

When the decoded symbols are converted to bits, the bit error probability P_b from (7.34) is

$$P_b = \frac{2^{k-1}}{2^k - 1} P_{MC} \tag{7.35}$$

The net error rate after decoding is often the quantity specified by the system operator and offered to the customers as a specification of the data service. Depending on the service type, this final error rate may be very low, and perhaps only achieved through the use of ARQ even with the improvements available through coding. For real-time services such as voice, ARQ may not be viable in which case the system will rely entirely on the coding for error correction with the expectation that the residual errors will be infrequent enough to go unnoticed by the end user.

With a final net error rate objective, the coding gain can then be used to find the necessary raw error rate, which in turn dictates the minimum SINR. For a fading link, and a given link availability objective, this minimum SINR (and error rate) can only occur a small fraction of the time. The error rate may then be specified as averaged over a 1-min, 5-min, 1-h or similar time frame that is relevant to service users. This leads to the idea of *adaptive coding*, which becomes more robust during fading events. This is similar to the adaptive modulation mentioned earlier in this chapter.

7.6.7 Space-time codes

STCs are codes constructed to exploit the uncorrelated channel response differences that exist when using multiple transmit antenna elements and optionally, multiple receive antenna elements. Multiple-input, multiple-output (MIMO) antenna systems are discussed in Section 6.7.1.

As stated above, coding is the technique of adding redundant symbols to the transmitted signal to improve the ability of the receiver to correctly detect the information data. As described above, conventional coding uses redundancy in the time domain by adding code bits to the time sequence of information bits. A STC uses both the time and space domains to send redundant information to the receiver, the space domain being represented by separate antenna elements distributed in space. This redundancy idea can also be extended to other dimensions of the wireless channel including frequency and polarization, for example.

With STCs, multiple transmit antennas are used to send separate data streams made up of symbols that are organized to achieve performance improvement when detected at the receiver. One of the earliest and simplest examples of an STC is found in the work of Alamouti [30]. This technique will be discussed in detail here for two reasons: (1) it is a straightforward example of link improvement through space-time coding, and (2) the Alamouti method is specified in the draft of the Institute of Electrical and Electronic Engineers (IEEE) 802.16a Standard for 2 to 11 GHz systems as an optional technique for link improvement.

Figure 7.19 shows the Alamouti method using two transmitting antennas and one receiving antenna. This was originally intended as a way to achieve diversity gain at a cellular handset without the burden of adding a second antenna with associated RF/intermediate frequency (IF) circuitry, or additional bandwidth and overhead, to achieve diversity gain. The signals from the two antennas are a sequence of two symbols or symbol blocks, shown as s_1 and s_2 in Figure 7.19. During the first-symbol period at time t, the symbols sent from Antennas 1 and 2, respectively, are s_1 and s_2. During the second-symbol period at time $t + T$, the symbols sent are $-s_2^*$ and s_1^*. The received signals at time periods t and $t + T$ are

$$\begin{aligned} r_1 &= r(t) = h_1 s_1 + h_2 s_2 + n_1 \\ r_2 &= r(t+T) = -h_1 s_2^* + h_2 s_1^* + n_2 \end{aligned} \tag{7.36}$$

where h_1 and h_2 are the channel transfer functions from each of the transmitting antennas to the single receive antenna.

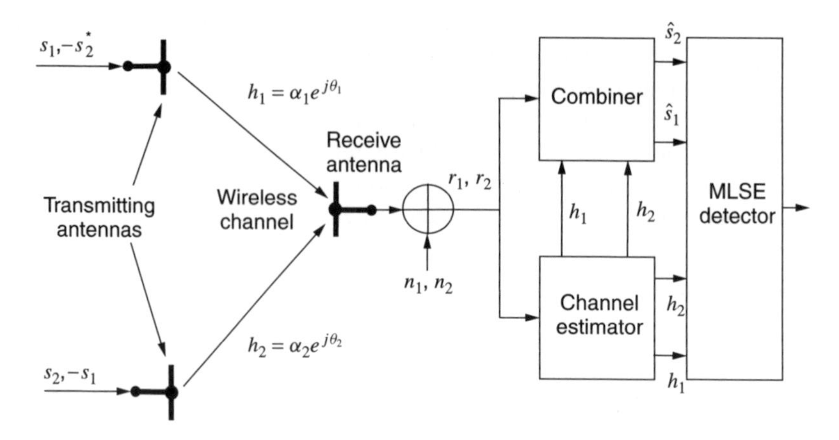

Figure 7.19 Alamouti two-branch transmit diversity with space–time coding.

The combiner produces two estimates of the symbols $\hat{s}_1$ and $\hat{s}_2$

$$\begin{aligned}\hat{s}_1 &= h_1^* r_1 + h_2 r_2^* \\ \hat{s}_2 &= h_2^* r_1 - h_1 r_2^*\end{aligned} \tag{7.37}$$

Inserting the received signals from (7.36) and the channel response from the Figure 7.19 yields

$$\begin{aligned}\hat{s}_1 &= (\alpha_1^2 + \alpha_2^2) s_1 + h_1^* n_1 + h_2 n_2^* \\ \hat{s}_2 &= (\alpha_1^2 + \alpha_2^2) s_2 - h_1 n_2^* - h_2^* n_1\end{aligned} \tag{7.38}$$

From [30], this is the same result as for a maximal ratio combiner (MRC) with two diversity antennas at the receiver, thus achieving the desired diversity gain result but by using only a single receiving antenna.

In this case the STC is the redundant transmission of the symbols s_1 and s_2 on the two transmitting antennas, but during the second-symbol transmission time, it is reversing the antennas that send the symbols, and sending the complex conjugates of each symbol. The sign of one symbol is also reversed. This STC achieves diversity gain, which in turn will substantially improve the BER in fading channels without additional bandwidth or power. In [30] the concept is extended to multiple receive antennas that provide smaller incremental diversity gain as might be expected. As with any block code, one drawback of the Alamouti scheme is that there is a delay of one symbol time in recovering the two symbols. For a single-channel symbol, this is not a problem, but if the symbols s_1 and s_2 represent long code blocks, this delay could be significant for real-time applications. Also, if the total transmit power remains the same (half to each transmit antenna), the Alamouti scheme also suffers a 3 dB penalty compared to MRC in a receiver using two diversity antennas for reception.

The Alamouti technique is an example of a space-time block code (STBC). The concept of space-time block codes can be generalized to more than two symbols and more than two transmitting antennas. The *space-time code word* for Q length symbol blocks and N transmitting antennas is

$$\mathbf{C}(k) = [\,\mathbf{c}_1(k) \quad \mathbf{c}_2(k) \quad \ldots \quad \mathbf{c}_Q(k)\,] = \begin{bmatrix} c_{11}(k) & c_{12}(k) & \ldots & c_{1Q}(k) \\ c_{21}(k) & c_{22}(k) & \ldots & c_{2Q}(k) \\ \vdots & \vdots & \ddots & \vdots \\ c_{N1}(k) & c_{N2}(k) & \ldots & c_{NQ}(k) \end{bmatrix} \tag{7.39}$$

Each row in (7.39) is a *space-time symbol* and each element a linear combination of possible channel symbols $\mathbf{s}$. Each space-time symbol in $\mathbf{C}$ can be sent in vertical layers as shown in (7.39), or it can be broken into subblocks so that one subblock is sent on each of the antennas. The latter approach is often called *diagonal space-time layering* since the subblocks of each space-time symbol lie on a diagonal across the matrix $\mathbf{C}$.

The received signal $\mathbf{y}(k)$ across a channel with response $\mathbf{H}$ can be written as

$$\mathbf{y}(k) = \sqrt{E_s}\mathbf{H}\mathbf{C}(k) + \mathbf{n}(k) \quad k = 1, 2, 3, \ldots, Q \tag{7.40}$$

where $\sqrt{E_s}$ is the energy per space-time symbol, so that $\sqrt{E_s/N}$ is the energy per constellation point within the code word.

Each symbol block may itself be separately encoded using a block code such as those described in Section 7.6.1. There are various approaches to designing the codes to create the space-time symbols, the most obvious being orthogonal block codes in which the space-time symbols (rows) are orthogonal. The Alamouti two-symbol code uses this approach. The maximum likelihood sequence estimator (MLSE) then decides which symbol was transmitted by calculating the Euclidean distances (analogous to the Hamming distance) between the received code word and the constellation of possible code words and chooses that code word with the smallest distance. From this, error rates can be determined by finding the probability of the incorrect code word being detected, which in turn depends on the Euclidean distance between code words. The derivation of this pairwise error probability is somewhat involved but can be found in [31] and several other publications. For code words $\mathbf{C}$ and $\mathbf{E}$, the probability, averaged over all channel realizations, that $\mathbf{E}$ will be detected when $\mathbf{C}$ is sent is given by an upper bound

$$\Pr(\mathbf{C} \to \mathbf{E}) \le \left(\frac{E_s}{4N_0}\right)^{-r(\mathbf{B}_{C,E})M} \prod_{i=0}^{r(\mathbf{B}_{C,E})-1} \lambda_i^M(\mathbf{B}_{C,E}) \tag{7.41}$$

where M is the number of receive antennas and

$$\mathbf{B}_{C,E} = (\mathbf{C} - \mathbf{E})^T(\mathbf{C} - \mathbf{E})^* \tag{7.42}$$

The term $r(\mathbf{B}_{C,E})$ denotes that the rank of matrix $\mathbf{B}_{C,E}$ and $\lambda_i(\mathbf{B}_{C,E})$ are the nonzero eigenvalues of $\mathbf{B}_{C,E}$. The term N_0 is the noise power so E_s/N_0 is the SNR. The superscript ^{T}

indicates the transpose of the matrix; the superscript * indicates the transpose with complex conjugates of each matrix element.

In addition to block codes, trellis coding can also be used as described in [31]. The design criteria for space-time (ST) trellis codes (STTC) are summarized in [31] as:

- *Rank Criterion*: Maximum diversity is achieved for N transmit and M receive antennas when the matrix $\mathbf{B}_{C,E}$ is full rank for every pair of distinct code words.
- *Determinant Criterion*: With diversity advantage of $N \times M$ as the objective, the minimum of the determinant of $\mathbf{B}_{C,E}$ taken over all distance code word pairs $\mathbf{C}, \mathbf{E}$ should be maximized.

Generally, trellis codes can outperform block codes in terms of error rate and diversity but require increased receiver complexity.

From [30], an original objective of STCs was to provide diversity improvement in narrowband flat-fading channels without adding receive antenna diversity. The best performance improvement is also achieved when the channels are uncorrelated. When these assumptions are not valid; that is, the channel is time-dispersive (frequency-selective rather than flat-fading), and/or correlated to a significant degree, the available performance improvement with STC is more limited. For this reason, ST codes have primarily been of interest when used with OFDM when the cyclic prefix (CP) or zero-padding (ZP) can be inserted to avoid the channel time where multipath signals arrive at the receiver. Both approaches reduce bandwidth efficiency. The percentage reduction can be made small by increasing the block length, but long block lengths are undesirable in a packet-based network with packet acknowledgement retransmission (see Chapter 9).

The ST code performance is described for a single user rather than multiple users in a multiple-access, interference-limited system. In such networks, the link performance will suffer because the number of antennas with interfering signals is increased by N when using STC (even though these antennas are close together and the same total power is distributed among them). As with any modulation/coding scheme, multiuser interference (MUI) can be reduced using beam-forming antennas as described in Section 6.6, but beam-forming requires multiple receive antennas that can only eliminate $M - 1$ interferers [32]. The use of STBC with multicarrier code division multiple access (CDMA) (see Section 8.7.1 for MC-CDMA) where the spreading codes are specifically designed to suppress interference is described in [33].

For fixed broadband NLOS networks, the use of OFDM with STC and MIMO technology is an approach that is being actively pursued. The success of this approach in single link situations has been demonstrated in the field tests (see references for Section 6.7), but at the time of this writing, tests of its performance in high-density MUI environments have not been reported.

7.7 CONCLUSION

There are a wide variety of modulation techniques available to a wireless system designer. Those discussed in this chapter including *M-ary* PSK, *M-ary* QAM, and OFDM are either the most widely used now or under active development for both LOS and NLOS fixed wireless systems. Several other modulation schemes such as *M-ary* FSK (frequency shift keying) are also available. The choice of a modulation constellation is driven by the required spectral efficiency in bps/Hz as derived from the available channel bandwidth and required channel data rate capacity. For a given link with power and noise limitations, this will also determine the link range for a given availability. Robust systems use adaptive modulation that maintains the link connection at acceptable error rates by reverting to lower level modulation formats and low rate codes when the link is stressed with rain fades, interference, or other impairments.

Errors due to ISI in time-dispersive channels can be effectively combated using both time domain and FDE. There are various trade-offs associated with the use of various equalizers in terms of expected channel delay spread and the severity of the ISI distortion that must be corrected. The appropriate choice of equalizer strategy will therefore depend on the nature of the scattering and multipath, which in turn depends on the nature of the propagation environment.

Coding schemes, especially TCM, provide a means of improving the net system error rate over the raw system error rate by several orders of magnitude. Coding can correct errors resulting from noise as well as interference. (STCs) are one of the more promising approaches that use multiple transmit antennas to exploit the spatial dimension along with the time dimension of the wireless channel to achieve diversity gain and improved error rate performance.

Modern fixed broadband wireless systems use many of these techniques in various combinations to achieve high reliability links. Understanding the parameters involved in modulation, equalizers and coding, and how they impact the system performance, is important for successful wireless system design.

7.8 REFERENCES

[1] J.G. Proakis. *Digital Communications*. New York: McGraw-Hill. 2nd Edition. 1989.

[2] J.G. Proakis and M. Salehi. *Communications System Engineering*. Upper Saddle River: Prentice-Hall. 1994. page 553.

[3] G.L. Stuber. *Principles of Mobile Communication*. Boston: Kluwer Academic Publishers. 1996.

[4] R. Van Nee and R. Prasad. *OFDM for Wireless Multimedia Communications*. Boston: Artech House Publishers. 2000.

[5] H. Sari, G. Karam, and I. Jeanclaude, "Frequency domain equalization of mobile radio and terrestrial broadcast channels," *IEEE Global Telecommunications Conference*, San Francisco, pp. 1–5, November, 1994.

[6] H. Sari, G. Karam, and I. Jeanclaude, "An analysis of orthogonal frequency-division multiplexing for mobile radio applications," *Proceedings of the 44th IEEE Vehicular Technology Conference*, Stockholm, pp. 1635–1639, June, 1994.

[7] R. van Nee and R. Prasad. *OFDM for Wireless Multimedia Communications*. Boston: Artech House Publishers. 2000. pp. 120 ff.

[8] T. Wilkinson and A. Jones, "Minimisation of the peak to mean envelope power ratio of multicarrier transmission schemes by block coding," *Proceedings of the 45th IEEE Vehicular Technology Conference*, Chicago, pp. 825–829, July, 1995.

[9] R. Van Nee. "OFDM codes for peal-to-average power reduction and error correction," *IEEE Global Telecommunications Conference*, London, pp. 740–744, November, 1996.
[10] J.G. Proakis. *Digital Communications*. New York: McGraw-Hill. 2nd Edition. 1989. p. 262.
[11] R.F. Pawula, S.O. Rice, and J.H. Roberts. "Distribution of the phase angle between two vectors perturbed by gaussian noise," *IEEE Transactions on Communications*, vol. COM-30, No. 8, pp. 1828–1841, August 1982.
[12] K. Murota and K. Hirade, "GMSK modulation for digital mobile radio telephony," *IEEE Transactions on Communications*, vol. COM-29, No. 7, pp. 1044–1050, July 1981.
[13] J.G. Proakis. *Digital Communications*. New York: McGraw-Hill. 2nd Edition. 1989. page 284.
[14] J.G. Proakis and J. Miller, "An adaptive receiver for digital signaling through channels with intersymbol interference," *IEEE Transactions on Information Theory*, vol. 15, pp. 484–497, July 1969.
[15] M. Schwartz, W.R. Bennett, and S. Stein. *Communication Systems and Techniques*. New York: McGraw-Hill, 1966. page 349.
[16] R.W. Lucky, "Automatic equalization for digital communications," *Bell System Technical Journal*, vol. 44, pp. 547–588, April 1965.
[17] G.L. Stuber. *Principles of Mobile Communication*. Boston: Kluwer Academic Publishers. 1996. page 265.
[18] D. Falconer, S.L. Ariyavisitakul, A. Benyayamin-Seeyar and B. Eidson, "Frequency domain equalization for single carrier broadband wireless systems," *IEEE Communications Magazine*, vol. 40, no. 4, pp. 58–66, April, 2002.
[19] J.G. Proakis. *Digital Communications*. New York: McGraw-Hill. 2nd Edition. 1989. p. 386 ff.
[20] R. Steele. *Mobile Radio Communications*. London: Pentech Press. 1992. page 424.
[21] A.J. Viterbi and J.K. Omura. *Principles of Digital Communications and Coding*. New York: 1979.
[22] G.L. Stuber. *Principles of Mobile Communication*. Boston: Kluwer Academic Publishers. 1996.
[23] J.G. Proakis and M. Salehi. *Communications System Engineering*. Upper Saddle River: Prentice-Hall. 1994.
[24] G. Ungerboeck, "Channel coding with multilevel phase signals," *IEEE Transactions on Information Theory*, vol. 28, pp. 55–67, January, 1982.
[25] G. Ungerboeck, "Trellis coded modulation with redundant signal sets- Part I: Introduction," *IEEE Communications Magazine*, vol. 25, pp. 5–11, February, 1987.
[26] G. Ungerboeck, "Trellis coded modulation with redundant signal sets- Part II: State of the art," *IEEE Communications Magazine*, vol. 25, pp. 12–21, February, 1987.
[27] G.L. Stuber. *Principles of Mobile Communication*. Boston: Kluwer Academic Publishers. 1996. page 367.
[28] P.Z. Peebles. *Probability, Random Variables, and Random Signal Principles*. New York: McGraw-Hill. 1980. page 72.
[29] J.G. Proakis. *Digital Communications*. New York: McGraw-Hill. 2nd Edition. 1989. page 430.
[30] S.M. Alamouti, "A simple transmit diversity technique for wireless communications," *IEEE Journal on Selected Areas in Communications*, vol. 16, no. 8, pp. 1451–1458, October, 1998.
[31] V. Tarokh, N. Seshadri, and A.R. Calderbank, "Space-time codes for high data rate wireless communications: performance criterion and code construction," *IEEE Transactions on Information Theory*, vol. 44, no. 2, pp. 744–765, March, 1998.
[32] A.F. Naguib, N. Seshadri, and A.R. Calderbank, "Applications of space-time codes and interference suppression for high capacity and high data rate wireless systems," *Proceedings of the 32nd Annual Asilomar Conference on Signals, Systems, and Computers*, Pacific Grove, CA, pp. 1803–1810, November, 1998.
[33] Z. Lui and G.B. Giannakis, "Space-time block-coded multiple access through frequency-selective fading channels," *IEEE Transactions on Communications*, vol. 49, no. 6, pp. 1033–1044, June, 2001.

8

Multiple-access techniques

8.1 INTRODUCTION

The wireless communication spectrum from approximately 150 kHz to 100 GHz is segmented into bands that have been designated for use by particular kinds of wireless services. These designations have been established by the International Telecommunications Union (ITU) on a worldwide basis, with further spectrum partitioning instituted by countries within these broader spectrum designations. Dividing the spectrum into blocks so that a multitude of users can have simultaneous access to it without conflict is the most basic form of frequency division multiple access (FDMA). While global spectrum partitioning is the mechanism for multiple access to the entire spectrum, for wireless systems operating in a particular geographic area in the same frequency band, multiple access means the simultaneous access to the same frequencies or channels within the same area by multiple users. In this context, 'simultaneous' means the system end user perceives the service to be continuous and immediate whether or not the wireless system is actually providing information to that user on a continuous and immediate basis. Service that is not perceived as continuous or immediate by the user (blocked or dropped calls in a cellular system, for example), or is otherwise flawed, is considered to have low quality of service (QoS). The number of simultaneous users that can be supported at a given QoS level is one measure of the capacity of the system and therefore directly tied to its success in achieving its service and commercial objectives. The capabilities of the system's multiple-access technology is therefore of fundamental importance in determining whether these objectives can be met.

While the multiple-access QoS delivered to the user is a function of the service type and acceptable service delay, the radio channel performance criteria for all multiple-access schemes can be reduced to the fundamental consideration of interference between systems (intersystem) and within systems (intrasystem). Controlling intersystem interference

Fixed Broadband Wireless System Design Harry R. Anderson
 ISBN: 0-470-84438-8

is usually the role of administrative rules or industry standards since the system operators competing for use of the spectrum will not necessarily conduct themselves in a cooperative way. The systems competing for use of the spectrum could be operating in adjacent geographic areas or could be operating in the same geographic area or market.

By contrast, intrasystem interference is caused by transmitting nodes within the network over which the operator has complete control. Within the broader context of avoiding intersystem interference and complying with any mandated modulation types, channelization formats, and so on, the system operator has considerable flexibility in choosing technology and multiple-access techniques to maximize system capacity and service levels. Since the greatest advantage can be gained for the operator through an intelligent system design to enhance capacity, intrasystem multiple access will be the primary subject of this chapter.

As noted in [1], the radio spectrum is a natural resource that is distinctly different from other natural resources such as forests, petroleum reserves, water sources, and so on. It is not depleted through use – extraordinarily, it is only wasted when it is not being used. The most efficient spectrum usage is achieved by maximizing the occupancy of time, frequency, and space within a designated frequency band.

8.1.1 Intersystem multiple access

Within a given frequency band in the same general geographical area, there may be several users who seek to operate their systems without destructive interference from other users in the same frequency band and area. The criteria for when destructive interference occurs are often written into the administrative rules for the spectrum such as those found in the Rules of the Federal Communications Commission (FCC). These rules can be quite detailed, such as those governing two-way multipoint, multi-channel distribution service (MMDS) systems in the United States where the specific propagation model, databases, system description, calculation methods, and interference thresholds are specified in detail. These specific methods must be used in order to be granted authorization to build a network in the MMDS spectrum (2.5–2.7 GHz) [2]. In contrast, the FCC Rules for intersystem interference for cellular and personal communications service (PCS) systems only describe power density limits beyond the boundaries of the system service limits.

Besides government statutes, there are also industry standards that have been established to accomplish harmonious sharing of spectrum among various users. Again, these can be quite detailed like the Telecommunication Industry Association (TIA) TSB-88-A for calculating interference and spectrum compatibility between mobile radio systems [3]. Other industry standards such as the interoperability standard for 10 to 66 GHz developed by the Institute of Electrical and Electronic Engineers (IEEE) 802.16 Working Group do not specify parameters for interference calculations but simply leave much of this to rely on 'good engineering practice' that can certainly lead to significant differences of opinion about how interference should be calculated and whether it is 'destructive' [4].

The spectrum administration rules and industry standards are all nonhardware mechanisms to achieve successful FDMA among various spectrum users. They are driven by the capabilities of the technology to the extent that the rules can allow tighter spectrum sharing when the technology employed by the operators is less susceptible to external interference. For example, at various times in the past the FCC has mandated the use of equipment and antennas that meet certain technical performance thresholds (out-of-band emissions, modulation efficiency, receiver bandwidth, antenna directionality, etc.) to increase the number of users and systems that can be accommodated within a spectrum band. Spectrum capacity and how technologies affect capacity is discussed further in Chapter 12.

8.1.2 Intrasystem multiple access

A given fixed broadband network may have one or more users all of whom require service of a given capacity, quality, and consistency. The capacity, quality, and consistency may be embodied in a contractual commitment between the service operator and the end user. The system capacity and the extent to which the system can accommodate the predicted traffic levels (and requisite service quality) are the primary design drivers. The way in which the available spectrum is partitioned for sharing by multiple users, and the multiple-access technologies that are then used, directly affects the traffic volume that can be supported. The capacity and quality issues ultimately determine whether the system can meet the commercial objectives of the system operator who seeks a system that is profitable, sustainable, and scalable to serve increasingly dense and dispersed demand.

In Chapters 7 and 12, the concept of adaptive modulation is discussed in which the modulation type is adjusted to respond to the conditions of the wireless channel. Multiple access can also be adapted in the same way to fulfill the requirements of service agreements with the end users. Wireless operators can construct several service levels with different price structures that offer different levels of access. These service levels vary from top-level service in which the end user is provided with immediate best access to the network when requested to lower-level access that is often described as 'best effort'. The Internet is an example of a network with a 'best effort' service objective. An adaptable multiple-access scheme is therefore not just a technical mechanism to increase capacity but is also a fundamental part of the system operator's business plan.

8.1.3 Duplexing

Duplexing is the term given to the method used to accomplish two-way communication between terminals of a point-to-point link or between a network hub and a remote terminal. Two methods to duplexing are used in fixed broadband wireless – frequency division duplexing (FDD) and time division duplexing (TDD). Unless mandated by administrative rules (which is the case for many services), the choice of the duplexing method is generally independent of the multiple access and modulation techniques used.

FDD is currently more widely used for fixed broadband wireless. With FDD a separate frequency is assigned for hub-to-remote communication (downstream or downlink) and for remote-to-hub (upstream or uplink) communication. The frequency separation between uplink and downlink channels is usually chosen to be great enough so that construction of suitable filters to separate the two signals at a given terminal is practical. FDD is the approach used with all 1G, 2G, and most 3G cellular systems, as well as most other mobile communications. However, when the volume of downlink and uplink traffic is not symmetric as may be the case with nonvoice data services, FDD results in an underutilized spectrum, usually on the uplink, which is inefficient. This has lead to an increased interest in TDD in recent years.

TDD uses the same frequency or circuit for both uplink and downlink communications between the network hub and remote terminal. In the broader wireless communications world, TDD is referred to as *simplex* transmission; the 'TDD' appellation has come into vogue only recently. TDD is actually the oldest form of duplexing, having been used for telegraph communications some 150 years ago, in which an operator would transmit a message and then listen for a response. Simplex or TDD is also used in many current-day communication systems such as aircraft communications in which all aircraft and the airport tower use the same frequency to send and receive voice messages. TDD potentially can make more constant and efficient use of a radio channel, especially for asymmetric uplink–downlink traffic flow. However, because both hub stations and remote transmitters can potentially use the same frequency, the interference considerations, and synchronization of transmissions to avoid interference, become more complex. This can impact the extent to which frequencies may be reused within a system.

The technical issues associated with FDD and TDD will be discussed in this chapter with particular consideration of spectrum efficiency and how interference calculations should be carried out.

It should also be mentioned that duplexing can be accomplished using multiple delivery technologies. For example, some proposed broadband systems use relatively high-speed satellite transmission or television broadcast channel transmission for the downlink and a relatively slow-speed telephone line for the uplink, again motivated by the assumption that the downlink traffic volume is greater than the uplink traffic volume. Because such hybrid systems are not entirely wireless, they will not be considered further in this book.

8.2 FREQUENCY DIVISION MULTIPLE ACCESS (FDMA)

With FDMA, the spectrum available to the system operator is segmented into frequency channels. A frequency channel or slot is a 'physical' channel that is identified by its center frequency and bandwidth. When a remote terminal is installed, or initiates communication, a frequency channel is assigned to support the communication between the system hub and that terminal. If the system uses FDD, both an uplink and downlink channel are assigned. If the system uses TDD, only a single channel is assigned for both uplink and downlink. The bandwidth of the frequency slots can be uniform or variable. In many cases, administrative regulations, and sometimes industry standards, mandate the channelization scheme so that all channels are of uniform width. Figure 8.1 shows a simple diagram of an FDMA with N channels that are occupied full time. Figure 8.2 shows the traditional view of the spectrum axis with the radio channels of varying bandwidths and a representation of the power spectral density (PSD) within each allocated channel.

The channel assignment process can be done statically as part of the planning process in which a given frequency is assigned to a terminal when the terminal is installed. The channel may also be assigned dynamically whenever the terminal initiates communication. This is the standard approach used in cellular systems. The assignment criteria for selecting a channel include interference avoidance and, in systems with bandwidth

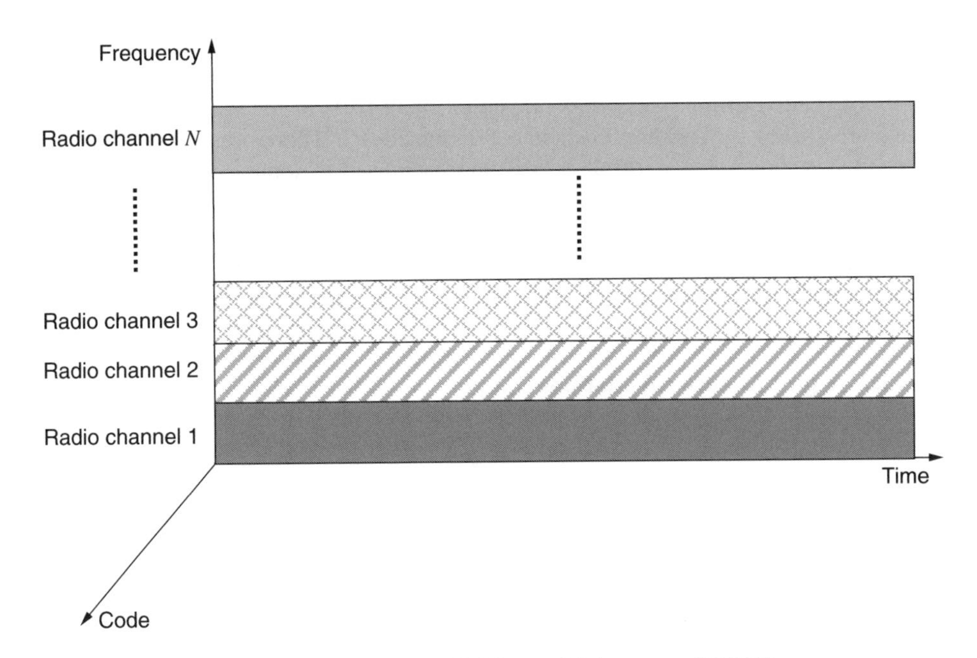

Figure 8.1 Frequency division multiple access (FDMA).

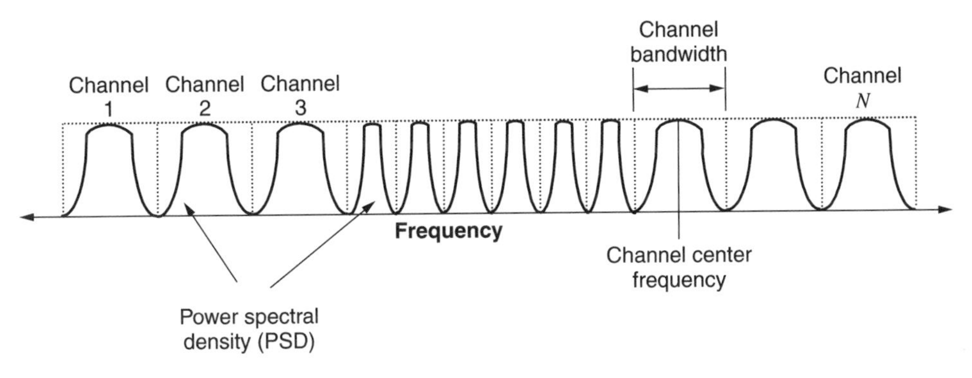

Figure 8.2 FDMA channelization of allocated spectrum with different occupied bandwidths.

flexibility, adequate bandwidth to support the data rates required by the terminal. An effective channel assignment process is fundamental for achieving efficient use of the spectrum. Accordingly, frequency assignment and planning will be discussed in some detail in Chapter 12 for particular network configurations.

Essentially all wireless systems use FDMA in some form, even systems that are normally referred to as *time division multiple access* (TDMA) systems. For example, the IS-136 TDMA cellular system used in the United States uses the same FDMA channelization scheme as the first generation advanced mobile phone service (AMPS) cellular system; it additionally allows multiple users per channel by dividing the channel into assigned time slots (TS). A similar approach is used for the global system for mobile communications (GSM) cellular system deployed in many parts of the world: 124 frequency channels are shared by dividing each into 8 assigned TS. Therefore, FDMA should be recognized as a 'first level' multiple-access approach used to some extent by all wireless technologies with further multiple access achieved through 'second level' spectrum partitioning of each FDMA channel using TDMA, code division multiple access (CDMA), or space division multiple access (SDMA). TDMA, CDMA, SDMA, and variations of these techniques are discussed in the following sections of this chapter.

8.2.1 FDMA interference calculations

The interference in an FDMA channel can come from three sources:

- Cochannel interference from other transmitters occupying some or all of the frequencies of the receiving channel currently used by the victim receiver.
- Adjacent channel interference from transmitters occupying some or all of the frequencies either above or below the receiving channel of the victim receiver.
- Spurious interference from intermodulation products and other, often unidentifiable, sources.

Spurious interference can usually be controlled to an acceptable level by proper installation techniques and equipment usage and is thus not a part of the design process. This section will focus on cochannel and adjacent interference.

The signal-to-interference + noise ratio or SINR, is the fundamental quantity that usually serves as a design objective for determining whether the interference level is acceptable. SINR is given by

$$\text{SINR} = \gamma = \frac{S}{I + N} \tag{8.1}$$

Each of the variables in (8.1) is a random process. Calculation of the noise power, N, and the interference power, I, in (8.1) is discussed in the following sections.

8.2.1.1 Noise power

The noise voltage N is usually assumed to be a zero-mean Gaussian random variable, with the power given by the variance of its amplitude distribution. The noise can come from a variety of sources but for frequencies above 2 GHz, the dominant noise source is thermal noise in the receiver circuitry and antenna system. The noise contribution from these sources is characterized by the *noise figure*. A perfect device that contributes no noise to the system has a noise figure of 1 or 0 dB. For other values of noise figure, the noise power density η (variance of the Gaussian distribution) is given by

$$\eta = (F - 1)K_b T_0 \quad \text{(Watts/Hz)} \tag{8.2}$$

where F is the noise figure, K_b is Boltzmann's constant $= 1.37 \times 10^{-23}$, and T_0 is the ambient temperature in degrees Kelvin of the environment in which the device operates; it is usually taken to be 290°. The noise of the device can also be characterized by its effective *noise temperature* T_e, which is related to noise figure by

$$T_e = (F - 1)T_0 \tag{8.3}$$

For very low noise devices such as cryogenically cooled amplifiers, the effective noise temperature is often the preferred way to characterize the noise contributed by the device.

The noise power density given by (8.2) can be used to find the total noise power N at the input of the receiver in a given bandwidth by simply multiplying by that bandwidth B

$$N = (F - 1)K_b T_0 B \tag{8.4}$$

The bandwidth used here is called the *equivalent noise bandwidth*, a uniform rectangular bandwidth that produces the same noise power as the actual receiver filter bandwidth. The noise figure in (8.4) is the noise figure for the entire receiving system, although the first stage of a receiver is usually the most significant contributor to the total system noise. Using the manufacturer's noise figure for the receiving equipment for the relevant bandwidth provides the system *noise floor* via (8.4). A 6-MHz MMDS system with a 4-dB noise figure will have a noise floor of −134.4 dBW or −104.4 dBmW. Link thermal fade margins are calculated by comparing the mean received signal power with the noise floor power.

The signal, interference, and noise powers in a receiver system, and the resulting SINR value, are normally referred to the input terminals or connector of the receiver. The transmission line loss is taken into account by appropriate attenuation of the signal and interference powers. Alternately, the SINR calculation could be referred to the antenna terminals in which case the transmission line loss adds directly to the noise figure and calculated noise power, but the signal and interference are not attenuated by the transmission line loss. The two approaches result in the same value of SINR.

Since the transmission line loss directly decreases the signal-to-noise ratio (SNR) [but not signal-to-interference ratio (SIR)], it is important to reduce its length as much as practically possible for traditional wireless systems in which tower-mounted antennas are some distance from the transmitter or receiver equipment. As described in Chapter 6, recent equipment developed for LMDS and other fixed broadband systems designed for building rooftop installation have outdoor units (ODUs) in which the transmitter power amplifier or receiver preamplifier modules are directly connected to the antenna, thus eliminating the radio frequency (RF) transmission line or waveguide entirely.

The noise power is just one moment (the variance) of the noise amplitude distribution. As shown in Section 7.4.1, errors in signal detection occur when the noise amplitude exceeds a relative signal level depending on the modulation type. It is the noise amplitude distribution, then, not just the variance that determines when errors occur. Thermal noise amplitude is modeled as a Gaussian distribution; atmospheric and manmade noise have different, more impulsive, amplitude distributions that result in different error values for the same average noise power.

8.2.1.2 Cochannel and adjacent channel interference

The interference term I in (8.1) is a random process that is the sum of the cochannel and adjacent channel interference

$$I = \sum_{n=1}^{N} I_{co,n} + \sum_{m=1}^{M} I_{adj,m} \tag{8.5}$$

where N is the number of cochannel interferers and M is the number adjacent channel interferers. The amplitude of the interferers I_n and I_m may be relatively constant most of the time as with LMDS systems using narrowbeam antennas, or they may be fading signals that are best described as compound Rician or Rayleigh fading processes with lognormal-distributed mean values (see Chapter 4). Such descriptions for interference are applicable to non-line-of-sight (NLOS) fixed broadband systems.

As mentioned in the introduction to this chapter, the interfering signals can occupy the same bandwidth slot as the desired signal S or a bandwidth slot that overlaps the desired signal bandwidth slot to some extent. This situation is illustrated in Figure 8.3 in which the receiver filter shape and interfering PSD are depicted. The amount of interference intercepted by any arbitrary receiver bandwidth from any arbitrary interfering PSD can be found by integrating the overlap area. This approach is sufficiently general that it can be used for either cochannel or adjacent channel interference; indeed these descriptions of interference become somewhat meaningless when considering the wider variety of possible desired receiver bandwidths, center frequencies, and interfering PSDs than may be used within a given network.

The mean power from each interferer is first calculated at the terminal of the victim receiver taking into account the interferer's radiated power in the direction of the victim receiver, the total mean propagation path loss, and the victim receiver antenna gain in the direction of the interferer. That power is then further reduced by the extent to which the receiver filter reduces the intercepted power based on the interferer's PSD. The net mean interference power P_i resulting from this process, is given by

$$P_i = P_T + G_T(\varphi_T, \theta_T) + G_R(\varphi_R, \theta_R) - L_p - A_F - A_S \text{ dBW} \tag{8.6}$$

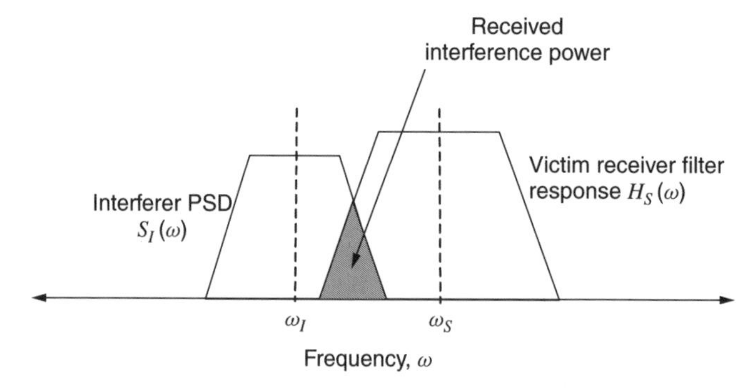

Figure 8.3 Received interference for arbitrary desired and interfering frequencies and occupied bandwidths.

where

P_T = transmitter power in decibel-watts.
$G_T(\varphi_T, \theta_T)$ = the gain in decibels of the transmitting antenna in the direction of the victim receiver where φ_T is the horizontal plane angle (azimuth) and θ_T is the vertical plane angle (elevation).
$G_R(\varphi_R, \theta_R)$ = the gain in decibels of the victim receiving antenna in the direction of the interferer where φ_R is the horizontal plane angle (azimuth) and θ_R is the vertical plane angle (elevation).
L_p = mean path loss in decibels, which includes free space loss, clutter loss, reflections, diffraction, and multipath.
A_F = the attenuation factor in decibels due to the receiver filter rejecting some or all of the interfering power.
A_S = the attenuation in decibels due to other system losses such as transmission line or waveguide losses, radome losses, and so on

The transmitting power level, antenna gains, and mean path loss will be specific to each interferer. The system losses A_S will be common to all received signals including the desired signal. From Figure 8.3, the receiver filter attenuation A_F is given by

$$A_F = 10\log\left[\frac{\int_{-\infty}^{+\infty} H_S(\omega-\omega_s)S_I(\omega-\omega_I)\,d\omega}{\int_{-\infty}^{+\infty} S_I(\omega)\,d\omega}\right] \tag{8.7}$$

where $H_S(\omega)$ is the desired signal receiver filter response center on frequency ω_s and $S_I(\omega)$ is the PSD of the interfering signal centered on frequency ω_I. As a practical matter, (8.7) is usually performed as a numerical summation using receiver filter response and transmitter PSD curves provided by the equipment manufacturer. Examples of some actual transmitter PSD and receiver filter response curves are shown in Figures 8.4 and 8.5, respectively.

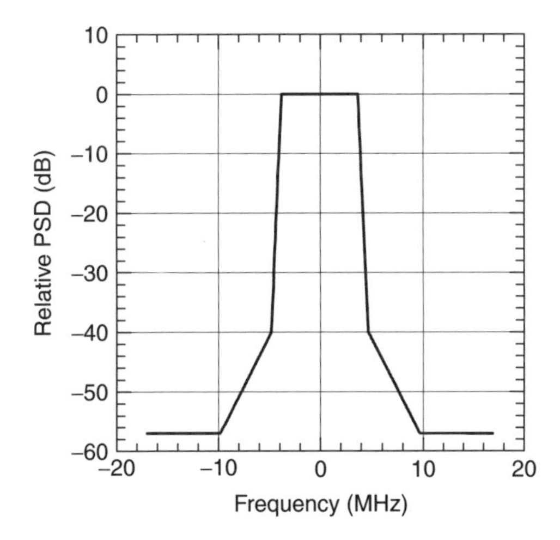

Figure 8.4 Example of a transmitter power spectral density (PSD) envelope.

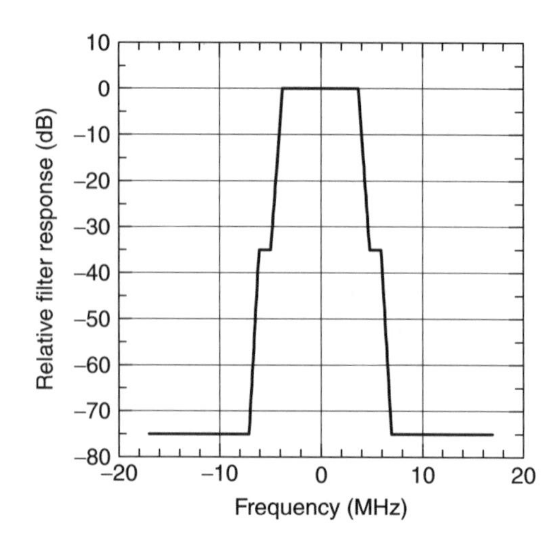

Figure 8.5 Example of a receive filter response envelope.

These curves are the envelopes of PSD and receive filter responses for LMDS equipment that are developed from the measured responses of sample equipment taken from the production process.

Since the receive filters and transmitter power spectra are not ideally band-limited, FDMA systems require a *guard band*, which is usually provided for by making the allocated channel bandwidth wider than that necessary to accommodate the intended signal spectrum width. *Spectrum masks*, similar to the PSD curve in Figure 8.4, define the required transmitter power limits as a function of frequency. If the administrative or industry standard spectrum masks are defined, they can also be used for interference calculations instead of measured equipment performance data.

8.2.1.3 Multiple interferers in LOS networks

As discussed in Chapter 7, the amplitude of the interference itself can actually be subject to fading due to changing propagation path conditions, or other time variations such as power increases resulting from transmitter power control action when the interfering link itself is being subjected to some fading event like rain. The statistics for the variations that might be expected are generally not known. The attempts to model variations in interfering signal amplitudes as Rayleigh or Rician are approximations based on observations in mobile radio systems that can be significantly wrong for line-of-sight (LOS) fixed broadband systems. Moreover, as indicated in (8.5), the total interference can actually be the sum of several interferers, each with different mean values from (8.6) and different statistics describing the amplitude variations.

Owing to the lack of specific information about the fading properties of the interferers, the standard approach is to simply consider the interferers as additional system noise and by doing so, approximate their envelopes as Rayleigh-distributed like Gaussian noise. For

M interferers of arbitrary center frequency and PSD, the denominator in (8.1) is then given by

$$I + N = \sum_{m=1}^{M} P_m + P_N \tag{8.8}$$

where P_m is given by (8.6) for each interferer. For power changes that may be predicted, such as power increases due to interferer transmitter automatic power control (APC) action, the mean interference power is increased accordingly and the SINR calculation in (8.1) is performed under these conditions. Of course, the desired system transmitter APC may also be capable of responding to the increased interference. *Link coupling* interactions of this type will be considered in Chapter 11.

The quality of the approximation in (8.8) where interferers are considered as additional noise will be increasingly poor as the interfering power is less and less 'noiselike'. For multiple fading interferers of comparable amplitude, modeling the combined interference as additional noise power is a reasonable approach. The worst case for this approximation occurs when the total interference consists of a single nonfading interferer. Even if the multiple interferers are present, if a single interferer dominates the others by several decibels, the envelope amplitude of the combination will be more Rician than Rayleigh.

The worst-case quality of the approximation in (8.8) can be examined using the analysis presented in Chapter 7 for the error rate performance of quadrature phase shift keying (QPSK) and 16QAM signals being detected in the presence of a single sine wave interferer plus Gaussian noise. Consider two cases for QPSK: one has noise and interference powers of equal amplitude and is 10 dB weaker than the desired signal and the second has the interfering signal removed and the noise power increased to 7 dB weaker than the desired signal. The second case is equivalent to considering the single interferer as additional noise as given in (8.8). The error probability P_e in the first case from Figure 7.13 is about 2.5×10^{-3} while in the second case the error probability from Figure 7.7 is about 1.6×10^{-3}. The results from a similar comparison for 16QAM at SNR = 15 dB and SIR = 15 dB level using Figures 7.10 and 7.8 are included in Table 8.1.

From this comparison, using the noise approximation for interference provides P_e values that are close to or higher than the correct P_e values for the single sine wave + noise case. Similar comparisons done numerically using the equations in Chapter 7 show that the 'all noise' approximation provides a worst-case estimate of P_e for SNR and SIR values of practical interest, that is, values that would be applied to realistic system designs. Therefore, using the noise approximation in (8.8) for combined interference constitutes a conservative approach to system design.

Table 8.1 P_e for the noise approximation of interference for C-QPSK and 16QAM

Modulation	SNR,SIR level	Single sine wave + noise	All noise approximation
C-QPSK	10 dB	2.5×10^{-3}	1.6×10^{-3}
16QAM	15 dB	1.0×10^{-4}	7.0×10^{-4}

8.2.2 Spectrum utilization

Pure FDMA systems have reserved channels for the various communication links in the network. This is efficient spectrum utilization only if the service on the link requires full-time access, as may be the case with 'real-time' services such as video or audio transmissions or building-to-building business data links. However, even these signals can be stored and delayed so that time-sharing the channel with another service is still possible. From the perspective of the end user, the service still appears to be uninterrupted. The memory capabilities of the transmitting and the receiving equipment accessing the channel therefore become an important factor in determining the flexibility in choosing a multiple-access scheme. The delay that can be tolerated, and the somewhat nebulous definition of 'real time', also play a role. Clearly, an entire sporting event could be stored and transmitted incrementally on a shared channel so that it appears as a continuous real-time transmission, but the event, or critical parts of the event, may have finished long ago, thus destroying the immediacy of what the end user experiences.

The issues associated with service type and when immediate transmission versus deferred transmission is acceptable are discussed in more detail in Chapter 9. For the present purpose, it is sufficient to say that with current technology, it is a very rare service that truly requires full-time continuous access to a communications channel. In recognition of this fact, alternative schemes such as TDMA and CDMA have been developed, which can achieve greater spectrum utilization for most services through shared multiple access to the frequency band without deteriorating the apparent real-time immediacy of the end user experience.

8.3 TIME DIVISION MULTIPLE ACCESS (TDMA)

Within a specific radio channel, multiple users can be provided with service if the times during which transmissions are sent and received by each user are controlled and coordinated by the system so that interference conflicts do not occur. With TDMA systems, access to the radio channel is typically divided into TS that are then assigned to individual users, much as full-time radio channels would be assigned to users in a purely FDMA system. The TDMA approach is illustrated in Figure 8.6. Each radio channel is divided into TS that are synchronized within a radio channel on a given base station or hub, but are not necessarily synchronized from channel to channel or from hub to hub on a given channel.

As a second-level multiple-access scheme, TDMA provides another degree of freedom that allow greater flexibility for accommodating service types. For example, a user that requests a service that has high data rate demands could be assigned multiple TS on a given channel, or across multiple channels, to increase the total data throughput for information transmitted to that user, up to full usage of all available TS with the concomitant denial of service to other users. Flexing the number of TS assigned to a user on the basis of requested service is the approach employed in General Packet Radio System (GPRS) and Enhanced Data rates for GSM Evolution (EDGE) cellular radio systems. It is another form of link adaptation mentioned in Chapter 7.

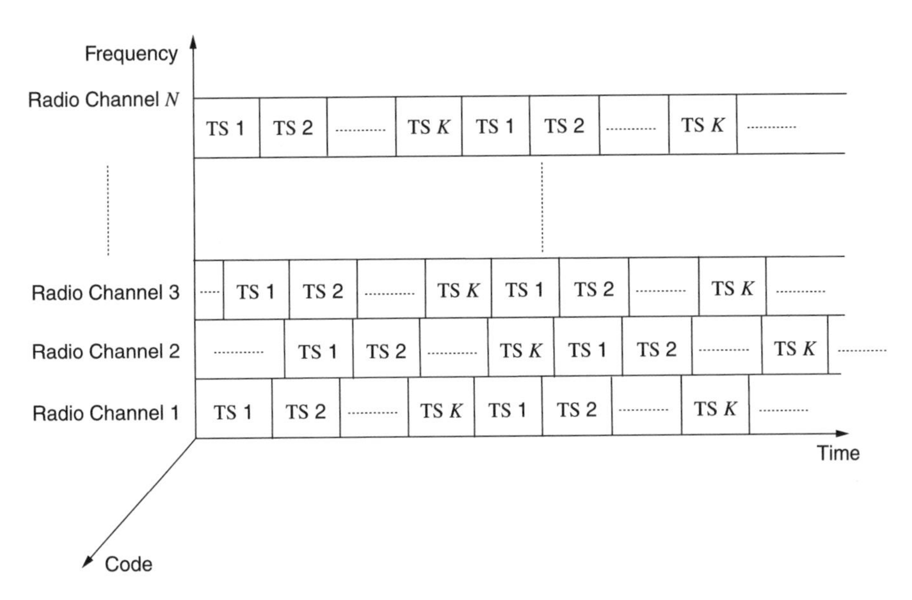

Figure 8.6 Time division multiple access (TDMA) with schematic of time slots (TS).

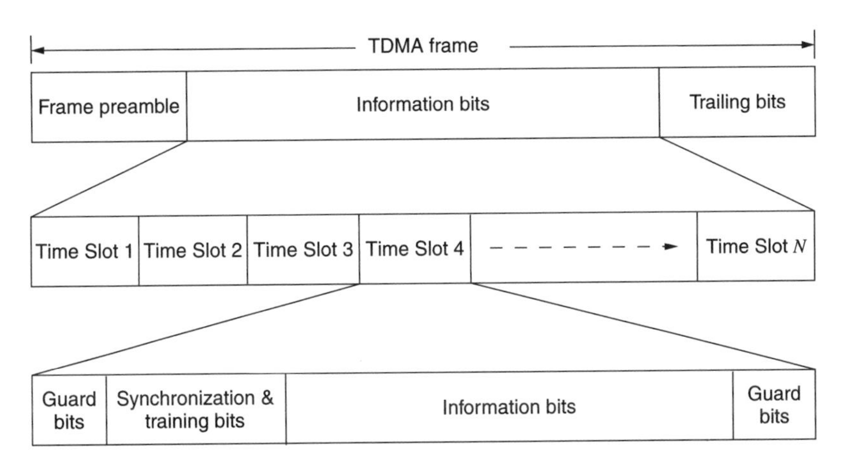

Figure 8.7 Generic TDMA frame and time slot structure.

The TS shown schematically in Figure 8.6 are actually grouped together in *frames* to ensure that they are coordinated with one another. Figure 8.7 shows a generic TDMA frame with preamble and trailing bits. The information frame is divided into TS that are self-contained *bursts* of data including training and synchronization bits. The training bit sequence is known to the terminal and used to train the equalizer as discussed in Section 7.4.4. Depending on the modulation type, these bits may also be used to establish a synchronous reference to facilitate data symbol detection.

Essentially any modulation type may be used with TDMA to provide the required data rate within the constraints of the link budget and design SINR. The GSM cellular system is a TDMA system that uses Gaussian minimum shift keying (GMSK) modulation, while EDGE, the higher data rate 2.5G version of GSM, uses 8PSK for a higher bits/symbol ratio. 16QAM or 64QAM are used in some forms of TDMA for fixed broadband systems.

The generic TDMA frame structure shown in Figure 8.7 may vary from system to system depending on what is to be accomplished, but all TDMA systems must dedicate a certain portion of the transmitted symbol stream to synchronization and timing and guard bits or symbols to prevent overlap of consecutive frames or consecutive TS. The overhead associated with these functions is in addition to the overhead for coding described in Section 7.5. The TDMA overhead bits will be system-specific, but can range from 10 to 30% of the total data throughput capacity. Therefore, overhead in TDMA systems is an important consideration in assessing the information payload that can be delivered.

The term 'channel' is most commonly applied to a radio channel with a particular center frequency and occupied bandwidth as with an FDMA system. However, a sequence of TS utilized by a terminal is also referred to as a *channel* in the rudimentary sense of the physical mechanism that conveys the communicated information. In system protocol nomenclature, the frequency and time physical channels are referred to as the physical channel layer or *PHY* layer.

Frequency and time physical channels are also filled with information that make up different 'logical' channels. Logical channels make up part of what is called the *medium-access control* or *MAC* layer. The MAC layer logical channels and protocols are properties of the system standard. While the MAC layer is important for messaging and services support, fixed broadband wireless system design and planning primarily involves the successful operation of the radio or PHY layer. For this reason, except in a few specific areas, most aspects of the MAC layer design will not be dealt with in this book.

8.3.1 TDMA intercell interference

From Figure 8.6 it is clear that within each radio channel the assignment and use of the TS by each terminal is coordinated so that the signals arriving from each terminal fall in the correct TS. This requires the hub to send a signal to each terminal on a given channel that assigns a *timing delay* or advance to that terminal, so the burst transmission from each arrives at the correct time. The required delay will be largely a function of the distance from the hub to the terminal, which can be deduced from the round-trip transmission delay. Multipath signals may introduce some uncertainty in this process. So that the signal from each terminal occupies its assigned time slot, the timing adjustments τ_k are used as shown in Figure 8.8 for a multicell system.

The time delay adjustments are made at each hub to prevent overlap of TS used by the terminals with which it is communicating; however, this synchronization does not extend to the timing of TS used on the same radio channel at nearby cells. The downlink interference from neighboring hubs at a remote terminal can be readily calculated since the hub position and path loss to a terminal location are known, and it is reasonable to assume that all TS on a radio channel are going to be occupied. For the uplink interference, however, the interfering signals on a radio channel are originating from remote terminals

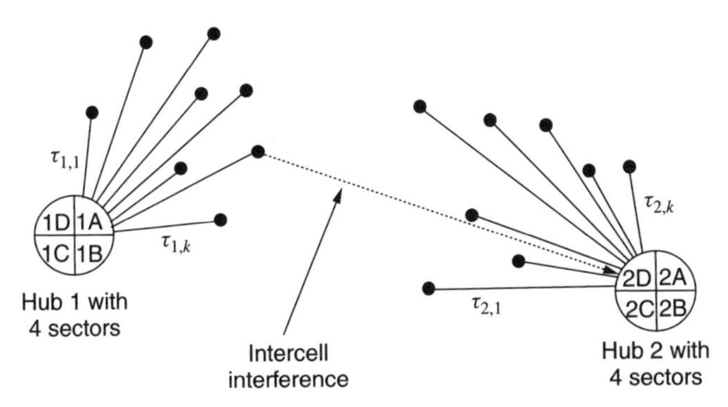

Figure 8.8 Remote terminals in a TDMA system with time delays τ_k adjusted so that signal arrives in assigned time slot at hub sector. Intercell uplink cochannel interference is approximated by full-time transmission from worst-case terminal interference from neighboring cell.

at different locations using assigned TS that are usually not fixed assignments but made as the traffic dictates. It is therefore not feasible to design the network, assuming the interference on a given uplink time slot will be originating from a particular terminal.

If the uplink radio channel assignment, but not time slot assignment, is known for terminals in a neighboring cell, the uplink interference on that channel could be calculated by averaging the interference from each terminal on that channel in each neighboring cell. This requires that the path loss from each terminal in each neighboring cell to the victim hub receiver be calculated as

$$I = \frac{1}{N_1}\sum_{i=1}^{N_1} P_{\mathrm{i}} + \frac{1}{N_2}\sum_{i=1}^{N_2} P_{\mathrm{i}} + \cdots + \frac{1}{N_m}\sum_{i=1}^{N_m} P_{\mathrm{i}} \tag{8.9}$$

where $N_{1\cdots m}$ are the number of TDMA terminals in each of the neighboring hubs 1 through m using the uplink radio channel under study. The power of each of the interference terms P_{i} are calculated using (8.6).

The weakness of this approach is that the terminal with the highest interfering power P_{i} conceivably could be actively transmitting for many or all of the channel TS, so worst-case interference essentially exists all the time. The resulting net interference is larger than the average. The conservative approach, therefore, is to use the interference from the worst-case terminal in each neighboring cell and assume that this terminal is actively transmitting during all TS.

$$I = P_{1,\max} + P_{2,\max} + \cdots + P_{m,\max} \tag{8.10}$$

where $P_{1,\max}$ is the uplink interfering power from the worst-case terminal operating in neighboring cell 1.

Interference in TDMA systems can be further mitigated by applying a technique known as *frequency hopping* (FH). FH is one method for creating a spread spectrum communications system that will be discussed in a later section. Its improvement value for TDMA systems can be appreciated from the foregoing discussion of interference levels on the uplink, and by extension, on the downlink. If the TDMA system uses a set of TS taken from different radio channels, the average interfering power on any given radio channel is reduced. FH with TDMA is an option that is available with GSM cellular systems that can potentially reduce the average interference levels by 2 to 3 dB.

8.4 CODE DIVISION MULTIPLE ACCESS (CDMA)

CDMA uses spread spectrum techniques and their inherent interference immunity to achieve access to the spectrum by multiple users. Spread spectrum systems take an information data stream, including coding and interleaving, and multiply it by a pseudorandom noise (PN) code sequence at a data rate that is much higher than the rate of the information data stream. This process, in effect, spreads the signal over a bandwidth that is much greater than the bandwidth of the information signal.

By coding the signal with a PN sequence and spreading the signal energy over a broad bandwidth, upon despreading and demodulation, both single carrier (SC) interference from outside the system and other spread spectrum signals from within the system can be effectively rejected. The result is a postdetection SINR that is adequate to achieve a low bit error rate (BER) even though the radio carrier SINR at the receiver antenna terminals is usually less than 0 dB. The bandwidth spreading can be achieved in two ways – frequency hopping FH and direct sequence (DS). Both approaches are discussed below.

Spread spectrum systems have recently seen more widespread interest and use for fixed broadband communications, especially for NLOS systems intended to serve residences and for license-exempt bands such as the IEEE 802.11b band at 2.4 GHz where spread spectrum techniques must be used according to FCC Rules. Note that 802.11b specifies direct sequence spread spectrum (DSSS) but this is not a CDMA system since there is no coordination among the codes used by the various network users. The same codes are used by all users and the channels are shared on a time domain basis [see Carrier Sense Multiple Access (CSMA) in Section 8.6].

CDMA is used for simple point-to-point links in both licensed and license-exempt bands. CDMA technology is also potentially well suited for point-to-multipoint NLOS systems that have system design issues that are quite similar to mobile cellular CDMA systems. With a variety of possible applications for CDMA, and with the differences in the CDMA systems that are now being deployed for fixed broadband wireless, the following discussion of CDMA systems is intended to be a generic treatment of the principles involved. However, certain CDMA system types have emerged at this time as leading contenders for use in NLOS networks such as the 3G UMTS W-CDMA standard, which includes both FDD and TDD implementations. The capabilities of W-CDMA as applied to the particular task of fixed broadband system construction will be covered in a following section.

8.4.1 Frequency-hopping spread spectrum (FHSS)

With frequency-hopping spread spectrum (FHSS), a pseudorandom sequence is used to select one of perhaps thousands of frequencies within the available spread spectrum bandwidth. At any given time only a few of the available frequencies are being modulated by the signal data. Since the spreading bandwidth is much greater than the signal bandwidth, the instantaneous occupied bandwidth of the hopped-to frequency is much less than the total spreading bandwidth. Simple frequency shift keying (FSK) modulation can be used to modulate the carrier. Like Orthogonal Frequency Division Multiplexing (OFDM), because the modulated rate of this carrier is low, it generally is within the coherence bandwidth of the channel and subject to narrowband rather than wideband fading. As such, it provides some immunity to multipath as well as providing frequency diversity. Also like OFDM, it is possible that a given hopped-to frequency will be in a narrowband fade so that the resulting detection of that transmitted symbol (or symbols) results in an error. Therefore, for similar reasons, it is essential to use interleaving and coding to achieve an adequately low error floor.

When the rate of hopping between frequencies occurs at a rate greater than the symbol transmission rate (multiple hops per symbol), the system is referred to as *fast frequency hopping*. If the system hops at a rate that dwells on a frequency for one or more symbol times, the system is referred to as *slow frequency hopping*.

For successful detection, of course, the receiver must have the same random code that determines the hopping sequence so that it can track the hops and correctly detect the signal. Other signals that do not follow this particular pattern are reduced to noise in the detection process. By using multiple random codes, multiple users can share the same spread spectrum channel with each being separately detected among the others by the particular code sequence it uses. In this way, FH-CDMA multiple access to spectrum is accomplished.

Compared to DS systems, FHSSs have not been as widely deployed, although their operation and potential advantage are essentially duals of the advantage available in DS CDMA (DS-CDMA). DS is the approach used in second-generation cellular CDMA systems (IS-95), and in third-generation cellular systems (cdma2000 and UMTS W-CDMA). The wealth of real operational experience and associated large production quantities of the necessary chip sets has made DS-CDMA much more popular than FH-CDMA.

8.4.2 Direct sequence (DS) spread spectrum

In DS-CDMA, the information signal is multiplied by a higher data rate (broader bandwidth) spreading signal so that there are multiple spreading symbols per data symbol. A block diagram of a DS spread spectrum transmitter and receiver is shown in Figure 8.10. The spreading signal is a PN code sequence that is unique to a particular user on a particular hub, thus allowing a separation of the signal at the receiver from other signals in the code domain by once again multiplying by the same unique PN code assigned to that user. The codes can be thought of as a dimension of the spectrum along with frequency and time as shown in Figures 8.1, 8.6, and 8.9. As shown in Figure 8.9, one or more frequency channels may be used for a DS-CDMA system, depending on the data requirements and the available channels. Dividing the data to be transmitted among several CDMA carriers with smaller bandwidth rather than using one CDMA carrier with a wider bandwidth may fit better when a large contiguous block of spectrum is not available.

The transmitted signal is usually binary phase shift keying (BPSK) or QPSK with a symbol rate of R. This signal is multiplied by the binary spreading code with symbol rate

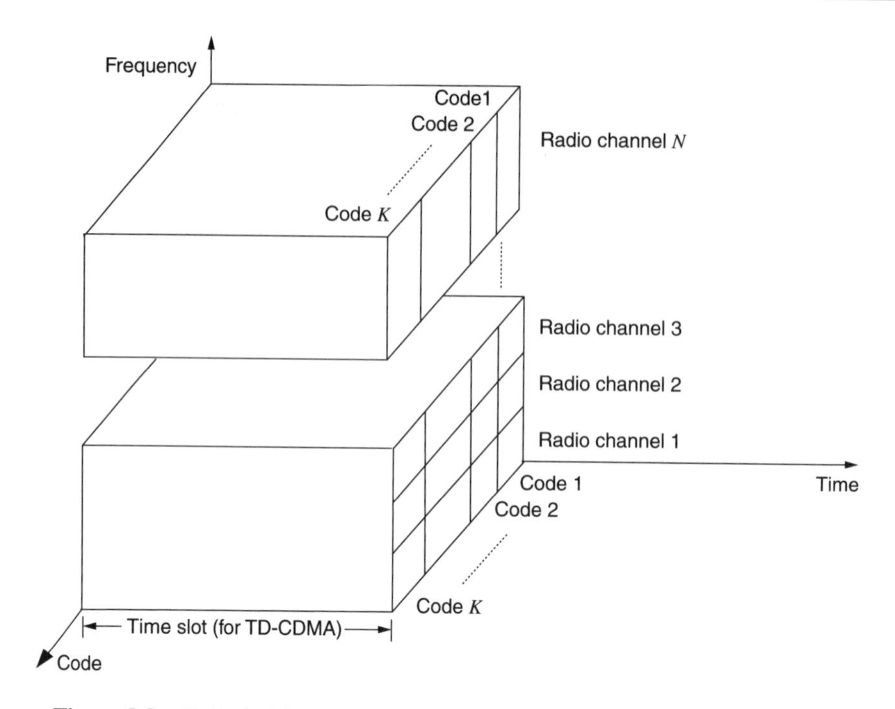

Figure 8.9 Code division multiple access (CDMA) on multiple radio channels.

W, resulting in a transmitted signal symbol rate at the spreading code rate and an occupied bandwidth essentially the same as the spreading code bandwidth (provided $W \gg R$). The transmitted symbols are known as *chips* with the rate of the spreading code called the *chipping rate*. The ratio of the chipping rate to the data rate, W/R, is called the *spreading factor* (also called *processing gain*). The spreading factor is one fundamental indication of the degree of processing gain in the system, and consequently, the degree of interference rejection (multiuser access) that can be achieved.

As stated in the introduction to this chapter, the core issue with all multiple-access schemes is their ability to operate in the presence of interference. The following sections will present standard interference calculations for CDMA systems. As will be shown, CDMA systems generally rely on careful control of the interference levels from all terminals communicating within a given hub on a radio channel; specifically, best (unbiased) detection of the uplink signals from all terminals connected to a hub is realized when all are received with equal power. This means that CDMA systems must have the ability to control the transmitting power of the remote terminals so that the signals received from terminals close to the hub do not dominate the signals received from more distant terminals. This is the so-called *near–far* problem in CDMA systems. It is dealt with using effective APC that may need a dynamic range of 80 dB or more to equalize the effects of wide path loss differences that can occur with near and far terminals connected to the same hub. The use of advanced techniques such as joint detection (JD) (see Section 8.4.5) can relax the demands on the uplink APC adjustments.

8.4.3 Downlink interference calculations

As with the other multiple-access techniques, the ability of CDMA to support multiple simultaneously communicating terminals depends on the impact of interference from other cells and from the home cell of a given terminal. The composite incoming signal consisting of the desired signal, the interfering signals, and the receiver noise is multiplied by the known spreading signal for the desired signal as shown in Figure 8.10. The signal that was spread by the desired code is recovered and the interfering signals remain as noiselike power across the spreading bandwidth. The SINR can then be calculating by assuming that the interfering signals are additional noise and the desired signal power has been increased by the processing gain or spreading factor W_d/R_d.

The codes used for the downlink and the uplink can be different, so the approach to finding the postdetection SINR is calculated somewhat differently. For the downlink or the forward channel, as it is often referred to in CDMA, the transmissions to all remote units can be synchronized since they are originating at one location and pass over the same channel to each remote unit. Consequently, the spreading codes used are orthogonal codes, sequences of binary pulses that are orthogonal (their inner product is zero). Two types of orthogonal codes are *Walsh functions* used in IS-95 CDMA and orthogonal variable spreading factor (OVSF) codes used in 3G W-CDMA. When the incoming composite signal plus interference is multiplied by the desired signal spreading code, the desired signal is recovered, and ideally, the product of the desired Walsh code with each of the interfering signals is zero. In practice, multipath distortion on the interfering signals to some extent destroys the orthogonality property of the interference, so the interference contribution to the final SINR is not zero.

To establish terminal connections to the network, the hubs typically transmit four kinds of signals – pilot, traffic, paging, and sync. These are referred to as *channels* in the sense of logical rather than physical channels. The pilot signal is detected by the remote terminal to establish a connection to the network, while the traffic channel actually carries the information to be communicated. The ability of the terminal to connect to the network depends on the SINR, or E_c/I_0, of the pilot channel. The quality of the received

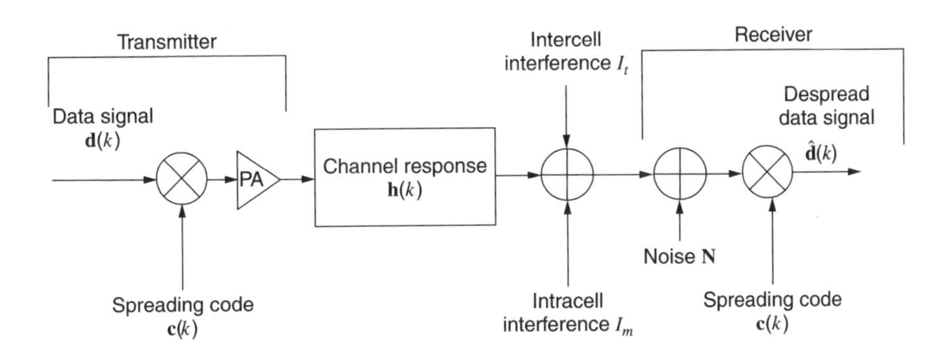

Figure 8.10 Discrete time diagram of a CDMA transmitter and receiver with transmission of block data vector $\mathbf{d}(k)$ for user k.

data depends on the SINR, or more specifically, the E_b/N_0, of the recovered traffic channel data.

8.4.3.1 Downlink pilot channel E_c/I_0

The pilot channel E_c/I_0 as received at a terminal is given by [5]

$$\frac{E_c}{I_0} = \frac{S_R}{I_h + I_o + I_m + I_t + I_n + N_t} \tag{8.11}$$

where

I_h = downlink overhead (sync and paging) power from home hub
I_o = downlink total overhead power (sync and paging) from other hubs
I_m = downlink total traffic channel power from the home hub
I_t = downlink total traffic channel power received from other hubs
I_n = external interference for all other sources such as adjacent channel interference or spurious interference from other transmitters
N_t = receiver thermal noise power
S_R = desired signal pilot channel power received at the remote terminal. This will be some (adjustable) percentage of the total power transmitted by the hub.

The 'home hub' refers to the hub with which the remote terminal is communicating. Since fixed broadband systems do not need to provide for handoff capabilities during transmission, the home hub will normally remain the same hub. The interference quantities listed above are calculated as the sum of the interfering powers as given by (8.6), taking into account the antenna gains and radiated powers in the directions from the hub to the terminal and vice versa. In CDMA systems, the pilot signal is not despread, so E_c/I_0 is usually less than 0 dB (by design, typically −12 to −16 dB for IS-95 CDMA cellular systems).

Like the desired signal S_R, the interference terms in (8.11) are subject to fading due to changing propagation conditions so that E_c/I_0 is also a fading quantity that is properly described in statistical terms. As discussed in preceding chapters, the extent of fading will depend on the propagation environment and the directivity of the antenna used at the remote terminal.

8.4.3.2 Downlink traffic channel E_b/N_0

The downlink traffic channel SINR per information bit, E_b/N_0, can be calculated in a similar fashion to the pilot channel E_c/I_0:

$$\frac{E_b}{N_0} = \frac{S_t}{(1-\alpha)I_h + I_o + (1-\alpha)I_m + I_t + I_n + N_t}\left(\frac{W_d}{R_d}\right) \tag{8.12}$$

where

α = the orthogonality factor (ranging between 0 and 1), which is designed to take into account the lack of perfect orthogonality due to multipath between downlink signals. If perfect orthogonality is preserved, $\alpha = 1$.

W_d = the downlink CDMA chipping rate
R_d = the downlink symbol transmission rate
S_t = desired signal traffic channel power received at the remote terminal for that terminal. The total percentage of hub power dedicated to traffic to all remote terminals is some (adjustable) percentage of the total power transmitted by the hub. The traffic power for a single terminal will normally be the total traffic power divided by the number of terminals being served by the home hub. For CDMA systems with downlink power control, the power allocated to each remote terminal traffic channel will also depend on distance and path loss to that remote terminal.

The definitions of the other interference terms I are the same as given above. The resulting traffic channel E_b/N_0 can be used with the BER equations in Chapter 7 for the data modulation method used to determine the raw error probability. When coding is used, error corrections at the terminal are made to further reduce the net error rate.

8.4.4 Uplink interference calculations

The uplink from the terminal to the hub does not require a pilot channel, although a sync channel may be used if the CDMA system also uses synchronous uplink transmission. For cellular systems, synchronous uplink transmission is problematic owing to the changing mobile locations and the resulting changes in time delays needed to maintain synchronization. The uplink SNR E_b/N_0 is given by

$$\frac{E_b}{N_0} = \frac{S_h}{(1-\beta)(1-\kappa)I'_m + I'_t + I'_n + N_b}\left(\frac{W_u}{R_u}\right)\nu \tag{8.13}$$

where

I'_m = uplink total traffic channel power from the home hub. This is the sum of the power from the other remote terminals connected to this hub and will therefore vary according to the number of these terminals. In general, the APC will adjust the transmitter power for each of these remote terminals so that the received power from each is the same, and is the same as the received power from the subject terminal. This term is also called *multiple-access interference* (MAI).
I'_t = uplink total traffic channel power received from remote terminals served by other hubs. Again, this will vary depending on the number of terminals being served by these other hubs, their locations (path loss to the home hub), and the APC-adjusted transmitter power levels of each of these remote terminals.
I'_n = external interference for all other sources such as adjacent channel interference or spurious interference from other transmitters.
β = the uplink improvement factor (ranging between 0 and 1), which is intended to take into account the reduction in intracell interference (MAI) from the relative degree of preserved orthogonality due to the use of synchronous uplink spreading codes. If perfect orthogonality is preserved, $\beta = 1$. Note that this factor only applies if the CDMA system uses synchronous uplink transmission and orthogonal spreading codes. The IS-95 CDMA system for cellular systems does not provide synchronous uplink transmission, so β is always 0.

κ = the uplink improvement factor (ranging between 0 and 1), which is intended to take into account the reduction in MAI due to the use of JD (see Section 8.4.5). If JD perfectly suppresses MAI, then $\kappa = 1$.

W_u = the uplink CDMA chipping rate

R_u = the uplink symbol transmission rate

S_h = desired signal traffic channel power received at the hub from the remote terminal. This power will vary according to the APC setting for the remote terminal.

ν = a general factor ranging from 0 to 1 that accounts for imperfect power control on the various terminals involved in the E_b/N_0 calculations. For perfect APC action, $\nu = 1$. For imperfect APC, $\nu < 1$ and the resulting E_b/N_0 is degraded.

The value of uplink E_b/N_0 can be used to find the uplink BER using the equations in Chapter 7.

It should be clear from (8.13) that the E_b/N_0 is primarily a function of how many remote terminals are being served by the hub on this frequency (as represented by I'_m) and secondarily by the number of remote terminals connected to neighboring hubs (as represented by I'_t). As the load on the system increases, the power the remote terminal has to transmit to achieve adequate E_b/N_0 increases until the point is reached when the transmit power cannot be increased further. Service to this remote terminal may then be interrupted. This will occur first in the most distant terminals (or those with the highest path loss to the hub), in effect, shrinking the service range of the hub. The dynamically changing coverage area that results from varying cell load is called *cell breathing*.

The increase in interference at the hub is commonly described by the *noise rise* that is essentially the increase in the interference + noise term in the (8.13) from the no-interference case. The noise rise is given by

$$R_N = \frac{(1-\beta)(1-\kappa)I'_m + I'_t + I'_n + N_b}{N_b} \tag{8.14}$$

The hub or cell *load factor* η is a measure of the extent to which the capacity of the cell and hence network is being utilized. The load factor ranging from 0 to 1 is given by

$$\eta = 1 - \left[\frac{N_b}{(1-\beta)(1-\kappa)I'_m + I'_t + I'_n}\right] \tag{8.15}$$

The load factor is normally given as a percentage. For example, given a particular traffic distribution, the system design (hub layout, channel assignments, antenna types, power, etc.) can be configured so that the load factor is 50 to 75%.

The uplink interference level is also a function of the power transmitted by remote terminals in neighboring cells or intercell interference. If the locations of these remote terminals are not known because they are mobile or at fixed but unknown locations, the path loss values cannot be calculated. The interference levels that are needed for summation of I'_t are not explicitly known. In this case the intercell interference is represented by a *frequency reuse factor* that is usually taken as a fraction of the home cell interference (55%, for example). The frequency reuse factor ξ is defined as

$$\xi = \frac{I'_t}{I'_m} \tag{8.16}$$

Calculating intercell interference using the home cell interference and the reuse factor is a highly approximate approach to finding total interference. However, there are alternatives. For mobile cellular systems in which mobile locations are not known, their location and communication traffic patterns can be simulated through the use of *Monte Carlo* simulation. A Monte Carlo simulation statistically represents the mobiles by guessing at their locations and analyzing the system based on these guesses. The hypothetical locations can be weighted to represent nonuniform traffic distributions throughout the system. This approach will be discussed further in Chapter 11. By using a Monte Carlo simulation for several different random mobile distributions, an overall view of how the system will perform when loaded can be developed.

For fixed broadband systems, the situation is somewhat different in that the location of the remote terminals in the system can be known, at least to the resolution of the address where the system end user has subscribed to the network. Therefore, the opportunity exists to directly calculate the sum of the intercell interference I'_t and therefore achieve a system analysis that is more accurate and that can be dynamically revised as new end users are added.

8.4.4.1 Rake receiver

CDMA terminals use a matched receiver architecture called a *Rake* receiver as illustrated in Figure 8.11. A Rake receiver is essentially a tapped delay line in which the tap delays are equal to one chip symbol time $T_s = 1/W$. As illustrated in Figure 3.6, the received signal includes multipath components to a greater or lesser extent. The objective of a Rake receiver is to isolate multipath energy into resolvable signals spaced at T_s and use those resolved signals as branches in a diversity-combining scheme. The received signal $r(t)$ is sent through a delay line with $L - 1$ delays with delays of T_s. After each delay, the signal at the tap or branch is multiplied by the reference signal $u^*(t)$ (a portion of the desired signal spreading code). This correlation process produces a result that is proportional to the amplitude of the desired signal multipath component. The interfering signals come out of this correlation process as noise. The signal at each Rake tap is then multiplied by a coefficient $C_l(t)$ that depends on the amplitude of the channel energy or SNR in that tap or branch, usually adjusted as a maximal ratio or equal gain combiner. The L branch signals are summed and integrated over the symbol spacing T to yield a decision variable. In this way the Rake receiver collects the multipath energy resulting from the propagation environment and uses it constructively.

Although a Rake receiver is not essential to the operation of CDMA, it does provide useful diversity gain. The amount of diversity gain depends on how multipath-rich the propagation environment is and the chip symbol duration time versus the multipath delay spread. For open environments with no multipath, a Rake receiver provides little diversity improvement. For heavily urban areas, a Rake receiver can provide useful diversity gain.

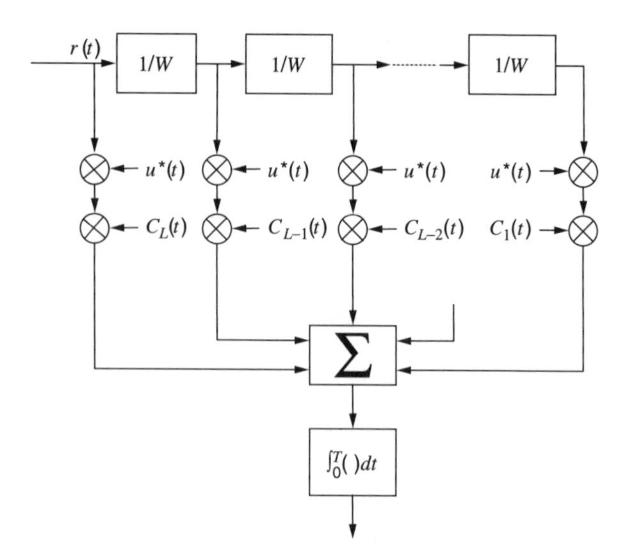

Figure 8.11 Basic rake receiver architecture.

8.4.5 Joint (multiuser) detection

The analysis of the CDMA uplink and downlink E_b/N_0 performance approximated the intracell interference I_m (MAI) and the intercell interference I_t as additional noise in (8.12 and 8.13). For MAI, however, the interference is not random noise but actually consists of signals from other remote terminals served by the home hub. The hub necessarily knows the spreading codes and estimated channel characteristics associated with each of these interfering remote terminals. This information can be exploited to reduce the magnitude of I_m and its impact on E_b/N_0 for the subject user signal. The process of taking advantage of all the uplink signals information arriving at the home hub is called *joint detection* (JD) or *multiuserdetection* (MUD).

The concept of JD as an optimum nonlinear detector for asynchronous CDMA systems was examined in [6] and for synchronous CDMA systems in [7]. These detectors are designed to eliminate MAI but not intersymbol interference (ISI) due to multipath. Since this early work, detectors that suppress both as MAI and ISI have been devised, with the main problem being computational complexity, especially as the number of simultaneous users K increases.

Joint detectors basically function by solving a set of simultaneous linear equations that increase in number as K increases. Joint detectors are therefore most effective (computationally reasonable) when the number of concurrent users is relatively small. It was therefore targeted at TD-CDMA systems such as UMTS terrestrial radio access (UTRA) TDD that are a combination of TDMA and CDMA in which the transmitted data is broken up into time domain frames that are then transmitted on the CDMA spread spectrum channel [8].

Ideally, JD can potentially reduce the term I_m in (8.12 and 8.13) to zero. The capacity improvement (improvement in E_b/N_0) is then given by the reduction in the denominator of (8.12 or 8.13). If the system is loaded to a reasonable level such that the noise and spurious interference terms, and I_n, are small compared to the system interference, then the upper bound on the capacity improvement ΔC, assuming that JD perfectly suppressed MAI, is given by

$$\Delta C = \frac{I_m + I_t}{I_t} \tag{8.17}$$

As mentioned, for rudimentary CDMA network planning it is often assumed that the intercell interference is some fraction of the intracell interference, for example, $I_t = 0.55 I_m$. The resulting potential improvement factor ΔC from (8.17) is thus 2.8. Unfortunately, the 0.55 factor is just a guess based on simple uniform cell service radii and hexagon cell layout patterns. This value is not supported by detailed simulation of realistic CDMA systems, especially those in highly nonhomogenous propagation environments such as urban areas and microcells where the cell service areas are not contiguous. Consequently, the resulting geographical distributions of interfering remote terminals in neighboring cells are not neatly constrained. The 0.55 factor in reality must be replaced with other values derived from simulations of the actual system. These values could range widely with the resulting improvement factor from (8.17) being equally variable. Whatever capacity improvement does result from JD, this greater capacity can be exploited as support for more users at a given data rate or the same number of uses at higher data rates.

In addition, JD can alleviate the tight accuracy requirements on automatic power control due to the fact that the BER is less dependent on the power settings of the individual remote terminals. The term ν in (8.13) can then be set close to one, also improving the net E_b/N_0.

Although JD is a relatively new technique, it is now seeing use in some commercial deployments. One of the first such deployments using TD-CDMA in the UTRA TDD specification has shown that JD can reduce the intracell interference term I_m to negligible levels [9].

8.4.6 CDMA broadband standards

The modulation and multiple-access schemes for fixed broadband wireless systems discussed here are generally not mandated by regulatory standards bodies leaving the industry free to develop technologies that best meet the perceived needs of the marketplace. One exception to this is third generation (3G) CDMA systems that offer data rates of 2 Mbps (higher in some cases) within the framework of worldwide standards. In particular, the 3G UMTS terrestrial radio access (UTRA) standard provides for FDD and TDD systems with chipping rates of 3.84 Mchips in a 5-MHz bandwidth and data rates up to ~2 Mbps (~4 Mbps for TDD). Basic parameters for these systems are shown in Table 8.2. Both FDD and TDD use a common fixed 15 time slot frame structure with each time slot being 2560 chips long.

For UTRA FDD, clearly a data rate of 2 Mbps cannot be achieved with a 3.84-Mcps chip rate, QPSK, coding overhead, and the lowest spreading factor of 4. A technique known as *multicode* transmission is employed in such cases. With multicode, two or more of the codes available on the hub sector are used to send data to a single remote,

Table 8.2 Basic parameters for 3G CDMA standard systems

	W-CDMA		cdma2000	
Parameter	FDD	TDD	1x EV-DO	3x EV-DO
Chip rate	3.84 Mcps	3.84 Mcps	1.2288 Mcps (downlink) 1.2288 Mcps (uplink)	3 × 1.2288 Mcps (downlink) 1 × 3.6864 Mcps (uplink)
Bandwidth	5 MHz	5 MHz	1.25 MHz	5 MHz
Frame length	10 ms	10 ms	5, 10, and 20 ms	5, 10, and 20 ms
Spreading modulation	QPSK (downlink) QPSK (uplink)	QPSK (downlink) QPSK (uplink)	QPSK (downlink) QPSK (uplink)	QPSK (downlink) QPSK (uplink)
Data modulation	QPSK (downlink) BPSK (uplink)	QPSK (downlink) BPSK (uplink)	QPSK (downlink) BPSK (uplink)	QPSK (downlink) BPSK (uplink)
Spreading factors	4–512 (downlink) 4–256 (uplink)	1, 16 (downlink) 1–16 (uplink)	4–256 (downlink) 4–256 (uplink)	4–256 (downlink) 4–256 (uplink)
Multirate	Variable spreading and multicode	Variable spreading and multicode	Variable spreading and multicode	Variable spreading and multicode
Handoff	Soft handoff	Hard handoff	Soft handoff	Soft handoff
Power control	1.6-kHz update rate	1.6-kHz update rate	0.8-kHz update rate	0.8-kHz update rate
Multiuser detection	No	Yes	No	No

thus multiplying the maximum throughput to a remote terminal by the number of codes used. Of course, if multiple codes are used to communicate with one terminal, fewer codes are available to communicate with other remote terminals at various data rates. Multicodes are also used in UTRA TDD.

For UTRA TDD, the TS are assigned as uplink or downlink slots. The number assigned to a particular direction can be varied in response to the traffic load. One of the 15 TS is used as a beacon channel, a second is used for downlink signaling/control information, and a third is used for uplink signaling/control information. The remaining 12 TS are used for data. At least one slot in each frame must be a downlink or an uplink slot. In order to make the JD interference suppression computationally feasible in UTRA TDD, the maximum number of simultaneous codes that can be transmitted in any time slot is 16. If all codes are devoted to one remote terminal using the multicode technique, the total maximum data rate is (3.84 Mchips/16) $\times$ 16(12/15) = 3.07 Msymbols/s. Since the data modulation is QPSK (2 bits/symbol), the raw maximum bit rate is 6.11 Mbps. The overhead due to coding, midambles data, and so on is about 35%, so the net throughput is about 4 Mbps. The FDD mode allows more simultaneous codes but does not allow for JD for intracell interference suppression. The increase in intracell interference will limit the overall throughput in the hub sector.

The UTRA FDD and TDD standards also anticipate higher chip rates that are multiples of 3.84 Mchips, that is, 7.68 Mchips in 10-MHz bandwidth, 11.52 Mchips in 15-MHz bandwidth, and 15.36 Mchips in 20-MHz bandwidth. Depending on the frequency band and channel allocation constraints where the system is to be deployed, these bandwidths and associated chip and data rates may be viable. The equations in (8.11, 8.12, and 8.13) are generally applicable for system design purposes with arbitrary chip rates, data rates, and spreading factors. Further detailed information on the FDD and TDD modes of W-CDMA systems can be found in published papers [8] and in the voluminous 3GPP specification documents [10].

Table 8.2 also shows the basic parameters for the cdma2000 standards. There are many similarities with W-CDMA but also some distinct differences that make W-CDMA and cdma2000 incompatible. The cdma2000 is more closely related to the 2G- CDMA standard IS-95, which was designed to work in the 1.25-MHz channel bandwidth (one-tenth of the uplink/downlink spectrum width allocated for 870-MHz cellular systems in the United States). The cdma2000 standard has chipping rates and channel bandwidths that continue to fit this channelization scheme. Further details on cdma2000 PHY and MAC layer specifications can be found in the documents of 3GPP2 [11].

All of these CDMA multiple-access technologies have been created to operate in an interference-rich environment. This means that the same frequency channel can be used on adjacent cells and adjacent sectors (a frequency reuse of 1). In comparing the total capacity of various multiple-access technologies and their ability to meet the traffic loads expected in a system service area, the interference immunity properties of a multiple-access technique under *loaded* system conditions must be considered to determine its true usable capacity. This issue is taken up again in Chapters 11 and 12 where specific multipoint systems design and channel assignment methods are discussed.

8.5 SPACE DIVISION MULTIPLE ACCESS (SDMA)

Through the use of high gain fixed directional antennas pointed at the intended service destination, SDMA has been the approach used for many years to share microwave radio channels in the same geographic area. The term SDMA has only recently been applied to make more apparent the parallel purpose of SDMA with TDMA and CDMA in achieving enhanced multiple-access capacity by exploiting all the available dimensions of the radio spectrum resource.

Figure 8.12 shows a sketch of a hub with multiple narrowbeam antennas, each serving a small fraction of the total remote terminals communicating with the hub. Since each antenna is providing separate signals for a set of remotes, it can be regarded as a *hub sector*. With sufficient directionality in the hub sector antennas (see Chapter 6) and in the remote terminal antennas, each hub sector can reuse the same frequencies for most terminals without the need for TDMA or CDMA schemes. The processing gain (relative interference rejection) realized is simply the antenna beamwidth at which sufficient suppression is achieved to prevent interference to a remote terminal on an adjacent hub sector so that its SIR is adequate to provide the service-required BER. For example, if the required SIR is 10 dB and the antenna has a 10 dB beamwidth of Φ, then the number of sectors M that can be situated at the hub is

$$M = \frac{360}{\Phi} \tag{8.18}$$

If the 10-dB beamwidth is 10°, M is 36 or a processing gain of about 15 dB without utilizing additional bandwidth as with CDMA or time-sharing the channel as with TDMA.

As the beamwidth of each sector radiator decreases so that it is only serving a single remote terminal, the processing gain increases still further. The simple LMDS antenna

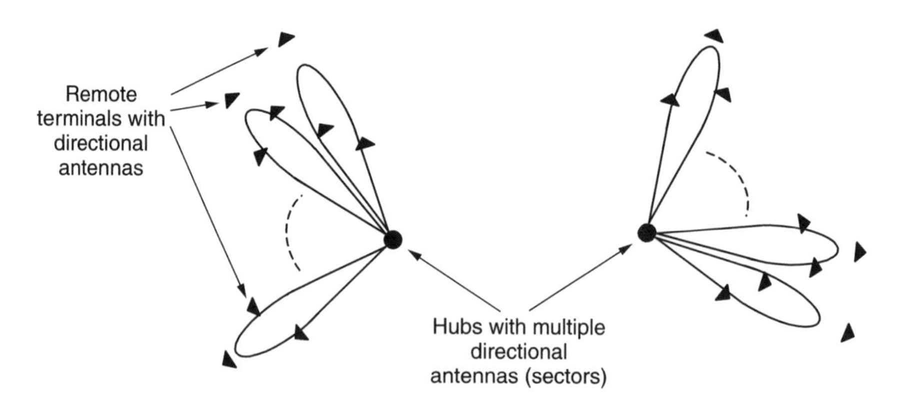

Figure 8.12 Space division multiple access (SDMA) using directional antennas in a point-to-multipoint system to achieve separation of transmissions to/from remote terminals using directional antennas.

whose radiation pattern is shown in Figure 6.12 has a 10-dB beamwidth of about 5°, increasing the processing gain to over 18 dB. The limit of this approach is perhaps represented by free space optics (FSO) in which the laser has a typical beamwidth of about 0.34°, resulting in a processing gain of 30 dB over the omnidirectional case.

These simple calculations are for the 2D horizontal plane (azimuth) case. If the antenna has directionality in the vertical plane as well, and the remote terminals are also distributed vertically as they are in a fixed broadband system in a densely urban area, the vertical plan directionality would make the SDMA processing gain even greater. From these examples, it is obvious that the potential capacity increases available with SDMA are substantial without increasing bandwidth or partitioning transmission time.

Many fixed broadband hubs deployed in the 24-GHz, 28-GHz, or 38-GHz bands are actually collections of point-to-point links using high gain antennas on each end. The 'hub' is simply a building roof or tower top where a portal to an optical fiber or other high-speed backbone link is available, and thus becomes the common terminal for a collection of point-to-point links. Since each antenna is directed to serve a single customer, these antennas are not really hub sector antennas.

The fixed sector antenna approach has been implemented in cellular radio systems as the so-called *switched beam* technology in which the signals from fixed narrow beam antennas are compared to find which has the best signal from the remote terminal (see Section 6.6). For mobile systems, this comparison is periodically done so that the mobile can essentially be tracked as it moves. Suppression of interference from other remote units is basically whatever incidental suppression is provided, given the fixed orientation of the antenna. The fixed beam approach can be replaced with adaptively beam-steered antennas that seek to maximize the signal level from the desired remote terminal. A steered antenna does not explicitly seek to maximize the SIR by suppressing interference as well.

The adaptive or smart antennas described in Section 6.6 use adaptive beam-forming to create a gain maximum toward the desired remote terminal while simultaneously forming pattern gain minima toward cochannel interference sources, thus maximizing SIR. From an achieved SIR and capacity perspective, this action can be modeled as the fixed narrow beam antenna continuously pointed at the other end of the link.

For fixed broadband systems, the problem of adapting the antenna to point at a moving target is eliminated compared to cellular systems. Nonetheless, the propagation environment is constantly changing, even from highly localized effects such as persons moving in a room around the antenna; so to achieve the processing gains possible with SDMA, the antennas must still be capable of adapting to this changing environment.

If the antenna used on the remote terminal end of each link is a low gain antenna that illuminates a wide extent of the propagation environment, the angular spread of the signal arriving at the hub may not be a single direction but span a range of directions due to scattering. The interference source can be scattered in a similar way. The angular spread of the desired and the interfering signals inhibits the ability of the adaptive antenna to maximize the SIR. However, when compared to a mobile system, it is possible to exploit the relatively lower variability of the angular spread in a fixed broadband system with longer sampling periods in the adaptive antennas to arrive at the best SIR results.

The number of discrete interferers that can be suppressed using an adaptive antenna is equal to $N - 1$, where N is the number of antenna elements. For the simple adaptive antennas being designed for use in remote terminals on NLOS system, only 2 or 3 elements are used. If 2 or more strong interferers exist, such simple adaptive antennas may not be able to suppress all the interference and the SIR will not be adequate. This scenario can only be investigated in real terms when a system is fully deployed and fully loaded with remote terminals. It has been the experience with cellular systems that they are designed with an over-abundance of downlink signals in certain service areas (called *pilot pollution* in CDMA systems). Similar design errors in a SDMA system, in which adaptive antenna directionality was relied upon to achieved adequate SIR, could lead to more strong downlink signals than the remote terminal can successfully suppress. Avoiding this situation is an important aspect of designing a fixed broadband system that relies on SDMA and adaptive antennas to achieve multiple access.

8.6 CARRIER SENSE MULTIPLE ACCESS (CSMA)

A straightforward means of achieving multiple access on a radio channel is to establish a protocol in which each remote terminal monitors the radio channel and does not attempt to transmit any data until it detects that the channel is currently idle (no other terminal is attempting to transmit). This approach is quite general and can be applied to form a wide variety of 'ad hoc' wireless networks consisting of user terminals and network access points. This technique is usually called carrier sense multiple access (CSMA) because the network nodes 'sense' the presence of other radio carriers before transmitting.

Of course there are limitations to this approach. The primary problem is that two or more terminals may attempt to transmit at the same time because neither sensed the other first, such that some part of their transmissions overlap. The resulting interference will cause the transmission to be received with errors. When such collisions are possible, the network can remedy the situation in several ways:

- *CSMA/CD*: The 'CD' appended to CSMA stands for *collision detection*. When a terminal detects a transmission collision, it terminates the transmission and waits for an open time to begin transmitting. This approach is used with wired Ethernet networks and is viable on other networks when each terminal can detect the transmissions of every other terminal.
- *CSMA/CA*: The 'CA' stands for *collision avoidance*. This approach is appropriate for wireless networks when all the terminals cannot necessarily detect the transmissions of other terminals. Collisions are avoided by a terminal first inquiring of the receiving terminal if it is acceptable to transmit using a short request to send (RTS) message. If the receive terminal sees a clear channel, it responds with a Clear to Send (CTS) message that also tells others terminals (within detection range) that the channel is busy. When a transmission is received without error, the receiving terminal can send an acknowledgement (ACK) of correct reception or, conversely, a NACK message (negative acknowledgement). This is the basic (though not the only)

approach for CSMA in the IEEE 802.11a and 802.11b wireless network standards. The timing and time-out restrictions in these standards also impose fundamental constraints on their usable range.

Unlike the other multiple-access techniques discussed above, CSMA does not provide for simultaneous use of the radio channels, although it is somewhat like TDMA without reserved time slots and thus, without reserved spectrum capacity for active users.

The drawback of CSMA is that terminals must wait for an open window in which to transmit. That wait gets longer as the number of connected terminals increases. The probability of collisions is directly related to the propagation delay between terminals and the detection delay at a given terminal, and, of course, the number of terminals. As the probability of collision increases, and retransmission (sometimes multiple retransmissions) is needed to ensure correct data receipt, the net data throughput of the network decreases dramatically and the turnaround delay experienced by the terminal user increases.

8.7 MULTIPLE ACCESS WITH OFDM

OFDM (or COFDM in normal use) as described in Chapter 7 by itself is not a multiple-access method. It is subject to cochannel interference from other users in the network that will result in detection errors depending on the type of modulation used for each OFDM subcarrier.

OFDM is used in *single frequency networks* (SFN) in which cochannel interference is anticipated and planned. Correct operation of an OFDM SFN requires exact synchronization of all transmitters in the network as well as appropriate selection of the guard interval so that echo signals (interference) from other transmitters occur within the guard interval and are thus ignored, or occur within the same transmitted symbol time so they can be used constructively in the receiver to improve system performance. However, an excessive guard interval to successfully achieve SFN operation represents idle transmission time that directly subtracts from the overall capacity of the channel. Also, such systems are really not multiple-access systems because the transmitted information from each station in the network is the same.

Since OFDM is only a modulation type, some other process must be included for it to be used with efficient and competitive capacity in a multiple-access network. Five main approaches have emerged to accomplish multiple access using OFDM as the modulation type:

- multicarrier CDMA
- orthogonal frequency division multiple access (OFDMA)
- OFDM with TDMA
- OFDM with CSMA/CA
- OFDM with SDMA.

Each of these will be discussed below.

8.7.1 Multicarrier CDMA (MC-CDMA)

Despite its name, multicarrier CDMA is actually a method of employing OFDM to multiple-access scenarios. With MC-CDMA, the spreading code is used to spread the data symbols across multiple OFDM carriers. The fraction of the data symbol corresponding to a chip in CDMA is transmitted via a single subcarrier. The entire data symbol is therefore transmitted at one time on the entire ensemble of OFDM subcarriers. This necessarily means that the symbol rate on any individual subcarrier is the same as the data rate. Successive data symbols are transmitted in the same way.

An important aspect of OFDM is that the symbol rate on the individual subcarriers must be slow enough, and hence their occupied spectrum must be narrow enough, such that frequency-selective fading does not occur across the spectrum of any subcarrier. If this is not the case, with MC-CDMA the data stream must be further subdivided using a serial-to-parallel converter, with each data symbol from the output of the converter further subdivided to modulate a set of OFDM subcarriers in the usual way as described in Section 7.3.3. A block diagram of this process is shown in Figure 8.13.

The number of subcarriers needed for MC-CDMA depends on the extent to which the serial-to-parallel converter must group data blocks for transmission. If the converter sets up P output bits for transmission, the number of subcarriers required is

$$N = PK_{MC} \tag{8.19}$$

where K_{MC} is the spreading code length. The total occupied bandwidth BW_{MC} is given by

$$BW_{MC} = \frac{(PK_{MC} - 1)}{(T_S - T_G)} + \frac{2}{T_S} \tag{8.20}$$

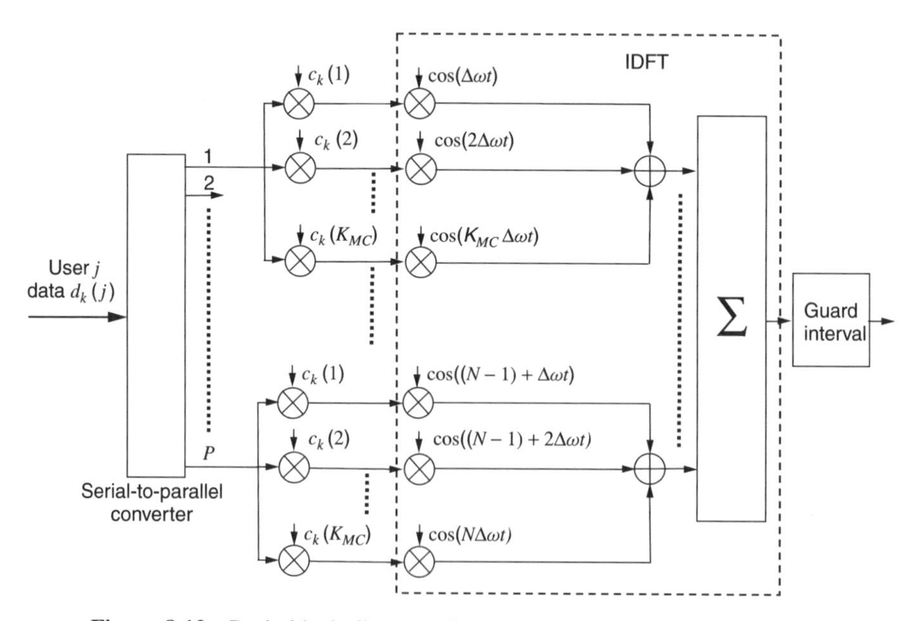

Figure 8.13 Basic block diagram of an MC-CDMA transmitter for user j.

where T_S is the subcarrier symbol duration and T_G is the guard interval. The subcarrier spacing is then $\Delta\omega = 2\pi\Delta f = 2\pi/(T_S + T_G)$.

An analysis of MC-CDMA as compared to DS-CDMA is presented in [12] for several different system designs and multipath scenarios. This comparison arrives at the conclusion that MC-CDMA has no major advantage over DS-CDMA in terms of required bandwidth to achieve the same BER performance. Its ability to reject interference and therefore provide capacity in a multiple-access system design is also the same as DS-CDMA. Like other OFDM systems, however, MC-CDMA does place greater linearity requirements on the power amplifier compared to DS-CDMA, as well as subcarrier stability constraints to avoid intercarrier interference (ICI). Using SC modulation, there obviously are no ICI problems with DS-CDMA.

While MC-CDMA has undergone significant analysis by a number of researchers (see the reference list in [12]), at present it is not a technology that is being pursued for use in fixed broadband wireless system deployment.

8.7.2 Orthogonal frequency division multiple access (OFDMA)

OFDMA is a logical extension of OFDM in which a subset of the available subcarriers is used to carry the data for a given remote terminal. The subcarrier universe is that established via OFDM, but the subset of subcarriers assigned to each user can be (randomly) distributed among the universe of subcarriers. This makes it somewhat similar to FDMA if a user's data were subdivided among multiple subcarriers; however, because of the orthogonal relationship of the carriers in OFDM, it is easier to filter one subcarrier from another, so guard bands and tight filtering as with FDMA are not needed.

From the use of multiple subcarriers, it is an obvious step to further randomize the channel usage and average the interference potential by employing FH among the OFDM subcarriers. In FH-CDMA, the frequencies used in the hopping sequence can be arbitrary as long as the receiver knows what the frequencies are and the order in which they are accessed. With OFDMA, the frequencies are established by the OFDM modulation with the FH sequence used to distinguish the signal of one terminal from another. The multiple-access concept can be further extended to subdivide the spectrum in time. Individual remote terminals are assigned TS as in TDMA during which they are permitted to access the subcarriers to transmit data via their uniquely assigned coded hopping sequence.

As with any FH system, the hopping patterns can themselves be orthogonal; that is, for N frequencies in the systems, N orthogonal hopping patterns can be constructed. OFDMA is one of the modulation/multiple access techniques specified in the draft of IEEE 802.16a (see Section 8.7.6). Because all the subcarriers used on a hub for communication with various remotes are orthogonal, intracell interference is eliminated, subject to phase/frequency distortions from the multipath. The only interference source will be from remotes and hub sector transmitters in other cells. This interference will experience spread spectrum interference averaging over the entire channel bandwidth as normally seen with spread spectrum (SS) systems in general and CDMA in particular.

8.7.3 OFDM with TDMA

Any of the modulation types described in Chapter 7 can be used with TDMA by simply defining a TDMA frame structure with TS and using the selected modulation during each time slot to communicate with separate remote terminals. Using TDMA with OFDM is one of the multiple access/modulation methods included in the draft of the IEEE 802.16a standard (see Section 8.7.6).

8.7.4 OFDM with CSMA/CA (IEEE 802.11a)

This combination of technologies is included here because it is the combination selected for IEEE 802.11a systems operating in the 5-GHz license-exempt band in the United States and Europe. Both OFDM and CDMA/CA were discussed in earlier sections.

The 802.11a standard currently provides for up to 12 nonoverlapping 20-MHz-wide channels, each with a raw data rate of up to 54 Mbps at short ranges. The allowed transmitter power output ranges from 50 to 250 mW for the 8 low band channels intended for indoor communications, and up to 1000 mW on the upper 4 channels intended for outdoor building-to-building communications. Each 20-MHz channel is occupied by 52 OFDM subcarriers resulting in a carrier spacing of about 300 kHz. To achieve the maximum transmission a rate of 54 Mbps, each of the OFDM subcarriers is modulated with a 64QAM signal as described in Chapter 7.

As with 2.4-GHz IEEE 802.11b that uses a DS spread spectrum air interface (not CDMA), the multiple-access scheme in IEEE 802.11a still relies on the relatively inefficient CDMA/CA approach. For a large number of remote terminals attempting to use the network, the net data throughput will decrease and the network response delay will increase.

8.7.5 OFDM with SDMA

Some systems currently being planned for NLOS operation in the licensed bands at 2.5 to 2.7 GHz and 3.5 GHz use a combination of OFDM and adaptive antennas to accomplish multiple access. This combination provides good multipath immunity and interference rejection properties that will potentially allow multiple users to operate on the same frequency on the same hub sector. The relative ability of OFDM to handle cochannel interference as already discussed is modest, so the multiple-access strengths of the scheme rely solely on the ability of the adaptive antennas to enhance the desired signal and suppress interference. As mentioned earlier, several strong interferers will present the greatest challenge to OFDM and SDMA systems. In such cases, with multiple strong interfering signals close by, intelligent assignment of different frequency channels is an appropriate solution. With an NLOS system and remote terminals intended to be self-installed by the end user, the frequency selection process must also be automatic, even with the potential close-by interferers not yet in operation. Frequency assignment strategies of this type are discussed in Chapter 12.

8.7.6 OFDM multiple-access standards

There are at least two standards in which OFDM is currently specified as a modulation type with associated multiple-access methods. The two most relevant to current fixed broadband wireless system design are the IEEE 802.11a standard for the 5- to 6-GHz band

Table 8.3 Draft 802.16a OFDMA parameters

Parameter	Value
FFT size, N_{FFT}	2048, 4096
Bandwidth, BW	Depends on band and country
Sampling frequency, F_S	$BW \cdot (8/7)$
Subcarrier spacing, Δf	F_S/N_{FFT}
Useful time, T_b	$1/\Delta f$
Cyclic prefix (CP) time, T_g	Based on T_b and allowed ratios
OFDM symbol time, T_S	$T_b + T_g$
Ratio, T_g/T_b	1/32, 1/16, 1/8, and 1/4
Sample time	$1/F_S$
Number of downlink/uplink channels	32
Number of data subcarriers per channel	48

and IEEE 802.16a standard for the 2- to 11-GHz band. The IEEE 802.11a is described in Section 8.7.4.

The draft IEEE 802.16a standard [13] for wireless access systems in the 2- to 11-GHz range is intended for much longer range metropolitan area networks (MAN) rather than local area networks (LAN), which is the target of IEEE 802.11a. The IEEE 802.16a standard has specifications for both licensed and license-exempt bands. IEEE 802.16a specifies the use of three different modulation/multiple access techniques:

- single carrier (SC) with TDMA;
- OFDM with TDMA: The fast Fourier transform (FFT) size can be 256 or 512, resulting in 212 or 424 active subcarriers, respectively;
- OFDMA with FFT sizes of 2048 or 4096, with 1696 or 3392 active subcarriers, respectively. For the 2048 case, the subcarriers are divided into 32 channels of 48 subcarriers each, with the remaining subcarriers used as pilots and other overhead functions.

Given the large number of bands in which the 802.16a standard can be applied worldwide, there is no fixed channel bandwidth. Instead, the OFDM subcarrier frequency spacings are scaled for the available channel bandwidth. Some of the basic parameters are shown in Table 8.3. Both FDD and TDD duplex methods are permitted by IEEE 802.16a. Given the various options available in this standard, transmissions contain preamble information that instructs the remote units as to the technology being used.

8.8 DUPLEXING METHODS

As discussed in the introduction to this chapter, duplexing is the mechanism by which two-way communication is accomplished between two points in a single link system or between a remote terminal and the network hub or backbone in point-to-multipoint networks. The duplexing method chosen for a system is essentially independent of the modulation type and the multiple-access-technology. The main issues that influence the duplexing method selection are the nature of the available spectrum allocations and the nature of the traffic that is to be communicated.

The two duplexing methods primarily used in fixed broadband wireless are FDD and TDD. Both are discussed in the following sections along with the lesser-used methods of polarization duplexing and space duplexing.

8.8.1 Frequency division duplexing (FDD)

Frequency division duplexing is the method most widely used in traditional fixed broadband point-to-point and point–to-multipoint systems. A drawing illustrating FDD is shown in Figure 8.14. The primary parameters with FDD are the channel bandwidths themselves for downlink and uplink and the duplex frequency separation between pairs of downlink and uplink frequencies. The downlink and uplink bandwidths are chosen to accommodate the required system data rate and modulation scheme, or they may also be mandated by administrative rules as in the case of first- and second-generation cellular systems. The total duplex bandwidth available to a user is $D_n + U_n$. If asymmetrical downlink and uplink traffic volume is anticipated, D_n does not necessarily have to be the same as U_n, although in conventional systems it is typical for the downlink and the uplink bandwidths to be equal.

The duplex frequency separation is chosen largely to make the design of duplexing waveguides or circuitry at a terminal practical. For a full duplex system, a terminal will be sending and receiving simultaneously so that both the downlink and the uplink channels are active. If a single antenna is used, which is the economical approach, the device known as a *duplexer* at the remote terminal must separate the signals being received on D_n from those being transmitted on U_n. The design of duplex filters to accomplish this task becomes more difficult and problematic as Δf decreases. Since the uplink and the downlink signals are different, signal leakage of one into the other represents interference and

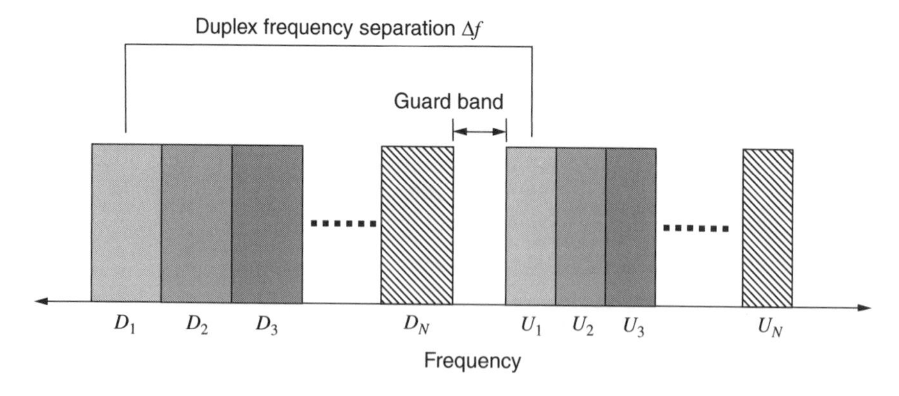

Figure 8.14 Frequency division duplexing (FDD) with downlink channels D_n and uplink channels U_n with duplex frequency separation Δf.

correspondingly lowers the fade margin. A typical minimum duplex frequency separation is 5% of the carrier frequency. For the 28-GHz LMDS band, this is about 1.4 GHz.

Anything that can increase the isolation between D_n and U_n will relax the filtering requirements of the duplexer. For microwave systems with fixed high gain antennas, a common approach for improving downlink–uplink isolation is to use one polarization for the downlink and the orthogonal polarization for the uplink. As illustrated in Figure 6.12, in the main lobe of antennas typically used in such systems, the order of 30 dB of isolation can be achieved between orthogonal polarizations. In effect, this is *polarization duplexing*. The required duplex frequency separation Δf can be significantly reduced with effective polarization duplexing.

From a spectrum efficiency standpoint, the main drawback to FDD is that it reserves spectrum in D_n or U_n that may be underutilized much of the time. This is especially true when the traffic is highly asymmetrical. The usual situation considered is the 'web-browsing' application in which the downlink transmission is a relatively high volume of data traffic including web pages, images, or video or audio files, while the uplink transmission consists simply of a series of mouse clicks or key strokes – relatively low traffic volume. If a full uplink channel is reserved for this terminal, it could well be idle most of the time, which is an inefficient use of the spectrum. This is the argument most often put forward in favor of TDD discussed in the next section.

Other data applications such as e-mail represent much more symmetrical downlink–uplink traffic volume distribution, and the potential benefits of TDD are not quite so apparent. Also, for traditional two-way point-to-point microwave links that carry high volume data traffic in both directions, or multiplexed voice traffic with a fixed number of channels in each direction (even if idle), TDD can have drawbacks. Such point-to-point links are commonly used in cellular backhaul (network interconnect) applications or corporate interbuilding or intercomplex connections. For such situations, FDD is the better approach for duplexing.

8.8.2 Time division duplexing (TDD)

Time division duplexing (TDD) uses the same frequency channel for both downlink and uplink transmission. For conventional analog wireless systems, the TDD operation is called *simplex*. In a typical simplex analog wireless system, the network user listens for an idle channel before transmitting a message. The user then monitors this same frequency for a response to the transmitted message. What distinguishes TDD from simplex systems is that modern digital hardware makes it possible to rapidly switch the channel from transmitting to receiving information, and interleave (in time) packets of transmitted and received data such that from the end user's viewpoint, the channel is perceived to be a full duplex channel that can transmit and receive information 'simultaneously'. A drawing illustrating TDD is shown in Figure 8.15. A guard time is inserted for transitions from transmit to receive mode, and vice versa, between TS.

The main advantage of TDD is that the time occupancy of the channel can be adjusted to accommodate the asymmetrical traffic volume on either the uplink or the downlink. The reduction in channel idle time leads to greater efficiency than FDD in which some channel idle time would occur with asymmetrical traffic volume. For an asymmetry factor of 2, the

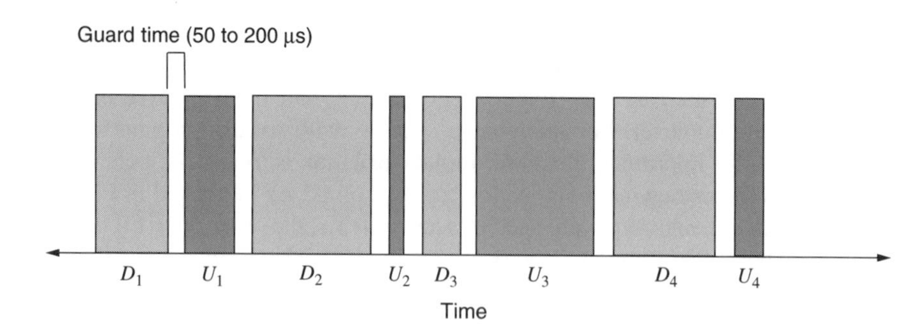

Figure 8.15 Time division duplexing (TDD) with downlink transmission intervals D_n and uplink transmission intervals U_n with guard time. Transmission interval length can vary with adaptive TDD.

increased efficiency is given at about 16% in a recent report. Because of basic overhead information, all TDD systems (like UTRA TDD, for example) use a fixed frame length and TS or 'burst' structure of some sort that becomes the quantum unit for uplink/downlink transmission. The flexibility to adjust the symmetry of the uplink/downlink flow is therefore not unlimited but restricted to possible ratios as discrete values. The limited number of discrete values still provide for considerable range in adjusting the uplink/downlink time slot allocations.

For multiple access or adaptive antenna techniques that use channel estimation, the fact that the uplink and the downlink channels are the same can be used to some advantage. TDD systems are also more flexible in administrative frequency allocations in which only a single rather than a pair of frequency channels are available. Of course, for equivalent full rate downlink and uplink transmission, the bandwidth of the TDD channel must be equal to the bandwidth of the sum of the downlink and uplink channel bandwidths used for FDD.

8.8.2.1 TDD interference calculations

Although TDD has many benefits, one drawback of a TDD system that is rarely pointed out is the doubling of the number of interference sources that must be dealt with as compared to an FDD system. This situation is illustrated in Figure 8.16(a) for FDD with a simple two-hub system with one remote terminal connected to each hub using the same downlink and uplink frequencies f_d and f_u in each case. The interference paths for the FDD interference analysis are shown by dotted lines for the uplink and dashed lines for the downlink. The interference analysis for the downlink (interference to remote terminals) is confined to interference from other system hubs only. Similarly, uplink interference calculations (interference to hubs) are confined to interference from other remote terminals only.

The TDD interference situation is shown in Figure 8.16(b). Because both the hub and remote terminals are transmitting on the channel f_d, both must be considered in the interference calculations at the hub and remote terminal (uplink and downlink) on

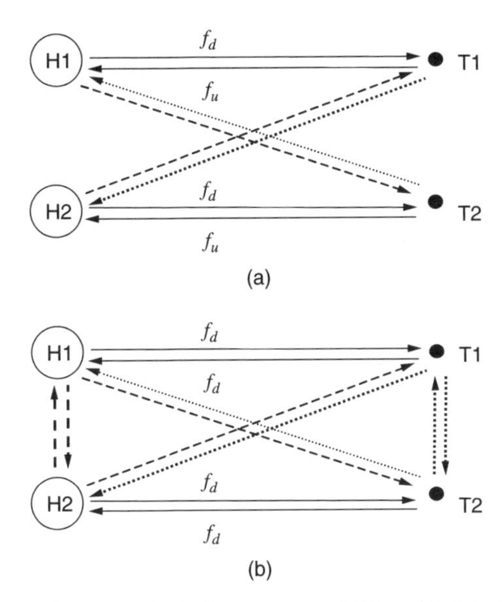

Figure 8.16 (a) FDD interference calculations paths and (b) TDD interference calculation paths.

the neighboring H2-T2 link. While the uplink/downlink time division allocation of the channel is synchronized so that the transmissions do not conflict between a single hub and its remote terminals, this uplink/downlink time division allocation does not generally extend to other hubs and communication with their respective terminals. Synchronizing the two-way uplink/downlink allocation is difficult because the allocation between a hub and remote terminal is traffic-dependent. Synchronizing them all with other links would impose severe uplink–downlink allocation restrictions on those links. Because the uplink–downlink transmission allocations are asynchronous, both the hub and the remote terminal must be considered as interference sources with some reduced average duty cycle. The interference paths for the equivalent two-hub TDD system are shown as dotted and dashed lines in Figure 8.16(b). For UTRA TDD, frame synchronization can assist in mitigating this interference. Some simulations of this problem for UTRA TDD are found in [11].

For systems using SDMA that rely on adaptive antenna technology for interference suppression, the increase in the number of potential interference source locations means these antennas have more interference sources they must suppress. For remote terminals with modest adaptive antennas made up of only a few elements, the increase in interference source locations (and corresponding different arrival angles) will increase the difficulty of adjusting the antenna to suppress interference and achieve the desired SIR.

8.9 CAPACITY

A measure of the success of any multiple-access technique is the overall capacity it can achieve in terms of bps/Hz/km^2 throughput in a two-dimensional service area. Fixed broadband systems in densely urban areas actually operate in a three-dimensional propagation environment, which can be exploited using SDMA or fixed broadband microwave systems using antennas with high gains in both the azimuth and elevation planes. Very narrow beam FSO technology can best exploit the 3D spectrum space. Capacity for such systems should really be recognized in three dimensions, or bps/Hz/km^3. While this wireless 'box of beams' is an accurate generalized view of a fixed broadband network deployment, for simplicity, capacity will be considered here in the more globally applicable two-dimensional framework.

Unfortunately, the raw capacity metric bps/Hz/km^2 is not very useful for assessing the quality of the communications process under a variety of single-user, multiuser, noise-limited, and interference-limited circumstances. Capacity can be assessed in a variety of ways depending on how that assessment will be used. A number of capacity calculation approaches are discussed below including

- Shannon theoretical channel capacity limit
- capacity in interference-limited, multiuser systems
- user capacity
- commercial capacity.

Capacity figures are often used to make relative comparisons of the success or strength of one technology versus another, but sometimes these comparisons are influenced by other nonengineering issues that are not relevant to a system operator trying to select a technology to meet service objectives and business plan milestones. With that in mind, the following discussion will describe capacity in terms of what a communication system operator can offer customers by way of data throughput, reliability, and service quality.

8.9.1 Shannon theoretical channel capacity

A presentation of Shannon's theory for the upper bound of channel capacity in bits per second for a channel corrupted by additive white Gaussian noise (AWGN) can be found in any modern communications engineering textbook, for example, [14,15]. The Shannon capacity bound is a rate at which information can be transmitted with arbitrarily small error rate (including zero) over an AWGN channel with the use of appropriate coding (which Shannon did not identify). That Shannon information capacity bound is given by

$$C = B_W \log_2\left(1 + \frac{S}{N}\right) \text{ bps} \tag{8.21}$$

where S is the signal power, N is the noise power in the channel, and B_W is the channel bandwidth in hertz. The normalized channel capacity bps/Hz is then

$$\frac{C}{B_W} = \log_2\left[1 + \frac{E_b}{N_0}\left(\frac{R}{B_W}\right)\right] \text{ bps/Hz} \tag{8.22}$$

where N_0 is the noise power density in hertz, E_b is the signal energy per bit, and R is the information rate in bps.

The Shannon capacity limit is not realizable with currently known coding schemes. More importantly, coding schemes that approach the limit require significant lengths and complexity; the resulting long delays tend to infinity as the Shannon limit is approached. While achieving the maximum information capacity from a channel may be desirable, from a user's perspective, infinite delays and the cost associated with high complexity are not. Short of this upper limit, it is possible to identify a transmission rate and particular coding scheme rate or set of rates, that achieve some nonzero (but acceptable) error level.

For most wireless channels, the SNR in (8.21 or 8.22) is a time-varying quantity with a statistical description. The capacity is also similarly varying with time and statistically described. Even though the Shannon capacity is not a very practical measure for wireless system design and planning, it can be a useful benchmark for those engaged in designing coding and multiple adaptive antennas technology for wireless systems.

The Shannon capacity is a channel capacity rather than a system capacity. It relies on some notion of a defined 'channel', which for Shannon in 1947 was better represented by a telephone wire than a complex wireless communications channel. The increased capacity results with multiple input, multiple output (MIMO) systems described in Chapter 6 leads to the view of a 'channel' that is not even a distinct entity but rather a continuum of decoupled transport mechanisms that are enabled in partial, time-varying ways by the complexity and fractured nature of the propagation environment.

8.9.2 Capacity in interference-limited, multiuser systems

Beyond consideration of a single channel, the more generalized capacity of interest is the overall capacity of the network when there are multiple channels and multiple users. More directly, the network capacity is the number of users whose communication objectives can be 'simultaneously' met, with the caveats on 'simultaneously' previously discussed in the introduction to this chapter. This approach to calculating capacity has been used extensively for cellular system technology comparisons. The basic weakness is that analytical studies require several assumptions, particularly, interference levels from other parts of the system and the channel response characteristics. The traditional hexagon cell layout with assumed cell service radii with associated intercell interference and reuse distances is only theoretical – no real cellular system has this topology. As the propagation environment gets less homogeneous, and directional antennas are used, the interference levels leading to BER calculations are not uniform. The system capacity thus becomes less dependent on the technology and more dependent on the environment where the technology is deployed. Environment-specific simulations with full-featured propagation, channel, and hardware models are the appropriate approach to assessing the potential capacity of competing technologies.

The factors in a multiuser system that affect the system capacity are as follows:

- Required net error rate for a given application after decoding and forward error correction (FEC)
- Modulation efficiency (bps/Hz)
- Coding overhead

- Multiple-access technique – FDMA, TDMA, CDMA, SDMA, and OFDMA
- Diversity (time, space, frequency, code, etc.)
- Overhead for guard bands, guard times, and so on
- Hard or soft handoff network overhead requirements
- Directional antennas
- Automatic power control (APC) on both uplink and downlink
- User service duty cycle or packet traffic statistics (voice activity factor in voice systems)
- System intelligence in terms of dynamic resource assignments to respond to changing service demand. Dynamic channel assignment (DCA) and dynamic packet assignment (DPA) are two examples of system intelligence that affects capacity.

As an example, consider a 1-MHz channel using 16QAM modulation. Theoretically 16QAM has an efficiency of 4 bps/Hz, but with coding overhead and imperfect filtering, the net actual efficiency is typically around 3.0 bps/Hz. In a channel with a 1-MHz bandwidth, the data rate furnished to the user will be 3.0 Mbps. If the channel is time-shared using TDMA among 10 users, and TDMA overhead is ignored, the data rate to each user is 300 kbps. If the system has 10 such channels using a single hub, the total hub capacity is 30 Mbps.

The above calculation assumes that the SNR at the user terminal is adequate to achieve an acceptable error rate after decoding and error correction. For a given transmitter power, receiver noise figure, and antenna gains, this implies there is a limit to the range or link pathloss at which service can be furnished. Choosing a more efficient modulation scheme such as 64QAM with 6 bps/Hz will increase the total data rate available in a channel but will reduce the transmission range over which this data rate is available because a higher SNR is required to achieve the required error rate. The more efficient modulation is also more susceptible to errors from interference as shown in Chapter 7, so the ability to reuse frequencies on nearby hubs is reduced, thus reducing the overall *network capacity*.

A multihub hypothetical hexagon system layout is shown in Figure 8.17. For a multihub system, each with multiple users, the performance of each link on each of the 10 channels is now degraded by intercell interference. The extent of interference will depend on the distance and resulting pathloss between the receivers (both hub and remote terminal) in a victim cell and the transmitters (both hub and remote terminal) in neighboring cells. The interference level will also depend on the directional antenna properties of all transmitters and receivers, and whether APC is used on transmitters. If the locations of all transmitters and receivers are explicitly known along with the characteristics of the antennas and RF equipment, a pathloss model can be used to find the interference level from every transmitter to every receiver. The capacity of the 10 channels used by the receivers can then be reassessed in light of the interference + noise level. A similar procedure can be used for each cell and the total capacity of the network explicitly calculated. Of course, when anything changes (more terminals, different modulations, etc.) the system capacity must be recalculated. Chapter 12 includes the mathematical details of the capacity calculation.

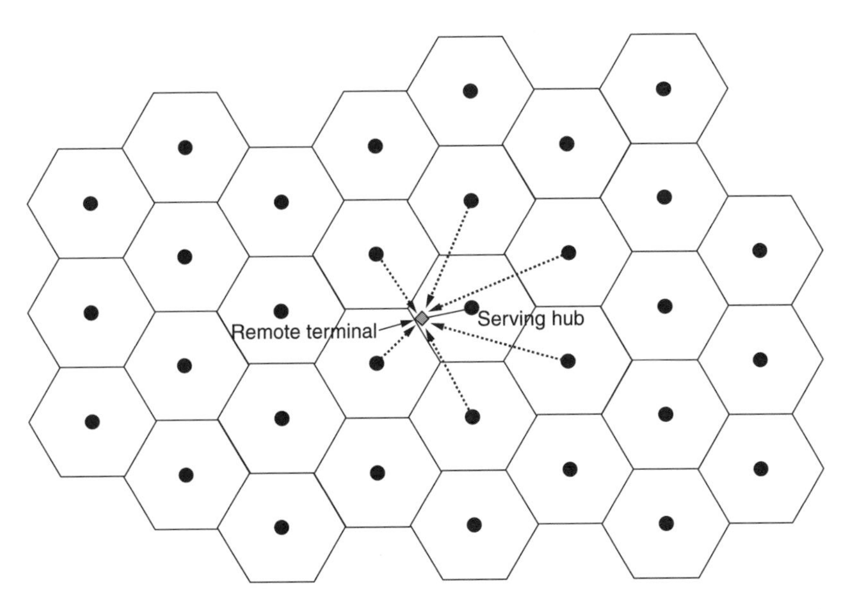

Figure 8.17 Hypothetical hexagon grid cell layout with interference paths to remote terminal from closest (first 'ring') interfering hubs shown as dotted lines.

The point of describing this example system is to demonstrate how dependent the system capacity is on all the bullet points listed above. In simple analytical analyses of networks, many of the direct calculations described above are replaced by assumptions. For example, the service range for a given modulation type is usually assumed to be a fixed distance from the hub, and the interference from other cells is summarized by a single number that is proportional to its distance from the victim cell. Even though these assumptions cloak the details and complexity of the problem and thus lowers the quality of the capacity estimate, this approach at least provides a simple, easily understood 'ballpark' capacity number.

In Chapter 11, specific link budget calculations will be presented that quantitatively take into account the factors cited above. Comprehensive, iterative use of such calculations, used in the context of wireless system planning software, can provide much more realistic estimates of the capacity of particular system design by taking into account all the particular parameters of that design.

As a general principle, network capacity will increase as the cell service area size decreases. This is simply a result of shifting the capacity from the wireless radio resource to a larger number of network access point resources. However, the infrastructure cost associated with this shift is significant, and includes the infrastructure access points themselves (including towers, etc.) and the cost of connecting these access points to the optical fiber backbone (backhaul). For mobile systems, there is also the issue of handing off service from one access point or hub (cell site sector) to another. Depending on the handoff scheme, some capacity may be sacrificed in maintaining multiple remote-hub connections. This cost trade-off is another aspect of assessing system capacity.

8.9.3 User capacity

The capacity calculations for the system in the preceding section considered service to be delivering a given data rate at a given BER. The concept of capacity can be extended to calculations of how many users are receiving 'acceptable' service that include several other QoS parameters. Such additional parameters include

- connection time overhead (how fast does it connect)
- turnaround delay (latency)
- jitter (the variation in latency)
- total uplink and downlink throughput
- throughput variation (service consistency).

A user's perception of the network is based on a qualitative assessment of these characteristics. Bridging the gap from the qualitative user perception to specific engineering design objectives is one of the challenges in modeling network traffic requirements. These aspects of service quality will be discussed in more detail in Chapter 9 on traffic and application modeling.

8.9.4 Commercial capacity

In most cases, the fixed broadband wireless system will be built as part of a commercial wireless business in which the system capacity represents the inventory of the business – the product that the wireless business has for sale. How the link capacity is packaged is important to how desirable it will be to system customers. A system that very efficiently provides a set of highly reliable slow speed 10-kbps data channels does not provide the operator with a product inventory that is competitive or marketable. From this perspective, the starting point in calculating system capacity is market research, which in turn defines the required service and application inventory. The engineering details of the system are then determined so that the required inventory is provided in the most cost-effective way.

The required inventory can be quite varied with different customers requiring different data rates, service quality, and availability. The terms of what the wireless operator is providing (and what the customer is paying for) are sometimes embodied in service level agreements (SLAs). Lower data rates, quality or availability carry a lower price. Technological flexibility that allows the service to be packaged in a wide variety of ways is another aspect of the commercial capacity that should be considered in network design.

Commercial capacity can be directly tied to business revenue potential and ultimately to whether there is a viable business case for building the system. If the business case is not initially attractive, modifications to the engineering elements that affect capacity and cost as listed in Section 8.9.2 can be made in an effort to create a system that is commercially feasible.

8.10 CONCLUSION

Fixed broadband wireless systems fall into two broad categories. The first category includes individual links used to connect two distinct points. The only access question is whether an interference-free frequency can be found to carry the service. In the second, more general, case, there will be multiple users contending for use of the same spectrum. These users will be both cooperative (within the control of the system operator), and uncooperative as in the case of a competing system in the same geographical area or in a neighboring area.

For the case of cooperative multiple users in the same system, this chapter has described a variety of approaches to harmoniously allowing many users to successfully access the spectrum simultaneously. The techniques described include FDMA, TDMA, CDMA, SDMA, OFDMA, and CSMA. Each is intended to exploit particular dimensions of the spectrum space, but more importantly, modern systems have evolved in which two or more of these approaches are used in combination to achieve system capacity that is well suited to the network requirements.

SDMA is particularly universal as it can be added in combination with any of the other described methods without the use of additional bandwidth or time. The cost associated with SDMA is hardware complexity; a cost which continually decreases with more efficient and powerful signal processing chips. SDMA also requires more physical space for the adaptive antenna elements. Nevertheless, of the multiple-access techniques described here, SDMA offers the greatest future potential for increasing system capacity when compared to the potential of the other technologies.

8.11 REFERENCES

[1] H.R. Anderson, "Digicast: toward a more efficient use of the radio spectrum resource," *Intelligent Machines Journal*, no. 10, page 12, June 1979.

[2] Federal Communications Commission (FCC) MM Docket 97-217, "Methods for predicting interference from response stations transmitters and to response station hubs and for supplying data on response station systems," April 21, 2000. (Appendix D).

[3] Telecommunications Industry Association (TIA/EIA), "Wireless Communications – Performance in Noise, Interference-Limited Situations – Recommended Methods for Technology-Independent Modeling, Simulation and Verification." TSB-88-1, June 1999, and Addendum 1, January, 2002.

[4] IEEE Computer Society Working Group 802.16, "IEEE Recommended Practice for Local and Metropolitan Area Networks – Coexistence of Fixed Broadband Wireless Access Systems." *Institute of Electrical and Electronic Engineers*, September, 2001.

[5] S.C. Yang. *CDMA RF System Engineering*. Boston: Artech House, 1998. Chapter 7.

[6] S. Verdu, "Minimum probability of error for asynchronous Gaussian multi-access channels," *IEEE Transactions on Information Theory*, vol. 32, pp. 85–96, January, 1986.

[7] R. Lupas and S. Verdu, "Linear multiuser detectors for synchronous code-division multiple-access channels," *IEEE Transactions on Information Theory*, vol. 35, pp. 123–136, January, 1989.

[8] M. Haardt, A. Klein, R. Koehn, S. Oestreich, M. Purat, V. Sommer and T. Ulrich, "The TD-CDMA based UTRA TDD mode," *IEEE Journal on Selected Areas in Communications*, vol. 18, no. 8, pp. 1375–1385, August, 2000.

[9] Tim Wilkinson. Private communication. June, 2002.
[10] 3GPP web site. *http://www.3gpp.org*.
[11] H. Holma, S. Heikkinen, O.A. Lehtinen, and A. Toskala, "Interference considerations for the time division duplex mode of the UMTS terrestrial radio access," *IEEE Journal on Selected Areas in Communications*, vol. 18, no. 8, pp. 1386–1393, August, 2000.
[12] R. van Nee and R. Prasad. *OFDM for Wireless Multimedia Communications.* Boston: Artech House, 1998. Chapter 8.
[13] IEEE P802.16.a/D4-2002. "Draft Amendment to IEEE Standard for Local and Metropolitan Area Networks. Part 16: Air Interface for Fixed Broadband Wireless Access Systems – Medium Access Control Modifications and Additional Physical Layer Specifications for 2–11 GHz, IEEE Standard for Local and Metropolitan Area Networks: Coexistence of Fixed Broadband Wireless Access Systems," *Institute of Electrical and Electronic Engineers*, May 27, 2002.
[14] V.K. Garg and J.E. Wilkes. *Wireless and Personal Communications Systems.* Upper Saddle River, N.J.: Prentice Hall PTR. 1996.
[15] B.P. Lathi. *Modern Digital and Analog Communication Systems*. New York: Holt, Reinhart and Winston. 1983.

9

Traffic and application mix models

9.1 INTRODUCTION

The objective of any wireless communication system is to convey information or communications traffic from one location to another. The design of a suitable system to accomplish this task depends to a great extent on the locations of the points between which the traffic is to be communicated. These points could be across the office, across the street, across the city, across the country, or around the world. Each of these circumstances will dictate a different type of fixed broadband wireless system.

Assessing communication traffic requires two main tasks:

- finding the geographic location of the traffic sources and destination;
- finding characteristics of the traffic: maximum data rate, arrival time and duration distribution, required throughput, latency, and so on.

Designing a successful broadband wireless system requires detailed knowledge so that these tasks can be completed. Sometimes the information is exactly known. For example, a fixed wireless link connecting two office buildings in a large company, or a fixed link connecting a cell base station site with a switching center usually have known traffic characteristics including the traffic source and destination locations and the type and magnitude of the traffic. In such cases, estimates and models of the traffic are not needed.

In the more general case, however, the traffic sources and destinations (terminals) will be potential, but not yet identified users of the wireless system. Models that describe the location and density of such traffic sources are the subject of the first part of this chapter. Several approaches to this problem are presented, relying primarily on a variety of geographic data sets, some of which were discussed in Chapter 5. For convenience, the type of users can be divided into classifications that can be used to predict the type of traffic that can be expected between their terminals and the network. The locations of traffic terminals,

Fixed Broadband Wireless System Design Harry R. Anderson
 ISBN: 0-470-84438-8

or *nodes*, as they will be referred to here, have a direct impact on where the access points to the network hubs are built and what capacity those access points must have.

The second part of this chapter will discuss the traffic characteristics themselves that vary greatly depending on the communication needs of the users at each node. These needs will include urgent messaging and e-mail, downloading commercial products (software, images, audio and video), secure financial transactions, nonurgent idle web-browsing for entertainment purposes, and real-time streaming audio and video applications. All these services or applications place different data requirements and quality-of-service (QoS) constraints on the wireless network, as well as on the wired or optical fiber to which the wireless network is connected.

For traffic that is known to be, or anticipated to be, continuous, a circuit must be continuously dedicated to carry the traffic. Even if the traffic is not continuous, a customer may be willing to pay to have guaranteed full-time access. Microwave links that serve as cell systems backhaul links, utility telemetry and control supervisory control and data acquisition (SCADA) circuits, corporate building interconnections, and broadcast station studio-transmitter links (STLs) are examples of systems in which the traffic is likely to be continuous with some maximum continuous data rate or required bandwidth.

Traffic that is intermittent (as most traffic is) can be handled using either circuit-switched or packet-switched sharing of network resources. Circuit-switched traffic is applicable to traditional transmission applications in which the user is assigned a circuit or *trunk* (radio, wire line, physical, logical, or other) and retains exclusive access to that circuit for the duration of the communication, even if the circuit is idle much of the time. Wire line telephone systems as well as first- and second-generation cellular systems were designed as circuit-switched systems. The quality of service is easily described by the probability of the user getting a circuit when one is requested (*blocking probability*), and the rate at which an unexpected termination of circuit access occurs (*dropped call probability*). While the call is in progress, there is also the issue of sound quality – fidelity, distortion, and background noise. The Bell system developed and maintained exacting standards for call sound quality throughout its system. Dropped calls were almost nonexistent, especially with trunk circuits on optical fiber, and trunk blocking was a predicable design parameter. Cellular systems have a long way to go to offer quality in these three areas that rival wire line quality.

Circuit-switched traffic will be briefly discussed in this chapter since it can be applicable to certain deployments of fixed broadband wireless systems. However, handling traffic from multiple sources by switching circuits is relatively inefficient when those circuits may be idle for much of the time. Consequently, the dominant trend with modern wireless communication systems is to switch communication resources to handle traffic on a packet-by-packet basis. A packet, which will be discussed in detail later, is simply a block of information bits like a time division multiple access (TDMA) frame described in Chapter 8. Even real-time services such as voice, audio, and video can be broken up into packets and can use storage mechanisms at various points in a system, presenting the data stream to the end user in such a way that it is perceived as continuous. Packet switching allows a much more adaptive, and hence, efficient, use of the radio spectrum resource. Packet switching is the approach used in 2.5G and 3G cellular systems as well as IEEE 802.16 and IEEE 802.11 standards in the form of *burst transmissions*.

9.2 TRAFFIC GEOGRAPHIC DISTRIBUTION MODELS

When a wireless system is being designed to provide service to an area where the specific customer locations are not known, some model must be devised for describing where the traffic is likely to originate. For simplicity, this discussion will refer to the distribution of traffic sources with the understanding that in a two-way communication system, each traffic source is also a traffic destination.

For fixed wireless systems, the location of traffic sources are assumed to be in fixed, defined locations rather than in arbitrary locations, or in motion. This makes the description of the geographic distribution of traffic distinctly different from that used for cellular systems. In cellular systems, considerable traffic is generated from handsets being used while people are driving along the road, walking around the streets, or in restaurants, offices, and other random locations. The geographic distribution of traffic can be weighted toward these geographic entities. They are less likely to use cell phones when there is a wire line phone available with comparable connection cost since the wire line phone quality is superior to cell phone service.

In contrast, fixed broadband wireless service will originate from terminals in large office buildings, small businesses, and residences. It will also originate in areas where business travelers are in transit, such as airports, train stations, hotel lobbies, and other places that do not have other high-speed data connections. In general, all these locations are structures of some type. The idealized notion of people on the beach or in a park making extensive use of high-speed wireless data connections is unlikely to occur. Deliberately planning the system for such a context would be misdirected.

The models that are appropriate to describe geographic distribution of traffic sources are those that provide the locations of these structures. Such distributions can be drawn from a number of data sources, some of which are also used for propagation modeling. Several sources of traffic distribution models are discussed in the following sections.

9.2.1 Residential demographic data

Demographic data is a rich source of information about where people live and work. It also provides useful economic and age distribution data that can be used to predict the probability of whether they will subscribe to a wireless service and what type of service would be most desirable. Demographic data include information on residences and on businesses. Business demographic data are discussed in Section 9.2.2.

Demographic data are available from several sources. Most commonly, countries will conduct a periodic census of the population, which, among other things, will establish the number of people, where they live, and other details of their lifestyle. In addition, private organizations will undertake substantially more focused studies of particular groups as a general part of market research on behalf of businesses that are developing and selling various products. It is not the intent here to survey this myriad of data sources, but rather to illustrate the use of some of the generic data types in formulating a geographic distribution model of traffic sources that can be effectively used in designing fixed broadband wireless networks.

Demographic data for residences usually include basic statistics of the population in different age groups and the number of housing or dwelling units. For US Census, these numbers are provided for several basic and summary geographical areas. The most basic (smallest) area is called a 'block', which has an average population of 50 to 60 persons in 20 to 30 dwelling units. The data statistics are geographically defined by attaching them as attributes to a block 'centroid' given in terms of latitude and longitude. Although this approach is specific to the US Census, the concept is generic enough that other government-created census databases or private data sets can be considered in the same way. The fundamental point is that the data consist of discrete geographic locations that have attached to them a number of attributes, some of which include attributes that can be used for determining wireless traffic distribution.

Figure 9.1 shows a map of the housing unit density distribution for Eugene, Oregon, as derived from the 2000 US Census. Superimposed on this map are hypothetical service radii for several four-sector hub sites. These simple boundaries, of course, are not realistic service areas but will serve the explanatory purposes here. The housing unit densities were calculated from discrete centroid census data by constructing a grid of tiles (in this case, $0.25\,\text{km} \times 0.25\,\text{km}$), totaling the number of people or housing units attributed to each centroid that falls within each grid tile, and dividing that number by the grid tile area ($0.0625\,\text{km}^2$).

Since fixed broadband wireless service will normally be provided by a service agreement to a household or a housing unit, the housing unit density is actually a better model for the number of potential customers than raw population density. By integrating the housing unit density over the service area for a particular hub sector, the total number of housing units that are potentially served by that hub sector is found. Given that the

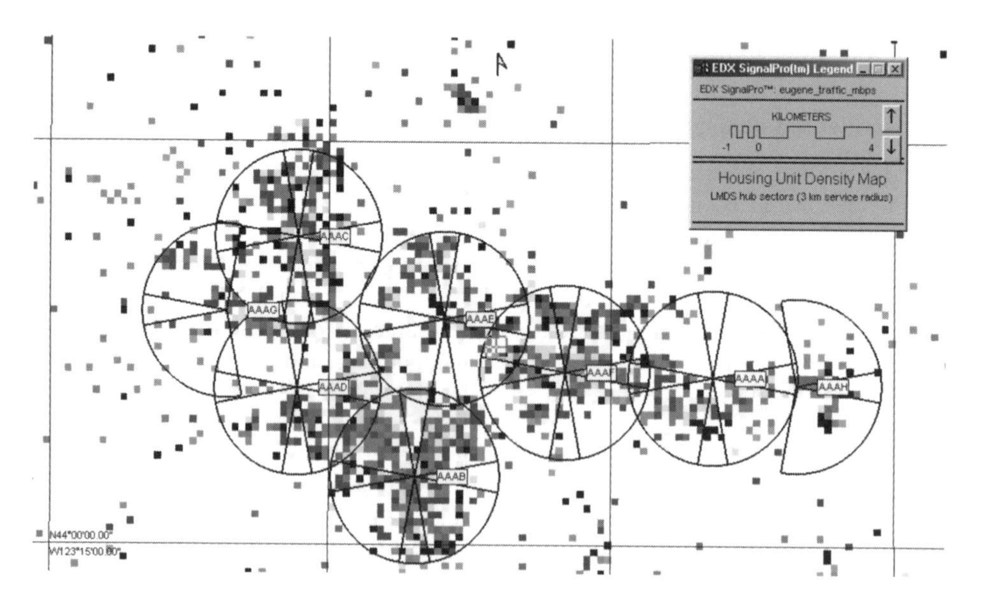

Figure 9.1 Housing unit density based on US 2000 Census data.

service area has been quantized into a regular grid already, this integration is logically done as a numerical summation. The total number of remote terminals (users, U_i) that are potentially served by hub sector i is

$$U_i = \sum_{n=-N}^{N} \sum_{m=-M}^{M} \rho(m,n)\, dA \text{ for } S_i > S_j, i \neq j \tag{9.1}$$

where $\rho(m,n)$ is the housing unit density in grid tile m, n, and S_i is the received signal level from hub sector i, which is assumed to be higher than the signals available from any other hub sector, and thus defines the basic service area of that hub. The term dA is the area of the grid tile.

Within the service areas of the hubs only a fraction of the potentially served housing units will actually subscribe to the wireless service. This subscriber percentage is called *the penetration rate*, a term that is generally applied to the rate at which any service is taken up by those to whom it is available whether it is wireless broadband, cable or digital subscriber line (DSL) broadband, or cellular phone service. The penetration rate is largely a function of nonengineering considerations such as the price, features, promotion, and support of the service, especially when compared with competitive services. It also depends on the extent to which customers have already subscribed to a competing service. It usually is harder to lure customers away from a competing service than to sign them up for first-time service. These issues and other similar ones are the core domain of business market research.

From a system design perspective, a projected penetration rate will normally be furnished to the design engineer from these market research activities. For a successful service, the penetration rate will increase over time, so a multiyear projection of the penetration rate will provide longer-range objectives for the total traffic the system must handle. The penetration rate is also a function of the QoS. The QoS of a data service is generally described by its throughput, latency, jitter (latency variation), and the consistency of performance. These engineering issues will be discussed later in this chapter. For now a penetration rate r_p can be used to modify U_i in (9.1) to yield the total number of residential connections J_i for hub sector i

$$J_i = r_p U_i \tag{9.2}$$

The penetration rate for broadband service in the United States as of April, 2002 was about 25%, or about 25.2 million households [1], with DSL and cable modems each making up about 1/3 of this number, with the remaining third divided among satellite, fiber, and wireless services. Broadband connections roughly doubled over December, 2000, largely driven by the rapid increase in home networking (including wireless networking), which increases the number of potential users inside the household and thereby provides a strong motivation to increase the speed of the Internet connection that everyone shares [2]. From these figures, a reasonable estimate for the initial penetration rate for fixed broadband wireless is 10%, with the understanding that this number carries the qualifications listed above.

Demographic databases are continuing to evolve. As discussed in the next section, a technique known as *address matching* can be used to find the latitude/longitude coordinates of any address with reasonable accuracy. This approach has been used to find the geographic location of every business in the United States. It is a logical next step to use the same approach to find the latitude/longitude coordinates of every dwelling unit along with the associated list of attributes for the dwelling unit, including the number of residents, income, and ultimately, Internet usage. With specific coordinate locations established for data traffic sources, and a projection of the likelihood that a dwelling unit will subscribe to a wireless data service, the process of estimating the expected traffic load on hub sectors in a wireless system design should become much more accurate.

9.2.2 Business demographic data

The housing unit data captures one segment of the potential traffic sources – residences. Population and housing unit counts are based on where people live, not where they work. However, businesses are an important traffic source for wireless systems because they represent a much higher average data usage per connection and potentially higher revenue. From a system resource viewpoint, a single high-usage connection is preferable to several lower speed connections dispersed throughout the section service areas with the same total data requirements. Small to medium size businesses are particularly good candidates for service since they are rarely directly served by optical fiber, whereas large businesses often do have a direct fiber optic connection. In the United States, only about 10%, or about 1.5 million, commercial buildings have a direct fiber optic connection. Cable TV systems are also rarely built in business areas, so the only remaining competitor for wireless broadband is DSL. All these reasons have lead to many broadband wireless system operators exclusively targeting small to medium size businesses with high-speed microwave line-of-sight services that provide data rates well beyond what DSL can provide.

Fortunately, the geographic data for businesses are actually more refined than the data for residences. The geographic coordinates (latitude and longitude) of essentially every individual business in the United States have been determined through address matching. Address matching uses the known coordinates of the street intersections that bracket the business street address. The address is used to interpolate between the bracketing intersections. For example, if the coordinates of the 400-block and 500-block intersections on a city street are known, and a business is located at address 415 along the street, it is assumed to be 15% of the distance along the street from the 400-block intersection to the 500-block intersection. For closely spaced regular street grids, this process can provide good accuracy with a few tens of meters. For rural roads, especially curved roads, where the cross-street intersections are more widely spaced, the resulting business address matching coordinates are less accurate.

The primary motivation for finding the coordinates for businesses is location-based commercial services. Common web-based queries that find all the Chinese restaurants within 5 km of a hotel are enabled by having the coordinates of the hotel and all the restaurants in a database that can be searched using the 5-km radius.

The data for businesses includes other numbers that are useful for establishing the extent to which it will be a traffic source for a fixed broadband wireless system. In the

United States, each business is assigned a classification code called a Standard Industry Classification (SIC), a relatively old system that is now being replaced with North American Industry Classification System (NAICS). The basic classifications for NAICS and SIC are shown in Table 9.1. Within these main classifications, there are a large number of subclassifications that identify particular business sectors. The NAICS has a systematic classification code hierarchy that has codes ranging from 2 to 8 digits, allowing a business type to be specified with some precision.

In addition to its classification code, the business information also includes the number of employees within several set ranges, as given in Table 9.2. One approach to assessing

Table 9.1 NAICS and SIC industry (business) classifications (major categories)

NAICS codes	NAICS sectors	SIC codes	SIC divisions
11	Agricultural, forestry, hunting, and fishing	01–09	Agriculture, forestry, and fishing
21	Mining	10–14	Mining
23	Construction	15–17	Construction
31–33	Manufacturing	20–39	Manufacturing
22	Utilities	41–49	Transportation, communications and public utilities
48–49	Transportation and warehousing	—	—
42	Wholesale trade	50–51	Wholesale trade
44–45	Retail trade	52–59	Retail trade
72	Accommodation and food services	—	—
52	Finance and insurance	60–67	Finance, insurance, and real estate
53	Real estate, rental, and leasing	—	—
51	Information	70–89	Service
54	Professional, scientific, and technical services	—	—
56	Administrative support, waste management, and remediation services	—	—
61	Education services	—	—
62	Health care and social assistance	—	—
71	Arts, entertainment, and recreation	—	—
81	Other services (except public administration)	—	—
92	Public administration	91–97	Public administration
55	Management of companies and enterprises	—	(parts of all divisions)

Table 9.2 Employees per location for U.S. business (1999 data from US Census)

Employee range	Number of companies	Number of locations	Total employees	Average employees per location
No employees	709,074	711,990	0	0.000
1 to 4	2,680,087	2,685,788	5,606,302	2.09
5 to 9	1,012,954	1,027,212	6,652,370	6.48
10 to 19	605,693	643,106	8,129,615	12.64
20 to 99	501,848	670,822	19,703,162	29.37
100 to 499	81,347	309,211	15,637,643	50.57
500 to 999	8,235	101,268	5,662,057	55.91
1,000 to 1,499	2,756	57,193	3,356,793	58.69
1,500 to 2,499	2236	70,753	4,272,562	60.39
2,500 to 4,999	1706	102,720	5,904,452	57.48
5,000 to 9,999	871	117,503	6,064,760	51.61
10,000 or more	936	510,878	29,715,945	58.17

the average amount of data traffic from a business is to determine the data requirement per employee for each business classification, and to multiply that number by the number of employees in that business. Taken together with the latitude/longitude location of each business, the total number of businesses in the service area of a hub sector, along with the traffic requirements of each can be summed. That total, again multiplied by the business penetration rate (which will be different from the residential penetration rate), will yield the data traffic carriage requirements of the hub sector. Other factors already discussed for the residential case also affect the business penetration rate and also change over time. However, it can be expected that the number of business customer connections will be less volatile than residential customer connections because the data connection is usually an integral and essential part of the business operation and not as price-sensitive as service to residential customers.

Ideally, an average data rate or system load could be associated with particular SIC codes so that for a particular business location, a direct calculation of the data load could be done. Within the expected service area of a hub sector, these loads could be summed to yield the total expected average business load on the sector. Associating a data load or requirement per employee for a given SIC business code is difficult. Clearly some businesses such as law firms (SIC = 81), banks (SIC = 60), and insurance companies (SIC = 63) will have higher expected data loads than other businesses such as manufacturing (SIC = 20–39), and food stores (SIC = 54), and restaurants (SIC = 58). One fixed broadband wireless equipment manufacturer, through working with wireless operators, devised tables of average data demand in Mbps per employee for each of the several SIC business classifications. The data demands fall into three categories:

- 0.192 Mbps per employee for high demand businesses
- 0.128 Mbps per employee for medium demand businesses
- 0.064 Mbps per employee for low demand businesses

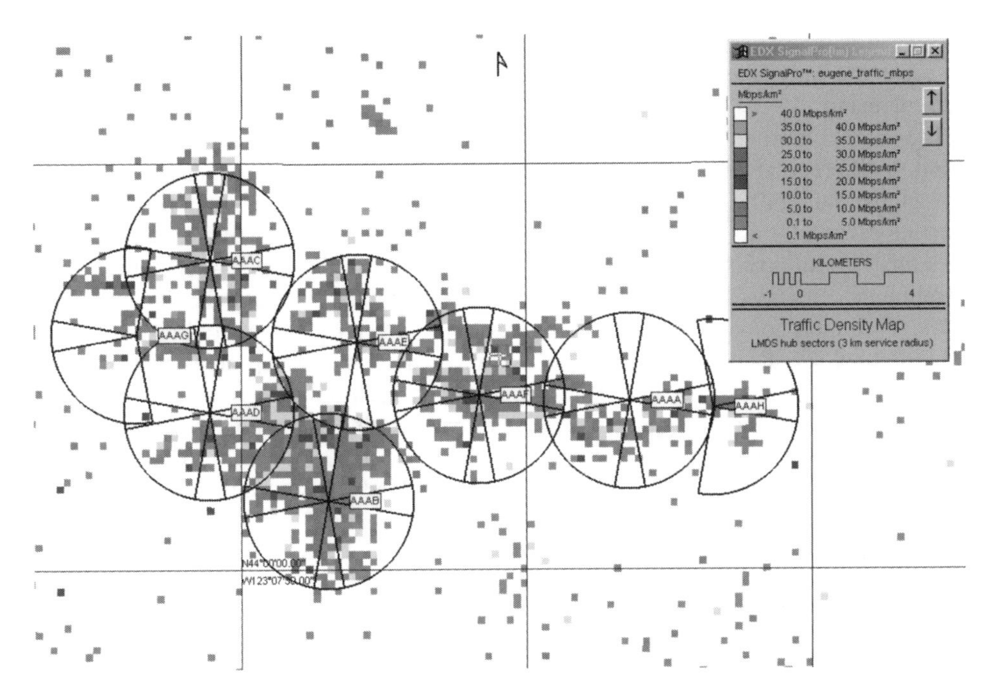

Figure 9.2 Example of an average data traffic density map based on business information.

Using these values, the business SIC code, and the number of employees at the given location as given by demographic and location data, an estimate of the total data requirements for the business can be obtained. Using the grid approach described in Section 9.2.1, the businesses in each grid tile can be found and their data requirements summed to find the data demand per grid tile. Figure 9.2 shows a map of average business data requirements for the Eugene, Oregon, area using this approach. As before, the total data demand can be integrated within the service areas of each hub sector to find the estimated business data load on the hub sector. An estimate of time-escalated penetration rate can also be applied. In particular, large businesses that have optical fiber connections can probably be excluded from the total data demand, although some businesses have used fixed broadband high capacity wireless links as a redundant connection (backup) to their optical fiber connections.

9.2.3 Land use data

Another coarser approach to finding the geographic distribution of traffic in a market is to use morphology (land use/land cover) data. This type of data was discussed in Section 5.4. The resolution of morphology data will usually be lower than demographic data, and the opportunity to use the location of specific businesses along with their expected data requirements per employee is not available. However, in cases in which such detailed demographic data is not available – which includes many parts of the world where fixed broadband wireless services might have the greatest application – morphology data may represent the only alternative to getting some idea of the location and magnitude of traffic demand.

The sample morphology map for Lisbon shown in Figure 5.11 has colors that correspond to different land use categories. Industrial and commercial areas can be expected to have higher data traffic densities, whereas agricultural and forest areas can be expected to generate essentially no data traffic. Of course, the challenge to effectively using this data to estimate traffic loads for a system is to find the data traffic rates associated with each morphology category. This is difficult enough for individual business classification (SIC) codes. With morphology data, such estimates become even more approximate. Nonetheless, morphology data can certainly be useful to identify built-up areas where service adoption will be most likely and hence, system hubs can be located to best serve these areas first. The transmission capacity needed to be available at these hubs can then be left to detailed market research surveys in achieved service areas. A specific survey of office and commercial buildings to establish possible traffic load is discussed in the next section.

9.2.4 Building data

Section 5.3 discussed vector and canopy building data models in the context of their use in propagation modeling. Vector building data has an additional use in identifying the locations of potential traffic sources for a wireless system. Basically, a ground survey of buildings in a service area can be done in which the surveyors proceed along the streets and record the building addresses, tenant lists, number of floors in the building, and any other data that could be useful to assessing the traffic demand from that building. Figure 9.3 shows an example of the footprints of buildings in a downtown area along with a query box that shows the addresses and the tenant list for the building that has been selected with the mouse at the cursor.

Even if the building footprint database is not available, the ground survey approach can still be used if the surveyors also carry global positioning system (GPS) units to establish the latitude/longitude coordinates of the building. This will be effective for traffic modeling, but a simple building coordinate will not be particularly useful for propagation modeling, especially at microwave frequency bands such as Local Multipoint Distribution Service (LMDS).

Once the basic building information is gathered through the survey, estimates of the number of employees working in that building can be obtained. Office spaces are typically designed to have an average of 200 square feet (18.5 m^2) per employee, which includes normally unoccupied areas such as hallways, common areas, stairwells, conference rooms, lobbies, and so on. The footprint of the building can be used to find the floor area of most floors, and together with the number of floors, can be used to find the total area of the building and the approximate number of employees who work there.

This manual approach to data collection can result in the most useful potential customer detail, but clearly is a labor-intensive and a daunting task for many cities. Large mixed urban/suburban areas may be several square miles and impractical to survey in this way. As a compromise approach, the survey can be limited to only the select

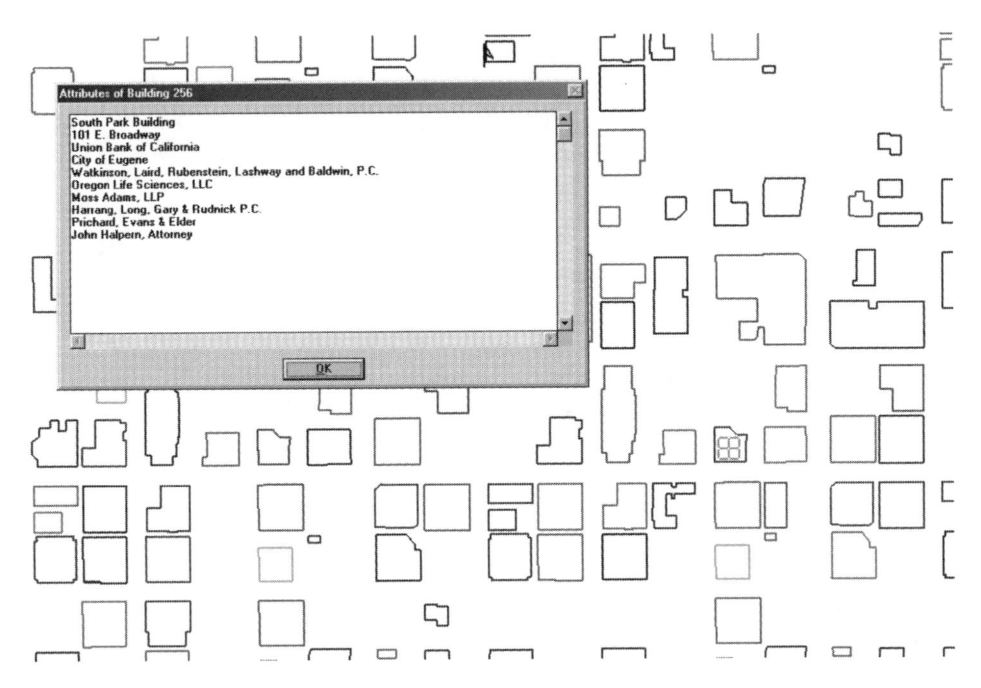

Figure 9.3 Building vector data with query results showing building tenant list.

highly urbanized clusters of buildings in which customer density is expected to be the highest.

9.2.5 Aerial photographs

Aerial photographs taken at appropriate altitudes and ortho-rectified can provide high-resolution information about the service area, geographic locations and actual shapes of buildings. The number of dwelling units in the service area can be found by counting all the structures on the photograph images. This straightforward, but labor-intensive, task may be justified for particular system deployments. Alternately, a count of the structures in a representative 1 km^2 area can be done and that structure density then applied to the entire service area to determine the number of buildings that are potential traffic nodes.

A typical aerial photograph image with an approximate resolution of 1.5 m is shown in Figure 9.4. A manual count of the identifiable structures inside the 1 km square box in the photo amounts to about 740 structures, mostly single storey homes. If an LMDS system was considered with a service radius of 5 km and 4 sectors, each sector would potentially serve 19.6 km^2 or about 14,500 structures, assuming that the density was uniform. Applying a penetration rate of 10%, the hub sector should be designed to handle traffic from 1,450 remote terminals. A similar calculation for hub service radius of 3 km (7.06 km^2 service area per hub sector) results in a total of 5,230 structures potentially served using 740 structures per km^2. Applying the 10% penetration factor leads to 523 remote terminals generating traffic that will have to be handled by each hub sector.

Figure 9.4 Aerial photograph of a residential area in Eugene, Oregon. By manual count, the structure (dwelling) density inside the 1-km square (black line) is approximately 740.

9.3 SERVICE AND APPLICATION TYPES

A number of service or application types are currently used or envisioned for data traffic on wireless systems. For many system users, especially those at home, the data traffic will result from Internet applications. For businesses, there will be additional private traffic between buildings on point-to-point link systems or consecutive point link systems, or mesh networks. Table 9.3 shows a list of Internet applications taken from [3] and various other sources, along with expected nominal data rates and an indication of whether the service is *real time* or *non-real time*. This distinction will be discussed further in the following sections.

As Table 9.3 shows, the data rates for these services vary substantially in terms of required data rates. The high rates listed for streaming video, even with motion picture experts group (MPEG) compression, is significant. If a wireless broadband network was to represent a competitive replacement for people renting videos or DVDs, the on-demand transmission capacity could potentially add up to several streaming video channels into each subscriber household with multiple televisions. This requirement clearly extends beyond the 1-Mbps rate set forth in the Preface for defining broadband wireless.

Table 9.3 Downstream/upstream nominal data rates for various applications

Application	Service type	Downstream nominal data rate	Upstream nominal data rate
E-mail	Non-real time	144 kbps	144 kbps
Web-browsing	Real-time block transfer	1 Mbps	64 kbps
File transfer (FTP) image transfer	Real-time block transfer	2 Mbps	2 Mbps
Corporate WAN*, remote office connect	Real-time block transfer	10 Mbps	10 Mbps
Interactive games	Real-time block transfer	1.5 Mbps	1.5 Mbps
CD-quality stereo	Real-time streaming	256 kbps	—
Broadcast quality stereo	Real-time streaming	64 kbps	—
Toll-quality telephone	Real-time streaming	64 kbps	64 kbps
Cellphone quality telephone	Real-time streaming	6.4 kbps	6.4 kbps
HDTV	Real-time streaming	20 Mbps	—
Video on demand, MPEG2	Real-time streaming	4–6 Mbps	—
Video on demand, MPEG1	Real-time streaming	1–2 Mbps	—
Videoconferencing	Real-time streaming	64 kbps – 2 Mbps	64 kbps – 2 Mbps
Low-quality videoconferencing	Real-time streaming	<28.8 kbps	<28.8 kbps
Computer-aided design (CAD), supercomputer connects	Non–real time	45 Mbps	45 Mbps

*WAN wide area network.

9.4 CIRCUIT-SWITCHED TRAFFIC MODELS

Every communication link between two points can be regarded as a 'circuit', regardless of the technology that is used to implement it. Circuit switching basically assigns a circuit to transfer information between two points. The circuit assignment stays in place until the information transfer is finished. The definition of 'finished' can be somewhat nebulous for data traffic and Internet applications. In traditional telephone technology, and cell phone systems, the initiation and termination of a call, and the circuit assignment, are well defined and directly determined by whether a handset was 'off-hook.' For terminals that have a virtual perpetual connection to the network, the initiation of communication by the terminal (or by the network) may occur at any time and consist of very short messages or very long messages. Assigning a circuit for the duration of the communication would require a guess at when the communication had been terminated, which perhaps could be reasonably derived from inactivity. Nonetheless, for the time that data transfer was not taking place, the wireless resource represented by the circuit would be idle and the resource wasted instead of being used to carry the traffic of some other user.

9.4.1 Circuit-switched quality of service (QoS)

The quality of service, or QoS, of a circuit-switched communications system is normally given by the blocking rate. The blocking rate is the probability that a circuit will not be available when a terminal or user attempts to make a call or initiate communication. The blocking rate basically depends on how much traffic is to be carried and how many circuits or channels are available to carry that traffic. The unit of circuit-switched traffic is the *Erlang,* which is defined as

$$A = \frac{QT_d}{60} \text{ Erlangs} \tag{9.3}$$

where Q is the number of call requests in an hour, and T_d is the average call duration in minutes. For example, for 2000 call requests per hour, and an average call duration of 1.76 min (a number commonly used in the cellular industry), the amount of traffic is 58.7 Erlangs.

If the system is being designed to achieve a certain blocking rate, the Erlang value can be used to find the number of channels needed to carry this traffic with the chosen blocking rate. Two formulas are commonly used for this calculation: the Erlang B and the Erlang C formulas, as described below. The Erlang formulas are based on a number of assumptions about the traffic such as the following:

- There are an infinite number of traffic sources.
- The sources are independent.
- The arrival times for calls are Poisson-distributed.
- The holding times are exponentially distributed (for Erlang C).

For a given set of circumstances, many of these assumptions may be invalid. A typical example is a sporting event or trade show where the call density is concentrated in a particular location and the arrival times of the calls being placed is not independent but highly correlated to event schedule (a burst of calls at the end of a football game, for example). The result is a blocking rate that increases dramatically and a QoS that decreases. For cellular systems, the *ad hoc* remedy is temporary cell base stations on mobile trailer units or built into cabinets that can be wheeled into the trade show exhibit hall.

Note that the notion of traffic and circuits is quite general even though voice traffic on telephone circuits is where it is normally applied. The same approach could be used to assess traffic requirements for 10-Mbps data transmissions or 6-MHz video transmissions, for example, where the transmissions have a defined duration and circuits are scaled to accommodate the wider bandwidth signals.

9.4.1.1 Erlang B blocking probability

The Erlang B formula determines the number of channels needed to achieve a given blocking rate for a given traffic load. The Erlang B formula is also called 'blocked calls cleared', which basically means that if the call attempt is blocked because there is no circuit available, the call is not retried. Blocked calls are almost always retried, whether cellular or wire line, so the Erlang B formula is optimistic rather than conservative in terms of the number of channels it predicts that are needed to carry the traffic. In spite of this, the Erlang B formula is the most widely used for calculating the number of channels needed to carry a predicted amount of traffic in a cellular phone system. The Erlang B blocking probability is given by

$$\Pr(\text{blocking}) = \frac{\dfrac{A^N}{N!}}{\displaystyle\sum_{n=0}^{N} \frac{A^n}{n!}} \tag{9.4}$$

where N is the number of channels and A is the traffic in Erlangs as given by (9.3). If the blocking probability is given and the number of required channels is to be found, (9.4) can be calculated iteratively to find the number of channels that result in a blocking probability below the required level. The number of channels, of course, is an integer value.

9.4.1.2 Erlang C blocking probability

The Erlang C formula calculates blocking probability assuming a block call request will be retried (delayed) rather than abandoned as with Erlang B. In this regard, it is a more realistic way to calculate the trunking service given how telephone customers actually use their wire line or cell phone service. The probability of a call being delayed using the Erlang C formula is given as

$$\Pr(\text{delayed}) = \frac{A^N}{A^N + N!\left(1 - \dfrac{A}{N}\right)\displaystyle\sum_{n=0}^{N-1} \frac{A^n}{n!}} \tag{9.5}$$

where the variable definitions are the same as given above.

As a practical matter, the difference in the number of circuits needed to achieve a desired QoS when calculated using the Erlang B and Erlang C formulas is small. When considered in the light of the approximate and variable nature of the traffic volume estimate, the difference is irrelevant so either formula may be used.

9.5 PACKET-SWITCHED TRAFFIC MODELS

To better utilize the wireless spectrum resource, modern communication systems are designed to apply or switch radio resources on a packet-by-packet basis, called *packet-switched traffic*. Data transmissions on the Internet are carried out using data packets of varying lengths having embedded header data that determines their destination and assembly order. The newer 2.5G and 3G systems cellular phone networks also use packet switching for data and packetized representation of real-time voice transmissions.

Because the total data to be communicated is broken up into packets, the data transmission as a function of time from any single node can be viewed as an ON/OFF string of packets. When a packet is being sent, it is sent at the maximum data rate of the communication link. Either a packet is being sent or not sent (the link is idle), so the average data rate over some defined period will always be less than the maximum data rate capability of the modulation scheme/coding for the link. If the network does not have the transmission capacity to accept the packets at a particular time, the packets are placed in a queue and delayed until such time as the network can accommodate them. Because of the ON/OFF nature of the packet transmission, packet data is inherently *bursty* even for a single source.

When multiple sources are randomly and independently transmitting packets, the total required maximum transmission rate is the sum of the maximum packet data rate of the individual sources. Obviously, this only occurs if all sources attempt to send packets at the same time, which can occur but has a very low probability. It represents the upper limit of the capacity required to serve a multitude of sources and assumes that no packets will be put in a queue and delayed. It is highly inefficient to provide this maximum capacity since almost all the time it will be unutilized. Consequently, a network is normally designed to provide capacity based on some other measure of the total data transmission rate and accepts that some packets will be delayed and put in a queue. This delay or latency is analogous to blocking in a circuit-switched system. Delay therefore becomes a QoS parameter with its own statistical description. Latency is the round-trip delay in getting a response to a transmitted packet or packets. It will, therefore, depend on the delay incurred in the wireless network, the delay in the wired or optical network, and the delay in the remote server that is responding to the transmission.

With a queuing process for packets that depends on network loading, the delay or latency will clearly vary considerably as the load changes. The variation in latency is known as jitter and can make it difficult to assemble packets at the receiving location within the constraint of the data buffer.

As the network load (requests to send packets) increases beyond its maximum capacity to transmit, data and packets must be delayed, the total average number of packets transmitted for a single user in a given time period decreases. The resulting net *throughput* is another measure of the network QoS. Depending on the service type such as those listed in Table 9.3, throughput may be more important than latency. For example, for network telephony services, or other voice traffic, large latency will be noticeable and annoying even though the average throughput requirements are modest. For file or image downloads, the average throughput will determine how long it takes to complete the download task. For streaming voice and video applications, both QoS metrics are important. For multimedia applications with both voice and video (and perhaps data), the differential delay between these data streams will be important so that the sound of a voice is reasonably well synchronized with lip movement, for example.

It should be pointed out that the latency and throughput experiences by a user are not just a function of the broadband wireless network connection but also a function of all the other linked elements of the Internet. The high capacity optical fiber links and routers

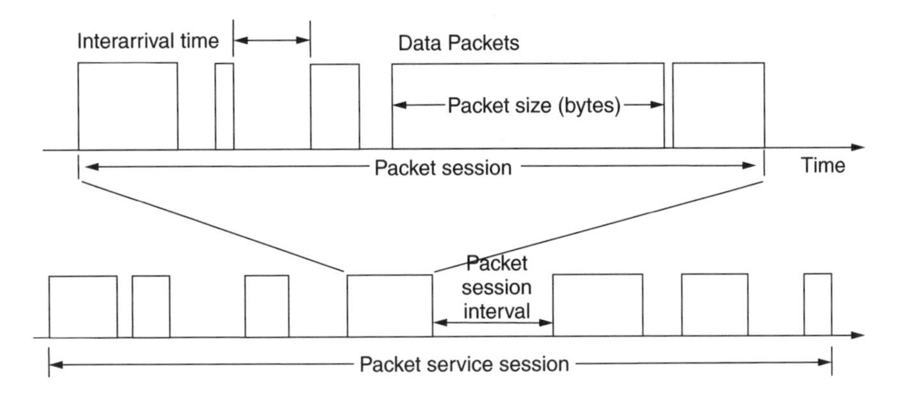

Figure 9.5 Schematic of data packet transmission sessions.

are usually not a large source of network QoS limitations. A significant factor can be the local connection to the Internet service provider (ISP) that hosts the Internet site, web pages, images, and so on, that a user requests. If there are limitations at that point, it will be reflected in the perceived performance, even those it has nothing to do with, the local wireless connection being accessed by the user. The larger questions associated with the QoS on the entire network is beyond the scope of this book. Instead, the focus here will be on design objectives for radio network capacity to carry packet traffic where the QoS metrics are determined for the fixed broadband wireless network alone.

Figure 9.5 shows a simple diagram of a collection of *packet sessions* or *packet calls* that consist of two-way communication of several packets. This diagram is a model for a single traffic source communicating with the network. This packet stream definition will be used to create models for aggregate traffic loads under various conditions.

9.5.1 Self-similar data characteristics

An interesting and somewhat surprising aspect of the data traffic on networks (Ethernet or the Internet) is its *self-similar* nature. Self-similar means that the statistical distributions that describe the data distribution (number of packets per second, per hour, per day, etc.) are very similar. Normally as quantity is observed over longer and longer periods of time, it tends to average or smooth out so that the statistical description changes. Network traffic is bursty over short periods (a packet is on or off), but continues to be bursty over longer periods indicating that the packets from multiple users tend to cluster together, and clusters tend to cluster together. This phenomenon was first reported in [4] after analyzing massive amounts of Ethernet traffic data.

As the system load increases and approaches the system capacity, however, more recent research [5] suggests that the traffic characteristics do tend to smooth out and more closely follow conventional Poisson distributions. From these somewhat divergent views of Internet traffic it is clear that reliable and consistent 'industry-accepted' models do not yet exist for network data traffic volume or time distributions. The modeling discussion and simulation results in this chapter follow the more commonly used packet traffic models for bursty data traffic distributions.

9.5.2 Packet probability distributions

On a fundamental level, network traffic can be viewed as the aggregate of the traffic from many different sources that may or may not be independent. Each of these sources are transmitting and receiving strings of packets. The packets have different lengths or sizes in bytes – the statistical distribution of the size is discussed below. The interarrival time of the packets (spacing between packet arrivals) is a second random variable that is not Poisson-distributed [6]. However, the arrival time of a packet session does reasonably follow a Poisson distribution [7] in a similar way to the calls in a circuit-switched network described above. This may be largely due to the packet interarrival distribution being determined by the network traffic control, whereas the initiation of a packet session is a human choice and therefore, more likely to be independent as required for a Poisson distribution.

A higher-level view of the network traffic is derived directly from the nature of the application itself. For streaming applications, a continuous stream of packets is presented for transmission by the source. In contrast, a web-browsing session may consist of multiple packet calls (the number being another random variable), with the time between packet calls depending on how long the user views the web page (the so-called 'reading' time).

In the following sections, some basic probability distributions are described that are currently regarded as viable choices for describing the many random variables involved in packet transmission.

9.5.2.1 Packet size distribution

The probability distribution most often used to describe the distribution of the packet size is the Pareto distribution given by

$$f(x) = \frac{\alpha k^{\alpha}}{x^{\alpha+1}} \quad x > k \tag{9.6}$$

where x is the packet size in bytes, k is the minimum packet size in bytes, and α is a constant that controls the shape of the distribution with $1 < \alpha < 2$. The cumulative distribution function (CDF) is

$$F(x) = \left\{ \begin{array}{ll} 1 - \left(\dfrac{k}{x}\right)^{\alpha} & \text{for } x \geq 0 \\ 0 & \text{otherwise} \end{array} \right\} \tag{9.7}$$

The Pareto distribution is similar to an exponential distribution except that the probability of large packet sizes is increased. While the Pareto distribution provides a convenient closed-form distribution, one set of observations of packet sizes showed that the packet sizes tend to have sizes that more closely follow a bimodal distribution [8]. In a bimodal distribution, the packets are either as large as allowed by the network (50% of the time) or the packets are of minimum size (40% of the time). The minimum sizes occur frequently because they carry acknowledgements (ACKs) of packet receipt and other data flow management functions. The remaining 10% of the packet sizes fall between these two extremes.

Because the Pareto is a single value monotonically decreasing function, for computer simulation purposes, values that occur following this distribution may be created using the inverse function method. This method uses a uniform random variable along with the inverse of the CDF to create a series of values x that occur with Pareto probability.

$$x = \frac{k}{U^{1/\alpha}} \tag{9.8}$$

where U is the random variable uniformly distributed from 0 to 1.

The minimum and maximum packet sizes depend on the network. In one model of web-browsing traffic [9] to be discussed in more detail below, the minimum and maximum packet sizes are given as 81 and 66,666 bytes, respectively, resulting in a mean value for the distribution of 480 bytes.

9.5.2.2 Packets and ADU's

An *application data unit* (ADU) is a block of data that is of arbitrary length and generated at an arbitrary time but is usually derived from the nature of the application itself. For example, an ADU may be a voice sample, a video frame, or a whole web page. Depending on the length of an ADU versus the length of the packets, the packet may consist of several ADUs or the ADU may consist of several packets [3]. This is also called packet fragmentation [10]. If an ADU or packet is corrupted so extensively that the forward error correction cannot fix it, an automatic request for retransmission (ARQ) is sent. The smaller the transmission block, the less likely the packet will be corrupted. However, smaller blocks, each bearing coding nonpayload bits, increase the total transmission overhead that decreases efficiency. This trade-off decision is usually designed into the medium access control (MAC) layer protocol for the system. For the discussion of traffic modeling here, a packet will be used as a general concept of a basic data transmission block.

9.5.2.3 Packet interarrival time distribution

As noted in [8], the packet interarrival times (the time between the end of one packet and the start of the next) for a single source are not Poisson-distributed. The nature of the distribution is a subject of current research with different proposed approaches. The web-browsing model in [9] suggests the geometric distribution to represent the packet interarrival times. The probability density function (pdf) of a geometric distribution describing a discrete random variable is given by

$$f(N = n) = p(1 - p)^{n-1} \quad \text{for } n = 1, 2, 3, \ldots \tag{9.9}$$

where p is the probability of the discrete time block time space between packets. From this it can be seen that the geometric distribution is the discrete form of the exponential distribution. The parameter $p = 1/\mu$, where μ is the mean value of the distribution. The value is clearly a function of traffic and thus a modeling choice. In [9] for a packet bit rate of 2.048 Mbps, the estimate of the average interarrival time is 0.00195 s.

The discrete (discontinuous) CDF is given by

$$F(n) = 1 - (1 - p)^{n+1} \tag{9.10}$$

From [9], the mean value is given in seconds as a function of the packet bit rate. The inverse function for generating random values with a geometric distribution is

$$N = \text{int}[\log(1 - U)/\log(1 - p)] \tag{9.11}$$

where U is a uniform random variable distributed from 0 to 1 and int[·] indicates taking the integer part of the argument.

An alternate distribution for interarrival time used is the Weibull distribution, which is a continuous distribution related to an exponential distribution. A Weibull distribution is given by

$$f(x) = \frac{1}{a}\left(-\frac{x}{a}\right)^{c-1} \exp\left[-\left(\frac{x}{a}\right)^{c}\right] \tag{9.12}$$

The CDF is given by

$$F(x) = 1 - \exp\left[-\left(\frac{x}{a}\right)^{c}\right] \tag{9.13}$$

The parameter c determines the shape of the distribution, while the parameter a can be used to shift the distribution range so that it has a specific mean value. If $c = 1$, the Weibull distribution becomes an exponential distribution. As the parameter c decreases, the traffic becomes more burstlike. Some studies of real traffic data indicate that a c value as low as 0.25 may be appropriate.

9.5.2.4 Distribution of the number of packets and the packet sessions

The number of packets in a packet session will be a random number with a particular distribution, as will the number of packet sessions in an entire user communication session. These are discrete (integer) values, which in [9] are assumed to be geometric distributions as described in Section 9.5.2.3 with particular average values. The geometric distribution for these two random quantities will be used for the simulations described later in this chapter.

9.5.2.5 Packet session interval distribution

A packet session consists of the transmission of several packets that may constitute the computer response to an action by the user to sent data, or by the network to respond to the request of a user. Between these packet sessions there may be a delay or interval that depends on the motivations of the user. A web-browsing model such as that found in [9] takes into account the fact that between packet sessions (sent or received), the user will pause to review the information received before sending any new information (called the 'reading time'). Certainly some services like web-browsing and e-mail will have intervals between packet sessions, while others like voice over Internet protocol (VoIP) and streaming audio or video will be a regular stream of packets with no user-introduced interval.

The average length of the user-introduced interval is highly variable given that it will be affected by the other tasks the user is engaged in. The probability distribution for the reading time from [9] is given as a geometric distribution as described in Section 9.5.2.3. A more general approach to packet session arrival times, and hence interarrival intervals, is discussed in the next section.

9.5.2.6 Packet session arrival distribution

Thus far the distributions have described the nature of the succession of packets that could originate from a traffic source; in this case, a remote terminal on a fixed broadband wireless network. The network will be required to handle a number of such independent sources so that the traffic distribution that will ultimately be used to define the traffic carrying capabilities of the wireless network will be determined by the distribution of this aggregate amount. For this purpose, a description of the arrival rate of packet sessions is needed.

The packet session arrival distribution is largely controlled by the actions of the human who is initializing communications. It is therefore likely to be more independent than packet arrival distributions; however, as noted in the discussion on circuit-switched traffic data, there will be events and schedules that result in highly correlated traffic creation. With that caveat, an appropriate model often used for independent arrivals is the Poisson distribution. The Poisson distribution is a discrete distribution that gives the probability that a certain number of (independent) packet sessions will arrive in a given time. The Poisson distribution is given by

$$f(N = n) = \frac{e^{-a}a^n}{n!} \quad \text{for } n = 0, 1, 2, 3, \ldots \tag{9.14}$$

where a is the mean value or the average number of packet session arrivals in a given time period. The CDF is given by

$$F(n) = \sum_{i=0}^{n} \frac{e^{-a}a^i}{i!} \tag{9.15}$$

The generating function for Poisson random values cannot be constructed using the simple inverse function method. Instead, a straightforward (but inefficient) method can be employed, which also works with any other discrete probability distribution. This method uses a uniform random number generator to first find a random value between 0 and 1. The cumulative distribution from (9.15) is then calculated incrementally for increasing values of n, $n = 1, 2, 3, \ldots$, until $F(n) \geq U$. The value of n where this occurs is the

Table 9.4 Parameters for ETSI web-browsing packet transmission model

Web-browsing data rate	Average number of packet calls within a session	Average reading time between packet calls within a session(s)	Average number of packets within a packet call	Average interarrival time between packets(s)	Parameters for packet size distribution
0.384 Mbps	5	412	25	0.01040	$k = 81.5$ $\alpha = 1.1$
2.048 Mbps	5	412	25	0.00195	$k = 81.5$ $\alpha = 1.1$

desired Poisson-distributed random variable. This process is repeated to generate a string of values that can be used for packet session arrival times.

9.5.3 ETSI web-browsing packet transmission model

Reference [9] presents a web-browsing model that has been commonly used to assess the traffic carriage requirements for 3G cellular systems. It postulates a basic packet transmission structure like that shown in Figure 9.5, including reading time between packet sessions or *packet calls* as it describes them. It also sets forth various parameter values to define each geometric and Pareto distribution needed to fully describe the packet transmissions. Those values are shown in Table 9.4.

This model can be extended to other data rates by adjusting the packet interarrival times. For the simulations used to create the traffic density results presented in Section 9.6, the European Telecommunication Standards Institute (ETSI) model is one of the two models used to create independent packet streams for each of the traffic sources in the amalgamation of traffic sources. The second model is the random packet cluster model described next.

9.5.4 Random packet cluster transmission model

The random packet cluster model is similar to the web-browsing model except that it eliminates the reading time transmission gap in the packet transmission scenario. As such, it represents a more continuous and thus intensive packet transmission as might be found with streaming data (audio and video), e-mail, and file transfer services.

9.6 MULTISOURCE TRAFFIC DENSITY MODELS

The core objective in considering the traffic geographic distributions and packet traffic densities in this chapter is to find a way of determining how much data capacity a fixed broadband wireless system must be designed to provide. The geographic traffic distribution model estimates where the traffic sources are located; the packet transmission models estimate how much traffic there will be from each traffic source.

The geographic traffic models are system-specific, so these distributions can only be invoked when a specific system design and its coverage are considered. However, regardless of where the cell or hub sector is located, it will still be serving multiple remote terminals (traffic sources), and the total traffic carriage requirements for that hub sector will be the aggregate of the traffic from these sources. As described at the beginning of Section 9.5, for a given number of traffic sources with a total requested transmission rate, the hub sector must provide sufficient capacity so that the QoS (throughput, latency, etc.) desired by the system operator is achieved. As a simple example, if 100 remote terminals on a sector are all seeking to simultaneously transmit data at 1 Mbps, the worst-case data rate capacity requirements for the hub sector would be 100 Mbps. If the system provides less that 100 Mbps, some packets from some remote terminals may have to be put in queues and delayed depending on the intervals between packets and packet sessions, and the length of the packets themselves. As the total provided capacity further decreases, the delays will increase and QoS will decline. As the delays increase, at some point the system may elect to drop packets.

For wireless system design, a quantitative description of this relationship is needed. The approach here is to use simulations of aggregate traffic demand with queuing to arrive at descriptions of this relationship under a variety of conditions. The description is normalized against the ratio of the total maximum user data rate and the channel transmission rate so that it can be applied to estimated traffic loads with delay for an arbitrary number of remote terminals connected to a hub sector. The simulation and the simulation results are presented in the following sections.

The simplest traffic flow is that which occurs from a single remote terminal and thus is viewed by the wireless system as a single service channel. In reality, a single remote terminal accessing the wireless network could actually be a single computer user with a wireless modem, or a home network with multiple computer users aggregating their data at a home hub, or an office where many computer users are aggregating their traffic through one wireless modem. The whole 'network' may thus be a hierarchy of several networks, each of which is aggregating and forwarding packets on to the next level (up or down) in the hierarchy.

The traffic generated by multiple remote terminals is the aggregate of the traffic from the individual remote terminals, each of which is generating packet sequences according to the probability distributions described in the previous sections. Given the complexity of the composite packet sequence distribution for the remote terminals, the most efficient approach to finding aggregate traffic is to construct a traffic simulator (software program) that contains a large, arbitrary number of independent packet sequence generators, each of which represents a single remote terminal in a wireless system. The packet lengths, packet intervals, session lengths, session arrival times, and so on can be controlled for each generator. The statistics of the aggregate data rate from these generators can then be examined over long periods of simulation time to arrive at an understanding of the capacity requirements of the radio access channel. The radio access 'channel' in this case is an all-encompassing term that includes all the possible multiple access/capacity dimensions including the modulation constellation, frequencies, time slots, code domains, polarizations, and colocated scattering channels, which the wireless infrastructure is designed to exploit. A traffic model is intended to predict the required capacity – how the wireless system provides the capacity may involve any or all the listed multiple access/capacity techniques.

The following sections provide the results and associated discussion for a number of simulation scenarios, along with the relevant simulation parameters.

9.6.1 Aggregate data rate statistics

The first useful value that can be developed from the simulation is the total aggregate data rate that the radio access channel must provide given that there is no packet queuing. The radio access channel permits transmission of every packet immediately whenever it is generated by a traffic source or sent to a traffic source (remote terminal). This value provides an upper bound on the wireless capacity.

For this simulation, three packet data rate mixes were used, as shown in Table 9.5. Other than the different data rate of the packets, the sources are assumed to be independent, identically distributed (i.i.d.) packet sources with the probability distribution parameters shown in Table 9.6.

The minimum and maximum packet lengths in Table 9.6 are taken from [9], which was prepared with wireless packet services in mind. However, general Internet packet traffic described in [3] indicates a minimum packet size of 40 bytes and a maximum packet size, or *maximum transfer unit* (MTU), of about 1,500 bytes. The minimum packet size corresponds to a simple acknowledgment or domain name system (DNS) packet with addressing and error control. For maximum packet length, there are some standards that proscribe a maximum packet length of 576 bytes. There are no imposed or agreed industry standards for upper and lower packet size limits. The wide variations in packet

Table 9.5 Data rate mixes

Data rate mix	64 kbps packets(%)	1-Mbps packets(%)	10-Mbps packets(%)
1	0	0	100
2	70	20	10
3	33	34	33

Table 9.6 Simulation probability distribution parameters

Probability distribution parameter	Parameter value
Minimum packet size	80 bytes
Maximum packet size	66,666 bytes
Pareto packet size α	1.1, 1.25, 1.40
Average number of packets in a packet session	25
Average packet interarrival spacing	500 bytes
Average packet session interarrival interval	0 s, 5 s

size parameters will affect the absolute value of the simulation results that follow, but the relative relationships of how parameters affect traffic loading are still valid.

The simulations that were carried out to produce the results presented here included 500 i.i.d. remote terminals. The simulation step times were set to equal the minimum packet length at the highest data rate. The results reported here were generated using a total simulation time of 10 min. Longer simulations of 30 and 60 min were also done with no significant change in the resulting statistics. The shorter time was therefore used for best simulation efficiency.

Figure 9.6 shows the CDF in percent for the normalized aggregate data rate for data rate mix 1 as defined in Table 9.5. The normalized aggregate data rate at time step k, $\overline{R}(k)$, is defined as

$$\overline{R}(k) = \frac{\sum_{n=1}^{N} R_n(k)}{\sum_{n=1}^{N} R_n} \tag{9.16}$$

where R_n is the packet data rate in bps for remote terminal n (64 kbps, 1 Mbps, or 10 Mbps from Table 9.5) and N is the number of remote terminals (500 in this case). Using the ON/OFF packet stream model discussed previously, the packet data rate at a time step k is either one of the three packet data rates (ON condition) or 0 bps (OFF condition) depending on the random values generated from the probability distributions. The normalized aggregate data rate was chosen as the parameter because it directly relates the capacity the wireless access channel must provide to the number of remote terminals in the hub sector service area and the data rate from each of those remote terminals. In this way it is tied to the geographic estimates of traffic distribution discussed in the first part of this chapter.

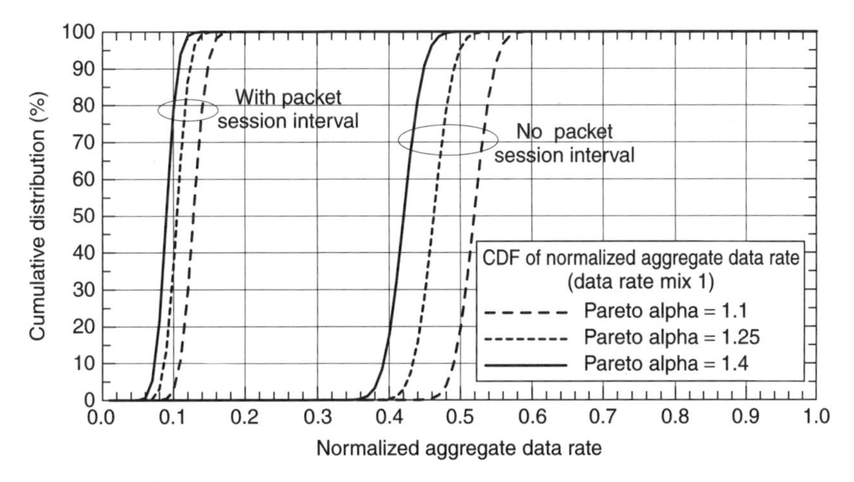

Figure 9.6 CDF for aggregate data rate for data rate mix 1.

Figure 9.6 shows the CDF for three different values of α that control the shape of the Pareto packet length distributions. Lower values of α result in a heavier tail on the distribution so that longer packet lengths are more probable. As expected, for a given number of packets, the longer packet lengths increase the relative overall time the packet ON condition exists from a remote terminal and thus raises the traffic demands for the wireless access channel. Even with $\alpha = 1.1$, 90% of the time the aggregate data rate required for the channel is less than 54% of the maximum possible total data rate from all remote terminals.

If a user-introduced interval between packet sessions is included with an average length of 5 s, the packet stream from each remote terminal has considerable time periods in which there is no packet being transmitted. The resulting three curves shown in Figure 9.6 reflect this fact with normalized aggregate data rates that are significantly less than those for the no-interval case. Again for $\alpha = 1.1$, 90% of the time the aggregate data rate required for the channel is less that 14% of the maximum possible total data rate from all remote terminals.

Figures 9.7 and 9.8 show similar results for data rate mixes 2 and 3 as shown in Table 9.5. The more spread out distribution results in Figure 9.7 reflect the low data rate packets that are sent from 70% of the remote terminals for data rate mix 2. Of course, in this case the maximum total rate from all remotes is less so the normalizing operation will also spread the distribution.

9.6.2 Aggregate data rate statistics with packet queuing (delay)

The wireless access channel capacity data rates indicated by Figures 9.6 through 9.8 assume that there is no multiple access delay due to the limitations in the capacity of the wireless access channels. The high access capacity required for this high level of service quality is generally not economical for the service provider to offer, nor is it

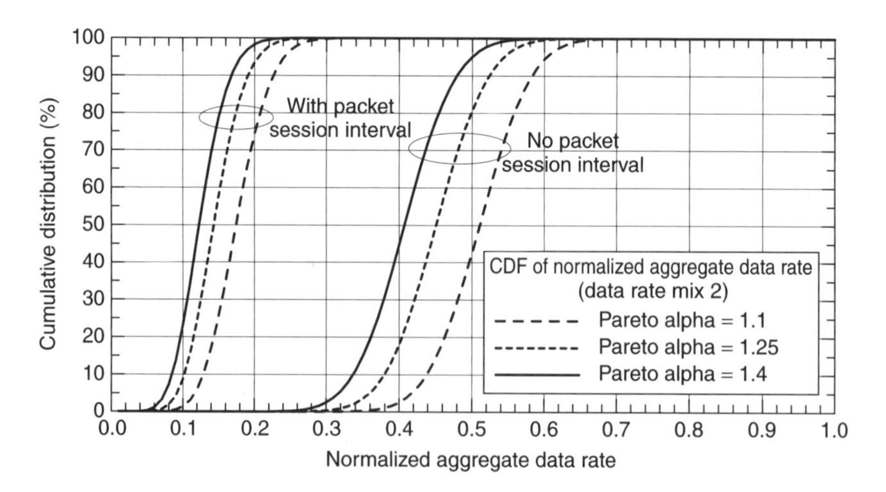

Figure 9.7 CDF for aggregate data rate for data rate mix 2.

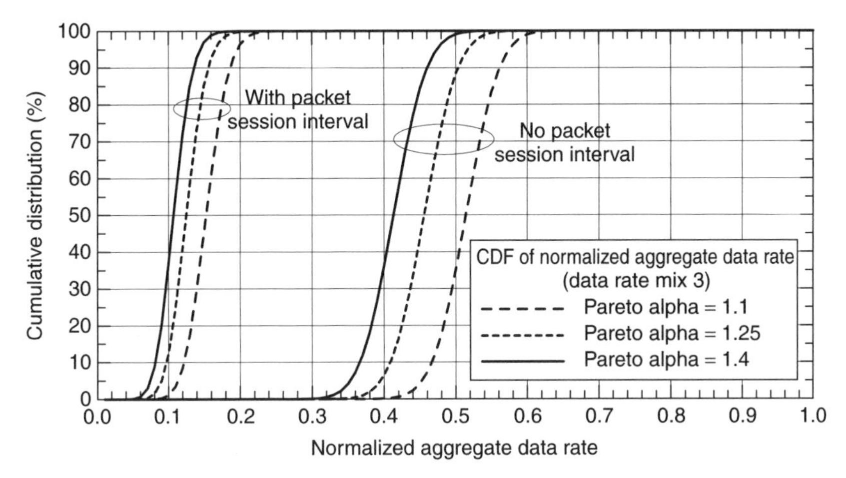

Figure 9.8 CDF for aggregate data rate for data rate mix 3.

sensible to provide such a 'no delay' radio network when significant delays or latency will be incurred in the Internet connection itself. A practical approach would be to choose a lower data rate than those indicated by Figures 9.6 through 9.8 incurring a delay that only slightly increases the total delay the user experiences, resulting from both the wireless system and the Internet. In short, the wireless infrastructure can be designed to impose no additional significant degradation in the QoS delay the user sees. Since the capability of the Internet itself with its myriad of possible connections and routings, and the remote server performance, cannot be controlled by wireless system design, the best that can be achieved with a wireless design is to significantly degrade the Internet-inherent QoS further.

Of course, for private networks (rather than the Internet) with a private multiple access wireless network, the latency and other QoS parameters for the data network may actually be under the control of the wireless network operator. In this case, an appropriate balance may be found between delay in the wireless access system and the rest of the network.

For either scenario, the relationship of how delay increases as the wireless access capacity decreases will provide the information needed to make these design decisions. For this purpose, the traffic simulator with i.i.d. remote terminals generating traffic was once again employed. However, in this case a maximum wireless channel access data rate threshold was set and the traffic from the remote terminals was regulated so that this threshold rate was never exceeded. When the number of remotes and their packet data rates exceed the threshold, the packets from some remotes are selectively delayed by putting them in queues to be sent at a later time.

The process of choosing which remote terminals can send data is usually referred to as *admission control*. The process of determining which packets are sent and when they are sent is known as *scheduling*. The admission control and scheduling algorithms can be adapted to achieve a variety of purposes that include giving higher priority to some customers or to certain high revenue services such as streaming video. This will be discussed further in Section 9.7. Other scheduling algorithms give preference to terminals with higher signal-to-noise ratios. Since these terminals are more likely to achieve error-free packet transmissions and thus require fewer retransmission requests, this scheduling strategy usually results in much higher total wireless access throughput than an algorithm that considers each terminal on an equal, unbiased basis.

For the first group of simulations shown here, the admission control and scheduling were done using a uniform random selection process so that no remote terminal or data rate level was favored over another. This is also the appropriate approach when assessing the total capacity requirements of the wireless channel for best-effort systems.

Figure 9.9 shows the one-way queuing delay percentage for data rate mix 1. This is the delay introduced by the wireless access network due to packet queuing. The delay percentage is the average number of time slots where a delay was required divided by the total number of time slots where packet transmissions were attempted. If the remote terminal had the wireless channel to itself (or the channel had the capacity equal to the maximum aggregate data rate), there would be no delays or queuing delay time slots, and percentage delay would be zero. A queuing delay percentage of 50% means that during the simulation, 50% of the time the remote unit attempted to transmit a packet, it was required to delay that transmission because no capacity in the wireless channel was available.

The average delay percentage (averaged over all 500 remote terminals) is shown as a function of the ratio of the wireless access data rate threshold to the maximum total data rate from all the remote terminals, as before. Here this is called *the normalized wireless access capacity:*

$$\text{Normalized wireless access capacity} = \frac{\text{Wireless access capacity (bps)}}{\text{Maximum aggregate data rate (bps)}} \tag{9.17}$$

For these simulations, the total simulation time was once again set to 10 min (600 s).

For these delay simulations, the Pareto packet size distribution parameter $\alpha = 1.1$ was used. The minimum normalized wireless threshold capacity of the wireless access systems was set to 0.01 (1%) of the maximum aggregate data rate. For 500 remote terminals, each

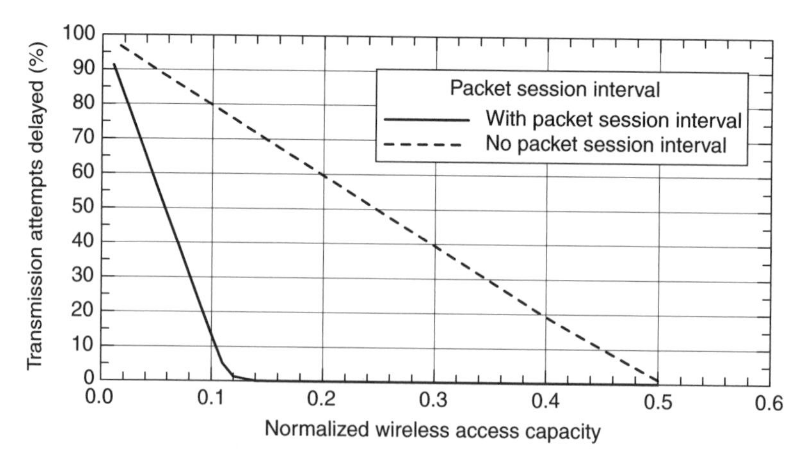

Figure 9.9 Wireless access-caused delay as a function of normalized access data rate.

capable of sending packets with a bit rate of 10 Mbps, the maximum aggregate data rate is 5 Gbps so the minimum wireless access data rate (1%) investigated with the simulations is 50 Mbps.

If intervals are inserted between packet sessions, the demand on the wireless channel is reduced as expected from the data in Figures 9.6 through 9.8. At a 10% normalized wireless access capacity, the percentage of transmission attempts delayed is about 13% as opposed to 80% for a continuous stream of packet sessions coming from each remote terminal.

As expected, as the normalized wireless access channel capacity is raised, the delay is reduced. If the maximum possible total data rate in bits per second from all remotes served by a sector is known, and the accepted wireless access-caused delay is known, the required channel capacity in bits per second to achieve a particular delay objective can be read directly from the curves in Figure 9.9.

The delays shown in Figure 9.9 are delays for one direction of transmission. If similar capacity ratios are provided in both directions as a function of maximum possible total data rate, the wireless network-caused round-trip delay will be twice this delay.

9.6.2.1 Internet latency

As mentioned, an important QoS parameter for the network is latency or absolute end-to-end transmission delay for a packet. Latency requirements only apply to real-time applications such as streaming audio or video and block transfer. It does not apply to non-real-time services such as e-mail or web-browsing interactions. Many studies have been done on latency on the Internet. Many of these studies have used data derived from a diagnostic technique known as *pinging* in which a packet is sent to a remote server and the time delay measured until a response is received. Continuing pinging studies are carried out for a long round-trip connection between the Stanford Linear Accelerator Center (SLAC) in California and CERN, the European Laboratory for Particle Physics near Geneva, Switzerland. Modeled as a Pareto random variable, the mean latency for data accumulated through 1999 is 188 ms [11]. The round-trip delay value used in the traffic model in [12] for a remote server is 300 ms. These delays are associated with the transmission of each packet since a packet must be sent and an ACK received before the next packet is sent.

While these latency figures are normally encountered, the latency requirements for what a user would perceive as 'good service' may be more restrictive. For voice communication, a round-trip delay should not exceed 500 ms, a delay level that telephone circuits using geostationary satellites struggle to meet. A response of 1 s for a service (like a movie) to begin after the user issues a 'start' command is considered acceptable, while a response time of 100 ms to any user-issued command is usually perceived as 'instantaneous' [3].

9.6.3 Throughput

Throughput is the amount of information bits that are *successfully* communicated from one location to another. Like delay, it is another QoS measure for systems handling packet traffic. The system throughput an individual remote terminal experiences is a function of several system characteristics:

- Maximum channel transmission rate.
- Multiple-access queuing delay.
- Overhead due to coding, preamble bits, synchronization bits, training symbols, and so on. Overhead can range from 10 to 50% of the total data bits transmitted.
- Packet receipt acknowledgment delay (ACK response).
- Retransmission of packets received with uncorrectable errors.
- Prioritized (biased) admission control and scheduling algorithms that give preferences to the transmission to/from one remote terminal over another.

The throughput can be investigated with the simulation, as was done with the delay multiple. Figure 9.10 shows the results of these studies as average throughput per remote terminal averaged over the simulation time. The maximum possible throughput will be the rate at which the sources actually generate data to send. With the packet length and inter-packet interval distribution, the average rate at which a remote terminal can generate data to send is about 5 Mbps, with no interval between packet sessions. If intervals are inserted between packet sessions, the maximum average traffic from a source is about 1.15 Mbps.

Figure 9.10 shows that with queuing, the packet scheduler essentially fills the channel completely so that the average throughput per remote is simply the wireless access channel capacity divided by the number of remote terminals accessing it. The curves for ratios below 0.12 shows this effect with the presence of the interval between packet sessions irrelevant since the channel is filled anyway.

Of the system characteristics listed as bullet points above, the throughput shown in Figure 9.10 only takes into account multiple access queuing delay. The throughput of actual information bits (the payload) can be linearly derated from the values in Figure 9.10 based on the percentage of transmitted bits devoted to overhead.

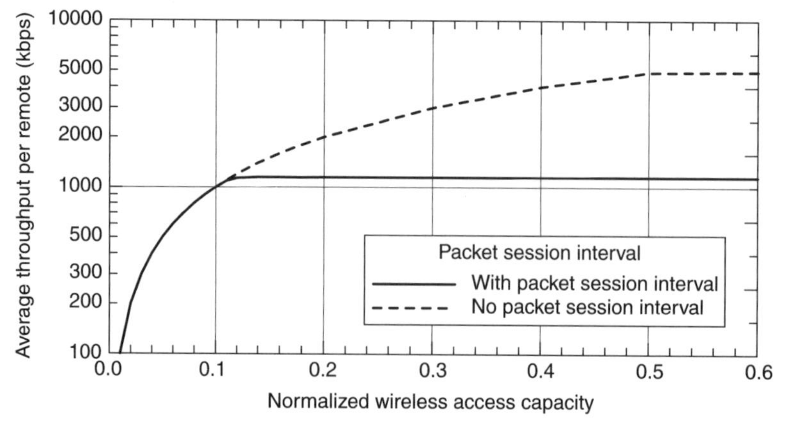

Figure 9.10 Average throughput per remote as a function of wireless access capacity.

The throughput limitations imposed by ACK, as noted above, will depend on the round-trip delay in whatever network the wireless access system is connected to, whether it is the Internet or a private network. It can be a significant limitation on throughput, and in fact, be the dominant limitation on throughput. The throughput limitations imposed by packet retransmission due to errors will be a function of the signal-to-interference + noise ratio (SINR) on the wireless channel from each remote terminal, assuming the errors introduced by the wired/optical network are substantially less. A typical design objective for the (uncorrectable) packet error rate requiring retransmission is 1 to 4% [13]. Retransmission of 1% of the packets will lower the throughput by this percentage and by the round-trip delays associated with the packet acknowledgement. The probability of packet retransmission can be a good measure for assessing system throughput.

9.7 APPLICATION MIX

The simulations done for most of the results in the preceding sections used uniform data rates and source distributions. In reality, each remote terminal will be generating a unique traffic pattern, with the peak packet data rate, packet length distribution, interval distributions, and so on, varying depending on the type of service. A useful approach to assessing the traffic load on a wireless hub sector is to postulate a mix of service types distributed across the population of remote terminals. Table 9.7 shows one possible application mix that is heavily weighted toward lower speed e-mail and web-browsing services. Lower weights are assigned to video and audio streaming services. These services consume more of the system resources and thus will probably carry premium usage rates and be self-regulated through price elasticity of demand. This type of weighted mix is indicative of current network usage, although for any particular population of users, this mix can vary widely from system to system and from sector to sector within a system. Determining an accurate estimate for the application mix for a particular service area and user population is the domain of current market research. This determination will be driven in part by the geographic distribution of the businesses and residences as described in the first part of this chapter.

For the real-time applications (telephony, streaming audio, and streaming video), the peak rates are set higher than the nominal rates in Table 9.3 so that the net throughput is approximately comparable to the rates in Table 9.3 using the peak packet-to-average throughput

Table 9.7 Sample application mix for simulations

Application	Percent of remote terminals (%)	Packet peak rate (kbps)	Average packet session interval(s)	Service priority
E-mail	50	64	30	5
Web-browsing	20	256	5	4
Telephony	20	64	10	3
Stereo audio	7	512	0	2
MPEG2 video	3	5	0	1

rates from Figure 9.10. For non-real-time applications (e-mail and web-browsing), the peak packet data rate is the nominal downstream rate from Table 9.3.

The packet session intervals are set to correspond to the type of delay that would occur because of human interaction. The telephony intervals assume a two-party conversation in which each party talks for 10 s and then listens for 10 s; it does not account for pauses during the speech of each party.

The real-time streaming services have no delays between packet sessions, although the packet lengths and inter-packet arrival times are determined by the size distributions and other parameters discussed earlier. Normally, a streaming service will be producing frames of data at some regular rate. For example, a movie with 30 frames/s, compressed with MPEG compression will have a data frame (30/s) of about 16.7 kB [3]. The frame content and packetization processing will cause the packets and interpacket intervals to vary, so a statistical description of some sort for these quantities is still appropriate.

The service priority factors in Table 9.7 are used to control the scheduling process. In the previously described simulations, random scheduling with uniform weights was used. The service priority in Table 9.7 biases the scheduling process so that higher priority applications are scheduled first (1 being the highest). A biased scheduling process also leads to the issue of a *reservation algorithm* in which initialization of a streaming video application, for example, would also reserve a certain wireless access capacity for this transmission so that its QoS in terms of delay and throughput could be maintained at a suitable level. Reservation algorithms can be complex and depend on *a priori* knowledge of the extent (data rate, duration, etc.) of the initiated transmission. Alternately, the first several seconds of the transmission can be used to create a forward estimate of the required capacity so that other non-real-time traffic is deferred as necessary. For the simulations here, no reservation scheme was used for the high-priority applications.

With these adjustments, the same simulation tool was used to evaluate wireless access delay and throughput as before. The results are shown in Figures 9.11 and 9.12 as a function of the normalized wireless access capacity as before.

The results show that a high level of normalized capacity is needed to support real-time streaming audio and video services. A reservation protocol of some kind is also necessary since packet transmission from lower rate services (e-mail, web-browsing) started during gaps in the streaming transmissions will introduce delays in those streaming transmissions because the lower rate packets cannot be interrupted while being transmitted even though the streaming services have a higher priority. For this service mix, a normalized capacity of 10% is a reasonable design point for the wireless access capacity if the streaming video services are not supported or supported at a very low level. If video is supported, the normalized capacity design point should be raised to about 25% so that the average throughput is about 500 kbps (16.7 kB/frame $\times$ 30 frames/s).

For the service mix in Table 9.7, the maximum aggregate data rate is 140.92 Mbps. At the 25% level, the capacity that the wireless channel must provide is 35.23 MB. If the wireless multiple access scheme (using all the various spectrum sharing techniques) could provide an average efficiency of about 3.5 bps/Hz, then 10 MHz of bandwidth could support service to this group of subscribers.

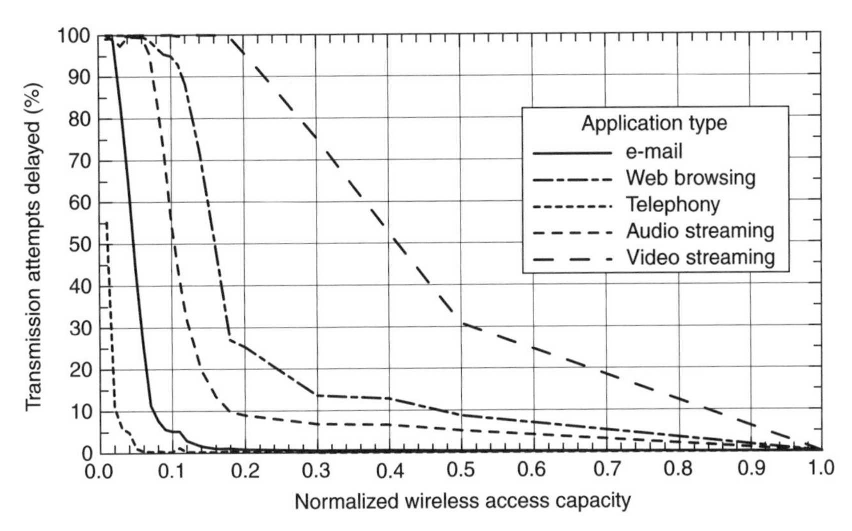

Figure 9.11 Percentage of packets delayed versus normalized wireless access capacity for application mix in Table 9.7.

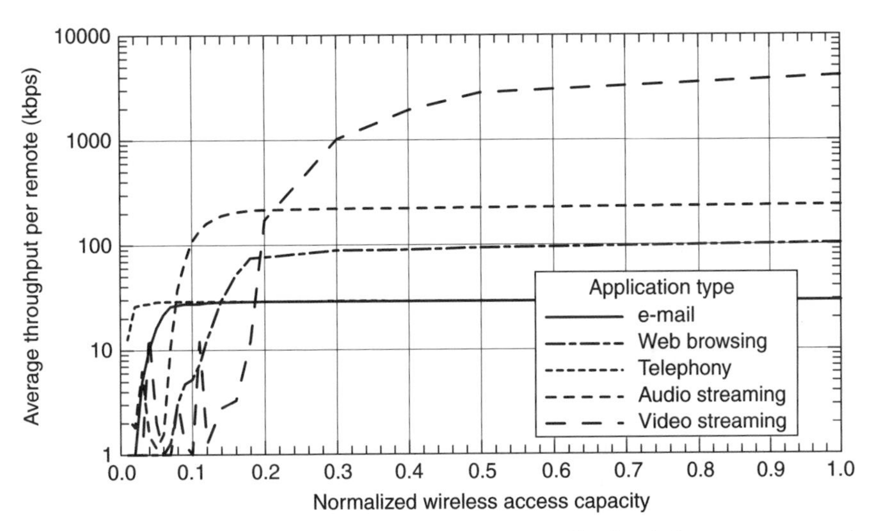

Figure 9.12 Average throughput per remote versus normalized wireless access capacity for application mix in Table 9.7.

9.8 BROADCAST AND ON-DEMAND VIDEO APPLICATIONS

One of the original ambitions of fixed broadband wireless services was to provide broadcast video to homes in direct competition with coaxial cable and satellite delivery systems.

The LMDS band in the United States at 28 GHz band and similar bands in other countries were set aside in anticipation of this type of service. A broadcast system of this type is primarily a one-way system in which a large number of television channels (10 or more) are available for selection at each remote terminal. The bandwidth needed to accommodate this system is about 6 MHz per channel assuming analog TV, normal definition with no compression. A system with a large number of channels could be accommodated in the multipoint, multi-channel distribution service (MMDS) and LMDS systems. In fact, the MMDS service was originally developed for television delivery. Several systems were deployed as one-way 'wireless cable' operations that successfully provided multichannel, cablelike service in low density and rural areas where building a coaxial cable system was not cost-effective.

The traffic requirements of such broadcast systems are significantly different from on-demand systems in which each remote terminal on a sector is receiving a potentially different and separate video program. This concept is essentially a 'wireless video store' in which a vast selection of videos could be delivered on demand to a customer's terminal with the convenience of not having to physically go to the video store. In this scenario, the wireless access system must provide for a separate video delivery channel to every customer who wants one. The potential demand here could be estimated from the rental activity at video stores. The number of households in a sector service area renting videos on a given day will be a small fraction of the 10% penetration rate discussed earlier.

A hybrid system that provides both the one-way video and a reasonable degree of on-demand video capability is certainly feasible with wireless and would potentially provide viable competition to cable and satellite services. Such systems would only require modest upstream capability. Cable currently has an average penetration rate of about 70% of television households. If the applications that a broadband wireless service is designed to provide are intended to compete for this very large audience, the traffic requirements of the wireless system may be substantially different from the two-way data-centric requirements presented in this chapter.

9.9 CONCLUSIONS

Modeling the traffic that a communication system will carry is an essential task in designing a successful fixed broadband wireless system. Unfortunately, it is often not given sufficient consideration relative to the hardware and software infrastructure aspects of the design.

Modeling traffic requires that the geographic location of traffic sources be determined along with the magnitude and the statistics of the traffic originating from and destined for each traffic source. The first part of this chapter presented several methods for estimating the geographic distribution of traffic sources. These include residential and business data from a government population census and commercially available sources, land use (morphology) data, specific building information, and aerial photographs. The choice of which of these resources is to be used will largely depend on what information can be conveniently (and economically) obtained and reduced to a form that can be used in a wireless system-planning tool.

The second part of this chapter described methods for estimating the capacity needed in the wireless access channel to accommodate a given amount of circuit-switched and packet-switched wireless traffic. While circuit-switched traffic is the traditional model used in cellular telephone systems, a fixed broadband wireless system using multiple access to serve wide areas will be handling packet-switched rather than circuit-switched traffic. The simulation results presented in this chapter show the relationship between the normalized wireless access capacity and the delay and throughput of the packet traffic for i.i.d. sources and for various mixes of sources with different data rates and statistics. These results can be used to estimate the wireless access capacity needed to achieve a given QoS objective in wireless system planning.

The actual traffic that must be accommodated with the desired QoS will potentially vary considerably from the traffic models presented here. Since no high-density fixed broadband wireless packet data systems currently exist, there is no operational data on which to base empirical traffic models. Moreover, the traffic demand will be driven by the desirability of the applications the system can support. Introduction of a new 'killer application' could cause traffic density to skyrocket and skew it in a direction that is not anticipated by the models presented here. The 'no packet session interval' curves in Figures 9.9 and 9.10 with a 10-Mbps packet data rate probably represent a reasonable upper bound of traffic load based on currently envisioned wireless applications.

The information in this chapter and Chapters 2 through 8 provides a broad wireless engineering foundation that will now be employed in Chapters 10 through 12 for the detailed design process of fixed broadband systems.

9.10 REFERENCES

[1] R. Greenspan, "Big Boosts in Broadband," CyberAtlas, INT Media Group, Incorporated. *http://www.cyberatlas.internet.com/markets/broadband/article* May 20, 2002.

[2] In-Stat/MDR, "Consumers adopt home networks to keep up with the Joneses," *http://www.instat.com/* April 10, 2002.

[3] T.C. Kwok, "Residential Broadband internet services and applications requirements," *IEEE Communications Magazine*, vol. 35, no. 6, pp. 76–83, June, 1997.

[4] W.E. Leland, M.S. Taqqu, W. Willinger, and D.V. Wilson, "On the self-similar nature of Ethernet traffic (extended version)," *IEEE/ACM Transactions in Networking*, vol. 2, no. 1, pp. 1–15, February, 1994.

[5] J. Cao, W.S. Cleveland, D. Lin, D.X. Sun, "Internet traffic tends *toward* Poisson and independent as the load increases." *Nonlinear Estimation and Classification*, eds. C. Holmes, D. Denison, M. Hansen, B. Yu, and B. Mallick. New York: Springer. 2002.

[6] V. Paxson and S. Floyd, "Wide area traffic: the failure of Poisson modeling," *IEEE/CM Transactions on Networking*, vol. 3, no. 3, pp. 226–244, June, 1995.

[7] M. Arlitt and C. Williamson, "Internet web servers: workload characterization and performance implications," *IEEE/ACM Transactions in Networking*, vol. 5, no. 5, pp. 815–826, Oct. 1997.

[8] C. Williamson, "Internet traffic measurement," *IEEE Internet Computing Magazine*, vol. 5, no. 6, pp. 70–74, November/December 2001.

[9] TR 101 112 V3.2.0, "Universal mobile telecommunications systems (UMTS): selection procedures for the choice of radio transmission technologies of the UMTS," *European Telecommunication Standards Institute (ETSI)*, pp. 33–37, (1998-04).

[10] D. Gesbert, L. Haumonte, H. Bolcskei, R. Krishnamoorthy, and A.J. Paulraj, "Technologies and performance for non line-of-sight broadband wireless access networks," *IEEE Communications Magazine*, vol. 40, no. 4, pp. 86–95, April, 2002.
[11] L. Cottrell "Overview and future directions of internet monitoring in EP," Presentation at the ICFA Standing Committee on Inter-regional Connectivity (SCIC) meeting, FNAL, Chicago, April, 1999.
[12] E. Anderlind and J. Zander, "A traffic model for non-real-time data users in a wireless radio network," *IEEE Communication Letters*, vol. 1, no. 2, pp. 37–39, March, 1997.
[13] J.C.-I. Chuang and N.R. Sollenberger, "Spectrum resource allocation for wireless packet access with application to advanced cellular internet service," *IEEE Journal on Selected Areas in Communications*, vol. 16, no. 6, pp. 820–829, August, 1998.

10

Single and multilink system design

10.1 INTRODUCTION

Whether the objective is a terrestrial point-to-point microwave link, an Earth-satellite transmission, a cellular network, a high frequency troposcatter system, or any other communication system, the elemental design component of each of the systems is the single wireless link that connects the transmitter with the receiver. As a fundamental building block, the design of single links will be examined in some detail.

Fixed broadband wireless links can be divided into three general categories:

- Long-range line-of-sight (LOS) links over mixed paths consisting of terrain, water, and buildings.
- Short-range LOS links over strictly urban paths with buildings of varying heights as the primary feature of the propagation environment.
- non-line-of-sight (NLOS) links of short to medium range in urban and suburban areas with the remote terminal antenna located either outdoors or indoors.

The design process consists of identifying the end points of the link where the transmitter and receiver are located, examining the details of the propagation environment along the circle path between those points, and on the basis of this information, postulating a link configuration (tower heights, antenna gains, transmitting and receiving equipment, etc.) and, finally, calculating the performance of this proposed design to determine whether it meets the required service and reliability objectives. The basic calculation of link performance is known as a *link budget* and takes into account all the gains and losses in the system, as well as noise and interference contributions. The fade margin derived from the link budget calculation is used to determine link availability under fading conditions.

Stand-alone single broadband links have several applications, including long-range connections between cities, cellular backhaul connections from cell sites to switching centers, inter-building high speed data connections, studio-transmitter links from broadcast

Fixed Broadband Wireless System Design Harry R. Anderson
© 2003 John Wiley & Sons, Ltd ISBN: 0-470-84438-8

stations, and telemetry links for utilities and other commercial operations. Broadband wireless links are often used in combinations or networks to achieve longer end-to-end range or to connect multiple points in some systematic way. Multilink systems fall into three categories:

- Multihop or tandem systems between transmitting-receiving locations that are too far apart, or not line-of-sight, so they cannot be connected with a single link.
- Consecutive point networks connecting multiple locations in 'ring' configurations.
- Mesh networks connecting multiple locations in 'branching' configurations.
- Point-to-multipoint (PMP) networks connecting many remote terminals to hubs in 'star' or 'spoke' configurations.

Multiple link systems, of course, are composed of a collection of single links, each of which can be designed and its performance assessed with a link budget as described above. Multiple link systems have some additional design consideration in that the overall system availability will be a function of the availability of all the links involved in the connections. Multiple link system design must also consider interference between links when the spectrum available for the system is limited. This may require careful selection of assigned link frequencies and directional antennas.

PMP networks are the most widely used topology for connecting large numbers of locations. Because of their significance, Chapter 11 is devoted entirely to the design of PMP wireless networks.

10.2 LONG-RANGE LOS LINKS OVER MIXED PATHS

A single link over a path consisting of a mix of physical features is the most general kind of link design problem. A typical link profile over a mixed propagation environment is shown in Figure 10.1. This profile was automatically drawn using a

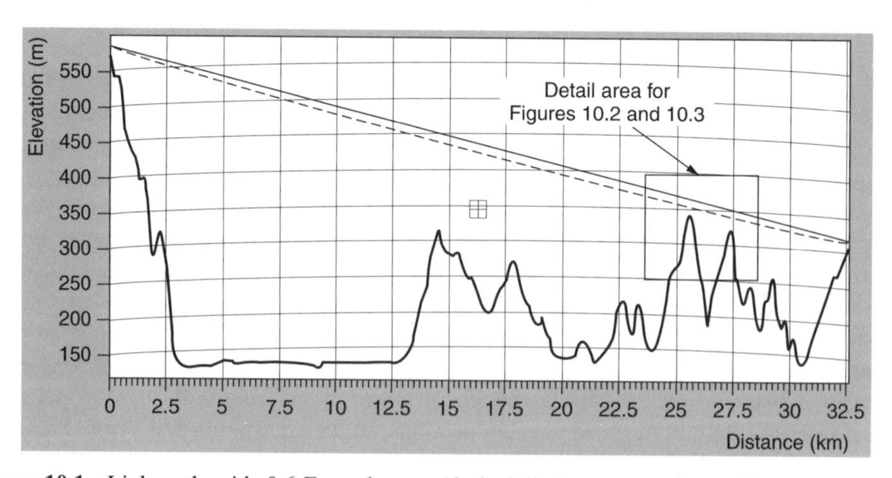

Figure 10.1 Link path with 0.6 Fresnel zone (dashed line) over terrain profile extracted from 30-m resolution terrain database (DEM).

wireless system-planning tool – a capability found in most modern planning tool software packages. The fundamental information to be gleaned from such profile displays is path clearance over intervening obstacles. While creating the profile, other link performance information is usually calculated.

The engineering issues associated with link path design are derived mainly from the transmitter-receiver link geometry relative to the propagation environment features along the path. The two most important path geometry considerations are path clearance and the reflection point location. These subjects are discussed in the following sections. Additional background information can be found in the classic monograph by White [1].

10.2.1 Path profile clearance analysis

The profile in Figure 10.1 is the fundamental tool for assessing path clearance. The typical design objective for an LOS path is to adjust the transmitting and receiving antenna centerline heights so that the 0.6 first Fresnel zone is free of obstructions. As discussed in Chapter 2, this value provides for essentially total power transmission in the propagating plane wave past the obstacle. The Fresnel zone is actually a 3-dimensional ellipsoid of revolution around the path whose width varies along the path (largest in the middle of the path) and decreases with frequency as given by (2.46).

In assessing path clearance, information that is not shown on the terrain profile must also be considered. In many areas, the heights of trees on top from the terrain high points are an obvious source of obstructions. Figure 10.2 shows a zoomed-in view of two obstacles inside the box in Figure 10.1. A tree obstruction has been added to the hills to represent the presence of trees on the actual mountain ridge. The tree shown in Figure 10.2 impinges on the 0.6 first Fresnel zone (shown by the dashed line), but not the radio path itself. This circumstance would still be considered an LOS path, although with an obstruction in the Fresnel zone. Figure 10.3 shows the same situation with a taller tree that actually does obstruct the path. For this geometry, knife-edge diffraction calculations,

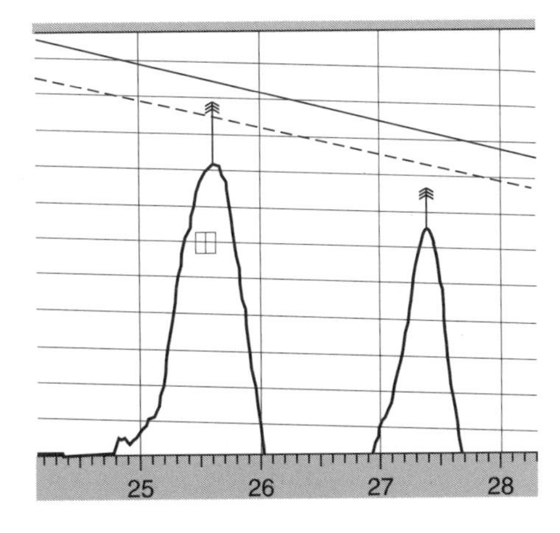

Figure 10.2 Representation of trees added to terrain profile in Figure 10.1.

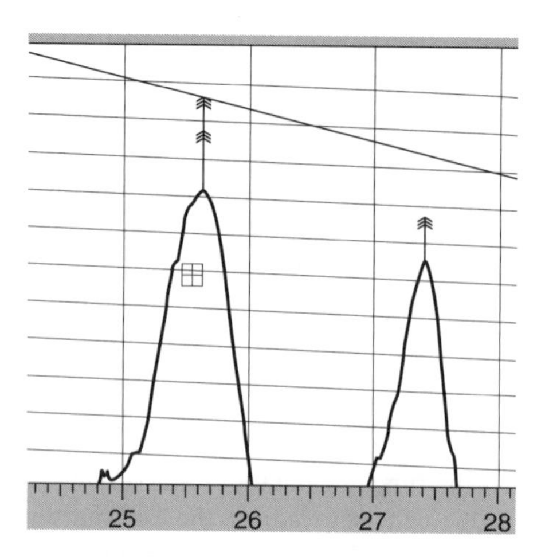

Figure 10.3 Tall tree obstructing radio path shown in Figure 10.1.

as discussed in Chapter 2, can be used to find the additional path loss introduced by the tree obstruction. Accounting for trees as opaque obstacles, like terrain, is generally a better and certainly more conservative approach to estimating path loss than treating the trees as partial absorbing/scattering screens.

The growth rate of trees must also be considered. Microwave links are often expected to function for 20 or 30 years. Over this period of time, the height of trees along the path can increase by several meters. If the initial clearance over trees is too tightly designed, at some point in the future the link performance could begin to degrade as trees partially obstruct the path.

Path clearance will vary with changes or variations in atmospheric refractivity gradients. Data for the distribution of atmospheric gradients from references cited in Chapter 2 can be used to set a range of k factors and the path clearance evaluated for each of these k factors. Typically three k factors are used – high, average, and low – with the actual numbers dependent on the specific location and climatic characteristics of the area in which the link will be built. Figure 2.23 shows a path profile clearance analysis with paths drawn as curved lines representing three refractivity gradients. A well-designed link will maintain adequate clearance under all these refractivity conditions.

A link that lacks adequate clearance can be corrected by increasing the antenna heights on either end or choosing a different link routing for the system if possible. If rerouting is not possible, increasing the antenna heights is the only option, but carries with it additional expense in terms of increased tower heights (if new towers are being built) and increased power loss due to longer waveguide lengths connecting the transmitter and its antenna, or the receiver and its antenna.

Depending on the location of the obstacle along the path, one end of the link will be the better candidate for increased antenna height. As a simple consideration of the geometry shows, increasing the antenna height on the end of the link closest to the obstacle will achieve adequate clearance with the least amount of antenna height increase.

This approach may be tempered by zoning and other environmental restrictions on tower heights.

Some rare physical situations may also require nearby lateral obstacles to be considered. Lateral obstacles can occur naturally as steep or vertical canyon walls. For such situations, a static analysis of path clearance is sufficient since significant variations in atmospheric refractivity occur vertically rather than laterally. The presence of nearby lateral obstacles is much more common in high-density urban (building) environments and will therefore be discussed in more detail in Section 10.3.

10.2.1.1 Path clearance validation

With high-resolution terrain databases and conservative estimates for trees and other supplemental obstacle heights, a computer analysis of the path should provide a very reliable indication of whether sufficient clearance has been achieved. However, when some aspect of the computer path analysis is questionable, either because the databases are known to be of low resolution or out-of-date, or the clearance is marginal, it is prudent to validate clearance by an on-site inspection of the link itself.

An on-site inspection uses one of several means to establish that *visual* LOS exists between the candidate link endpoints. This can be done by using telescopes and binoculars, with the addition of light sources or mirror flashes to aid in identifying the intended far link endpoint. A theodolite may also be employed that will not only establish LOS conditions, but azimuth and elevation point angles from one site to the other if the link is sufficiently short so there is little divergence between the visual and radio path. If the contemplated antenna placement is some height above the site on a tower yet to be constructed, balloons or helicopters are sometimes employed to simulate the actual location of the link antennas in their final elevated mounting positions.

The visual validation of path clearance is not necessarily an extraordinary expense. In completing the design of a link system, it usually is necessary in any event to visit each site to obtain construction-specific information such as power location of the power mains, any access restrictions, and other similar details.

10.2.2 Reflection point analysis

The location of the reflection point along the path can have an impact on the signal fading statistics present at the receiver. As discussed in Chapters 2 and 4, a single specular reflection adding out of phase with the directly received signal can cause nearly complete cancellation of the resultant signal envelope amplitude, as well as increase the notch depths of frequency-selective fading in wideband systems. The location of the reflection point(s) is the point (or set of points) where the angle of incidence and angle of reflection are equal. Path geometries in which the reflection point falls on a body of water or a flat field are more likely to result in high amplitude specular reflections that can increase the probability of deep signal fades. 'Flat' in this context is typically regarded as areas where the RMS variation in the surface height is less than a quarter of the wavelength and the size of the area is at least as big as the first Fresnel zone projected as an ellipse on the reflecting surface (see Section 2.6.2).

The location of bodies of water and flat terrain can often be effectively derived from land use or morphology databases as described in Section 5.4. Bodies of water are usually clearly delineated in such databases. Areas with flat ground can be classified in a variety of ways such as 'agricultural', 'rangeland', or 'open'. Considering the morphology class in conjunction with high-resolution terrain information can further refine the determination of whether a reflection point is on a surface that is potentially a strong specular reflection source.

The reflection point location can be adjusted by changing the antenna heights at the end of the link, as illustrated by the drawing in Figure 10.4. In this case, the antenna height of the transmitting end has been increased to move the reflection from the water to the land farther along the profile. As shown in Section 2.6.3, reflections from rough surfaces result in diffuse scattering with a suppressed specular component. This generally improves the fading statistics with deep fades less probable.

Controlling the location of a reflection point is more important for links in areas with relatively static atmospheric conditions such as flat agricultural areas or coastal regions. Fading results from the variations in the lengths of the direct and reflected path lengths so that the absolute phase delay of each will vary independently and at times will be completely or partially out of phase. For a static atmosphere, a particular refractivity gradient and direct/reflected path lengths can result in the out-of-phase condition existing from an extended period. Atmospheric conditions with substantial mixing, such as those in mountainous areas, are less prone to this phenomenon. The fade outage models presented in Chapter 4 and used for calculations in the following sections have parameters that depend on the link environment with the purpose of taking this static atmosphere into account.

For digital system path geometries in which the (mean) absolute lengths of direct and reflected paths are substantially different, the reflected digital signals can arrive sufficiently delayed to cause intersymbol interference (ISI) as discussed in Section 4.2.4. This is a factor to consider in link routing but normally nothing can be done with antenna heights to appreciably change the absolute direct and reflected path lengths so that the delay time is less. Instead, broadband digital systems rely on equalizers to suppress reflection components that are sufficiently delayed to cause ISI (see Section 7.5).

Given the path geometry, it may not be possible to move the location of the reflection point. However, if other factors are equal, this additional effort in the path design can be worthwhile for improving link reliability. Quantifying the amount of improvement is difficult since changes of this sort in operating links have not been the subject of extensive empirical research.

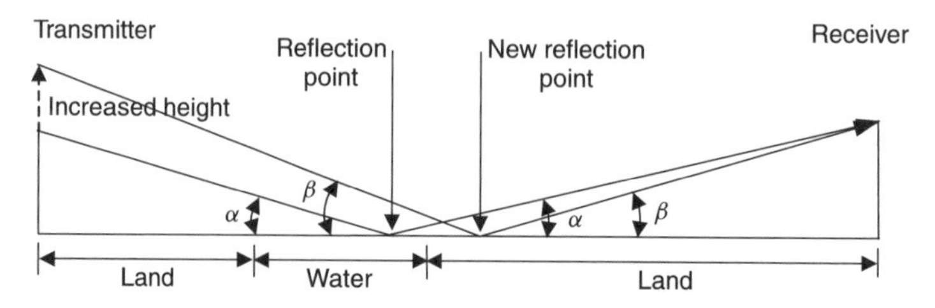

Figure 10.4 Moving a reflection point by increasing the antenna height on one end of the link.

10.2.3 Link budget

The link budget is a tabulation of all the gains and losses for the link that are added in order to arrive at the mean signal level at the receiver. This number can be used to assess the availability of the link under a variety of fading mechanisms. The link availability is the percent of time averaged over a month or a year when the link performance in terms of bit error rate (BER) or frame error rate (FER) meets or exceeds an acceptance criterion.

Table 10.1 shows a link budget for a 6-GHz link on a 50-km path. Each of the elements in the link budget is discussed below.

Table 10.1 Basic link budget

System element	Gain
Transmitter output power (dBmW)	35.00
Transmitter circulator/multiplexer loss (dB)	−0.50
Transmitter waveguide loss (dB)	−2.00
Transmitter antenna radome loss-dry(dB)	−0.50
Transmitter antenna gain (dBi)	35.00
Effective radiated power (dBmW)	**67.00**
Frequency (MHz)	6000.00
Path length (km)	50.00
Free space pathloss (dB)	−141.97
Multiple diffraction loss from terrain and buildings (dB)	0.00
Atmospheric absorption/water vapor loss (dB/km)	−0.01
Atmospheric absorption loss for path (dB)	−0.45
Partial Fresnel obstruction loss (dB)	0.00
Building clutter loss (dB)	0.00
Foliage loss (dB)	0.00
Building penetration loss (dB)	0.00
Total median path loss (dB)	**−142.43**
Receiver antenna gain (dBi)	35.00
Receiver antenna radome loss-dry (dB)	−0.50
Receiver waveguide loss (dB)	−2.00
Receiver circulator/multiplexer loss (dB)	−0.50
Median signal level at receiver input (dBmW)	**−43.43**
Receiver noise figure (dB)	5.00
Receiver equivalent noise bandwidth (MHz)	10.00
Receiver noise threshold (dBmW)	**−100.66**
Total external interference level (dBmW)	**−105.00**
Required signal-to-noise ratio, SNR (dB)	16.00
Required signal-to-interference ratio, SIR (dB)	16.00
Thermal noise fade margin (dB)	**41.23**
Interference fade margin (dB)	**45.57**

Transmitter output power. This is the time-average power of the link transmitter on the transmission channel. By convention, the power level is given in dB relative to one milliwatt, dBmW, or simply dBm. Normally the link is operated with the maximum possible transmitter power. However, automatic transmit power control (APC) is also used on links so that a power reserve is available to respond to rain fades or interference from other links. With APC, the nominal transmit power is automatically adjusted so that the BER or FER on the link meets some desired threshold.

Transmitter circulator/multiplexer loss. Normally the output of a transmitter is combined with the signal from other transmitters operating on other channels, or with the receiver input when the transmitter and receiver share the same antenna. This multiplexing/splitting device (called a circulator for combining/splitting waveguide signals) has some loss associated with it. The loss value in dB is obtained from the equipment manufacturer.

Transmitter waveguide/transmission line loss. This is the loss in dB from the waveguide or transmission line connecting the transmitter (or circulator) with the antenna. The loss depends on the waveguide type, waveguide length, and frequency. The loss as a function of length for several waveguide types can be read from the chart in Figure 6.22. The loss for lengths other than 100 m can be found by linear interpolation. For transmission lines, Table 6.1 can be used to find the loss for some common fixed broadband wireless bands.

Transmitter antenna radome loss. The radome cover on the antenna introduces some loss that varies with radome material type and whether it is wet or dry. Sections 4.3.4 and 6.9 discuss radome losses. If the radome loss when wet is significantly higher than the value when dry, the wet value should be used for conservative link design. This loss value is obtained from the antenna manufacturer who normally supplies the necessary radome.

Transmitter antenna gain. The transmitter antenna gain value depends on the antenna type (mostly its cross section or aperture size) and is obtained from the antenna manufacturer. It is assumed that the transmit antenna will be oriented so that its maximum gain direction is pointed along the path toward the receiver antenna, so the maximum gain value from the manufacturer is used here. The antenna type and its gain is one of the link system elements the design engineer can change to improve link performance if the initially selected antenna does not provide sufficient link fade margin. By convention, the gain of antennas used in microwave and other fixed broadband wireless systems are usually given in dB relative to an isotropic radiator, denoted as dBi.

Effective radiated power (dBmW). The effective radiated power (ERP) is the sum of the transmitter power and transmit antenna gain minus the losses from the multiplexer/circulator, waveguide, and radome. Since the radiated power is calculated using the antenna gain in dBi, the ERP value is denoted as 'ERPi'. It has the same units as the transmitter power, or dBmW.

Frequency. This is the operating frequency of the link. This is the nominal center frequency of the channel, which is only used in the link budget to find pathloss so it is not necessary to specify it with a great deal of precision.

Path length. The distance from the transmitting antenna to the receiving antenna along the great circle path is the path length. The great circle path length can be found using spherical trigonometric calculations.

Free space path loss. The total path loss is the free space path loss, L_f, calculated using the Friis space loss formula (Section 2.5.1):

$$L_f = 32.44 + 20 \log f + 20 \log d \text{ dB} \tag{10.1}$$

where f is the frequency in megahertz and d is the path length in kilometers.

Atmospheric absorption/water vapor loss. The atmospheric absorption (dry O_2) and water vapor losses in dB/km are read from the curves in Figure 2.27 for the link operating frequency.

Atmospheric absorption loss for the path. This is the total atmospheric absorption/water vapor loss along the link path, which is found by multiplying the path length by the atmospheric absorption/water vapor loss in dB/km.

Partial Fresnel obstruction loss. For a properly designed LOS link with adequate designed path clearance, there should be no partial obstruction of the 0.6 first Fresnel zone, so this loss should usually be zero. However, for special circumstances at low frequencies and long paths in which the Fresnel zone is large, it may not be possible to achieve such clearance. This loss varies from 0 to 6 dB proportionally to the degree to which the Fresnel zone is obstructed. The 6 dB loss occurs at grazing incidence (see Section 2.7.2).

Building clutter attenuation. For LOS paths, the path clearance is set so that there are no intervening building obstructions and this parameter is 0 dB. For NLOS paths, the loss is calculated using the models in Section 3.5.6.1 of Chapter 3 and using other models found later in this chapter.

Foliage attenuation. For LOS paths, this value is set to 0 dB since the clearance is set so that there are no trees or bushes obstructing the path. For NLOS links discussed later in this chapter, the loss becomes an important factor.

Building penetration attenuation. For LOS paths, the clearance is set so that the link terminals are outside with no building or structure obstructions attenuating the signal.

Accordingly, for LOS paths this parameter is set to 0 dB. For NLOS links discussed later in this chapter, building penetration is an important loss factor.

Total median path loss. This is the sum of the free space and other loss factors along the path. This is normally considered as either the median or mean path loss. For typical propagation conditions, the median and mean of the path loss variability distribution are within about 1 dB of each other.

Receiver antenna gain. See the discussion for the transmitter antenna gain mentioned earlier.

Receiver antenna radome loss. See the discussion for the transmitter radome loss mentioned earlier.

Receiver waveguide/transmission line loss. See the discussion for the transmitter waveguide/transmission line loss mentioned earlier.

Receiver circulator/multiplexer loss. This is the loss of the circulator/multiplexer used at the receiver. For two-way links, the equipment at each end of the link may be symmetrical so this multiplexer may be the same as transmit multiplexer.

Median signal level at receiver input. The signal level exceeds 50% of the time (locations) at the receiver input. This is the sum of the transmitter ERPi and receiver antenna gain minus the losses from the receiver multiplexer/circulator, waveguide, radome, and, of course, the total median path loss.

Receiver noise figure. The receiver noise figure in dB is a function of the receiver equipment design. The noise figure value is obtained from the receiver equipment manufacturer. Low noise amplifiers (LNAs) can use special components or cryogenic (cooling) techniques to achieve very low noise figures. For links with substandard or inadequate fade margins, using an LNA is another approach for improving link performance. Noise figures for standard receivers range from 3 dB at 2.5 GHz to 8 dB at 38 GHz [2,3].

Receiver equivalent noise bandwidth. The equivalent noise bandwidth of the receiver is the bandwidth necessary to accommodate the digital signal symbol spectrum and symbol rate. The receive filter is designed to provide bandwidth that maximizes the received SNR. For noise power purposes, the receiver filter is modeled as a perfectly band-limited filter with the same noise power characteristics as the actual filter. The bandwidth of this perfectly band-limited filter is the equivalent noise bandwidth. This value is obtained from the receiver equipment manufacturer.

Receiver noise threshold. The receiver noise threshold is found from the noise figure and equivalent noise bandwidth. The noise threshold is given as

$$P_N = (F - 1)K_b T_0 B \tag{10.2}$$

where F is the noise figure (as a real number, not dB), K_b is Boltzmann's constant = 1.37×10^{-23}, T_0 is the ambient temperature in degrees Kelvin of the environment, usually taken to be 290°, and B is the equivalent noise bandwidth in Hertz give above. See Section 8.2.1.1 of Chapter 8 for a detailed discussion.

Total external interference level. For a single link design, the external interference power may be zero. In high-density systems with many links where the performance is limited by interference rather than noise, this factor in the link budget may be very significant. This value can be calculated from specifications (location, power, antennas types, etc.) for the other potential interfering links. This will be discussed in more detail in Chapters 11 and 12.

Required signal-to-noise ratio, SNR. The required SNR value is the SNR value needed at the input of the receiver to achieve a specific signal quality at the output of the receiver. For an analog system, the output quality measure will be an analog SNR. For a digital system, the fundamental measure of output quality is the BER or FER. This value is the design objective, which the link must maintain for a specific percentage of time (link availability). This required SNR is found on, or can be derived from, the equipment specifications. Some typical values taken from [2,3] are shown in Table 10.2.

Required signal-to-interference ratio, SIR. The required SIR value is the SIR value needed at the input terminals of the receiver to achieve a specific signal quality at the output of the receiver. This may differ from the SNR value although normally the noise power and the interference power are added together to arrive at a signal-to-interference + noise ratio (SINR) value, which is then used with the required SNR objective to find the fade margin. During some multipath and rain fades, the interference level may be varying along with the desired signal so that the SIR is not changing even though the SNR is changing. For this reason, it is appropriate to calculate the SIR separately from the SNR.

Table 10.2 Nominal required SNR values by modulation type

Modulation	Net efficiency (bps/Hz)	Required SNR (dB)
QPSK	1.8 (10% overhead)	10
16QAM	3.4 (15% overhead)	15
64QAM	4.8 (20% overhead)	21
256QAM	6.0 (25% overhead)	26

Note: QPSK is quadrature phase shift keying and QAM is quadrature amplitude modulation.

Thermal noise fade margin. The thermal noise fade margin is the difference between the median signal level at the receiver input and the sum of the receiver noise threshold plus the required SNR. Under nominal link conditions, this difference (fade margin) is present. When the link fades, the probability that the fade depth exceeds the fade margin is the probability that the BER or FER will fall below the minimum value that is considered acceptable performance. In this circumstance, an outage is considered to have occurred. The fade margin is the ultimate result of the link budget calculation. It will be used for link availability calculations that are discussed in the following sections.

Interference fade margin. The interference fade margin is the difference between the median signal level at the receiver input and the sum of the interference plus the required SIR. Under nominal link conditions and interference levels, this difference is present. When the link fades or interference levels increase, the probability that the fade depth exceeds the fade margin is the probability that the BER or FER will fall below the minimum value that is considered acceptable performance. As with the thermal fade margin, this happens when an outage has occurred.

The formulas needed to calculate the values in the link budget use very basic mathematical functions and, as such, are very straightforward to implement in commonly available spreadsheet programs. A link budget spreadsheet provides a convenient way of trying different parameter values (antenna gain, waveguide losses, interference, levels, etc.) to determine their impact on the resulting fade margin.

The term 'link budget' is occasionally used to indicate the total path loss that can be accommodated before an outage occurs. Such a link budget value can be found using the same values in Table 10.1, but inverting the calculation by specifying the fade margin rather than calculating it. With the fade margin specified, the unknown quantity to be calculated is the total path loss that can be tolerated, or the 'link budget'. This approach is somewhat more convenient for NLOS links in which the path loss cannot be calculated with the same certainty as it can be with LOS links. However, it does require an assumption about the necessary fade margin, which, in turn, requires an estimate for the fading – a statically described quantity like varying path loss.

10.2.4 Fade margin

The link budget yields the thermal noise fade margin, A_T, and the interference fade margin, A_I. The flat fade margin, A_F, is found as

$$A_F = -10\log\left[10^{(-A_T/10)} + 10^{(-A_I/10)}\right] \text{ dB} \tag{10.3}$$

The fade margin is used to find the link availability as discussed below. As the flat fade margin is given, (10.3) is the fade margin assuming the signal bandwidth is less than the coherence bandwidth of the channel (see Section 4.2.4.1 of Chapter 4). In simple terms, it is the occupied bandwidth that is narrow enough so that essentially the same degree of fading occurs at all frequencies within the signal bandwidth. As the delay spread in the channel increases, the coherence bandwidth decreases and frequency-selective or dispersive fading will be more likely. When the coherence bandwidth is smaller than the signal bandwidth, the *dispersive fade margin* (*DFM*) is a better way to consider link outage performance. DFM is discussed in Section 10.2.8 below.

10.2.5 Link availability (reliability)

Many fixed broadband wireless links are designed to be available essentially at all times. 'Available' means that the BER or FER is at or below a given quality threshold level. Conversely, an 'outage' is the time when the link is not available; for example, the BER/FER level is above the quality threshold level. An outage of only 53 min a year is an availability of 99.99%. This link availability level is also referred to by the shorthand term *four nines*; an availability of 99.999% is *five nines*. The availability percentage is usually based on an annual average, although link outage due to fades is normally calculated on a 'worst-month' basis as discussed below.

The annual outage time is simply related to percentage availability by

$$\text{outage time} = \left[1.0 - \left(\frac{\text{percent availability}}{100}\right)\right] \cdot 525,600\ \text{min} \tag{10.4}$$

An outage can occur for a variety of reasons, including multipath fades, rain fades, dispersive (wideband) fades, and equipment failures. Calculating the probability that fades of a particular magnitude occur, or equipment failures occur, will lead directly to the probability of an outage and hence the link availability probability. The outage mechanisms and the methods for calculating their respective probabilities of occurrence are discussed in the following sections.

10.2.6 Multipath fade outage

Multipath fades occur as a result of reflected, diffracted, and other signals adding in and out of phase with the directly received signal, as discussed in Chapter 4, causing variations in the amplitude of the signal envelope at the receiver. The probability of a decrease or fade of a certain depth can be found by a number of methods. For conventional point-to-point LOS microwave systems, two largely empirical methods are used as described in Chapter 4 – the Vigants–Barnett and the ITU-R 530-8 methods. These methods or models are intended to find fade depth probabilities for links operating above a certain frequency and over a range of link lengths. The Vigants–Barnett model is also limited to modeling fade depth probabilities for fades deeper than about 15 dB, the fades of greatest interest for high availability links. For fade depths less than 15 dB, some other probability distribution must be employed such as Rayleigh, Rician or lognormal.

The details of the Vigants–Barnett and ITU-R fade probability prediction methods are given in Sections 4.2.2 and 4.2.3, respectively. The Vigants–Barnett section includes a discussion of supplementing the fade probability predictions with a lognormal distribution for fade depths less than 15 dB. The fade probabilities predicted using the Vigants–Barnett and ITU-R methods for a 6-GHz link 50-km long are shown in Figure 10.5. A

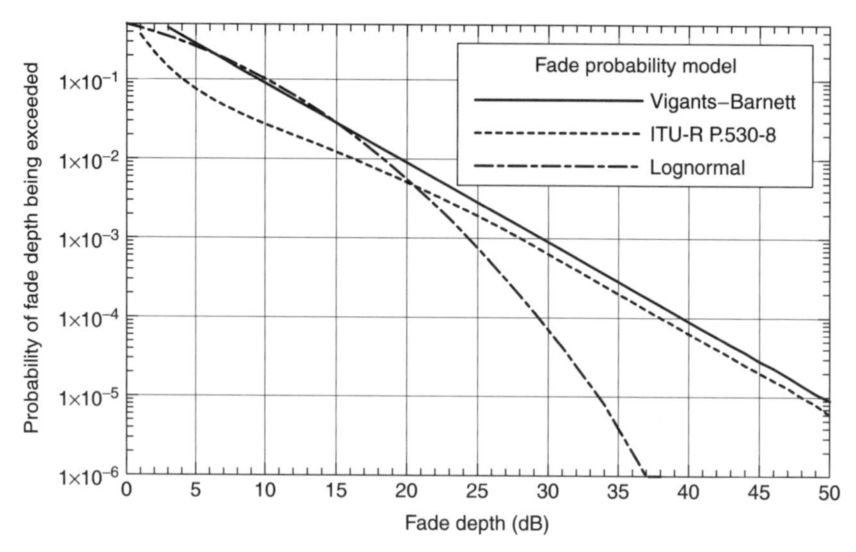

Figure 10.5 Fade probability versus fade depth for various multipath fade models.

Table 10.3 Parameters for ITU-R fade probability calculations in Figure 10.5

Parameter	Value
C_0	1.7 dB
C_{Lat}	0.0 dB
C_{Lon}	−3.0 dB
p_L	15.0%
ε_p	0.0
f	6 GHz
d	50 km

Vigants–Barnett parameter $C = 2.0$ was used for the Vigants–Barnett curve. The parameters in Table 10.3 were used for the ITU-R curve.

The models provide a probability that a fade of certain depth will occur. The relevant depth is the composite fade margin calculated from (10.3). As an example using the 6-GHz path cited above, if the fade margin is the thermal noise fade margin of 40 dB, or $A = A_T = 40$ dB, and the probability of a fade of this magnitude from (4.4) is 0.0090% averaged for the worst month. This is the probability of an outage, yielding link availability during the worst month of 99.9910%. If the usual assumption is made that the 'worst month' conditions occur during 3 months of the year (usually the summer months since the multipath fading is induced by heating of the Earth), the average annual fade probability $P_{F,A}$ is

$$P_{F,A} = 0.25 \times P_F \tag{10.5}$$

For this example, the annual fade probability is 0.00225% resulting in an annual average availability of 99.99775%.

10.2.7 Diversity improvement in flat fading links

Since multipath fading depends on relatively rare out-of-phase additions of direct and scattered signal vectors arriving at the receiver, it is highly dependent on a particular point location in space. Under these conditions, space, frequency, angle, and polarization diversity are effective techniques for mitigating multipath fading. From a spectrum conservation perspective, using another channel to achieve frequency diversity is an expensive option, so the much more commonly used techniques are space and polarization diversity. Space diversity is accomplished by using two or more antennas positioned so that the fading of the signals at the terminals of these antennas are as uncorrelated as possible. Normally, the greater the spacing between the antennas, the less correlated the fading signals are and the greater the diversity improvement. Space diversity in point-to-point link systems is normally achieved by spacing the antennas vertically since this is more practical (lower wind loads) for vertical towers and masts. Angle diversity also uses two antennas pointed at slightly different vertical angles.

Polarization diversity relies on the fact that the fading for horizontally and vertically polarized signals will be uncorrelated. This is a reasonable expectation since the phase and amplitude of the reflected horizontally and vertically polarized signals is different, depending on the angle of incidence to the reflecting surface as discussed in Chapter 2. Polarization diversity has an advantage over space diversity in that it only requires one antenna capable of receiving horizontally and vertically polarized signals, whereas space diversity requires two antennas displaced by some amount. However, it is common in system design to use the two independent polarizations for separate information transmission; for example, the uplink transmission on one polarization and the downlink transmission on the other. In this case, the second polarization is not available for carrying the redundant signal needed for diversity operation.

Generically, for two diversity branches, with independent fading at each branch, the probability P_D that both branches will fade below the fade margin is the product of their individual fade probabilities:

$$P_D(A < A_F) = P_1(A < A_F)P_2(A < A_F) \tag{10.6}$$

If the $P_1(A < A_F) = P_2(A < A_F) = 10^{-4}$, then $P_D(A < A_F) = 10^{-8}$ – a very significant improvement. In practice, such large diversity gains are not realized, suggesting that fading on the diversity branches is correlated to some degree. The practical industry approaches for calculating diversity gain for various diversity methods are discussed in the next sections.

10.2.7.1 Space diversity

Like the empirical fading models described in Chapter 4, empirical models of space diversity improvement have also been devised. These were discussed in Section 6.5.1, but for convenience are restated here. Diversity combining can be done in one of several ways. The most straightforward simply monitors the signal levels (or BER/FER of the demodulated signal) and switches to the diversity branch (antenna) that has the best signal. Such switched diversity combining requires completely separate receivers for each diversity branch. For baseband switched diversity, the improvement factor is given by [4]

$$I_{SD} = 1.2 \times 10^{-3} \left(\frac{f}{d}\right) s^2 \nu^2 10^{A/10} \tag{10.7}$$

where:

d = the path length in km
f = the frequency in GHz
s = the vertical antenna separation in m
ν = the difference in the amplitudes of the main and diversity branches. In most cases this will be the difference between the main and diversity antenna gains. Note that ν in (10.7) is a ratio as a real number, not dB.

It is evident from (10.7) that the diversity improvement increases with increased antenna spacing, s. The net fade probability with diversity is then

$$P_{FD} = \frac{P_F}{I_{SD}} \tag{10.8}$$

For the previous 6-GHz, 50-km link example, and using a spacing of 3 m between equal gain antennas, the diversity improvement factor from (10.7) is 12.96 at a fade margin of 40 dB, or roughly an order of magnitude. This is equivalent to increasing the fade margin by about 11.3 dB; diversity improvement is often given as 'diversity gain' in dB. At a fade margin of $A = 30$ dB, the diversity gain from (10.7) is only 1.13 dB, so the application of space diversity techniques is only appropriate for high reliability links that have already been designed with high link fade margins.

Instead of achieving diversity through baseband switching between two received signals, it is also possible to combine the signals at the intermediate frequency (IF) in a constructive way. Such combining techniques and the resulting combining gain were discussed in Section 6.5. This approach is particularly well suited to microwave repeater or relay sites where it is not necessary to demodulate the signal to baseband because there is no terminal at the site. As with switched combining, empirical formulas have been developed for the improvement from IF combining.

To calculate the improvement in IF combining, the first step is to calculate the increase in the fade margin:

$$A_C = A_T + 2.6 + 20 \log \left(\frac{1+\nu}{2}\right) \tag{10.9}$$

Using this value for increased thermal noise fade margin, the new composite fade margin is adjusted

$$A = -10 \log \left[10^{(-A_C/10)} + 10^{(-A_I/10)}\right] \tag{10.10}$$

Using this value, the diversity improvement factor is found as

$$I_{SD} = 1.2 \times 10^{-3} \left(\frac{f}{d}\right) s^2 \frac{(16\nu^2)}{(1+\nu)^4} 10^{A/10} \tag{10.11}$$

where the various variables have the same definitions as given above. Using the same example 6-GHz, 50-km link with 40 dB of thermal fade margin, the diversity improvement with IF combining is 377.3, significantly better than the switched diversity case. Part of the improvement is due to the increase in SNR resulting from the signal combining in (10.11).

In [5], a more elaborate formula for space diversity improvement is described

$$I_{SD} = [1 - \exp(-3.34 \times 10^{-4} s^{0.87} f^{-0.12} d^{0.48} P_0^{-1.04})] 10^{(A-V)/10} \tag{10.12}$$

where the variables have the meaning as used above with the following changes:

$$P_0 = P_F \cdot 10^{(A/10)} \tag{10.13}$$

$$V = |G_1 - G_2| \text{ dB} \tag{10.14}$$

where $G_{1,2}$ are the gains in dB for the two antennas used in the diversity system. P_F is the probability (as a fraction) that the fade depth A is exceeded. In [5], this is given as a percentage so the equivalent of (10.13) in [5] has an additional factor of 1/100. For the example system used above, the diversity gain from (10.12) is 50.8 (17 dB) assuming the fade margin is 40 dB and $P_F = 0.000090$ (0.0090%).

In addition to these basic formulas, there are other empirical formulas that have been developed by equipment manufacturers using their own performance data and equipment capabilities. The manufacturers can be consulted for specific information in their diversity improvement calculations.

10.2.7.2 Polarization diversity

The degree to which polarization diversity is effective depends on the extent of decorrelation between fading on the horizontal and vertical polarization signals. Theoretical studies [6] using ray-tracing propagation in urban environments at 1900 MHz with broad beam antennas show that the fading in the horizontally polarized (HP) and vertically polarized (VP) signals is highly uncorrelated, so that the diversity gain achieved is similar to that for space diversity. However, for the typical point-to-point microwave link, high-gain directional antennas are used so the scattering properties the HP and VP signals encounter are significantly different from those modeled in [6]. The reflections are created from the ground as well as from refractive and ducting layers in the atmosphere. The difference in ground reflections for the HP and VP cases can be drawn from the graphs and equations in Section 2.6.1. The corresponding differences for reflections in the atmosphere have not been reported in available published research. A useful assumption, however, is that when the multipath scattering/reflecting environment is sufficiently complex to permit the effective space diversity gain, the environment is also sufficiently complex to produce decorrelated fading on the HP and VP signals on the link.

10.2.8 Dispersive (frequency-selective) fade margin

Like the flat fade margin, the dispersive fade margin is the fading depth that must occur for the link signal quality level to decrease to the point where the signal is unacceptable and an outage occurs. The flat fade margin is a function of the mean signal level and the noise and interference levels, and the required SNR/SIR performance levels. The DFM is also a measure of the receiving equipment's ability to produce a useable signal, which depends largely on the design and capabilities of the equalizer in the receiver. The fading that occurs in a dispersive, frequency-selective link occurs at particular frequencies – the total signal power averaged over the link operating bandwidth may not fluctuate significantly during a frequency-selective fade. Measurements and calculations of the DFM are described in Section 4.2.4.2 of Chapter 4. The resulting value is an equipment specification, which should be available for all modern digital radio systems. The dispersive fade margin, A_D or DFM as used in Section 4.2.4.2 of Chapter 4, is incorporated in the overall fade margin as follows:

$$A_D = 17.6 - \log\left(\frac{S_w}{158.4}\right) \text{ dB} \tag{10.15}$$

The composite fade margin is then the average of the flat and the dispersive fade margins as given by

$$A_C = -10\log\left[10^{(-A_F/10)} + R_D \cdot 10^{(-A_D/10)}\right] \text{ dB} \tag{10.16}$$

The factor R_D is called the dispersive fade factor, a number related to the environment where the link is operating and is a relative indicator of the likelihood of dispersive (high time delay) fading. As such it is an arbitrary weighting factor that ranges from 1 upwards. Unfortunately, no published studies are available showing how to select or calculate R_D as a function of other environment factors such as terrain or atmospheric conditions. As such, setting R_D to some value other than 1 will result in a highly uncertain modification to the fade margin yielded by (10.16).

ITU-R P.530-8 [5] provides an alternate approach to finding the probability of a link outage in a channel with frequency-selective fading. The basic outage probability P_S is given by

$$P_S = 2.15\eta\left(W_M \times 10^{(-B_M/20)}\frac{\tau_m^2}{|\tau_{r,M}|} + W_{NM} \times 10^{(-B_{NM}/20)}\frac{\tau_m^2}{|\tau_{r,NM}|}\right) \tag{10.17}$$

where:

$$\eta = 1 - \exp(-0.2P_0^{0.75}) \tag{10.18}$$

$$\tau_m = 0.7\left(\frac{d}{50}\right)^{1.3} \text{ nS} \tag{10.19}$$

and

W_x = the minimum (M) or nonminimum (NM) phase dispersive fade measurement signature width in GHz (see Section 4.2.4.2 of Chapter 4)

B_x = minimum (M) or nonminimum (NM) phase average signature depth (see Section 4.2.4.2 of Chapter 4)

$\tau_{r,x}$ = the reference delay used to obtain the minimum (M) or nonminimum (NM) phase dispersive fade measurement signature width in GHz (see Section 4.2.4.2 of Chapter 4)

The basic signature measurement methodology described in Section 4.2.4.2 of Chapter 4 is used to obtain the parameters for (10.17) as well, however the specific ITU-R definitions can be found in [7].

10.2.9 Diversity improvement for dispersive (frequency-selective) channels

Since the primary mechanism for combating wideband fades in frequency-selective channels is the equalizer in the receiver, the link availability improvement that can be achieved through the use of space, frequency, angle, and other diversity techniques is less clear than in the flat fading case. Even so, in [5] the ITU-R has developed some methods set for diversity improvement for digital systems in frequency-selective fading channels. The outage probability P_D is given by

$$P_D = \left(P_{ds}^{0.75} + P_{dns}^{0.75}\right)^{1.33} \tag{10.20}$$

where

$$P_{ds} = \frac{P_S^2}{\eta(1 - k_s^2)} \tag{10.21}$$

and

$$P_{dns} = \frac{(P_F/100)}{I_{SD}} \tag{10.22}$$

with I_{SD} given by (10.11), P_F given by (4.6), P_S given by (10.17) and η given by (10.18). The remaining term k_s^2 is found as follows:

$$k_s^2 = \begin{cases} 0.8238 & \text{for } r_w \le 0.5 \\ 1 - 0.195(1 - r_w)^{0.109-0.13\log(1-r_w)} & \text{for } 0.5 < r_w \le 0.9628 \\ 1 - 0.3957(1 - r_w)^{0.5136} & \text{for } r_w > 0.9628 \end{cases} \tag{10.23}$$

with the correlation coefficient r_w given by

$$r_w = \begin{cases} 1 - 0.97469(1 - k_{ns}^2)^{2.170} & \text{for } k_{ns}^2 \le 0.26 \\ 1 - 0.6921(1 - k_{ns}^2)^{1.034} & \text{for } k_{ns}^2 > 0.26 \end{cases} \tag{10.24}$$

The nonselective correlation coefficient is calculated by

$$k_{ns}^2 = 1 - \frac{I_{SD} \cdot P_{ns}}{\eta} \tag{10.25}$$

The degree of diversity improvement is not intuitive from (10.20 to 10.25). As with the unprotected dispersive fade margin, the calculation also requires detailed knowledge of the dispersive fade measurement signature of the equipment and the measurement delay.

10.2.9.1 Frequency diversity

Although frequency diversity is a spectrally inefficient technique, in areas with low wireless system density (rural areas, developing countries, etc.) it can be a viable alternative. The empirically derived formula for frequency diversity improvement in frequency-selective digital systems can be found using the same approach as in (10.20 to 10.25), but using a different value of I_{SD}

$$I_{SD} = I_{FD} = \frac{80}{fd}\left(\frac{\Delta f}{f}\right)10^{A/10} \tag{10.26}$$

where f and d are the frequency in GHz and path length in km and Δf is the frequency difference in GHz between the two frequencies in the diversity system. If $\Delta f > 0.5$ GHz, then $\Delta f = 0.5$ GHz is used in (10.26).

10.2.9.2 Angle diversity

Angle diversity is achieved by pointing two antennas in slightly different angles in the vertical plane under the assumption that the signals arriving at different angles are uncorrelated, with the degree of correlation a function of the angular difference. When tower or mast space is at a premium, angle diversity can sometimes be an attractive alternative. The calculation of angle diversity improvement involves several steps and knowledge of the half-power beamwidths of the antennas. Refer to [5] for the calculation details.

10.2.10 Rain fade outage

Rain fades occur because of scattering from raindrops as discussed in Section 2.12. The depth of the fade is proportional to the intensity of the rain, which is generally given as rain rate in millimeters per hour (mm/hr). As referenced in Section 5.5.2, rain rate information has been collected for locations throughout the world and tabulated as the probability that a given rain rate is exceeded. By calculating the fade depth as a function of the rain rate, the probability of a rain fade of a certain depth can be found. Using this fade depth with the fade margin, as was done for the multipath fading case, yields the probability of an outage due to rain.

Two methods are primarily used to calculate outage due to rain in wireless communication systems. One was developed by Crane (see Section 4.3.1) and the other is contained in the ITU-R recommendations (see Section 4.3.2). The background details concerning these fading models can be found in the referenced sections. The use of these models for link design is explained below.

Unlike multipath fading, the fading due to rain cannot be mitigated by standard diversity techniques since rain will reduce the amplitude of the directly received and scattered multipath signal vectors equally. The only diversity technique that can be effective is route diversity using an alternate microwave link or links over a path sufficiently removed from the primary path so that it is not subjected to the same rain intensity at the same time. This can be a prohibitively expensive alternative in most situations, although multilink consecutive point and mesh systems, discussed later in this chapter, can potentially provide alternate routing that avoids rain-stressed links and can therefore mitigate rain outages to some extent.

Because rain attenuation cannot be effectively dealt with using diversity techniques, it has become the main factor in limiting the range of fixed broadband links operating at frequencies above 8 GHz.

10.2.10.1 Link availability with Crane rain fade model

The details of the Crane rain attenuation calculation method are described in Section 4.3.1. The probability of rain attenuation calculated using (4.32) through (4.37) being greater than or equal to the flat fade margin in (10.3) is the probability that a link outage due to rain will occur.

To use the Crane method, an iterative approach is required in which successive values of the rain rates from Table A.1 in Appendix A for the appropriate zone are inserted into (4.32 or 4.33), depending on distance, and into the constituent (4.34) through (4.37). The resulting attenuation A_R at this rain rate is calculated and compared with the flat fade margin A_F. If the iteration loop index is i, then the point where $A_R(i) \leq A_F \leq A_R(i+1)$ with rain rates $R_p(i)$ and $R_p(i+1)$ determines the rain fade outage probability for the flat fade margin A_F. This is done using linear interpolation between the bracketing values.

$$P_R = P_R(i) - [P_R(i) - P_R(i+1)]\left[\frac{A_F - A_R(i)}{A_R(i+1) - A_R(i)}\right] \tag{10.27}$$

where $P_R(i)$ are the probabilities of rain rates $R(i)$ occurring. The iterative approach is somewhat inefficient from a computational viewpoint when large numbers of outage calculations are required. However, the advantage is that it allows an arbitrarily tabulated rain rate probability distribution to be used. Where specific local rain rate distribution information is available, this can provide more precise rain outage predictions.

10.2.10.2 Link availability with the ITU-R rain fade model

The details of the ITU-R rain attenuation calculation method are described in Section 4.3.2. Equation (4.39) and the following constituent (4.40) through (4.46) are used to find the probability of rain attenuation that is equal to or greater than the flat fade margin.

Since the value of attenuation due to rain, A_R, depends on the value of attenuation ($A_{0.01}$) for the rain rate that occurs 0.01% of the time ($R_{0.01}$), it is actually not necessary to have a tabulated function of rain rates and the probability that they are exceeded, as with the Crane method. The probability distribution of rain attenuation is implicit in (4.45 and 4.46). Nonetheless, the ITU-R method is still most conveniently implemented using an iterative approach in which incremental values of p are tried, the associated attenuation calculated, and linear interpolation used as in (10.27) when the attenuation values bracket the flat fade margin; for example, $A_R(i) \leq A_F \leq A_R(i+1)$.

Figure 10.6 shows a comparison of the outage probabilities for the Crane and ITU-R rain fade models as a function of fade depth for a 28-GHz, 5-km link. For these curves it was assumed a link was operating north of 30° latitude using Crane region C rain rates from Table A.1. The rain rate for 0.01% of time for the ITU-R method is set at 29.5 mm/hr, as taken from Table A.1. The curves in Figure 10.6 show that the two methods provide comparable rain outage predictions in this case. Additional rain attenuation curves are shown in Figures 4.8 and 4.9.

10.2.11 Composite link availability

The combined link availability is the availability taking into account outages from multipath and rain fades. These outage mechanisms are statistically independent and, as such, uncorrelated. For statistically independent events, the probability of an outage is the sum of the probabilities of the independent events occurring. For a multipath fade outage probability P_{FD} (with or without diversity) and a rain fade outage probability P_R, the probability of a link outage is

$$P_C = P_{FD} + P_R \tag{10.28}$$

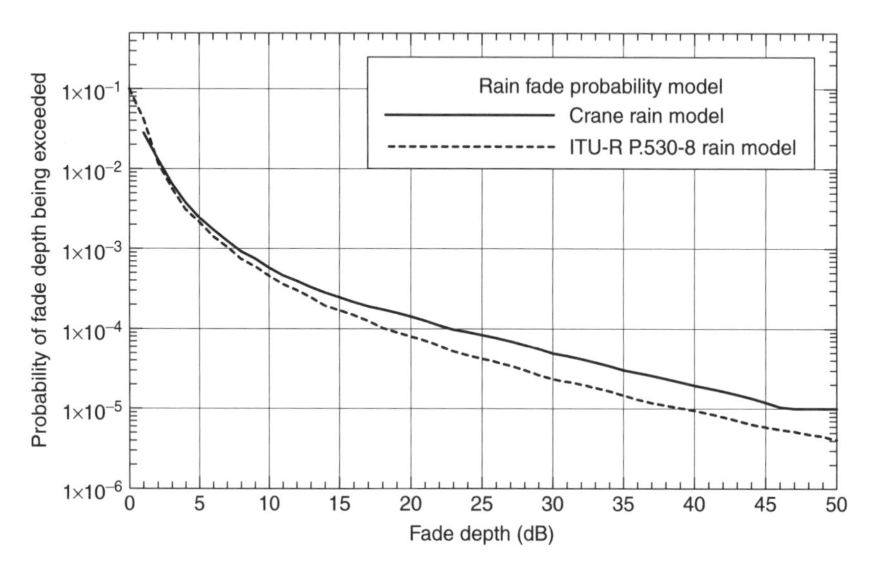

Figure 10.6 Rain fade depth probability using crane and ITU-R fade models for a 5-km path at 28 GHz with an ITU-R 0.01% rain rate of 29.5 mm/hr.

The combined link availability is then

$$\text{Annual link availability}(\%) = 100 \cdot (1.0 - P_C) \tag{10.29}$$

The value from (10.29) is usually the design objective set for the link by the system operator which, in turn, is derived from the requirements for the service and the data application. Starting with the link availability and working back through the link budget in Table 10.1, a maximum range (path length) can be found, which is used to determine the service radius of PMP systems. This will be discussed further in Chapter 11.

10.2.12 Equipment failures

For high reliability link systems that must have outages of only a few minutes a year, equipment failure can play an important part in achieving this reliability. Equipment failure can come from component failure in the equipment itself or physical damage to the equipment from violent weather or vandalism. If a microwave relay site is on a remote mountaintop, access by road or even by helicopter to repair a problem can take several hours or longer. A single such failure alone may violate the reliability objective of the system.

The equipment reliability is usually given as the *mean time between failures* or MTBF. This is a published equipment specification that varies from 50,000 h (5.7 years) to about 300,000 h (34.2 years) for currently available microwave link equipment. The *mean time to repair* (MTTR) must also be considered in looking at the overall probability of an outage due to equipment failure. Using an MTBF of 100,000 h and an MTTR of 8 h, the probability that a single link terminal will fail is

$$\text{Terminal outage probability} = 1.0 - \left[\frac{\text{MTBF}}{(\text{MTBF} + \text{MTTR})}\right] \tag{10.30}$$

or 0.00799% for the sample values. The link fails when the terminal at either end fails, so the link failure probability due to equipment failure is 0.0159%, or a link availability of 99.984%. For a link with a high fade margin, the equipment outage probability can dominate the overall link availability.

A link that has a single terminal radio (transmitter/receiver) is often referred to as *unprotected*. Note that this term also applies to links with no diversity. The *unprotected* availability because of equipment failure may not be acceptable when considering the availability requirements. For this reason, systems intended for high reliability applications usually employ redundant equipment at each terminal. Redundant radio equipment is usually referred to as *hot standby* equipment, indicating that the equipment is turned on and at operating temperature. It may then be immediately and automatically put into service in the event of failure in the primary radio units. A rapid switching process between primary and secondary units interrupts the signal for 20 to 50 ms. With a hot standby terminal, and assuming the failures in the two redundancy units are independent, the increased hot standby MTBF_{HS} can be calculated from the MTBF_{S} for a single radio as shown in [1]

$$\text{MTBF}_{\text{HS}} = \frac{(\text{MTBF}_{\text{S}})^2}{\text{MTTR}} \tag{10.31}$$

For the example with the numbers used above, the outage probability of the link is now $1.6 \times 10^{-7}\%$ or a link availability of 99.99999984%. This availability is substantially higher than can be expected from multipath and rain fade outages, thus removing equipment failure as a significant factor in determining overall link availability.

10.3 SHORT-RANGE LOS LINKS IN URBAN ENVIRONMENTS

In general, the design information, link budget calculations, and link availability predictions for long range LOS links in Section 10.2 also apply to short-range urban links. However, there are simplifications and also some complexities in dealing with short urban links. For the purpose here, 'short' links mean links with path lengths ranging from as little as 50 m up to 5 km. A 50-m link results when two buildings across the street from each other, or within a campus or office compound, are to be connected. If a cable or optical fiber connection is impractical or prohibitively expensive because of the required excavation, a wireless link (either radio or optical) is an attractive alternative. On the other end of the scale, a 5 km link is a nominal practical range limit for a local multipoint distribution service (LMDS) system in typical rain fade conditions. It is also a reasonable estimate of the range where atmospheric refractivity along the path can be reasonably ignored because of the small amount of radio ray bending that will occur over this short distance.

10.3.1 Building path profiles

For short urban links of this type, the main obstacles to be considered in assessing path clearance are buildings, trees, and local hills or slopes. In general, the environment would be a composite mix of these features with buildings constructed on the tops and the slopes of hills and trees and other foliage interspersed among them. Figure 10.7 illustrates a typical urban link path profile with buildings and low hills.

The path clearance objective is the same as for long-range links – 0.6 of the first Fresnel zone radius around the link path. An example of that zone is shown as the dashed line in Figure 10.7. For path clearance analysis, the Earth radius k factor should usually be set to 1 for such short paths indicating that there is no significant ray bending due to atmospheric refractivity.

The 'urban canyons' created by tall buildings also require lateral Fresnel zone clearance to be considered. A lateral move of 1 or 2 m can potentially eliminate a building as an obstacle on a path profile like Figure 10.7 even though the path is close enough to the vertical edge or side of the building to experience loss due to partial obstruction of the Fresnel zone. There is also the issue of database accuracy, which, at its best is of the order of 1 m. Engineering a wireless link that has very close spacing to the vertical edge of the face of a building requires a careful evaluation of the exact path geometry relative to that building. This evaluation can be done with the aid of a sophisticated fixed broadband wireless planning tool and a precise building database.

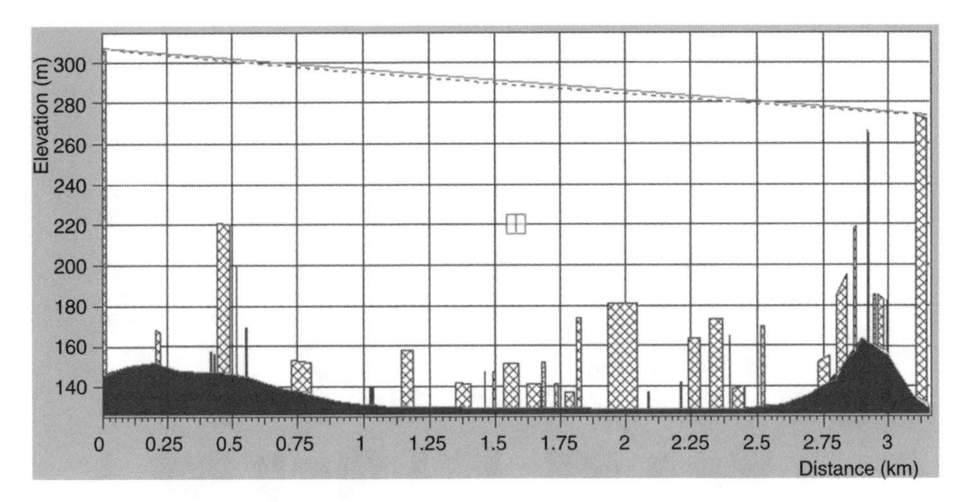

Figure 10.7 A 3.1 km urban link over buildings on low hills.

The overhead map view of the building outlines in Figure 10.8 can be used to judge whether the link path passes very close to a building edge or surface. A second perspective view in Figure 10.9 of the same view can give a further indication of clearance since it accurately depicts the building heights as well. The small dish antenna in the foreground of Figure 10.9 is located just 1 m above the building roof – a typical mounting situation that reduces visual impact of the installation, reduces the wind load, and simplifies the support structure construction. Often the antenna is positioned just high enough to provide clearance over the parapet or railing along the outside perimeter of the roof. The parapet is not shown in Figure 10.9.

10.3.2 Short-range fading

The Vigants–Barnett and ITU-R fading probability models in Section 10.2.6 were developed from measurement data for relatively long paths. In the case of ITU-R P.530-8, it specifically notes that the equations were developed for paths with lengths ranging from 7 to 95 km. Similarly, the Vigants–Barnett model was formulated using performance results primarily from long-haul telephone microwave circuits. Given this foundation, these models are not well suited to predicting fade outage probability for short links in urban areas.

The physical link geometry suggests the fading distributions will be different for three main reasons:

1. Antennas will generally be mounted on rooftops placing them high above the reflecting ground or street surface and (for short paths) increasing the reflection incident angle.
2. High-gain antennas with narrow beamwidths will more strongly suppress ground reflections and other multipath components because their angles of arrival are further removed from the antenna maximum pointing angle.

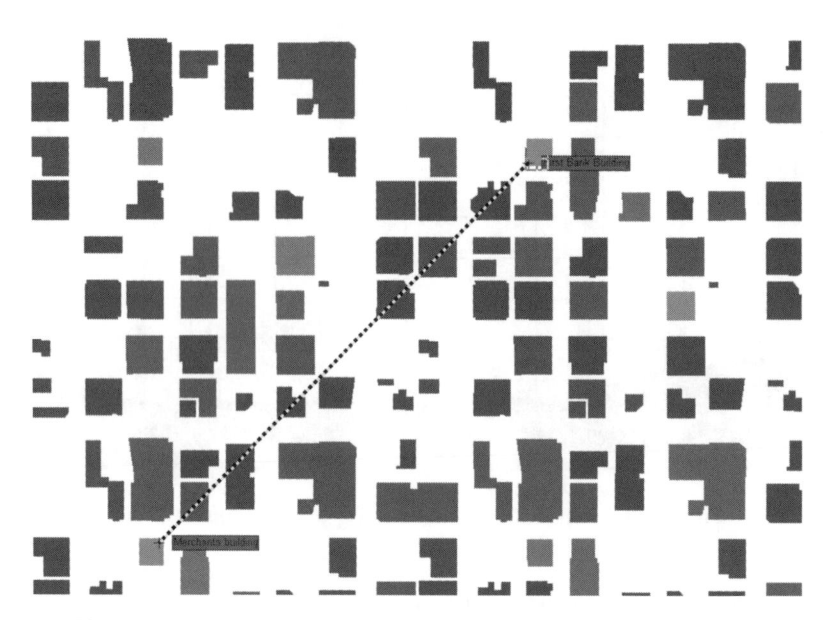

Figure 10.8 Overhead map of short urban link with building shapes.

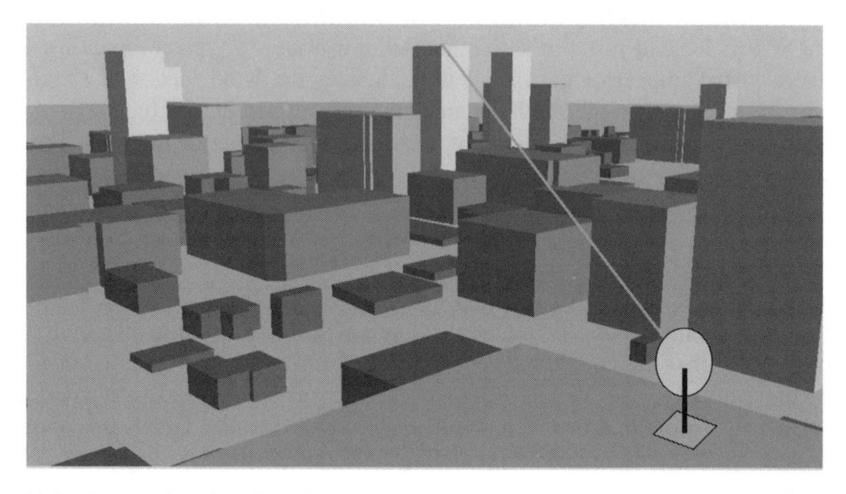

Figure 10.9 Perspective view from first bank building to merchant building for link in Figure 10.8.

3. Because the paths are short, the degree of power illumination of higher atmospheric layers is reduced which, in turn, reduces fading induced by reflections from atmospheric layers.

A recent study of short range 20 to 50 GHz paths whose results are included in [8] essentially dismisses multipath fading as a factor in such links when compared with other factors like rain, citing high-gain antennas as the rationale. Anecdotal evidence from short-range link systems deployed in several US cities in the 24-, 28-, and 38-GHz bands also indicate that multipath fading is not a significant link outage mechanism.

While multipath fading is a secondary outage mechanism, for high reliability systems it would be imprudent to exclude it entirely. For a strong directly received signal component, and relatively weak, random-phased scattered signals, a distribution that can potentially be applied to describing the fading is the Rician distribution discussed in Section 4.6.1.2 of Chapter 4.

For short-range urban links, a Rician distribution with a large k factor value from 12 to 20 dB could be applied to take into account a modest degree of multipath fading. For digital systems that have FER performance curves that change significantly with a few dB changing in SNR or SIR, the few dB of fading from the Rician distribution in conjunction with rain fading could be a determining factor in adequate link fade margin design. A graph showing Rician fade depth probabilities for various k factors is found in Figure 4.15.

As mentioned before, a Rician fading distribution is applicable to describing narrow-band fading. For short urban links where high-gain antennas are usually employed, any strong reflections that do arrive at the receiver will generally be coming from obstacles (buildings) that are close to the direct propagation path. Because of their amplitude and the short wavelength, these reflections could still add out of phase with the directly received signal, even though their arrival time delay relative to the arrival time of the direct signal is small. The relationship between higher antenna gain resulting in lower delay spread for LMDS systems in urban areas was demonstrated in [9]. From this reasoning, the Rician fade models are an appropriate choice for describing the multipath fading for such links. Unfortunately, there is no strong measurement data set that can assist in selecting an appropriate Rician k factor for such a model. In the absence of such empirical guideline information, a nominal value of $k = 9$ dB can be used.

10.3.3 Short-range urban rain fading

Unlike the multipath fade probability models that were developed for long links, the rain fade models from Crane and ITU-R described in Section 10.2.9 provide rain attenuation in terms of dB/km so that the attenuation figures in principle may be applied to arbitrarily short paths. The difficulty here is that rain cells and the relative changing geographic distribution of rain intensity can result in the entire length of a short link being inundated instead of being subjected to some average rain intensity over the length of the path.

Over short ranges, the structure of rain cells and clusters becomes more significant. From the information in [10], rain cells are composed of *volume cells* that typically have an area of 5 km^2 (average radius of 1.26 km) and a lifetime of less than 15 min. Volume cells can be thought of as the basic building block of larger convective structures. A closely spaced collection of volume cells is a *small cell cluster*. A small cell cluster contains an average of 3 volume cells and has an average area of 19 km^2 (an average radius of 2.45 km). A small cell cluster is what is normally referred to by radar meteorologists as a *rain cell*. Single detected clusters are isolated showers, whereas multiple (closely spaced) clusters represent wide areas of showers or squall lines. For links <5 km in length, the entire link could be within the boundaries of a single rain cell (volume cell cluster).

The distribution of the rain intensity within the cell is also a variable but unfortunately there is almost no measured data to characterize intensity distribution. Instead, rain data is generally long term distributions of the rain rate at selected measurement points exceeded for various percentages averaged over a year. The sample period for the data to construct these averages may be as short as 1 min, but more typically 5 or 10 min. For high reliability links, the acceptable outage time per year may be only a few sample periods so the outage probability predictions must be considered in light of this relatively crude rain rate sampling.

Even though there are questions as to their applicability for short-range links given the issues discussed above, in the absence of any better models, the Crane or ITU-R models remain the only currently viable choices for predicting link outage from rain events.

10.3.4 Interference diffraction paths over building edges

For urban broadband wireless networks with high link densities, the link performance will generally be limited by interference from other links rather than from receiver noise – interference-limited rather than noise-limited performance. While the desired link paths by design are LOS, the interference paths from other link transmitters may be LOS or obstructed by intervening buildings. The buildings themselves and the path attenuation they represent to interfering paths thus become an important element in the system design. Where a choice of locations on a roof are available, a link receiving antenna can be deliberately positioned so that it is shielded from an interfering transmitter by an intervening building along the path to the interferer. Exploiting the nonhomogeneous nature of the propagation environment in this way can make more links viable and increase the overall system capacity. It is perhaps another justification for the expense of having a high-resolution building database for this type of system design.

The additional path loss introduced on the path by an intervening building or buildings can be found by considering the buildings as a collection of diffracting edges. In special cases, such as steel framework buildings with large glass wall areas, the 'through building' path may actually be a lower loss path than the diffraction paths over and around the building. These situations are illustrated in Figure 10.10.

The path loss for the diffracting paths can be found from the formulas in Chapter 2. The wedge diffraction coefficient or loss can be calculated using the wedge diffraction coefficients in equations in Section 2.7.1. A simpler approach that is commonly used for the roof edge diffraction is simply to model the roof or parapet edge (which is most commonly a wedge with a 90° interior angle) as a knife-edge and use the simple diffraction loss formulas in Section 3.5.2. At the microwave frequencies used for urban systems of this type, this approximation is valid especially for large bending angles over the diffraction edge.

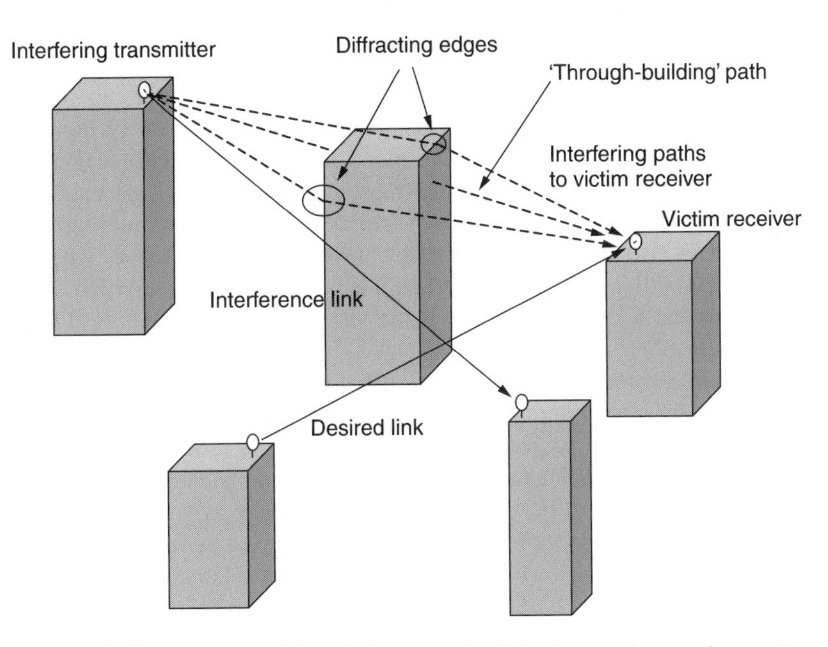

Figure 10.10 Desired and interfering paths for links in an urban environment.

The total power calculated from interferers arriving via LOS or diffracted paths is inserted in the link budget in Table 10.1 as 'Total external interference level' and incorporated into the flat fade margin as given in (10.3). If the interfering signal is not cochannel with the desired signal, or its occupied bandwidth is different, the techniques described in Section 8.2.1.2 of Chapter 8 can be used to further adjust the amount of incident interference that is accepted by the receiver. Of course, the 3D gain patterns of all the antennas must also be taken into account in calculating the interference level. These subjects will be discussed in more detail in Chapter 11 on PMP networks.

10.3.5 Urban link availability

The total link availability can be found by jointly considering the independent probabilities of a rain fade or a multipath fade that are greater than the designed link margin. Given the infrequency of these events, as before, the outage probability will be the sum of the probability of an outage due to rain and due to multipath fading.

$$P_{\text{outage}} = P_R(A < A_F) + P_{FD}(A < A_F) \tag{10.32}$$

As discussed, the dominant outage mechanism for these links will be fades due to rain. In practice, the multipath fading contribution to link availability is routinely ignored for the special case of short-range urban links using high-gain antennas ($G_{T,R} > 20$ dBi).

10.3.6 Free space optic (FSO) link design

The link geometry for short-range urban links, shown as examples in the preceding sections, is also applicable to FSO systems. Because of the extremely high frequency, issues regarding Fresnel zone clearance are moot. Because the electromagnetic energy is light, if visual clearance exists between the transmitting and receiving locations, then the link is viable if its length is not excessive. Using a high-resolution building database for system planning is also a valuable approach; however, database errors where the LOS analysis shows the path is clear when in effect it is not, are definitive. At FSO frequencies, there is no viable 'near LOS' capability. If the visual path is blocked, the FSO link will not work.

As discussed in Sections 2.13 and 4.5, if an FSO path is clear, then its performance is limited by fog, haze, atmospheric scintillation, and other natural effects that scatter and diffuse the laser beam, thus weakening its amplitude at the receiver. A useful guideline for a high reliability (99.999% availability) FSO system range is 200 to 400 m assuming typical fog and haze density probabilities. As a simple approximation for site survey purposes, an effective FSO system range will be about twice as far as a human can see.

FSO links also require that very careful alignment be maintained between the transmitter and receiver. Links set high in tall buildings can be affected by building sway during high winds, during seismic activity, and from thermal expansion and contraction. Modern FSO systems use two approaches to mitigate building sway effects.

First, a broader beam divergence angle can be used (see Figure 2.30). With a broader divergence, any movement of the detector relative to the beam will still fall within the cone of the beam and the link will remain connected. The trade-off of this approach is that for a fixed amount of power, broadening the beam weakens the signal so the effective system range is reduced.

A second, more expensive, approach is to use active tracking to keep the beam aligned. Modern FSO systems use multiple lasers (one is a beacon beam), motorized gimbal mounts, and quad detectors, which can provide sufficient beam redundancy and tracking to overcome building movement problems. The automatic tracking approach also eases installation since the system can automatically search for and acquire the beam. For challenging tall building installation situations, the choice of an FSO system should be made with this building movement in mind.

Another less significant performance factor is glass reflectivity deflecting the laser beam at certain geometries. Similarly, at certain angles sunlight can sometimes temporarily 'blind' the laser detector – essentially adding high amplitude noise signals at the receiver. Ideally, the FSO terminal units should be installed so the detectors are not pointed directly at the sun during any part of the day or season. This is only a factor for units mounted outside or on rooftops. For units used from indoor locations and transmitting/receiving through a windowpane, the sun will never be positioned directly in the aperture of the detector.

Dirty glass can also impair performance, although this is easily remedied by keeping the windows clean. Low probability events like birds flying through the laser beam are essentially nonproblems since they are of such short duration and the bird is unlikely to be large enough to block the entire beam cross section (see Figure 2.30).

FSO systems also depend on what amounts to a 'visual easement' between the transmitter and the receiver. Construction of a new structure that blocks the beam path can render the FSO system inoperable. Obtaining contractual rights to this easement is almost never done because the cost of prohibiting new construction would probably be exorbitant. Further details on FSO system design can be found in [11].

10.3.7 'Riser' and FSO backup links

Microwave frequencies above 60 GHz have attracted much greater interest in the past few years. In general, these frequencies are not yet regulated. Recognizing their potential use, in the United States the Federal Communications Commission (FCC) has identified three bands at 71 to 76 GHz, 81 to 86 GHz, and 92 to 97 GHz for new wireless services. The two main communication applications that have been discussed for these frequencies are

- Backup links for FSO systems when there is fog but no rain.
- 'Riser' links to move data from tall building rooftops to lower floors. This requires the lower antenna to be pointed straight up potentially causing interference to satellite services. However, atmospheric absorption (O_2) provides sufficient attenuation to prevent interference problems at the altitude of orbiting satellites.

It is expected that other very short-range uses will also be found for these bands to provide supplemental connections to support other wireless or wired communication systems.

10.4 NLOS LINKS IN URBAN AND RESIDENTIAL ENVIRONMENTS

The link budget in Table 10.1 applies to NLOS links as well as LOS links. The main difference is that the path loss will be based on a propagation model that is applicable to NLOS situations rather than the simple Friis transmission loss in (10.1). There will be path losses and scattering due to local foliage and buildings near the receiver. For a system model where the terminal unit is operating inside an office or house, the loss associated with the signal penetrating from the outside to the inside of the structure must also be included. These additional losses are explicitly identified in the link budget in Table 10.1.

Figure 10.11 (also Figure 3.19) is a vertical profile drawing of these loss mechanisms for NLOS links. There are two scenarios that fall into this category of links:

1. *Outdoor remote terminal*: The first scenario assumes that the remote antenna will be mounted outside the structure (whether office building or house) so that the NLOS losses are local foliage and clutter loss from nearby buildings obstructing the path. It also assumes the antenna will be pointed in some way, either through expert installation or automatic beam steering for smart antennas, which may permit nonexpert user installation. Even for a user-installed smart antenna, enough intelligence must still be applied to the installation to mount the antenna on the side of the structure toward the serving hub sector.

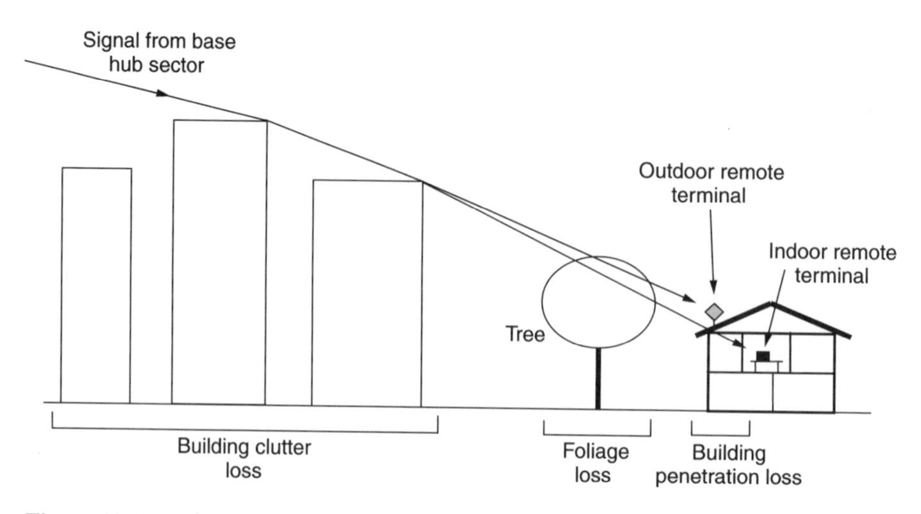

Figure 10.11 Path loss mechanisms for NLOS paths links in urban and residential areas.

2. *Indoor remote terminal*: The second scenario assumes the remote terminal is indoors on a desktop or similar user-convenient location. It assumes the terminal can be moved around (portable) but will not be operating while in motion. Given its portability within the structure, the building penetration loss can vary considerably. Locations that are at the outside wall of the structure next to a window will have much lower path losses than locations that are in the center of the structure with a large number of intervening interior walls in addition to the exterior walls.

These scenarios are not mutually exclusive within a system. A system could be generally designed to provide a link budget that is adequate to serve indoor terminals in a large majority of cases, with the outdoor terminal setup employed for locations that are on the fringes of the hub sector service areas. The losses associated with building clutter, foliage, and building penetration generally increase with frequency, sometimes substantially, so that the viability of NLOS paths of a given length will decrease with increasing frequency. Because of this, the indoor system configurations are primarily being considered for the frequency bands between 2 and 6 GHz; specifically, the 2.4 license-exempt band, the 2.5 to 2.7 GHz MMDS/ITFS band, the 3.5-GHz band, and the bands between 5 to 6 GHz allocated for U-NII and HiperLAN services (see Chapter 1).

Even with the additional path loss mechanisms, the basic link design approach still centers around the link budget calculation in Table 10.1 that provides an adequate signal level at the receiver to maintain performance and link availability during fading events. The additional path loss factors take into account building clutter, foliage, and building penetration losses; they also introduce additional statistical variations in the link budget. Even though the statistical uncertainty associated with building penetration loss, for example, is not a physical fading phenomenon such as rain or multipath fading, *per se*, the statistical variation must be accommodated in the same way as nature-induced fading in designing the link. This also means that as the information about these loss factors becomes more refined, the effective variance of the distribution describing them is reduced. The required link margin design objective is correspondingly reduced along with real benefits, like a reduction in required transmitter power or extended range. Such are the tangible benefits of increased propagation model accuracy.

10.4.1 Basic NLOS path loss

For the LOS case, a physical model of path loss such as the Free space + RMD (see Chapter 3) model described in Sections 3.5.1 and 3.5.2 produces the same results as the Friis transmission loss model in (10.1). For the NLOS case, the general-purpose fixed broadband model in Section 3.5.6 combines this free space loss with path losses due to path obstructions from terrain, buildings, foliage, and wall penetration. The total loss L_T is calculated as:

$$L_T = L_b(p_t, p_r) + C + F + B \text{ dB} \tag{10.33}$$

where L_b is the basic free space path loss plus diffraction loss given by

$$L_b = 32.44 + 20\log f + 20\log d + \sum_{n=1}^{N} A_n(\nu, \rho) \tag{10.34}$$

$A_n(\nu, \rho)$ is the diffraction loss associated with obstacle n as given by (3.46). This includes diffraction loss over buildings that can be specifically identified from buildings in a building database. The term C is the loss associated with structures along the path that are not explicitly identified as obstacles. Clutter loss is modeled as a lognormal random variable (Gaussian-distributed using decibel values). The term F is the loss associated with foliage, which is also a lognormal-distributed random variable. Finally, the term B is the loss associated with building penetration and also modeled as a lognormal random variable.

The free space and diffraction losses are calculated using models from Chapter 3. Estimates for the mean value of the losses for F and B can be found from the information in Sections 2.9.1 and 2.9.2. The values for C are found using the methods in Section 3.5.6.1 of Chapter 3.

Figure 10.12 shows the percent location availability versus distance for adequate signal performance for links operating at 2.6, 5.7, and 28 GHz using the parameters shown in Table 10.4. The equipment power levels, antenna gains, noise figures, and required SNR values in Table 10.4 are derived from various equipment manufacturer specifications. The indoor antenna gains are based on the assumption that simple two-branch diversity antennas of some sort are used. Antenna gains in a scattering environment are discussed in Section 10.4.2.

The deviation $\sigma = 8$ dB for the lognormal distribution that describes the standard deviation of the sum of the building clutter, foliage, and building penetration losses is an estimate; there is little substantive data on these loss factors to support a determination of σ for each of them separately. The resulting variation in path loss of different points at the same distance from the hub is known as *location variability*. Location variability is discussed further in Section 10.4.3. For the study in Figure 10.12, it is also

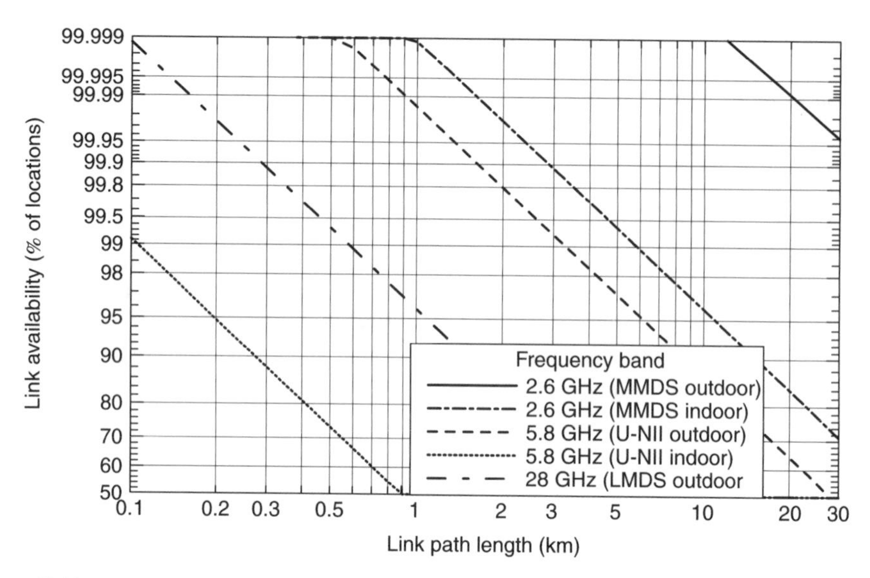

Figure 10.12 NLOS downlink availability for outdoor and indoor remote terminal antennas. Availability indicates the percent of locations served at the indicated distance.

Table 10.4 Parameters used for downlink availability curves in Figure 10.12

System component	2.6 GHz (MMDS)	5.7 GHz (U-NII)	28 GHz (LMDS)
Hub transmitter power (dBmW)	50.0	30.0	27.0
Hub antenna gain (dBi)	16.0	16.0	21.0
Average building clutter loss (dB)	15.0	17.5	20.0
Average foliage loss (dB)	2.5	5.6	20.0
Average building penetration loss (dB)	9.0	12.0	n/a
Path loss lognormal distribution σ (dB)	8.0	8.0	8.0
Outdoor receive antenna gain (dBi)	19.0	24.0	38.0
Indoor receive antenna gain (dBi)	6.0	6.0	6.0
Receiver noise figure (dB)	3.0	4.0	5.0
System bandwidth (MHz)	5.0	5.0	5.0
BER = 10^{-3} SNR threshold (dB)	16.0	16.0	16.0
ITU-R 0.01% rain rate (mm/hr)	n/a	n/a	42.0

assumed that multipath fading, which is usually described with a Rician or Rayleigh distribution, can largely be mitigated with antenna space diversity at the remote terminal. Accordingly, Figure 10.12 does not take into account any time variability due to changing multipath fading.

In addition to the model in (10.33), Chapter 3 described several types of propagation models including, among others, empirical path loss models developed for the 2.5 to

2.7 GHz band included in the IEEE 802.16.2 standard. These models predict signal levels and other channel attributes using very little specific information about the propagation environment. As such, the best use for these models is system dimensioning; that is, estimating the number of base stations or hubs that are required to serve an area. The cell service boundaries can be overlaid on a map of the service area and a simple count performed of the number of cells required. System dimensioning can be used to estimate the cost and capacity of the system, or by vendors preparing a price bid on furnishing equipment necessary to construct a system.

Because these empirical models do not consider specific propagation environment information and assume signal strength decreases monotonically with distance, these empirical models are not well suited to designing wireless systems in specific locations where the local factors such as terrain features, building clusters, and so on will dramatically affect the signal levels. Instead of simple coverage ranges, the coverage areas of the hub sectors will be nonhomogeneous and often discontiguous. For actual system hub design, a physically based model such as in (10.33) is required.

The results in Figure 10.12 are for the downlink. By reciprocity, the uplink path loss will be the same, as will the antenna gains. The main difference between downlink and uplink is that the remote transmitter will have lower power than the hub transmitter in the MMDS case. To some extent this is made-up, for if the uplink bandwidth is narrower than the downlink (because of a lower uplink data rate) the noise level is lower. In the U-NII band and LMDS band, the remote units could have the same powers as the hub transmitters (1 W in this case). The results in Figure 10.12 can therefore be interpreted as applicable to both downlink and uplink.

The results in Figure 10.12 show that MMDS links are viable for both outdoor and indoor situations as modeled here. Links in the 5.7 GHz U-NII band are also viable, although the additional building penetration loss for indoor links makes them usable over very short ranges of a few hundred meters. Figure 10.12 also shows that 28-GHz links can be successful over short ranges of 0.5 to 1 km or less, even with foliage and building clutter attenuation.

10.4.2 Antenna gain in scattering environments

For NLOS scattering environments, the power arriving at the remote terminal antenna comes via the (obstructed) direct path to the hub sector antenna as well as via the long-range multiple paths as shown in Figure 10.14 as well as via very localized scattering objects including the walls and other elements of the structure where the terminal is located. This set of scattering objects causes the angular dispersion of the power arriving at the remote antennas to be very broad. Because of this broad angular dispersion, an omnidirectional antenna is more successful at collecting the power than a highly directional antenna that only collects power arriving from a narrow angular range. The net effect is that the directional antenna appears to have less effective gain relative to the omnidirectional antenna.

This phenomenon has been measured in [12] and characterized as a *gain reduction factor* (GRF). From [12], GRF can vary from 0 dB (for the omni case) to 6 or 7 dB for an antenna with about 20° of beamwidth. More recent research [13] suggests that there is no significant difference in the effective gain between omnidirectional and directional remote terminal antennas for NLOS paths.

The actual effect here is really not a gain reduction for the directional case, but rather a lower path loss for the omnidirectional or broader beamwidth case because these broader beamwidth antennas are more efficient at collecting the scattered energy. From the link budget in Table 10.1, the antenna gains and direct path loss are still calculated as described above; however, for scattering environments the path loss can be decreased by a GRF, if desired, which is calculated on the basis of the remote terminal antenna beamwidth. Including this effect in the 'path loss' part of the link budget ledger is more appropriate than including it as a modified antenna characteristic since it actually is a characteristic of propagation environment scattering. The difference in the results reported in [12,13] show that this factor may be highly variable or not significant at all.

For indoor LOS situations, it is reasonable to expect the scattering to be more pronounced since an indoor location can be expected to have a greater degree of scattering and a more heavily attenuated direct signal by building penetration loss. With several signals arriving from various directions, the concept of using an adaptive antenna at the remote terminal becomes more attractive. An adaptive antenna will adjust itself so as to maximize the SNR (or SINR) regardless of the pattern shape and gain pattern that may result. If adaptively forming a pattern that is essentially omnidirectional results in maximized SNR, then the antenna is responding advantageously to the scattered power coming from the surrounding environment. Recent studies of antenna pattern shapes using an adaptive antenna indoors show this to be the case [14].

10.4.3 Location variability

As discussed in Chapter 4, fading models are used to describe a wide variety of uncertainty or unpredictability in wireless systems, not just physical phenomena like rain that defy deterministic analysis. The lognormal-distributed losses associated with building clutter, foliage, and building penetration represent location variability. For example, moving from room to room in a house will result in different values of building penetration loss – low losses may exist near windows on the side of the house toward the hub transmitter while the highest losses may occur in the center or opposite side of the house where the number of walls the signal must pass through is the greatest. The average of these losses is used in Table 10.4. In a similar way, the average losses associated with building clutter and foliage are listed in Table 10.4 with the understanding that these values are also modeled by lognormal distributions with the standard deviations indicated.

Considering the location variability of the attenuation along the entire path, the path loss includes the sum of these lognormal-distributed variables. There is no closed-form solution for this sum, but several approximate techniques have been developed that represent the distribution of the sum of lognormal variates as another lognormal variate with a mean and standard deviation set using one of several methods. One such method is described in Section 4.6.2. This approach was incorporated in the availability-range calculation results shown in Figure 10.12.

10.4.4 Time variability (narrowband fading)

Although the example used above assumed that spatial diversity at the remote unit would mitigate multipath fading, for narrowband systems without this technology, or as an indirect way to gauge the degree of channel scattering, it is useful to have some estimate of multipath fading statistics as an indicator of scattering.

A Rician distribution to model fading statistics for short-range urban links was described above, in Section 10.3.2. The Rician distribution has also been used successfully to model fading statistics in NLOS suburban paths based on measurements at 1.9 and 2.4 GHz as reported in [15]. The Rician distribution is described in Section 4.6.1.2 of Chapter 4. The pertinent parameter that controls the shape of the distribution and hence the fade statistics is the k factor. The model in [15] (converted to dB) proposes a statistical description of the k factor:

$$k = 10\log(F_s) + 10\log(F_h) + 10\log(F_b) + 10\log(k_0 d^{\gamma}) + u \text{ dB} \tag{10.35}$$

where

- F_s is the seasonal factor equal to 1.0 for summers (leaves present) and 2.5 for winter (no leaves).
- F_h is the receive antenna height factor, $F_h = (h_m/3)^{0.46}$ where h_m is the height of the remote antenna in meters.
- F_b is the beamwidth factor, $F_b = (b/17)^{-0.62}$ where b is the remote antenna beamwidth in degrees. k_0 is the seasonal factor equal to 1.0 for summers (leaves present) and 2.5 for winter (no leaves).
- u is the seasonal factor equal to 1.0 for summers (leaves present) and 2.5 for winter (no leaves).
- d is the distance from the cell site hub sector to the remote terminal.
- k_0 and γ are regression coefficients, $k_0 = 10$ and $\gamma = -0.5$.
- u is a lognormal variable with a mean of 0 dB and a standard variation of 8 dB

Figure 10.13 shows k factors as a function of the distance calculated using (10.35). The k factors can be used to estimate the probability that a narrowband fade will exceed the link fade margin using the curves in Figure 4.15 and assuming no space diversity. The k factor decreases with distance and increases with narrower remote terminal antenna beamwidths, consistent with the model of narrower beamwidth antenna suppressing multipath signals and thus reducing the fading. The full leaves of summer also decrease the k factor probably because the trees are more significant scatterers and thereby create a higher degree of multipath.

10.4.5 Time dispersion and arrival angles

Figure 10.14 shows a two-dimensional view of the signal path between the hub sector and the remote terminal. For a directional antenna (fixed or adaptable) at the remote terminal that limits the terminal's exposure to multipath or interference signals that are off the antenna's pointing angle, the direct path will usually be the dominant propagation mechanism for signals between the hub and remote terminal. For omnidirectional or broadbeam

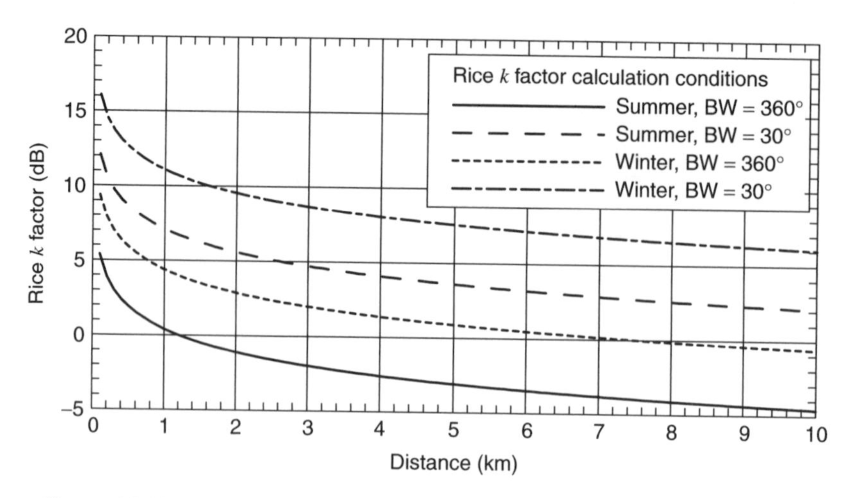

Figure 10.13 Rice k factor as a function of distance calculated from (10.28).

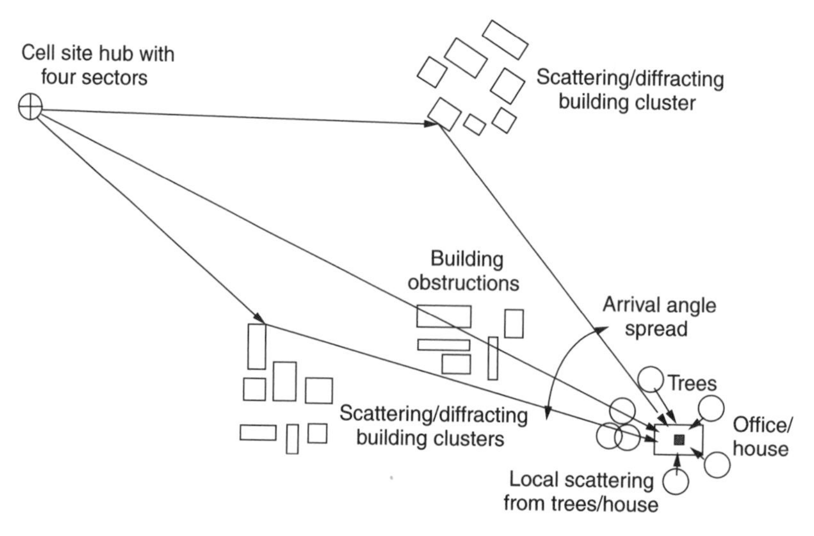

Figure 10.14 Overhead map perspective of scattering for NLOS link.

antennas, including smart antennas designed to suppress interference in selected directions, the viable propagation paths from the hub sector to the remote terminals will include this direct route as well as the multiple other paths via reflection, diffraction, and scattering. This will affect the mean path loss as well as other characteristics of the channel such as fading statistics (Rice k factor), delay spread, angular dispersion, and multiantenna channel correlation as exploited in multiple input, multiple output (MIMO) systems (see Section 6.7). All these factors affect the performance of the single NLOS link.

Figure 10.14 illustrates the geometry of the situation from an overhead map perspective. The various paths from the hub to the remote terminal will have different amplitudes depending on the propagation mechanisms along their individual paths and time delays based on the longer path lengths. This ensemble of paths can be formulated into a mathematical channel model as explained in Section 3.3.

The time dispersion in terms of delay spread depends on the frequency band and antenna gain. The study for the 2.5-GHz MMDS band in [16] shows delay spreads that vary from 20 ns to 5260 ns (average of 140 ns) with a directional antenna and 20 ns to 7060 ns (average of 370 ns) with an omnidirectional antenna. Even though these were NLOS paths, the antennas were still mounted outside, so these numbers do not take into account issues associated with an indoor antenna.

10.4.6 Channel spatial correlation

Characterizing spatial correlation in NLOS systems is important for assessing the operation of systems using diversity to mitigate fading and for MIMO systems that achieve increased channel capacity through the uncorrelated nature of the signals transmitted from and received on multiple antenna elements. As discussed in Chapter 6, the degree of correlation among signals on multiple closely spaced antennas ($\sim\lambda/2$ apart) is proportional to the degree of scattering that the signals undergo in transit from the transmitter to the receiver. This in turn depends on the number, type, and distribution of the elements of the propagation environment. A truly free space environment will have no scattering and perfectly correlated signals on multiple antennas (as practical link distances). Similarly, an open rural environment will see very little decorrelation among signals on closely spaced antennas. Conversely, so-called 'rich scattering' environments such as indoor situations and dense urban situations can have highly uncorrelated signals and thus benefit from diversity and MIMO techniques.

Because it depends on the detailed features of the propagation environment, spatial correlation will be highly site-specific (even varying considerably within the same indoor environment), and therefore be difficult to predict and quantify for the purposes of designing an NLOS fixed broadband system. The information currently available is derived from field measurements done with specific prototype systems. In the case of diversity systems, the results are characterized as diversity gain or improved performance resilience in a range of fading environments rather than statistical descriptions of correlation coefficients. Similarly, for the MIMO system, the results are given as capacity or data rate efficiency gain for the specific prototype system being evaluated. This makes it somewhat difficult to work backward to derive characterizations of spatial correlation properties of the channel, which can be generally applied to a variety of system configurations.

From a system capacity perspective, the expected channel spatial correlation is an ingredient that goes into assessing the amount of bandwidth that is required to accommodate a certain packet traffic load and QoS. Cumulative statistics like those in [17] suggest that throughout a hub sector service area, the average capacity may be increased through the use of MIMO by a factor of 4 to 8 times thus lowering the required access bandwidth by a corresponding factor, all else being equal. This factor can be incorporated in wireless access capacity calculations described in Section 9.8.2.

10.5 LINK ADAPTATION

When a fade occurs, the link can detect the fade by a change in the BER or FER at the receiver. If the transmitter is equipped with automatic power control APC and it nominally does not operate at its maximum power level, the link can respond to the fade by automatically raising the transmit power to restore the nominal performance. If the transmit power level is at maximum, and the fade is deep enough to continue to degrade performance, the link can still be maintained by changing the modulation constellation to a more robust, lower-rate scheme that can produce acceptable FER values with a lower SNR or SIR.

Such changes to the link to maintain a viable connection is called *link adaptation* (LA). Link adaptation is a common technique used in LOS link systems, which is now finding increased interest for use in fixed NLOS networks as well as cellular/mobile networks. Basically, LA monitors the state of the link channel and makes decisions to adjust the transmit power or chooses the best combination of modulation and coding for the current channel conditions. As mentioned, LOS microwave systems that are subject to rain fading are commonly equipped with the ability to select lower speed, more robust modulation schemes when rain attenuation degrades the quality of the link. For a family of related modulation types such as 256QAM, 64QAM, 16QAM, and QPSK described in Chapter 7, changing among modulation constellations is straightforward since they all share a common modulator/demodulator structure. In a similar way, the code rate can be changed as the channel quality degrades, employing lower-rate codes with more redundancy when the channel is poor so that the FEC-corrected BER remains at about the same acceptable level.

Different combinations of modulation and coding are referred to as *modes* in cellular GPRS and EDGE technologies. The choice of which mode to employ at a given time depends on some estimate or measurement of the channel state information (CSI). Some common approaches are to monitor the SINR during some time/bandwidth window and use the realized mean value to select modes. The time length of the window is chosen on the basis of the time variability of the channel, with a time window shorter than the coherence time of the channel. For fixed broadband channels, it is reasonable to expect the time variability of the channel to be more stable than a mobile channel so a longer averaging window will provide worthwhile estimates. Similarly, the frequency window (bandwidth) of the measurement should be adjusted so that it is indicative of the frequency-selective fading that the signal will experience. If multiple antennas are used, the SINR can also be sampled in space so that a three-dimensional sampling of the CSI can be used to set the mode.

Once a decision on where to set the mode has been reached by the receiver, it must be communicated back to the transmitter, so some overhead is involved when using LA. If rapid mode changing occurs because the LA channel state sampling is too aggressive, sending the mode change information could represent a significant overhead burden. Depending on the modulation types available in each mode, there may also be timing issues associated with reacquiring synchronization. For LOS microwave systems, rain fading is a very slow event compared to changing the desired signal and interference power levels in NLOS systems when rapid multipath fading can occur even with transmit or receive diversity. The appropriate responsiveness of LA must be selected as a trade-off of these factors.

Adjusting the parameters of an adaptive antenna is another form of link adaptation. Adaptive antennas are discussed in Section 6.6. Beam-steering antennas direct their maximum gain direction toward the other end of the link. Increasing the gain in selected directions to overcome signal loss due to multipath or rain fades is another way to maintain the link during these events. In a similar way, optimum-combining adaptive antennas respond to changing link conditions to make whatever adjustments are necessary in the array to maximize the SINR.

Link adaptation is an important network management technique that can also increase spectrum utilization as discussed in Chapter 12.

10.6 MULTIHOP (TANDEM) LINK SYSTEMS

Two terminals that are too far apart for a single link, or NLOS, can be connected using multiple wireless links arranged in an end-to-end configuration. This arrangement is illustrated in Figure 10.15. The intermediate link transmitter/receiver sites (*repeaters*) are usually located in isolated, unpopulated locations where no traffic demand or service requirements exist. The traffic flow is thus from the terminals with the same capacity provided at each link or hop in the system. The system can include both active and passive repeaters, although passive repeaters are only a suitable choice under particular circumstances, as discussed in Section 10.6.1.

In a multihop system of this type, each of the links between the terminal and first repeater, and between repeaters, is designed as described in the preceding sections of this chapter using a link budget to achieve a fade margin that provides the desired system availability under multipath and rain fading conditions. However, in the case of a multihop system, the primary interest is the end-to-end performance through all the links. The end-to-end performance, of course, will depend on the performance of the individual links.

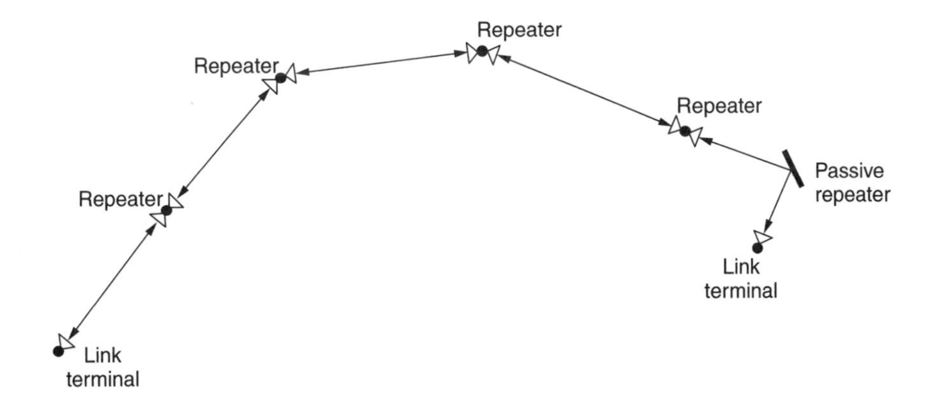

Figure 10.15 Multihop (tandem) link system with four active repeaters and one passive repeater.

Specifically, the commonly made and reasonably valid assumption is that the fading on the individual links is independent. The end-to-end availability can then be found from the outage probability as in (10.28). The sum of the individual link outage probabilities is given by

$$P_{MH} = K \sum_{n=1}^{N} P_{C,n} \tag{10.36}$$

where $P_{C,n}$ is the composite link outage probability given by (10.21) and K is a factor that can be used to adjust the composite outage probability if the individual link outage probabilities are correlated to some extent. Correlated outages can occur, especially for short links in which the link lengths are such that the same rain cells cause outages on both links.

The end-to-end availability is then

$$\text{Multihop link availability}(\%) = 100 \cdot (1.0 - P_{MH}) \tag{10.37}$$

As expected, the multihop system availability is worse than that of a single hop. For long routes that cross hundreds of kilometers before arriving at a terminal, the overall availability can challenge a high availability goal such as 99.999%. A particularly weak link in the system can also disproportionately erode system performance since it is not possible to make up for it through more reliable link design elsewhere in the multihop system.

10.6.1 Passive repeaters

Passive repeaters are normally used to redirect a microwave signal around an isolated obstacle such as a mountain or building. Passive repeaters have several advantages over active repeaters:

- They require no power (power line, generator, batteries, and solar panels).
- They do not require access roads.
- They can be installed in rough terrain (via helicopter) that is otherwise inaccessible.
- They are inherently broadband and may be used for any system type operating in the bands from 1.7 to 40 GHz.
- They require very low maintenance compared with active repeaters.

The primary drawback of passive repeaters is that they are less efficient in terms of spectrum conservation. The microwave signal is redirected such that interference can be caused to other systems in two directions instead of one. Because of this, passive repeaters are banded in some countries (such as the United Kingdom) where high-density spectrum reuse is a particular concern.

A passive repeater can be included in the link budget in Table 10.1 by calculating its insertion loss. The insertion loss for a plane reflector passive repeater is given by

$$L_I = L_T - (L_1 + L_2) + G_P \text{ dB} \tag{10.38}$$

where:

- L_T is the total free space path loss from the active link end points that illuminate the reflector
- L_1 is the total free space path loss from terminal 1 to the reflector
- L_2 is the total free space path loss from terminal 2 to the reflector
- G_P is the passive reflector gain

The plane reflector gain is given by

$$G_P = 42.8 + 40\log f + 20\log A_a + 20\log\left[\cos\left(\frac{\alpha}{2}\right)\right] \text{ dBi} \tag{10.39}$$

where f is the frequency in GHz, A_a is the area of the reflector in square meters, and α is the total angle (in three dimensions) between the two paths arriving at the reflector (see Figure 10.16). From (10.38) it is clear that the insertion loss is reduced when the reflector is placed near one of the terminals ($L_1 \gg L_2$ or vice versa). The reflector size is chosen on the basis of the path geometry and whether the reflector is in the near field of either terminal antenna. Reflector sizes range from a few meters on a side up to 12×18 m.

Normally the manufacturer of the repeater can perform the 3D angle calculations to set the angle of the reflector surface. The reflector and mounting structure are then manufactured with the required orientation and tilt as the center point on an adjustment range. Once the reflector is installed, small adjustments can be used in the field to maximize the signal level at the active receiving terminals or repeater.

Passive repeaters can also consist of two high-gain (large effective area) antennas connected back to back via a short length of waveguide. Such a repeater is analyzed in a similar way to the plane reflector using an insertion loss calculation. The insertion loss for a back-to-back antenna passive repeater is given by

$$L_I = L_T - (L_1 + L_2) + G_{P1} + G_{P2} \text{ dB} \tag{10.40}$$

where G_{P1} and G_{P2} are the gains of the two antennas used in the repeater.

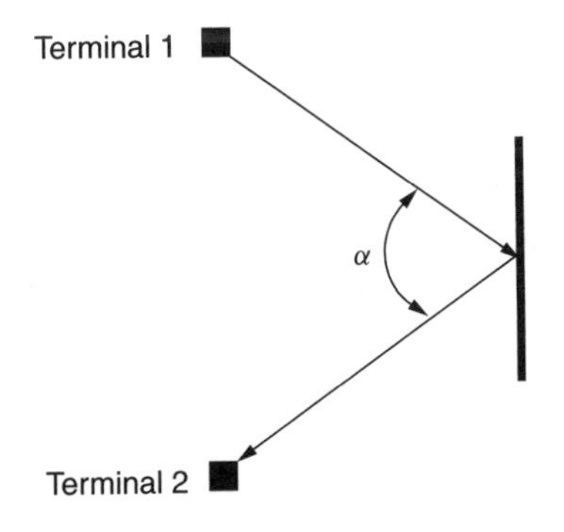

Figure 10.16 Passive repeater geometry.

For this configuration, the insertion loss can be very large so that a back-to-back passive repeater to turn the radio signal around an obstacle only makes sense for very short links. This use may be appropriate in short urban link situations to redirect a signal around a building though such use has been rare.

Passive repeaters are also applicable to FSO systems using mirrors as the reflector structure. With either mirrors for FSO or metal reflectors for microwave signals, it is important that the mounting structure be robust and stable since the angular orientation of the reflecting plane is critical to the reliability of such a redirected link.

10.7 CONSECUTIVE POINT NETWORKS

Consecutive point networks (CPNs) are well suited to high rate data transmission in urban areas where several buildings are to be connected to a fiber optic portal of some sort. As shown in Figure 10.17, CPNs are usually formed in rings, as optical fiber networks use rings in metropolitan areas. The advantage of the ring architecture is that if any link in the ring fails, the nodes in the ring can be automatically reconfigured to route traffic in the opposite direction away from the failed link, similar to the way it is now done with optical fiber rings.

The individual links in the CPN are designed as described in the preceding sections of this chapter. Section 10.3 on short-range urban links is particularly applicable. The interference, automatic power control, coupled fading, and other issues associated with a multilink system will be treated in Chapter 11 which discusses the design of PMP networks.

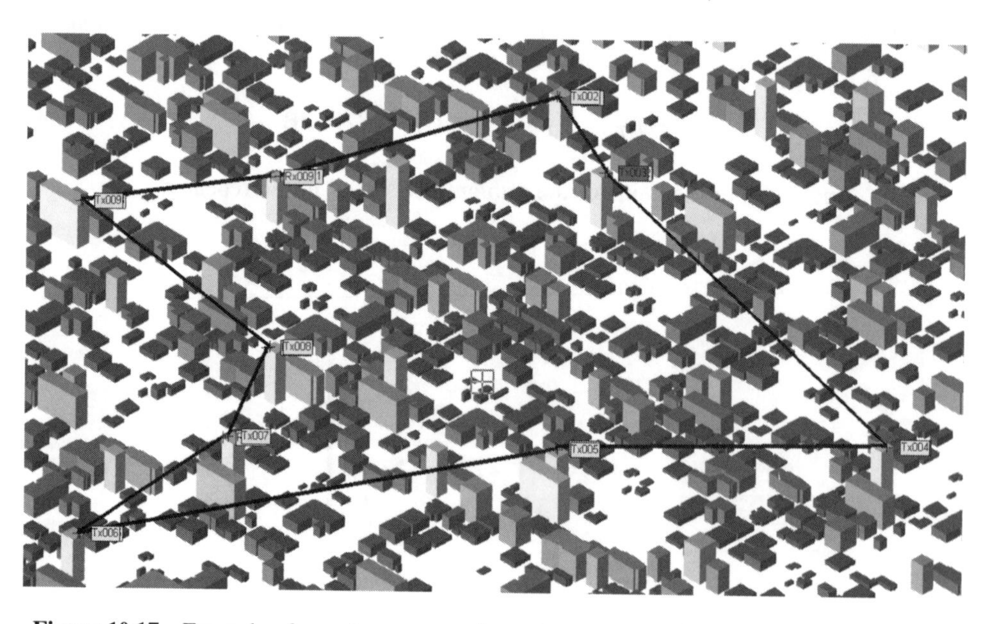

Figure 10.17 Example of an urban consecutive point network (CPN) in a ring configuration.

Consecutive point networks do have some drawbacks. To add a new node in the ring, the ring has to be broken and antennas reoriented to serve the new location. This is a manual, time-consuming process that requires expert technicians on site. One way to add a node and avoid breaking the link is to create a link 'spur' from one of the nodes on the ring. A collection of such spurs creates a multipoint network like those in Chapter 11.

The traffic flow on CPNs must also be carefully considered. As the number of nodes increases, the traffic flowing toward and away from the fiber optic portal increases. The links with the heaviest traffic are the links that connect directly to the portal. It may be appropriate to use higher data capacity equipment on these links as compared with the other links in the ring.

10.8 MESH NETWORKS

Wireless mesh networks have a topology that addresses some of the limitations associated with consecutive point and point-to-multipoint networks. Mesh network topology is a well-known approach that is currently used for the telephone trunk systems and the myriad of optical fiber, cable, and wireless connections that are used for routing Internet traffic.

A schematic diagram for a portion of a fixed broadband wireless mesh network is shown in Figure 10.18. The optical portal network (or other type of high speed portal) is the source point for serving signals to distribution nodes by single point-to-point links. The

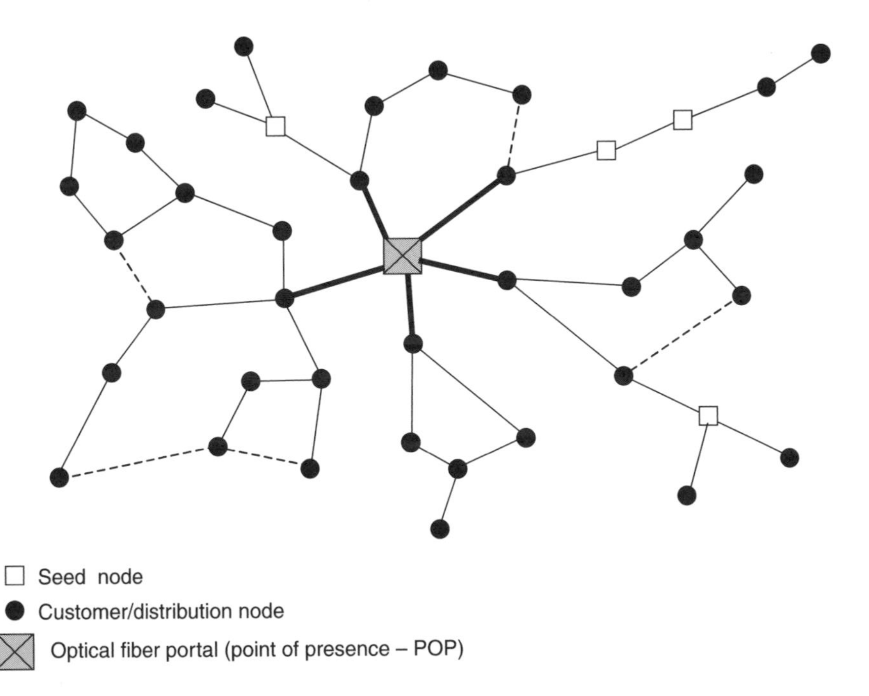

Figure 10.18 Portion of a mesh network. Dashed lines are planned alternate connections.

distribution nodes can also be customer nodes. Each customer/distribution node can route signals in several directions depending on the number of antennas that are used at that node. A standard node from a current manufacturer of wireless mesh uses four steerable high-gain antennas at each node. The antennas are steerable in the azimuth (horizontal) plane with relatively broad elevation plane radiation patterns to allow a reasonable range of height differences between nodes to be successfully connected. The seed nodes are placed at locations that are not customer premises but are simply used to extend the network to areas that are currently not served. The existing and planned alternate interconnections between nodes for microwave systems include only those paths that are LOS. NLOS mesh networks at lower frequencies such as the 2.6-GHz MMDS or 5.8-GHz U-NII bands are also possible as discussed below.

The individual short-range links in a mesh network are designed using the techniques described in the preceding sections of this chapter, especially Section 10.3. Since mesh works are intended to grow organically as the service demand increases, the link evaluation and design process, and the high resolution topographic/building database used to support the design and modeling activity, become integral parts of the network deployment and operation rather than being used only in the planning phase.

The main feature of this type of network is that there are multiple routes that can be used to transmit traffic from one terminal to another. This inherently gives the system the possibility of being more robust than a system in which communication relies on a single link. The diversity techniques discussed in the preceding sections show the value of having multiple mechanisms for achieving highly reliable communications. Although mesh networks can provide multiple routings, some routings are more efficient than others. The problem of choosing the most efficient routings for a large number of traffic sources over a mesh network has been the subject of research for many years. Optimum routing is primarily an ongoing network operational issue rather than a network or system design issue, although the number and type of routing options available is a result of the network design, so to some extent the layout of nodes in the network must be cognizant of the traffic routing issues.

Some of the advantages offered by a mesh network approach compared to the more commonly used PMP networks are:

- Alternate routing to bypass failed links or equipment. This is done by steering antennas to point toward alternate operating nodes.
- Alternate (dual) routing can potentially be designed into the network to avoid rain fades. Such dual routings must be designed because the antennas cannot be rapidly steered to respond to rain fades.
- As traffic load increases or changes, the network can be reconfigured to avoid traffic bottlenecks at particular nodes.
- For a given subscriber location, the ensemble of nearby nodes offers a wider selection of candidate LOS paths to connect to the network than does the typical PMP system.
- Since there are no large number of remote terminals directing signals at a given hub, as in a PMP system, the frequency planning problem is reduced and potentially more efficient use of spectrum can be achieved.

There are also disadvantages with the mesh network approach when compared to the PMP network:

- Because a customer's connection most probably transits through another customer node location that is not under the control of the network operator, what happens at that location can affect the service to nodes farther downstream. If a customer unilaterally disconnects his/her node (or something else happens to make the node stop functioning), a multitude of other nodes and customers could be affected.
- To achieve a small vertical stack of high-gain steerable antennas at each node, LOS mesh networks are limited to higher microwave frequencies above 20 GHz where such antennas are physically small. Larger nodes with physically larger antennas are conceivable but may be less practical because of the sizes involved for high-gain antennas at the lower frequencies.
- Because a node includes multiple antennas and electronics for both transit traffic and drop-to-customer traffic, it is more expensive than a customer remote terminal at an equivalent frequency and data rate in a PMP system that does not carry transit traffic and has just one fixed antenna and associated RF hardware. Since such a node is used for each customer location, the infrastructure cost multiple grows more rapidly than in a PMP system.
- Because of the expense of the nodes and their required LOS mounting requirements, mesh networks are not suited to providing service to typical residential customers or home office customers. For the most part, they are limited to business customers in traditional office buildings.
- Customers disconnecting and reconnecting to the network (network churn), can cause potentially significant disruptions in the network that require reconfiguration. The network may have to be reconfigured (by reorienting the antennas) to achieve alternate successful routing. High churn rates aggravate this problem.
- To accomplish optimum routing, plan alternate connections, and efficiently respond to customer churn, elaborate network management software and the expertise to use it is required. Such management software is an inherent part of what a mesh networks vendor should provide along with the wireless nodes and data routing hardware.

One of the early applications of this type of network is as a flexible backbone or backhaul network that provides signal delivery to PMP hubs or access points for several IEEE 802.11 wireless LAN WLAN systems. In this application, the mesh network does not replace a PMP network but rather provides support for it. Given the advantages and disadvantages cited above, this type of backbone application rather than a PMP replacement is an appropriate use of mesh networks.

10.8.1 NLOS mesh networks

The concept of a mesh network for NLOS paths is actually quite similar to packet radio networks that have been widely used by everybody from amateur radio operators to the military for data communications. A data packet with an address attached is passed through the system from the source terminal to the destination terminal. Along the way, the packet may pass through several nodes in the system that are at customer locations or are at noncustomer repeater sites (like the seed nodes described above). A mesh network is somewhat different in that the traffic is destined for, and originates from, the portal in the optical fiber network. General packet networks can be *ad hoc* 'peer-to-peer' networks that pass traffic between any pair of nodes.

Although flexible and robust, these packet networks are generally very inefficient unless the traffic flow and routing is closely managed, as described for LOS mesh networks. This can certainly be done in the NLOS case, however, the well-confined point-to-point links that the LOS system relies upon are not usually found. The relatively wide dispersion of energy from multiple nodes trying to pass traffic through the network will create aggravated interference problems that must be addressed via advanced smart antennas at the nodes that are capable of simultaneously suppressing multiple interference sources.

From a spectrum utilization standpoint, every time a wireless channel is used to pass the message, spectrum capacity is used. Mesh networks may pass a single message multiple times from node to node, occupying wireless spectrum each time. Even assuming that lower transmit power can be used, the extent of spectrum occupancy with mesh networks is much higher than in PMP or other systems that must only occupy the wireless spectrum once to transmit a message. Spectrum occupancy is discussed further in Chapter 12.

The range of the peer-to-peer links is also constrained since the path loss between each of the nodes will suffer twice from building clutter, foliage, and building penetration loss (if both nodes are indoors). The link budget in Table 10.1 can be modified to increase the amount of loss in each category so that the expected received signal level for a peer-to-peer link can be calculated.

10.9 CONCLUSIONS

The wireless link between two terminals is the fundamental building block for wireless networks. Whether the link is LOS or NLOS, a successful link design requires detailed evaluation of the propagation environment in which the link must operate.

The link performance is primarily determined by the fade margin, the difference between the median signal strength at the receiver versus the signal level the equipment requires for adequate receiver output error rates or signal-to-noise ratios. Using propagation models that can predict path loss with reasonable accuracy, as discussed in Chapter 3, are essential to this process. One or more propagation or channel modeling techniques may be applicable to designing a link.

The performance of the link is also determined by the nature of the fading on the link that occurs from multipath signal variations and from rain events at frequencies above 8 GHz. Several models of these fading phenomena were presented in this chapter that can be used in conjunction with the fade margin to predict link outage and its inverse, link availability. The space, frequency, and other diversity techniques covered in this chapter are effective methods for improving link availability. The nature of the propagation environment and link performance in various LOS and NLOS situations point to wireless hardware, like smart antennas and MIMO systems, which can exploit these characteristics to achieve higher capacity and more robust availability.

For terminals that are located too far apart for successful communication via a single link, a multihop or tandem link system was described. A multihop system uses active repeaters (and occasionally passive repeaters) to relay signals from one terminal to another over long distances or over difficult terrain that prevents a direct LOS terminal-to-terminal connection.

Consecutive point and mesh networks are two approaches for distribution of high rate data communications to a limited set of discrete locations. Each approach has its advantages and disadvantages depending on the service objectives (customer profile) and the business plan of the service operator. The more commonly used approach for mass communication distribution to a large number of arbitrary points, in a service area, is the PMP network topology. PMP networks are the subject of Chapter 11.

10.10 REFERENCES

[1] R.F. White. *Engineering Considerations for Microwave Communications Systems*. San Carlos, California: Lenkurt Electric Co., Inc., 1975.

[2] Harris Corporation. Various equipment technical equipment specification sheets.

[3] P-Com Corporation. Various equipment technical equipment specification sheets.

[4] A. Vigants, "Space-Diversity Engineering," *Bell System Technical Journal*, vol. 54, no. 1, pp. 103–142, January, 1975.

[5] International Telecommunications Union, "Propagation data and prediction methods required for the design of terrestrial line-of-sight systems," Recommendation ITU-R P.530-8, 1999.

[6] H.R. Anderson, et al., "Theoretical polarization diversity studies in an urban microcell environment," *Proceedings of the Fourth International Symposium on Personal, Indoor and Mobile Communications*, Yokohoma, Japan, pp. 35–59, September, 1993.

[7] International Telecommunications Union, "Propagation data and prediction methods required for the design of terrestrial line-of-sight systems," ITU-R Recommendation F.1093, 1999.

[8] International Telecommunications Union, "Propagation data and prediction methods required for the design of terrestrial broadband millimeter radio access systems operating in a frequency range of about 20–50 GHz," Recommendation ITU-R P.1410-1, 2001.

[9] H.R. Anderson, "Estimating 28 GHz LMDS Channel Dispersion in Urban Areas Using a Ray-Tracing Propagation Model," *Digest of the 1999 IEEE MTT-S International Topical Symposium on Technology for Wireless Applications*, Vancouver, pp. 111–116, February, 1999.

[10] R.K. Crane. *Electromagnetic Wave Propagation through Rain*. New York: John Wiley & Sons. 1996. pp. 66–67.

[11] H. Willebrand and B.S. Ghuman. *Free-Space Optics: Enabling Optical Connectivity in Today's Networks*. Indianapolis: SAMS Publishing, 2002.

[12] L.J. Greenstein and V. Erceg, "Gain reductions due to scatter on wireless paths with directional antennas," *IEEE Communication Letters*, vol. 3, no. 6, pp. 169–171, June, 1999.

[13] M.J. Gans, et al., "Propagation measurements for fixed wireless loops (FWL) in a suburban region with foliage and terrain blockages," *IEEE Transactions in Wireless Communications*, vol. 1, no. 2, pp. 302–310, April, 2002.

[14] G. Xu, "Unwiring broadband: the challenges and solutions for delivering non-line-of-sight, zero-install, nomadic wireless broadband," *Proceedings of the 8th Annual WCA Annual Technical Symposium*, San Jose, CA, slideshow presentation, January, 2002.

[15] IEEE 802.16 Working Group. "Channel models for fixed wireless applications," Document 802.16.3c-01/29r4.

[16] J.W. Porter and J.A. Thweatt, "Microwave propagation conditions in the MMDS frequency band," *Proceedings of the 2000 IEEE International Conference on Communication*, New Orleans, pp. 1578–1582, May, 2000.
[17] V. Erceg, P. Soma, D.S. Baum and A.J. Paulraj, "Capacity obtained from multiple-input multiple-output channel measurements in fixed wireless environments at 2.5 GHz," *Proceedings of the 2002 IEEE International Conference on Communications*, New York, CD-ROM, April, 2002.

11

Point-to-multipoint (PMP) network design

11.1 INTRODUCTION

Point-to-multipoint (PMP) network architecture is the most widely used approach for electronic communication with a large number of destinations or nodes. It is used for wired telephone systems in which twisted wire pairs from each phone terminal are routed to a central office (CO) where calls are placed on shared trunk circuits for onward routing. It is also used for cable television systems in which signals are received at, or originated at, a 'headend' and from there distributed via trunk and feeder cables to individual homes. Cellular telephone networks, commercial and government mobile radio systems, and one-way television and radio broadcast systems are all examples of PMP networks. The wired PMP networks (telephone and cable) provide service to discrete locations, whereas the wireless systems listed are designed to provide service to an area where the specific locations of the service users are usually unknown. For cellular and mobile radio systems the users are assumed to be in motion.

In the context of fixed broadband wireless networks, for the line-of-sight (LOS) systems the terminal locations are assumed to be at known discrete locations. Installation of LOS remote terminals at the customer or subscriber location requires the expertise of system technicians; the precise position and engineering circumstances of the terminal are known. For non-line-of-sight (NLOS) systems, the general location of the customer terminal may be known; that is, at the office or residence of the subscriber. They may also be at *ad hoc* temporary locations such as airports, hotels, and so on. The objective of fixed wireless PMP network design is to provide service to customers at all these locations whether they are permanent and known or temporary and unknown.

Fixed Broadband Wireless System Design Harry R. Anderson
 ISBN: 0-470-84438-8

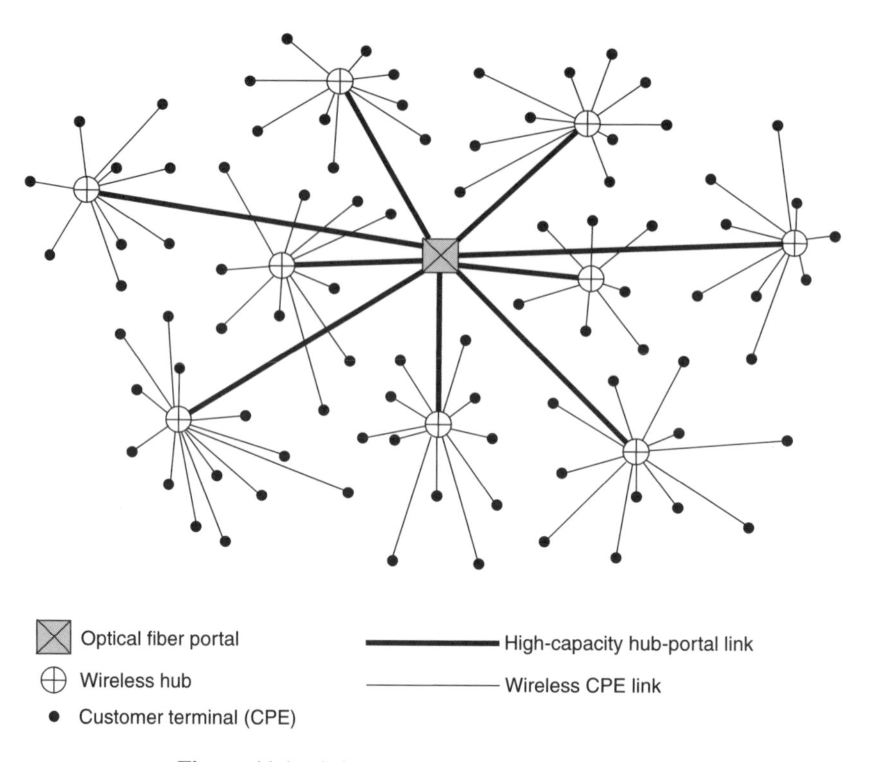

Figure 11.1 Schematic diagram of a PMP system.

Figure 11.1 shows the general topology of a wireless PMP network. In the industry nomenclature, the individual customer remote terminals have come to be known as *customer premise equipment* (CPE), a term used for many decades by the wired telephone industry. The CPE's are directly connected via wireless links to network nodes that are called cell sites, base stations, or hubs, depending on the type of wireless system. The term *hub* will be adopted here. At each hub, there may be one or more 'sectors' installed. A sector is simply a separate set of antennas for transmitting and receiving signals from the several CPEs. The sector antennas are relatively broad beam when compared to the high-gain narrow beam antennas used for multihop, consecutive point, and mesh networks described in Chapter 10. The network hubs may be connected directly to an optical fiber portal of some kind or the traffic passing through the hub may be put on another high-capacity wireless link of some kind [including a Free Space Optics (FSO) link] that connects it to an optical fiber portal. Such links are a common approach used in cellular wireless networks in which they are referred to as *backhauls* since they carry cellular phone traffic back to, and from, the mobile switching center (MSC).

The design process for a PMP network involves several steps, some of which have already been discussed. Traffic assessment and suitable propagation environment model databases have been discussed in Chapters 9 and 5, respectively. The methods for assessing the performance of the link between a hub and a CPE are found in Chapter 10. With these fundamental techniques and methods in hand, the network design process consists of three basic steps:

1. Perform the initial network design in which the hub layout is established using the traffic and the propagation environment model information. Tower heights, antenna types and pointing angles, power levels, and approximate hub capacity allocations are chosen at this stage.
2. Analyze the predicted performance of the design to evaluate whether the system operator service objectives have been achieved. For fixed LOS networks that serve discrete locations, this will require CPE locations to be postulated on rooftops or elsewhere to make a system performance simulation possible. For NLOS systems, the service objective will be defined over areas with particular traffic densities.
3. On the basis of the results of the performance analysis, revise the design as needed to correct deficiencies in the predicted performance.

When the system is ultimately built and tested, it is important that deficiencies, or excesses, that occur with the real system be evaluated in terms of the paper design prediction so that this feedback can be used to further refine the design process – for expansions of the initial system deployment or for the design of entirely new systems. Unlike mobile networks where the exact location of a mobile unit is not known (although this is an objective of future technologies), with fixed broadband wireless systems the locations are known and the quality of service (QoS) experienced at that location can be precisely logged. This unprecedented level of information about the real-world success of the design can lead to significant improvements to the system design process, especially in elements of the design process in which there is a scarcity of solid information – clutter, foliage, and building penetration loss, for example.

The sections of this chapter describe the methods and procedures that can be used to accomplish the three steps listed above. For large systems, many of the methods are algorithms that are best implemented in computer software, especially where efficient access to complex, high-resolution terrain and building databases is required for the path clearance, interference, and other engineering analysis.

Using the approach in Chapter 10, LOS and NLOS PMP networks will be treated separately since the engineering criteria used to assess coverage and availability are different.

11.2 LOS NETWORK DESIGN

The performance of the links in an LOS system is determined from the link budget, formulas, and methods described in Chapter 10. The primary performance difference in a network is that the presence of other links causes the interference term in the link budget in Table 10.1 to become the significant, and often limiting, factor in the link performance. The design of the network starts with hub site selection. A load (set of connected CPEs) for each hub sector in the system is then postulated and the resulting interference levels calculated.

11.2.1 Hub site selection

The objective of a wireless broadband PMP system is usually to provide service to the maximum number of CPEs with the minimum or lowest cost infrastructure. Deploying hubs is one of the main expenses associated with building a network, so hub locations are chosen so that the greatest number of potential customer locations are within their service ranges. For an LOS system, the service range (excluding interference) is usually determined by the existence of LOS path clearance between the hub and the potential CPE locations and whether the CPE is close enough to the hub so that the signal levels at the CPE and at the hub for the uplink are high enough to achieve adequate fade margin.

The potential locations for hubs are not unlimited. For LOS systems, high locations such as the tallest building rooftops, the existing radio cellular towers, the mountaintops, and so on represent a set of *candidate hub locations*. The set of candidate locations for hubs is further reduced to those places where legal access can be obtained through a lease option or some other contractual instrument. The set of potential hub sites can then be ordered on the basis of desirability, with the most desirable being those sites that serve the largest number of CPE locations, or as will be discussed below, the locations that represent the highest potential revenue in terms of business or other possible service users that are located in the served buildings.

11.2.1.1 Visibility/shadowing analysis

A visibility or shadowing analysis projects lines of sight from a candidate hub site to a grid of points in the area surrounding the hub to determine which points are within LOS. An example of such an analysis is shown in Figure 11.2. As illustrated by Figure 10.7, an elevation profile of points along the path from the hub to each point in the service area is constructed by extracting height values from the terrain and the building databases. The elevation angles to the terrain and building features points along each path are calculated. Comparing the elevation angles to the points on the path to the elevation angle to the path endpoint (postulated CPE location) determines whether the path is obstructed to that CPE location. The visibility analysis is converted or mapped into a uniform grid of analysis points for display as in Figure 11.2. This type of visibility analysis is routine in most wireless system planning tools and in some Geographic Information System (GIS) software packages.

The accuracy of the visibility or 'shadow map' is directly dependent on the accuracy of the underlying terrain and building databases that comprise the propagation environment model. Both vector and canopy building databases (see Section 5.3) can be used for this analysis with the caveats associated with canopy databases discussed in Section 5.3.3. It also depends on the spacing of the points used to construct the path. Normally the path is made up of points spaced at regular intervals and then connected together in piecewise linear fashion to form the elevation profile. The spacing of the points should be comparable to the resolution of the underlying database – a closer spacing does not yield new information for the profile, just additional points lying on a linear slope created by the underlying database grid.

Where building vector data is used, as in the case of Figure 11.2, the points at which the profile path crosses the vector representing a building wall must be captured as a

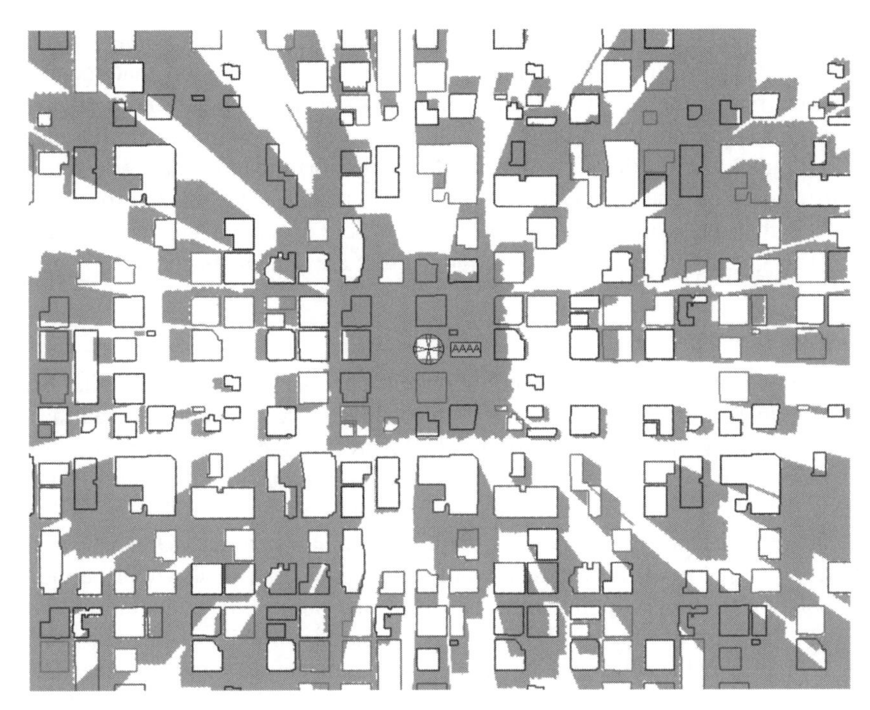

Figure 11.2 Shadow map for a rooftop site in an urban area with a 2-m study grid spacing.

profile elevation point, even though it may lie at an arbitrary distance from the hub transmitter. The ensemble of points describing the profile in this case will include uniformly spaced points and points at arbitrary distances along the path that represent building wall crossings.

Given the resolution limitations on building databases, the shadowing analysis should include some error margin that assumes the buildings may actually be taller and wider than the database depicts them. When LOS interference paths are considered, the opposite strategy is used.

There are several approaches described below for creating a shadow map, as shown in Figure 11.2. They vary in computational burden and accuracy:

1. *Direct grid calculation*: This approach constructs and evaluates a path to each of the uniform grid points in the map display grid. This is the most straightforward and exact approach to creating a shadow study but computationally takes vastly longer than the other approaches discussed below. For example, in a study area of 1 km by 1 km with a 2-m grid spacing, there are 25,000 paths to construct and analyze. If it is known that the only places where CPEs may be located are on rooftops, the analysis can be made substantially more efficient by only analyzing paths to grid points that are located on rooftops and eliminate calculations to points located at street level.

2. *Radial line*: With this approach, radial lines are constructed out from the hub at some small angular spacing like 0.25° or 0.5° and the LOS or shadowed determination

made for each point along each radial path. As the analysis is done for points progressively farther from the hub, the analysis already done to closer points on that radial can be utilized. This makes the elevation data extraction and analysis much more efficient than the direct grid approach. When the LOS/shadowed determination is made for each point along each radial, the results are mapped into the uniform map display grid for viewing and further analysis. The map in Figure 11.2 was created using this approach. The one drawback is that the radial lines become more widely spaced at greater distances from the hub. The spacing in meters between radials that are ϕ degrees apart is given by

$$s = d \sin \phi \tag{11.1}$$

where d is the distance along the radial in meters. The mapping from the radial lines to a point in the uniform grid is done using a simple linear interpolation between the closest points on the bracketing radials as shown in Figure 11.3. At a distance of 3 km and a radial spacing of 0.25°, the radial spacing is 13 m, so the maximum distance from a grid point to a radial line is about 6 m. If the point spacing along the radial is smaller, such as 2 m, then the distance from the grid point to a point where an LOS/shadow analysis was done is about 6 m. This is indicative of the amount of error that will be introduced by interpolating to the grid point from the radial line points. At points closer than 3 km to the hub, the error is smaller. While any error is undesirable, in this case the errors are small or reasonable, outweighed by the computational advantage of the radial line approach over the direct grid approach, which can be several orders of magnitude.

3. *Selected rooftop points*: An even more efficient approach is to carry out the LOS/shadow analysis at a single point or a set of points on each building or structure rooftop. For the single point approach, the point usually chosen is the centroid of the polygon representing the building rooftop, or *top print* as it is sometimes called. For unusually shaped top prints, the top print centroid may not be representative of the overall rooftop, and in some cases, such as U-shaped or open interior courtyard buildings, the centroid may not be on the roof at all. Although efficient, the centroid approach can lead to erroneous conclusions about whether a building rooftop is LOS or shadowed. There are a few examples in Figure 11.2 in which the rooftop centroid is shadowed but some other section of the rooftop is LOS – a *false negative* result. Only database errors can result in a *false positive* result. This can be mitigated to some extent by using several points on

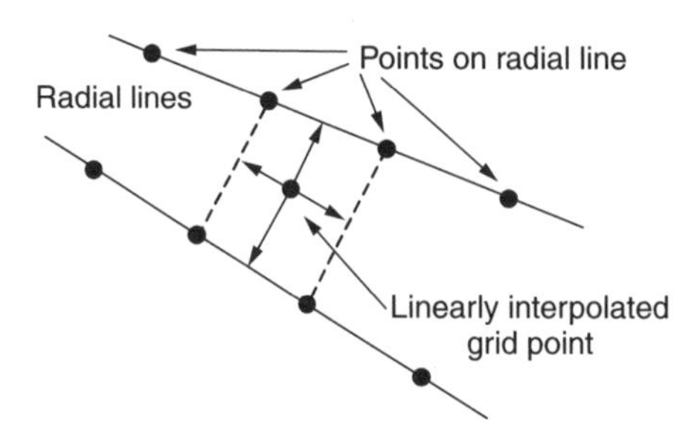

Figure 11.3 Interpolation from radial line points to grid point.

the roof – the logical choice for this set being the vertices of the top print polygon plus the centroid. The cases when this will result in a false negative are rare. Using multiple points for the visibility analysis decreases the efficiency, especially when a large number of buildings with many top print vertices are considered.

4. *Geometric shadowing*: Instead of analyzing visibility by considering the path to potential CPE service locations, the problem can be reversed and the visibility analyzed by assessing the shadows cast by each building. The grid points that fall in a projected arc behind each building are considered to see whether the elevation angle to the grid point is greater or smaller than the elevation to the rooftop edges of the building being evaluated as a shadower. This elevation angle calculation must take into account the fact that the grid point in the potential shadow region may be the roof of another building and elevated by some amount above ground level.

Another important consideration in assessing visibility from a potential hub location is taking into account the shadows cast by the building on which the hub itself is located. The example in Figure 11.2 uses an antenna tower or mast located in the center of the rooftop so that the hub antennas are 20 m above the rooftop. The square-shaped shadow region that surrounds the base of the building is caused by the edges of the building roof itself. For fixed broadband wireless PMP hubs, mounting antennas on a tower like this is undesirable and rarely done. It is also unnecessary. Instead, the four sector antennas (in this case) can be located at the edges of the roof (0.5 m from the edges) on much shorter masts (2 m in this case) as illustrated in Figure 11.4. The shadow map resulting from this configuration is shown in Figure 11.5. This example also illustrates the importance of submeter resolution and precision in the database building positions and the hub sector antenna locations, and the ability of software planning tools to accurately handle these small dimensions. Cellular system design tools intended for macrocells do not require, and often cannot successfully perform, calculations that correctly take into account this level of submeter accuracy.

11.2.1.2 Algorithms for efficient multiple hub site selections

When multiple hubs are being used, the objective is to find the set of the fewest number of hubs that will illuminate the rooftops of the required percentage of buildings, 90%,

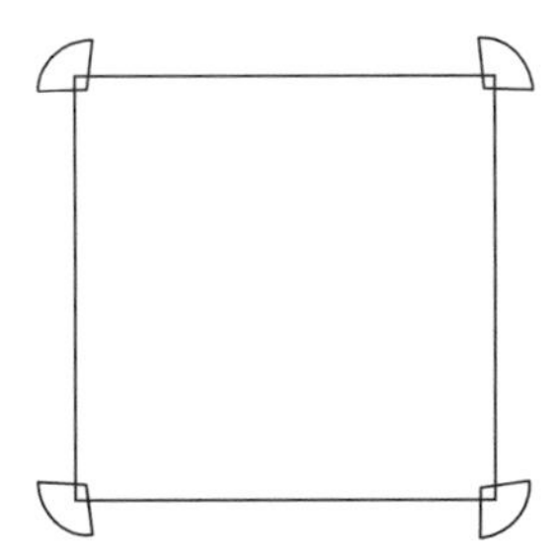

Figure 11.4 Hub sector antennas at edge of rooftop.

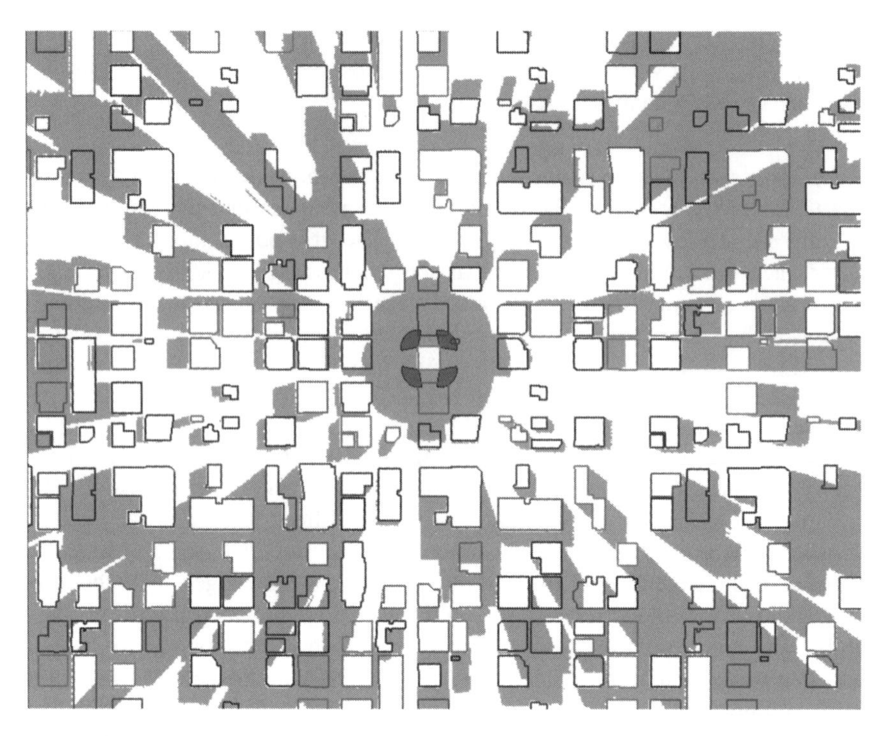

Figure 11.5 Shadow map with split hub sectors at the rooftop corners.

for example. A quasi-optimum set can be found by considering the visibility capabilities of candidate hub locations, starting with the hub location that is expected to provide the most visibility. In most cases this will be tallest candidate hub location, a candidate being a building or tower where a contractual option is in place to use the structure for the purpose of constructing a hub. Using a shadowing study as described above, the number of illuminated rooftops can be assessed. This will be some fraction of the total number of buildings in the intended service area.

After evaluating the tallest building or tower, the next candidate hub location to be evaluated can be selected in several ways:

- Choose the next tallest building – the most straightforward thing to do. This approach can be extended to select the third, fourth, etc. hub locations. The drawback to this approach is that the tallest buildings are often clustered so that the second or third tallest building is usually located very close to the tallest. The additional number of rooftops illuminated by the second, third, etc. hubs are not as great as they could be by choosing a different hub location.
- At some distance from the original first hub location, choose the tallest candidate building. The third hub is chosen in a similar way to be outside some distance ring from the others, where the radius of the rings is somewhat indicative of the service radius of a hub. This approach addresses the building cluster issue.

- Choose the building that shadows the most buildings from the first hub site. During the shadow analysis from the first hub, the number of buildings that are shadowed by other structures can be tallied. The building that creates the most shadows is a reasonable choice for a good location to place another hub, with the prospect that it can illuminate the rooftops of many of the buildings that were shadowed from the first (tallest) hub.

Each of these methods is a viable approach for determining the order in which a set of hub locations is selected, the first being the most straightforward. As each hub is added, a list of the number of illuminated rooftops is compiled. That number is compared to the percentage of rooftops that the system operator is seeking. The process of identifying hub locations continues until the required percentage is achieved. The list of hub locations can also be viewed as a priority list. As such, it can be used to create a phased deployment schedule in which the hubs that are positioned to connect the most customers are built first.

11.2.1.3 Hub traffic/revenue potential assessment

The use of a visibility analysis to select hubs is fundamental for building any system that uses LOS-enabled technology. Once visibility has been achieved, however, an appropriate second-level analysis is to determine the revenue potential of providing service to the building. Some buildings may already be served by optical fiber, so they are very unlikely to want a wireless connection unless as a backup to the fiber. The potential revenue from such a building is probably zero, so the value of illuminating it is zero.

The number of occupants in a building is another factor that will affect its revenue potential. This can be estimated by the floor space in the building, which can be approximated by the footprint area multiplied by the number of floors. The number of floors can be estimated as the building height in meters divided by 3.3; for example, a typical 10-story building is approximately 33 m high. The value of illuminating this rooftop will be greater than illuminating the rooftop of a four-story building, given the footprint is the same size. Rather than a simple tally of illuminated rooftops, then, a better measure of the value of constructing a hub in a given location is the revenue value V_H of that hub location estimated by

$$V_H = \sum_{n=1}^{N} U_S(0,1)\left(\frac{h_n}{3.3}\right) A_n \tag{11.2}$$

where h_n is the building height in meters, A_n is the footprint area in square meters, and $U_S(0, 1)$ is a binary function that is 1 if the building is visible from the hub location and 0 otherwise. This approach to assess the value of a hub is related to the traffic assessment discussed in Section 9.2.4. It necessarily requires the building outlines to be geographically defined. This is normally not the case with canopy building databases (see Section 5.3.2).

The valuation in (11.2) can be further refined if information about the building occupants is available. As discussed in Section 9.2.2, different business types have different data communication needs. Providing LOS service to a building occupied by lawyers or accountants is more valuable than a building of equivalent floor space, which is a warehouse.

The potential revenue assessment can also be used to judge whether a hub is worth building at all. The service value determined from (11.2) can be related to the traffic potential from that building if some information can also be obtained about the type of businesses occupying the buildings. An estimate of traffic volume in megabits per second for this building is the result. Assigning a revenue figure to this traffic volume and using (11.3) leads to the revenue potential for the hub. Given the cost of constructing the hub and its backhaul and the cost of the lease for the hub rooftop, this potential revenue may be too low to justify building the hub. Lower equipment costs and new building construction in the visibility areas of this hub could change the outcome of the analysis so that the hub construction is feasible. The decision not to build a hub must also be weighed against the wireless service marketing strategy that may promise ubiquitous service throughout the city.

11.2.2 Hub sector configuration

The example hub used for the visibility analysis above consisted of four hub sectors, each using an antenna with approximately a 90° beamwidth. The numbers of sectors and antenna beamwidths are determined by the area to be served from the hub and the required capacity. Early microwave band PMP systems used a collection of narrow beam antennas with each only serving a one of two CPEs. The antennas to serve these CPEs were often not put in place until the customer had signed up for service rather than in anticipation of service to multiple CPEs. These hubs are more accurately considered as the terminal node for multiple point-to-point (PTP) link systems rather than as true PMP hubs.

For a true PMP hub in which the antennas are intended to serve (as yet unsubscribed) multiple CPEs for the LOS systems in a service range/arc, the common approach is to use sectors with alternating polarizations. Figure 11.6 shows examples of a 4-sector and a 16-sector hub with an alternating sector polarization assignment pattern for both uplink and downlink. For LOS links in which high-gain antennas are used and scattering from the environment can be minimized, polarization discrimination can be an effective means of increasing capacity.

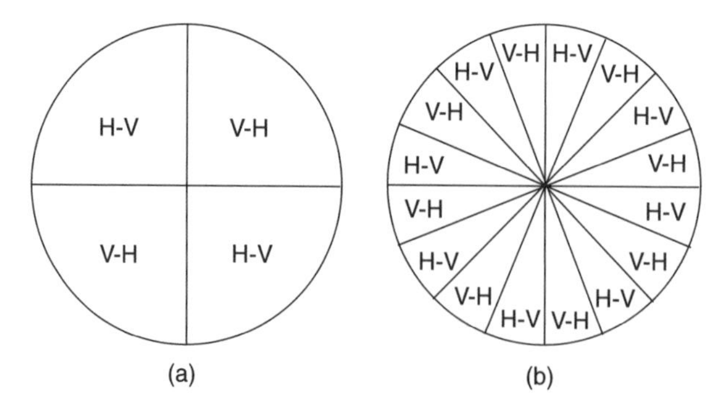

Figure 11.6 An example polarization assignments for a 4-sector and a 16-sector hub. The first letter in each case is the downlink polarization; the second is the uplink polarization.

The use of higher numbers of sectors (sectorization) with increasingly narrow antenna beamwidths is actually an example of the application of space division multiple access (SDMA) techniques discussed in Section 8.5. Sectorization of this type is common in the cellular industry to increase system capacity; however, the practical number of sectors for cellular systems is usually limited to three or six sectors. Theoretically, if the number of sectors was made arbitrarily large (with arbitrarily narrow antenna beamwidths), the connection from hub to CPE would approach a wireless 'string' with an infinitesimally small Fresnel zone and associated spectrum occupancy. Free space optic systems with very narrow angular beam spreading can most closely approach this theoretical boundary. Spectrum occupancy and efficiency will be discussed further in Chapter 12.

The practical number of sectors is limited by the performance characteristics of the antennas. The cross-polarization discrimination is usually highest along the boresight direction of the antenna and decreases with increasing azimuth and elevation angles away from the boresight. Figures 11.7 and 11.8 show the measured pattern responses for a 90° hub sector Local Multipoint Distribution Service (LMDS) horn antenna with horizontal and vertical polarization as the selected polarizations, respectively. For accurate interference calculation, the measured copolarized and cross-polarized pattern values for the intended antennas must be used. This is especially critical when tight sectorization as with 8, 16, or more sectors is contemplated for use at the same hub. In particular circumstances, physical features of the rooftop itself, such as penthouse structures at the top of elevator shafts, can aid in providing isolation between hub sector antennas.

The downlink S/I ratio seen by a CPE when just considering the signals radiated from a single hub with four sectors will be determined in the difference between the copolarized and cross-polarized pattern values in graphs as shown in Figures 11.7 and 11.8. The C/I ratio is minimum at the 3-dB crossover azimuth between sectors. Using alternating sector polarizations reduces this to the cross-polarization discrimination value of the antenna pattern at the crossover azimuth – in this case, S/I is about 30 dB at the crossover azimuth (±45° off of boresight). For hubs that are designed to use all available frequencies on every hub sector, this S/I ratio is the *maximum* S/I ratio that a CPE at the hub sector transition azimuth can expect to have. In a multiple hub network, this maximum value will be degraded by interference from other hubs as well as decreased cross-polarization discrimination during rain fades (see Section 11.3.1). There are frequency assignment techniques to mitigate this effect as discussed in Chapter 12.

The use of opposite polarizations for the uplink and the downlink on a sector is primarily intended to make the design of the CPE easier. For a system using frequency division duplexing (FDD), the downlink and the uplink (receive and transmit) signals must be separated at the CPE. The design of the duplexer that separates them becomes more difficult with smaller uplink/downlink frequency separations. Using separate polarizations for uplink and downlink adds additional isolation between the two, so the filtering action required from the duplexer is reduced. Time division duplexing (TDD) systems use a switch to alternate between uplink and downlink modes, so the duplexer design is not an issue.

In addition to the azimuth orientation of the hub sector antennas, the vertical orientation or beam tilt is also important. Figure 11.9 shows the relative LMDS received power levels for a remote terminal 10 m above the ground (3-story building) as function of distance.

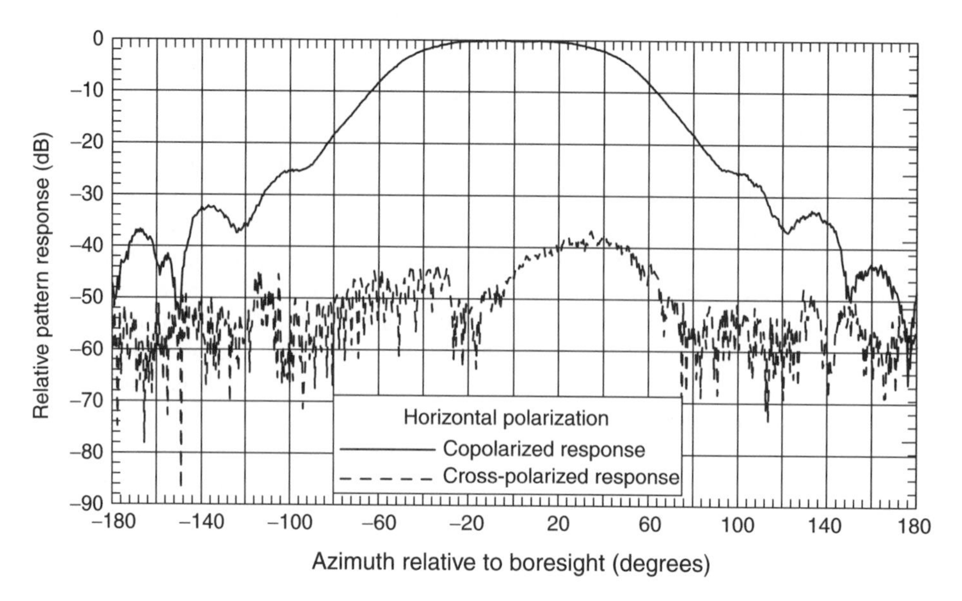

Figure 11.7 Copolarized and cross-polarized response for horizontal polarization from a 90° sector LMDS antenna. Data from [5].

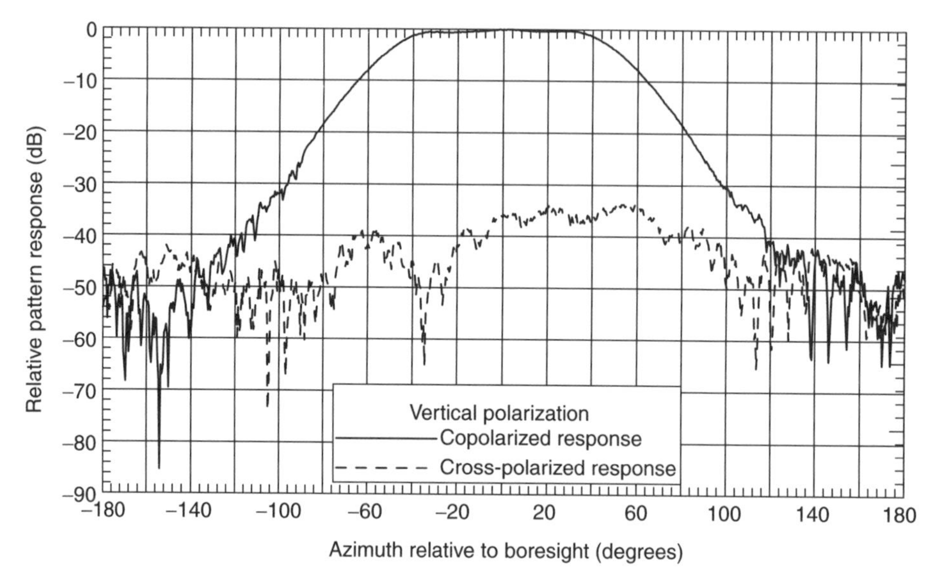

Figure 11.8 Copolarized and cross-polarized response for vertical polarization from a 90° sector LMDS antenna. Data from [5].

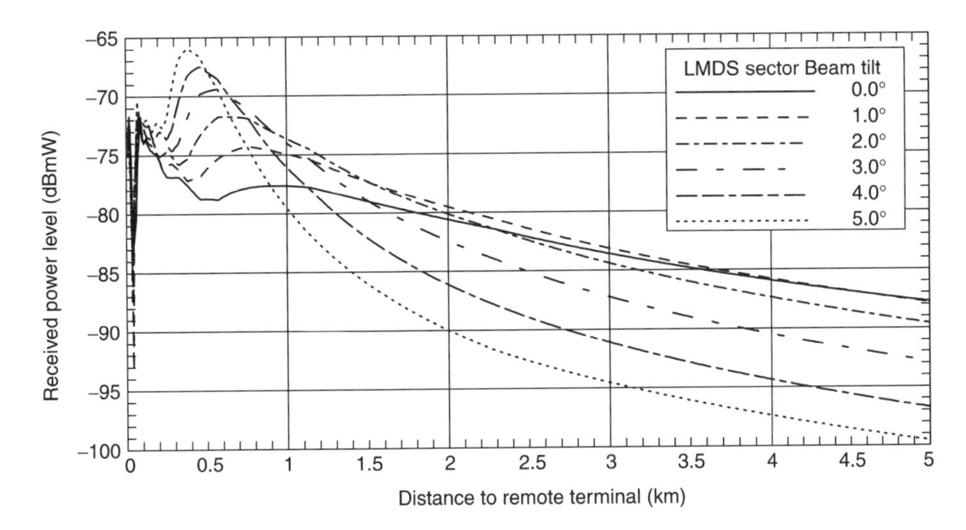

Figure 11.9 Received signal level for different hub sector antenna beam tilt angles. Hub antenna height = 50 m, remote antenna height = 10 m. Free space path loss. Hub antenna gain = 18 dBi, remote antenna gain = 0.0 dBi.

The tilt can be used to concentrate the radiated power close to the hub if the CPEs it will serve are expected to be close by.

The above discussion deals with fixed antennas with associated fixed pointing angles and beamwidths. Adaptive or smart antennas are also feasible for use at LOS hubs. Depending on the number of elements used and the practical frequencies, smart antennas can create independent beams that are pointed toward a given CPE when the CPE is transmitting. The main virtue of the adaptive antenna in this application is the flexibility of directing the beam as needed and its potential to achieve 3-dB higher power level (3-dB higher margin) on links to CPEs that are located near the half-power beamwidth crossover azimuth between sector antennas. Except for this difference, an adaptive antenna for each sector offers no capacity improvement over a spread of separate fixed high-gain antennas designed to serve the same azimuth arc. Moreover, since the CPE locations are presumed to be fixed in location and LOS so that the propagation channel is relatively stable, the flexibility advantage of adaptive antennas is small. However, for NLOS systems, discussed later in this chapter, an adaptive antenna is a much more attractive alternative.

Beyond the symmetrical hub sector configurations described above, there is obviously a limitless number of combinations of sector antenna beamwidths and pointing angles that can be used at a hub site to custom-tailor its coverage and capabilities to rooftops and potential CPE locations it is intended to serve. The decision on how many sectors are required is driven by the aggregate traffic requirements from the CPEs around the hub sector (see Chapter 9) and the channel capacity available for use on each sector as derived from the frequency bands available to the system operator, the efficiency in bits per second per hertz of the modulation/coding technique, and whether capacity enhancement techniques such as multiple input, multiple output (MIMO) are employed.

In a dense network with multiple hubs, the link performance will generally be limited by interference rather than noise, which in turn will affect link availability and traffic capacity. The straightforward approach is to configure sector antennas for each hub on the basis of the traffic to be served from that hub and assuming that the link performance is only noise- and fade-limited (multipath and rain). When the network is assembled and frequency channels assigned, the overall performance with interference can be analyzed. Adjustments can then be made to the hub sector configurations in which problems exist. LOS network performance evaluation is discussed in Section 11.3.

11.2.3 CPE best server hub sector assignments

With a set of hubs configured, the configuration of the second part of the network – the CPEs – can be done. The objective is to find a hub sector ('wireless server' or in WLAN nomenclature, an 'access point') that ideally provides the best, or at least adequate, service to each CPE. This is not necessarily the closest hub sector, and as will be shown, in highly nonhomogeneous propagation environments, the best server is rarely the closest hub sector.

For existing systems, the selection of the serving hub sector for a CPE can be done with a planning tool by evaluating the links from that CPE to each hub sector that is within link budget range and choosing the one that provides the highest link margin. This can be done with postulated CPE locations for initial system design or on a CPE-by-CPE basis as customers sign up for service.

For system designs in which there are no specific CPE locations yet, for the purpose of analyzing the quality of the design, it is important to project a distribution of potential CPE locations. There are a variety of ways to do this:

1. Distribute CPEs at the centroid location on each building rooftop. This approach can also qualify the selected buildings in some way. For example, all the buildings below two stories could be eliminated since their revenue potential can be assumed to be low.
2. Distribute CPEs randomly inside the service boundary.
3. Distribute CPEs randomly inside the service boundary but weight the random distribution according to the traffic density by region types found in a land use (morphology) database.
4. Distribute CPEs at traffic centroids found in a traffic density database.
5. Distribute CPEs at population centroids found in a demographic database.
6. Distribute CPEs at points on a uniform grid.

The best of these choices is clearly the first, since it most closely represents where the CPEs may actually be placed. However, it does require a vector-building database that may not be available, especially at the early stages of the network design when the commercial feasibility of the network is still being evaluated. The other methods have their uses, with a purely random uniform distribution being simple to implement and statistically unbiased.

These methods can provide a means of constructing hypothetical hub-CPE links for evaluating the system design. Using the downlink link budget, the assignment of CPE to hubs can be done for each CPE using a straightforward process of evaluating the path (and the path loss) from each CPE to each hub. The number of CPEs that are assigned to each hub sector can be used to gauge the traffic volume that the sector must accommodate. At this stage, adjustments to the sectorization of the hub can be made.

Figure 11.10 is a portion of a map showing the hub sector-CPE best server connection lines for a hypothetical LMDS system in the Eugene, Oregon, metropolitan area. One thousand CPEs have been randomly distributed inside the service boundary with six four-sector hubs located to serve them. The lines show the connection from each CPE to its best server based on the strongest downlink signal. In almost all cases, the CPE is assigned to the closest hub. Those CPEs in which this did not occur have a hill or some other terrain feature obstructing the path to the nearest hub sector. The propagation environment database in this case was a 30-m terrain elevation database with no clutter heights or building features included.

Figure 11.11 shows the best server connections for about 1600 CPEs located at the rooftop (top print) centroids of buildings in a hypothetical moderately dense urban area. The hub sectors were positioned automatically using the highest building method discussed in Section 11.2.1.2 in this chapter. This approach achieved LOS visibility to more than

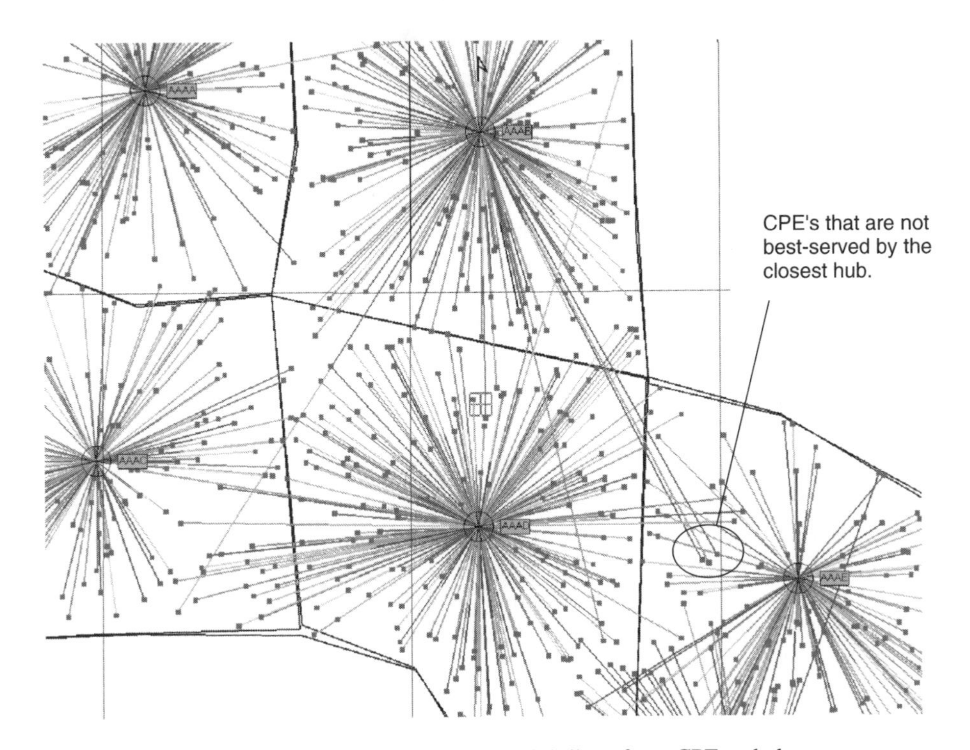

Figure 11.10 LMDS best server link lines from CPE to hub.

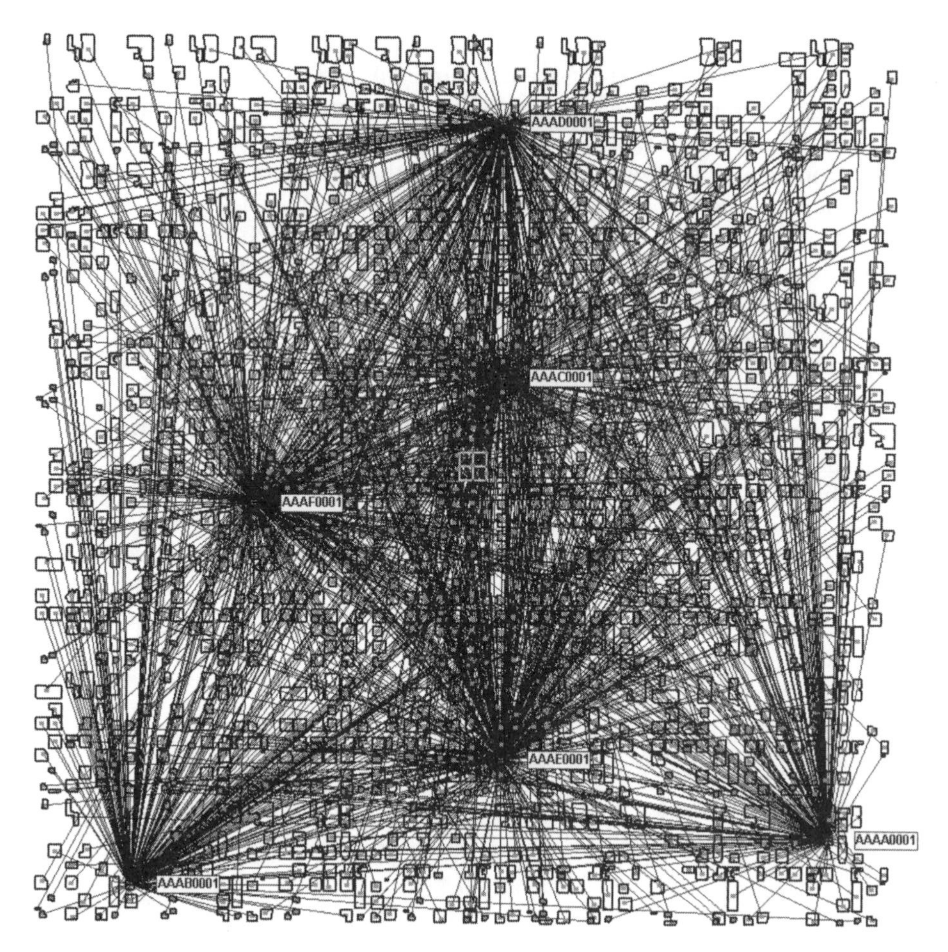

Figure 11.11 Best server link lines for 1600 CPEs and 24 hub sectors (six rooftop sites) in a moderately dense urban environment.

95% of the 1600 building rooftops in the service area using just six hub sites. From the complexity of the overlapping link connections, especially compared to Figure 11.10, the best server for a CPE clearly has very little to do with which hub is closest. The best server may not be the one that is a block away but rather one that is 3 km away. The complexity of the best server connections in this highly nonhomogeneous propagation environment shows that the notion of a hub service boundary, and frequency reuse patterns based on sector service boundaries, are really meaningless in this circumstance. It also indicates that selecting hub locations based on visibility criteria rather than from a uniform hexagon or rectangular grid based on a service range radius can provide a diversity of possible hub-CPE service connections. The network link assignment diversity realized with multiple viable hub-CPE connections achieves one of the advantages normally associated with mesh networks.

11.2.4 Signal distribution from a rooftop

One of the more vexing problems that has arisen with the deployment of LOS fixed broadband wireless systems is distributing the signal from the rooftop CPE location to the various locations in the building where it can be used. For large office buildings, the customer may be located several floors below the roof. The cable risers, conduits, or other cable chases where existing telephone, data network, and other communication cables run in the building may have a different set of access problems from those caused by using the roof. The room containing the telephone circuit terminal blocks for all the telephone distribution in the building is usually located in a vault in the building where the local telephone service provider has proprietary rights. Moreover, the twisted pair cable to the various offices or business located throughout the building may prove to be a data rate bottleneck. It is a case of the 'last mile' problem being solved by the wireless connection to the rooftop, but the 'last 100 feet' problem still remains. While access to these cable runs and other resources in the building can usually be obtained eventually, it represents a further obstacle that NLOS PMP systems are intended to overcome.

Wireless LAN technologies such as 802.11b at 11 Mbps or 802.11a at 54 Mbps may provide a solution in limited circumstances. The use of wireless connections at 60 GHz and above for 'riser' links is another option for signal distribution that is now being explored (see Section 10.3.7). The lack of dedicated capacity from the roof to an office may represent a potential limitation in the performance of an otherwise well-designed fixed broadband LOS wireless system. It should be noted that other wireless LOS systems such as mesh systems discussed in Section 10.8 will also be afflicted with this internal signal distribution problem.

11.3 LOS NETWORK PERFORMANCE ANALYSIS

In Section 11.2 some engineering guidelines and objectives were described that could be used to design an LOS system to provide broadband service to a designated service area. Once the design is developed, the next step is to analyze its performance to determine whether it meets the design goals. For a high density system in which spectrum resources are efficiently utilized, the system performance will be limited primarily by interference and secondarily by fading mechanisms, including rain at frequencies above 8 GHz. Analyzing interference requires the additional design step of assigning channels (frequencies, codes, time slots, etc.) to the uplink and the downlink between the hub sector and the CPE. Assigning channels to these links for both static and adaptive cases is a significant task; Chapter 12 has been devoted solely to this subject. The performance analysis here shall deal with interference conflicts in cochannel and adjacent channel assignments in a generic form via the generic signal-to-interference ratio (SIR), with the method for determining a particular SIR being left for discussion in Chapter 12.

The interference in a system depends not only on the multiple-access channel assignments but also on the interference management techniques that are employed – primarily automatic power control (APC) on both the uplink and the downlink. The analysis that follows includes the effects of APC, especially during rain fades and potential coupling interactions between cochannel links.

The best measure of the performance of links in a network is the same as for individual links discussed in Chapter 10 – the link availability that takes into account the interference and noise fade margins. The statistics of the link performance figures can be formed into a cumulative distribution function (CDF) that can be used as an overall assessment of the success of the network design.

11.3.1 Interference analysis

The interference that is received on a link (referred to here as the victim receiver) can be calculated using the formulas in Section 8.2.1. The interference power that a link receives from another link transmitter is given by the basic (8.6).

For fixed broadband LOS systems in which polarization discrimination is a useful tool for increasing system capacity, the antenna gains in (8.6) must take into account the cross-polarized radiation pattern; that is, the pattern of the radiation with a polarization that is orthogonal to the intended antenna polarization. The transmit and receive antenna gain terms in (8.6) have two components – the copolarized gain and the cross-polarized gain. By convention and for mounting convenience, the antennas are normally oriented so that the desired polarization is parallel to either the horizontal or the vertical plane, although this is not essential as long as the antennas are installed so that the employed system polarizations are orthogonal. Using polarizations that are rotated 45° from the horizontal or the vertical axis have been suggested by some as a way to balance out the advantage (lower attenuation) that vertical polarization has over horizontal polarization during rain fades (see Figures 4.8 and 4.9). For convenience, the orthogonal polarizations will continue to be referred to here as vertical and horizontal.

The total interference I_T at the receiver for link n is the sum of the interference from all the other link transmitters in the system given by

$$I_T(n) = \sum_{j=1, i \neq n}^{J} P_j \tag{11.3}$$

where P_j is the interfering power from transmitter j given by

$$P = P_H + P_V \tag{11.4}$$

with

$$P_H = P_T + G_{T,H}(\varphi_T, \theta_T) + G_{R,H}(\varphi_R, \theta_R) - L_p - A_F - A_S \text{ dBW} \tag{11.5}$$

and

$$P_V = P_T + G_{T,V}(\varphi_T, \theta_T) + G_{R,V}(\varphi_R, \theta_R) - L_p - A_F - A_S \text{ dBW} \tag{11.6}$$

where the various parameters in (11.5 and 11.6) have the same definitions as given for (8.6), except that the antenna gain values $G_{T,H}$, $G_{T,V}$, $G_{R,H}$ and $G_{R,V}$ are the transmit and the receive antenna gains for the horizontal and the vertical polarizations, respectively.

Of course, these values will depend on the intended polarizations of the interference transmitter and the desired receiver. For an interfering transmitter that is using an antenna that is cross-polarized to the desired receiving antenna, $G_{T,H} \gg G_{T,V}$ and $G_{R,V} \gg G_{R,H}$, or $G_{T,V} \gg G_{T,H}$ and $G_{R,H} \gg G_{T,V}$, resulting in a small value for the interfering power in (11.4). For the summation in (11.3 and 11.4), the decibel watt values in (11.5 and 11.6) are first converted to watts. The interference total $I_T(n)$ is converted back to decibel milliwatt and then used in the 'external interference' line in the link budget in Table 10.1.

The interference power calculated in (11.5 and 11.6) by way of (8.7) takes into account the offset between the channel center frequencies and potentially different occupied bandwidths of the interfering transmitter and the victim receiver.

The interference sources J that must be taken into account in (11.3) are different depending on whether FDD or TDD is used as the duplexing method, and whether time division multiple access (TDMA) synchronization can eliminate the need to consider interference from transmissions on links connected to the same hub sector. These subjects are discussed in Section 8.8.2.1 and in Section 8.3.1.

11.3.1.1 Reduced cross-polarization discrimination during rain fades

One aspect of the scattering that occurs during rain events in addition to the attenuation of the desired signal polarization is the scattering of interference signals at other polarizations into the desired signal polarization. The next effect is to increase the amount of interfering power in the desired signal polarization and thus degrade the S/I ratio, even if the rain event is occurring on the interfering signal path only. The amount of increase in the signal is given by the decrease in the cross-polarization discrimination, or XPD, as shown in Figure 4.11 as a function of rain fade depth on the interfering path.

If the system has been planned to take advantage of polarization discrimination to achieve the desired S/I ratio at a receiver, then that S/I ratio will be degraded during a rain event on the interfering signal path. This can be taken into account in the link budget by increasing the interference contribution from cross-polarized interferers in the external interference item in the link budget in Table 10.1. The amount of the increase is proportional to the rain depth and will therefore also be subject to the probability of that rain fade depth. When applied in this way, SIR during rain decreases more rapidly than if cross-polarization were not used to achieve the nominal clear sky SIR. Because it decreases more rapidly, it will reach the point where an outage occurs more often and thus lowers the link availability.

11.3.1.2 Correlated rain fades

In general, as the strength of the desired link signal at the receiver decreases during a rain fade, the fade margin above the noise decreases as does the fade margin above the interference. However, if the path of the interference arriving at the victim receiver is under the same rain cell, it too will be decreased in strength. The net effect is that the fade margin above the interference *for this particular interferer* does not decrease, or does not decrease as much as it would have if the interfering signal were not subject to the rain attenuation.

As illustrated in Figure 4.11, and suggested by (4.50), it is difficult to predict the degree to which the decrease in the strength of an interfering signal will track the decrease in the strength of the desired signal, and hence, the change in the fade margin. Clearly, if the interfering path lies along the same path as the desired signal, the attenuation caused to the interfering signal will be comparable to or greater if the path is longer than the desired signal path. This is also the maximum gain direction for the victim receiver antenna. At interference arrival angles that are removed from the maximum gain direction of the victim receiver antenna, the probability of correlated fading is reduced, but so is the gain of the victim receive antenna; so this interferer is less likely to be significant in limiting the link performance even under clear sky conditions.

With no usable empirical data for estimating the degree of fading correlation on interference paths, the following formula can be used as a first estimate:

$$\mathrm{SIR} = 10\log\left(\frac{P_D A_R}{\sum_{j=1}^{J} W_j P_j}\right)\ \mathrm{dB} \tag{11.7}$$

where the weighting factor W_j on interferer j is given by

$$\begin{aligned} W_j &= 1 - \cos[a(\phi_\mathrm{D} - \phi_j)] + A_R\cos[a(\phi_\mathrm{D} - \phi_j)] \text{ for } -90^\circ < a(\phi_\mathrm{D} - \phi_j) < +90^\circ, \\ &= 1 \text{ otherwise} \end{aligned} \tag{11.8}$$

with the desired and the interferer arrival azimuths at the victim receiver given by ϕ_D and ϕ_j, respectively. The factor a can be varied from 1 to higher values depending on how angularly broad a rain cell is expected to be. If $a = 2$, it means that an interfering signal on a path that is 45° different from the victim receiver antenna pointing angle (boresight) experiences rain fading that is completely uncorrelated with the rain fades on the desired path.

The parameter A_R is the rain path attenuation (as a fraction, not decibel) from (10.28, 10.35, or 10.36). As A_R increases for increasing rain rates, the SIR decreases until it reaches the point where it is below the required SIR for adequate link performance and an outage occurs. Equations (11.7 and 11.8) provide a simple way to weight interferences arriving along path azimuths close to the desired path azimuth.

11.3.1.3 Uplink interference calculations

A useful uplink interference calculation requires that the distribution of CPEs be a realistic representation of where CPEs will actually be. For this purpose, only options 1 and 2 in Section 11.2.3 are really viable. Distributing CPE's in a regular grid can be used to assess areas where the downlink signal strength is inadequate. However, distributing CPEs on a uniform grid results in CPEs being at locations that are highly unlikely to be used as actual CPE locations such as the streets, bodies of water, parks, and so on. The grid also represents a set of CPEs that could be greater than is representative of the actual number of CPEs that would be deployed. Under these circumstances, the sum of the uplink interference will not be representative of the interference that will occur with a real system.

With a realistic distribution of CPEs in place, the uplink interference to a received signal from a CPE is calculated in the same way as for the downlink case in which the CPEs included in the summation in (11.3) are limited to those that are on the same or nearby frequencies. The uplink interference calculation can also be made more efficient by excluding CPEs whose antenna boresight directions are far removed from the direction to the victim hub. A similarly simple filter approach can be used to eliminate consideration of CPEs that are sufficiently distant such that they will not represent significant interference under any conditions.

As with the downlink, the interference sources J that must be taken into account in (11.4) are different depending on whether FDD or TDD is used as the duplexing method and whether TDMA synchronization can eliminate the need to consider interference from transmissions on links connected to the same hub sector.

11.3.1.4 Impact of automatic power control (APC)

A standard feature of modern link radio systems is automatic power control or APC. Automatic power control is a feedback mechanism that monitors some quality parameter, such as frame error rate (FER) or block error rate (BLER), that characterizes the received signal and instructs the transmitter on the link to adjust power up or down to maintain signal quality. The main reason for adjusting power is to respond to multipath or rain fades that temporarily attenuate the received signal and to respond to increased interference levels resulting from the APC actions on other link transmitters. Because of these interference actions, analyzing the performance of a network using links with APC capability necessarily requires an iterative approach in which the nominal link powers are set to achieve adequate fade margin in the absence of interference. With the initial power setting established, the interlink interference can then be calculated as described above and new APC settings made to maintain link performance in each case. This sequence of adjustments continues until a steady state is reached; that is, no further adjustments on any links are needed to achieve adequate fade margin and the resulting FER or BLER performance is adequate.

From a capacity and spectrum utilization perspective, APC is desirable in any wireless system. Using the least amount of power necessary to achieve the desired link performance level reduces the interference level everywhere else in the system. The unused APC range or headroom (maximum transmit power minus the nominal clear sky transmit power) can also be thought of as additional fade margin that can be used to respond to rain and other fading events.

11.3.1.5 Coupled links

The coupled interaction of two links in a network can occur when each is causing interference to the other. Changes in the transmitting power level on one link will increase the interference on the other link, which, in turn, must respond by raising its transmitting

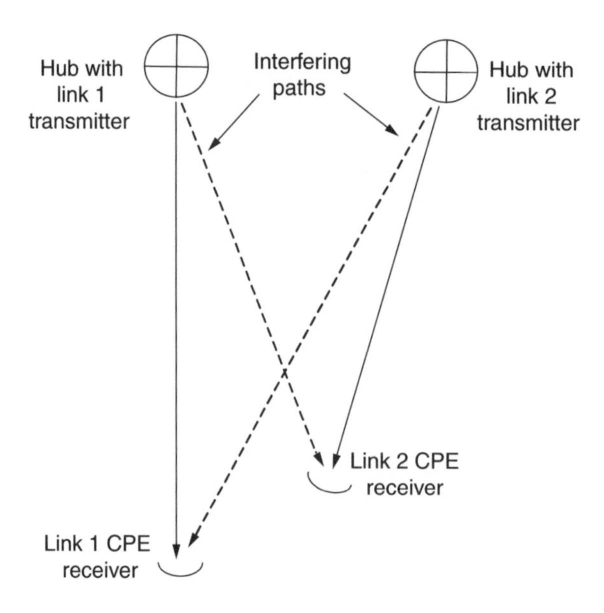

Figure 11.12 Coupled interactive links.

power via its APC. This situation is illustrated in Figure 11.12. Even links that do not interfere directly with each other may, through coupled intermediary links, have some degree of coupling. A rain fade on the desired link will cause its transmitter to raise power, which, in turn, increases the interference level at the Link 2 receiver, which lowers its fade margin causing the Link 2 transmitter to increase power. Theoretically, this sequence can continue until both transmitters are at maximum power and stay there. Since the links are bidirectional (not illustrated in Figure 11.12 for clarity), the same interaction is occurring on the reverse links. This can occur in any multilink system including a PMP system in which the links are between the hub sectors and the CPEs.

From Figure 11.12, it is clear that the degree of coupling is strongly affected by the antenna gains and pointing angles and frequency assignments. Indeed, it requires a fairly rare geometry for tight coupling to occur. The remedy, of course, is to assign different frequencies if possible to links that exhibit a high degree of coupling. If this is not possible, then seeking a different server assignment and thus changing the link pointing angle can also mitigate the problem.

11.3.2 Estimating hub sector capacity requirements

Estimating the capacity required on a hub sector is helpful in determining the system resources (frequency channels, time slots, polarizations, etc.) that should be allocated to that sector. The hub and CPE layout techniques described above affect the way in which the capacity requirements are estimated. If the CPEs were placed on every building, a tally of the CPEs connected to each hub sector as determined by the best server selection described above can be directly used to calculate the total traffic load on the sector.

In order to do this, some estimate of the traffic mix from each CPE can be used. The traffic volume can be estimated from a business or a collection of businesses occupying a building using the methods in Chapter 9. Of course, this capacity requirement is reduced by the penetration factor since it is unlikely that every business in every illuminated building will sign up for service. Answering these questions is the domain of market research and not an engineering task, even though they impact the engineering decisions.

The traffic volume and thus the capacity C_S that a sector must provide to serve K CPEs is then given by

$$C_S = \alpha_p \sum_{k=1}^{K} \sum_{m=1}^{M} p(k, m) R(m) \text{ bps} \tag{11.9}$$

where $R(m)$ is the maximum data rate (packet ON) for application type m and $p(k, m)$ is the probability that CPE k is using application type m. The term α_p is the aggregate traffic factor that is a fraction that sets the multiple-access delay and throughput on the wireless radio layer. This factor is the normalized wireless access capacity that can be estimated from the graphs in Figures 9.6 or 9.12.

If the CPEs were deployed so that they represent demographic or traffic nodes of some kind, the total number of nodes N served by a sector is found by simply counting the nodes that have the sector as the best server based on the preceding analysis. The weighting of traffic volume at the node can be used with a tally to find the total traffic the hub sector must accommodate. For example, if the collection of CPEs connected to a sector represents residential population taken from a demographic database and the expected average traffic per person C_p is 5000 bps (0.005 Mbps), then

$$C_S = \alpha_p \sum_{n=1}^{N} W(n) C_p \text{ bps} \tag{11.10}$$

For this example, the weighting factor $W(n)$ in (11.10) would be the number of people at the census node. For census block data in which the block centroid is the demographic node, this is the population in the block. If the information is available, the average data rate per person can be further adjusted using economic or household income data. Households with higher income will be more likely to have computers and have higher Internet usage.

Note that if the capacity that a sector must provide is higher than what the bandwidth and technology can provide, then some other sector configuration or additional hubs should be deployed so that the traffic load is divided among more hub sectors.

The initial system design should provide an ample hub deployment with hub sectors initially lightly loaded in anticipation of service growth. The traffic estimates should also be adjusted on the basis of projections of data rate demand and penetration rates over 5 years so that the network design is viable for at least that period of time. For a fixed LOS network, adding new hubs and reassigning CPEs to new locations requires the CPE antennas to be reoriented by a technician. Given the expense of such a 'truck roll' to send a technician to the CPE location, such events should be avoided wherever possible.

With an estimate of the required capacity that each hub sector must provide, technology selection decisions can be made that will provide this capacity. The first and fundamental choice is the number and the bandwidth of channels that are needed on each hub sector. This rarely is an open choice; it is usually highly restricted by regulatory considerations or by the licenses that the system operator has obtained for use in deploying the system.

Within these restrictions, the traffic handling capabilities of a given channel bandwidth will be a function of the radio frequency (RF) hardware operating in that bandwidth. Selecting the technology and the system hardware components to provide the required capacity can be complicated because there are many parameters and design approaches that affect the capacity capabilities of the technology. Some of these parameters include the following:

- *Modulation Type*: The theoretical efficiency of various modulation types are discussed in Chapter 7. The modulation type is an equipment parameter. Note that the modulation type may also be adaptable with lower rate, more robust schemes used for links in which the fade margin is substandard. This can be taken into account in the link budget in Table 10.1 by adjusting the required signal-to-noise ratio (SNR) and SIR values. The modulation types specified in IEEE standard 802.16–2001 [1] covering 11 to 66 GHz are limited to quadrature phase shift keying (QPSK), 16QAM and 64QAM for both the uplink and the downlink.
- *Coding and Other Overhead*: A certain amount of the transmitted data is not information data (the payload) but other data for error correction, system synchronization, equalizer training, and so on. Coding and other overheads can consume from 10 to 50% of the raw data rate capacity of a channel. To some extent this is already incorporated in the required SNR/SIR values used in the link budget in Table 10.1.
- *FDD or TDD*: The virtue of TDD is that it does not require a separate channel for uplink transmissions. Instead, time slots on the same frequency are used. This provides more flexibility in handling asymmetrical uplink/downlink traffic, but to some extent this is offset by the increased interference potential as noted in Section 8.8.2.1 in Chapter 8.

As an example, if a 5-MHz channel is available, and 16QAM modulation is used with an overhead of 20% (net efficiency of 3.2 bps/Hz), the data rate payload capacity of this channel is 16 Mbps. If the required sector capacity from (11.9 or 11.10) is 50 Mbps, then approximately three 5-MHz channels are required to accommodate the traffic volume.

For FDD, the aggregate downlink and uplink capacities can be calculated separately in (11.9 or 11.10) and the required sector resources found as described above. Asymmetrical traffic (downlink greater than uplink) will indicate that fewer channels and less bandwidth can be used to accommodate the uplink traffic. For TDD systems, the downlink and the uplink traffic requirements can be added together to determine the number and the capacity of the RF channels that are needed on a hub sector.

The actual process of assigning channel resources in a network to achieve the required capacities while controlling interference-induced performance limitations will be discussed in Chapter 12.

11.3.3 LOS network performance statistics

The overall performance of the wireless network can be divided into two categories:

1. The service availability of each hub-CPE link, taking into account fading and other outage mechanisms.
2. The QoS (multiple-access delay and throughput) of the traffic flow between the CPE and the network.

With the best server CPE assignments, the link connections, antenna pointing angles, and interference levels calculated, the availability percentage of each of the links in the network can be determined from the link budget in Table 10.1. This determination takes into account the multipath and rain fading mechanisms. A cumulative distribution can be used to character the overall network performance as shown in Figure 11.13. This figure shows that 89% of the links in the network have a link availability of 99.9% or greater. Similar cumulative distributions graphs based on simulations for a variety of network configurations can be found in [2]. Once the network is built, monitoring the availability of the links as a routine part of managing the network can provide similar statistics. Comparing the realized link availability statistics with the predicted availability statistics provides an assessment of the success of the network design techniques and methods.

Traffic delay and throughput is largely a function of how heavily loaded the hub sector is compared to its allocated capacity. This can be estimated from the maximum aggregate packet data rate on the hub sector (from the number of connected CPEs) versus the available hub sector capacity. With these two numbers, a ratio is constructed that can be used with the curves in Chapter 9 to estimate throughput. This data can also be graphed for sectors in a similar way as shown in Figure 11.11 in the network to judge performance.

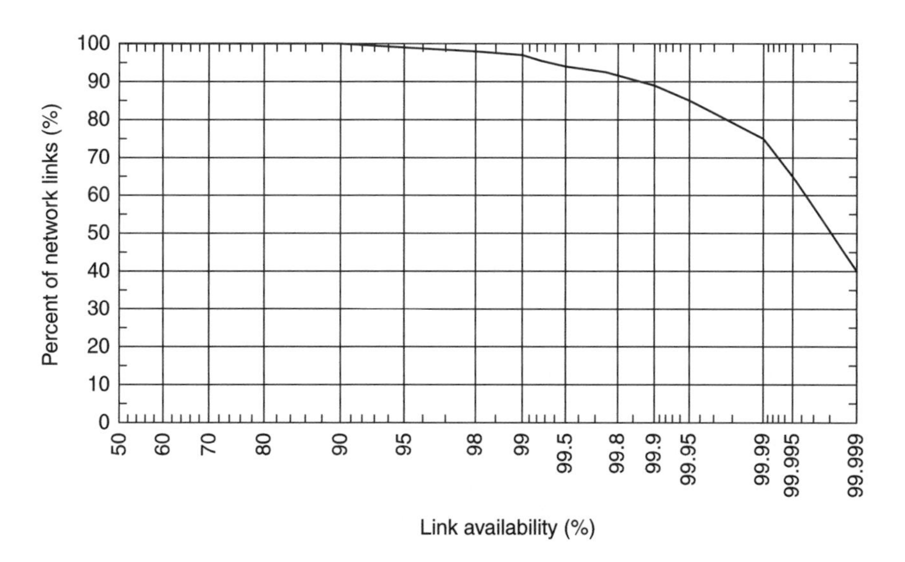

Figure 11.13 Cumulative distribution of network link availability.

Fading events that result in the link adapting to be lower efficiency modulation/coding schemes will lower the normalized wireless access capacity. The delay will increase and the throughput will decrease accordingly.

Analysis of the delay and the throughput performance for specific individual CPEs requires a traffic model for each and a time-based simulation that takes into account the network loading and the interference on a packet-by-packet basis. This type of dynamic Monte Carlo analysis is described in Section 11.5.2.

11.4 NLOS NETWORK DESIGN

The steps in designing an NLOS network are basically the same as for an LOS network but the engineering calculations used at each step are different. For example, the service areas of the hub sectors are determined by areas with adequate signal levels rather than areas that are LOS to a given hub sector. The specific association of a CPE with a building and the businesses within it cannot be done in the same way as was done with LOS systems. Instead, the evaluation of a system design can be done with Monte Carlo simulation techniques in which a random distribution of CPEs with different traffic loading is used to simulate both the downlink and the uplink performance in terms of link availability and traffic flow QoS. Monte Carlo simulations are currently widely used for simulating cellular phone systems in which the specific location of remote terminals is unknown.

11.4.1 NLOS hub site selection

The candidate list for hub sites in an NLOS system will include existing towers, buildings, and hilltops. The construction of a large number of towers to support cellular and PCS (Personal Communication Service) systems both in the United States and throughout the world has created a large number of possible hub locations that, by virtue of their original purpose, may already be well placed for the NLOS network application.

The process of initially selecting hub site locations can make use of the simple empirical propagation models such as the IEEE 802.16 model (see Section 3.4.1) for system dimensioning to arrive at approximate hub service radii and hub spacing. With this radius established, existing tower sites can be evaluated for coverage of the service area by positioning a circle of that radius on each tower site and assessing the extent to which the service area is covered. This is the basic system dimensioning process that leads to an approximate count of the number of hub sites required to deploy the system.

For situations in which the locations of existing towers are unknown or known to be unavailable, most wireless planning tools can provide a uniform hexagonal layout or a rectangular grid layout showing approximate hub site coverage.

11.4.1.1 Coverage/service area calculations

The expectation on link availability for fixed broadband services is substantially higher than the link availability that normally suffices for cellular and PCS system design, especially considering that NLOS fixed wireless systems must compete with wireline services such as DSL and broadband cable.

There are several predicted parameters that can be used to assess NLOS network coverage, among them the following:

- Received power level
- S/I ratio
- Percent availability
- Available data rate at a given percentage of the time and locations. This applies to systems that can adapt data rates based on link performance.

The received power is the most straightforward approach, while the S/I ratio needs frequency assignments in order to calculate interference. Percent availability is a very useful approach because it can take into account fading events such as multipath and rain fades. It can be calculated with or without knowing the interference. For adaptive links, the data rate will be lower for links with poorer link performance. The coverage showing available data rates can provide a more direct indication of what can be provided to customers.

As an example, a study of the coverage area of a 2.6-GHz multipoint, multi-channel distribution service (MMDS) network designed to serve indoor remote terminals was done. This network consists of six 4-sector hubs with 90° beamwidth antennas. The coverage was analyzed at points on a uniform study with a grid spacing of 100 m. Figure 11.14 is the resulting map showing those areas where link availability will be 95% or greater, or 99% or greater. Only noise and not interference was considered for this study. The parameters used for this study are shown in Table 11.1. They are largely drawn from the link availability discussion in Section 10.4.1. The propagation model in (10.33) was also used with terrain obstructions for the 30-m terrain elevation database explicitly taken into account as diffracting obstacles in calculating the path loss.

For the availability map in Figure 11.14, the assumption is made that the channel is flat fading channel signal location variability described by a lognormal distribution with

Table 11.1 Parameters for example MMDS coverage study

System element	Value
Hub sector transmitter power	50.0 dBmW
Hub sector transmit antenna gain	16.0 dBi
Hub transmit antenna height	25.0 m
Average building clutter loss	15.0 dB
Average foliage loss	2.5 dB
Average building penetration loss	9.0 dB
CPE receive antenna gain (indoor)	6.0 dBi
CPE receive antenna height	2.0 m
CPE receiver noise figure	4.0 dB
Channel bandwidth	5.0 MHz
SNR threshold for uncorrected BER = 10^{-3}	16.0 dB
Fade margin for 99% link availability (lognormal distribution σ)	13.5 dB
Fade margin for 95% link availability (lognormal distribution σ)	9.0 dB

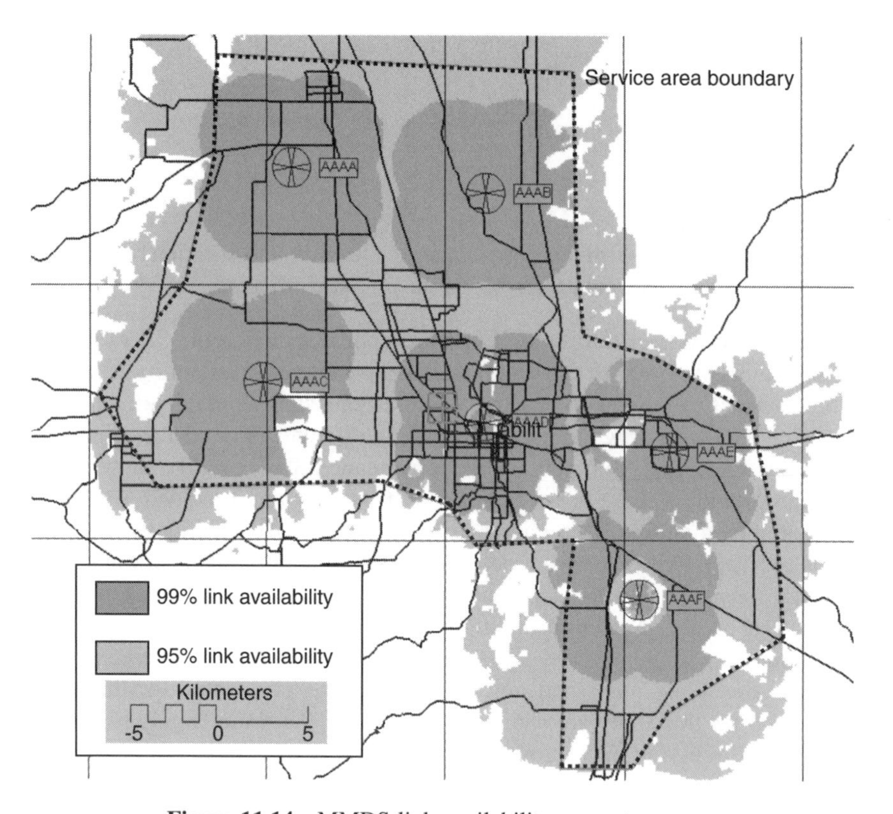

Figure 11.14 MMDS link availability percentage map.

$\sigma = 8\,\text{dB}$. It is also assumed that multipath fading is mitigated using diversity antennas. Even with diversity antennas there will be some multipath fading effects that probably will be wideband (frequency-selective) fading due to the scattering characteristics of an indoor environment. The map in Figure 11.14 provides a good indication of the extent of network coverage and locations where coverage difficulties or 'holes' are likely to occur because of terrain and other obstructions. The white areas inside the service area boundary in Figure 11.14 are examples. If these white areas represent significant sources of traffic, the hub locations can be modified, or additional hubs added, to correct these coverage deficiencies.

The map in Figure 11.14 shows the composite coverage for the overall network. Within this area, the individual hub sector service areas can be determined by simply assessing which hub sector furnishes the strongest signal at a given point in the coverage analysis grid. Those areas can be used for assessing the traffic load on each hub sector.

11.4.1.2 Automatic algorithms for hub site selections

Given the number of detailed aspects of the propagation environment that affect the hub site coverage, researchers have developed automated methods for choosing optimum hub site locations that provide the required coverage. One example in [3] uses the general optimization technique, known as simulated annealing, to find optimum cell site locations in a complex urban environment. A similar approach is applicable to finding optimum NLOS hub site locations, particularly if there is a defined candidate set of possible tower or building sites, a subset of which represents an optimum solution to the problem of choosing the hub sites.

11.4.2 CPE locations

With NLOS systems the expectation is that the CPEs will not typically be located on rooftops of significant buildings that are occupied by businesses, but rather be located inside offices and residences. This assumption was used for the coverage map in Figure 11.14. Some of the techniques for postulating a model distribution of CPEs listed in Section 11.2.3 are therefore not applicable. Instead, the most plausible approach is a random distribution of CPEs, which is weighted according to demographics or land use. Using the weighting, the total number of CPEs can be adjusted to represent the total loading that is expected on the network, including projections of future growth in demand.

11.5 NLOS NETWORK PERFORMANCE ANALYSIS

The techniques used for analyzing the performance of NLOS systems are generally different from those for LOS systems. Instead of discrete CPE points where service must be provided, the service objective is defined by providing service inside a service boundary line that the system operator has defined. Within this service boundary are the service areas of the individual hub sectors that can be calculated after the hub site locations have been selected as described earlier.

The performance analysis can be done in two ways with different levels of complexity and a different depth in the performance prediction results:

1. *Downlink/uplink signal analysis*: With this approach the downlink signal level, the availability, and the S/I ratio analysis are calculated for each study grid point in each sector service area. Traffic loads on each hub sector are estimated from the coverage areas and quantity of traffic sources within the hub coverage area. Uplink interference can also be assessed with this approach; however, it requires that some hypothetical geographic distribution of CPEs be postulated throughout the service area and that all the CPEs are presumed to be transmitting at the same time. This is a straightforward approach but is limited by the assumptions that the downlink and the uplink transmitters are all transmitting at the same time. Because of this, the number of CPEs must be carefully chosen so that the uplink interference calculation is a reasonable representative of the interference that will actually be present.

2. *Dynamic Monte Carlo simulation*: A dynamic Monte Carlo simulation (described in the sections below) is similar in that it also uses various hypothetical random distributions of CPEs. However, in addition, it uses traffic pattern generation at each CPE to assess network performance under a variety of CPE distribution and traffic loading conditions. This

allows computation of uplink and downlink interference for each simulation realization taken into account when packets are actually being transmitted. It therefore provides a more accurate and complete network performance assessment than the downlink/uplink signal analysis method above. It also allows different traffic distribution patterns and applications to be used in the performance analysis by incorporating simulation techniques such as those used in Chapter 9. The drawback is that this approach is more computationally intensive than the downlink coverage assessment method, especially when a large number of simulation realizations are used to assess network performance. It requires a wireless planning tool specifically equipped with this simulation capability.

In the context of the various network performance issues (coverage, interference, and capacity), both of these methods will be discussed in the sections that follow.

11.5.1 Downlink signals for basic NLOS interference analysis

As with LOS systems, the interference analysis relies on the channel assignments (frequency, time slots, codes) for each hub and each CPE. The techniques for making these assignments and their impact on interference are discussed in Chapter 12. These assignments can be made statically or dynamically in response to traffic demand or by assessing which spectrum resources (channels) are already in use.

For interference calculations, the primary difference between LOS and NLOS systems is that fixed high-gain directional antennas are not used at the CPE in an NLOS system. Directional antennas at the CPE increase the link margin, but more importantly in high density systems, they suppress received interference from other transmitters and transmit less interference toward other receivers. The processing gain achieved with directional antennas significantly increases the spectrum utilization efficiency and associated capacity of LOS systems when compared to NLOS systems. The trade-off for this greater efficiency is a system that is more difficult to deploy since it usually requires a technician to install and correctly orient each directional CPE antenna.

Because of relatively low gain antennas used at the CPE in NLOS systems (even in the adaptable antenna case), the interference conflicts (S/I ratio) cannot be assessed on a straightforward link-by-link basis as with LOS systems. The most general kind of analysis is an S/I ratio map that shows the areas where the ratio results in a link margin that is acceptable and not acceptable.

11.5.1.1 Downlink interference analysis

The total downlink interference at the CPE can be calculated by adding the interfering power from all the hub sectors:

$$I_T = \sum_{j=1,i\neq n}^{J} P_j \tag{11.11}$$

where J is the number of hub sectors and P_j is found from (8.6). The hub sectors that must be included in total number J can be adjusted to exclude those that may be irrelevant, given the channels (frequencies, time slots, codes) that are being used on that hub sector. Owing to scattering in the propagation environment in an NLOS system, polarization discrimination cannot be used for interference suppression, although it can be exploited for capacity in a MIMO-type system. Therefore, separate consideration of horizontally and vertically polarized interference is not applicable here. The resulting SIR is found as

$$\text{SIR} = \frac{P_D}{I_T} \tag{11.12}$$

where P_D is the desired signal power level as before. The interference level in (11.11) converted to decibel milliwatt can be used in the link budget in Table 10.1 for the assessment of link availability. If the system is using downlink APC, the downlink interference calculations can be done assuming all transmitters are operating at maximum power, even though this will rarely be the case. If a distribution of hypothetical CPEs is used throughout the service area, then the downlink APC levels can be calculated for each, using an iterative approach that accounts for changing interference levels from other downlink transmitters. This iterative APC adjustment method to model APC action is described in Section 11.3.1.4 in this chapter.

The use of adaptive antennas at the CPE for NLOS has two beneficial effects: (1) it can reorient the pattern to enhance the desired signal regardless of the scattering directions from which it may be arriving and (2) it can suppress interference in one or more directions depending on the number of elements used in the CPE adaptive antenna as discussed in Section 6.6. Both actions increase the SIR in (11.12). The increase in the desired power level can be included in the network performance model by simply increasing the effective gain used for the CPE antenna in Table 11.1.

Adaptive antenna interference suppression can be individually applied to each interfering signal arriving at the CPE receiver

$$I_T = \sum_{j=1,i\neq n}^{J} \alpha_j P_j \tag{11.13}$$

where α_j is the adaptive antenna suppression of the signal from interferer j. For a simple adaptive antenna with two elements, only one interferer can be explicitly suppressed (see Section 6.6), which is logically the strongest interferer of the maximum of the P_j in (11.13). The suppression factor α_j is some fraction that is really dependent on the specific capabilities of the adaptive antenna. For example, a suppression factor $\alpha_j = 0.1$ (10 dB) can be applied to the strongest interferer in (11.13) and the total interference recalculated accordingly. Simulations in many networks show the contributions to interference, and the resulting limit on performance at a remote unit is primarily coming from one or two strong interferers. When this is the case, an adaptive antenna that suppresses interference by 10 dB can allow much more intensive reuse of downlink channels and thereby increase the overall network capacity. On the basis of the field tests with prototype systems, one manufacturer of a CPE with an adaptive antenna claims that the frequency reuse of 1 can be achieved [4]. A frequency reuse of 1 means that every frequency can be used on every hub sector and still maintain adequate S/I rates at the CPEs (see Chapter 12).

11.5.1.2 Uplink interference analysis

As discussed above, when using simple downlink signal level predictions to model network performance, as in Figure 11.14, if no individual CPE locations are postulated, there is no practical way to calculate a summation of uplink interference since the location of the transmitters and their APC power levels are unknown. When this type of analysis has been applied to 1G and 2G cellular systems in which FDD and separate frequency channel reuse planning is employed, the downlink interference analysis is done as described here and the uplink interference is assumed to track the downlink interference because of the fixed relationship between the downlink and the uplink channel plans. For fixed broadband NLOS systems, a similar assumption can be used so that the assessment of the uplink network performance rests solely on the downlink performance predictions.

For Code Division Multiple Access (CDMA) systems, the aggregate interference on the uplink depends on how many CPEs are communicating with each hub sector and their APC transmit power level settings. If the normal assumption is used that the APC adjustments are done so that all CPEs on a hub sector are adjusted so that they have the same uplink received signal levels at the hub sector, the intracell interference power can be calculated from a simple total of the CPEs connected to that hub (both as the primary or the secondary server). The number of CPEs on a hub sector can be estimated from the traffic load (see Section 11.5.3).

As discussed in Section 8.4.4, the intercell interference can be estimated from a frequency reuse factor that assumes the total interference coming from all neighboring cells is some fraction of the total interference coming from intracell sources. This fraction can be derived from an assumption about the distribution of CPEs in the neighboring cells and their distances from the hub sector cell being studied. A value of 0.45 is commonly used.

The weakness of this approach, of course, is that it relies on a number of assumptions about the signals from CPEs operating in neighboring hub sector cells. Without taking into account specific CPE locations, it ignores the propagation conditions from each CPE location to each hub sector where uplink interference might be caused. For highly nonhomogeneous propagation environments, this can substantially affect the amount of interference received from neighboring cells.

This limitation can be addressed by evaluating network performance using a variety of possible random CPE location distributions. An example is shown in Figure 11.15. If a random distribution of CPEs is used, then the downlink and the uplink APC levels can be calculated for each link direction and adjusted using an iterative approach that takes into account the changing power levels at nearby interfering transmitters. However, it still requires some assumptions to be made about each link. Rather than explicitly account for when packets are being sent, the conventional method is to apply an *activity factor* that is found from the average amount of time the transmitter is actually sending data (a fraction).

A more complete analysis that resolves these limitations and assumptions will include traffic generating distributions and different geographical distributions of CPEs. The resulting network performance statistics are then derived from a wide range of CPE distributions, APC and interference levels, and traffic loading scenarios. This type of Monte Carlo analysis is discussed in the next section.

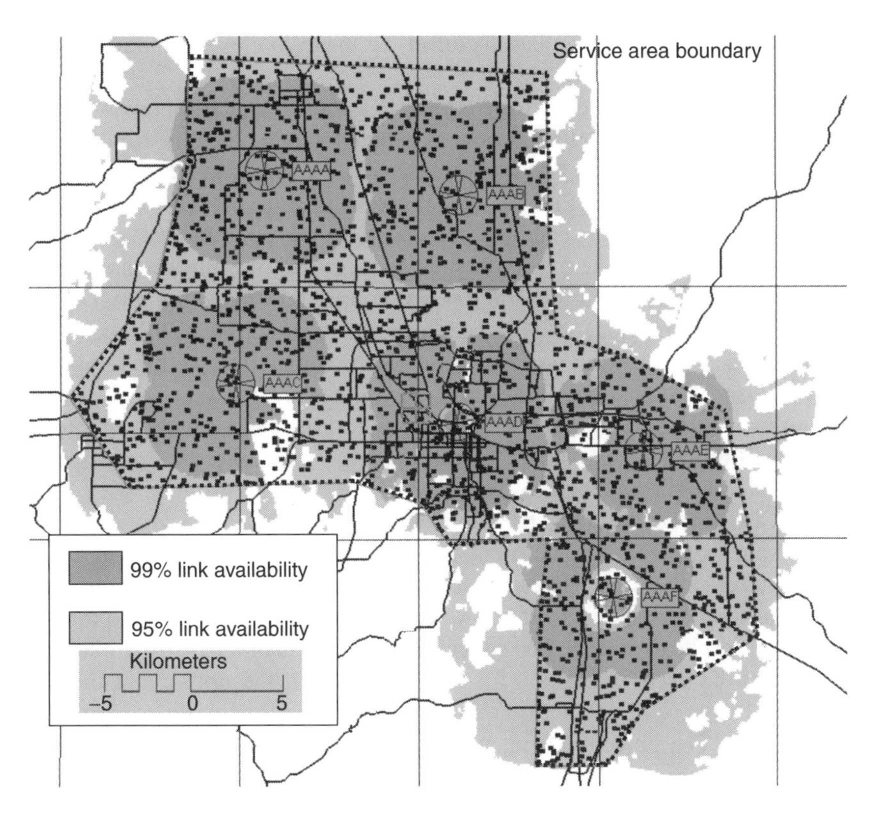

Figure 11.15 MMDS NLOS network with random distribution of 2100 CPE locations.

11.5.2 Dynamic Monte Carlo interference simulation

Even though the service objective is defined by areas rather than by discrete points, in order to analyze uplink and downlink interference, especially with dynamic channel assignments, some discrete point representation of CPE locations must be postulated. By analyzing the system performance for a large number of randomly weighted CPE distributions, an overall view of the system performance can be obtained. Because a large number of random CPE distributions are used, this analysis approach is usually called a Monte Carlo simulation. Monte Carlo simulations of this type are routinely used to analyze the performance of CDMA and W-CDMA cellular systems in which the uplink interference at cell sectors is strongly dependent on the location of the mobile units and which sectors serve them. The difference with NLOS fixed broadband is that the remote units are not mobile; however, they are still randomly distributed at locations that are not specifically known.

A dynamic Monte Carlo simulation uses a simulation clock with some defined time step. At each time step the state of the network and its components (hubs, CPEs) is calculated using an algorithm or mechanism that replicates the action of those that control the actual system. The simulation can therefore be made to closely mimic the operation of the real system once it is built.

After the hub site layout and sectorization is determined as described above, the next step is to make an initial CPE distribution. A random geographical distribution of CPEs is done to simulate where the traffic is expected to come from. Figure 11.15 shows a distribution of about 2100 CPEs for the example Eugene MMDS system. The total number of CPEs should be selected to represent the penetration rate and how many CPEs are expected to be in active communication during the busiest hour. The 2100 number used here is based on a total of 104,000 households inside the service boundary, 10% penetration (subscription) rate, and 20% of those subscribed CPEs being engaged in communication. The distribution can be weighted so that areas of high population, or high expected user density, have more CPEs than other areas. Areas where very few or no CPEs are expected, such as bodies of water or forests, can be underweighted so that essentially no CPEs fall in such areas. These weighting functions require a geographical definition of some kind to define those areas.

For each CPE in the initial distribution, a traffic pattern or packet session is then established using the probability distribution for packet size, packet spacing, number of packets in a packet call, number of packet calls in a packet session, and so on. This set of packets is what the CPE intends to send. A similar approach can be used to represent the downlink (CPE received) traffic. The techniques and probability distributions described in Chapter 9 can be used as a guide for how to establish the uplink and the downlink traffic patterns for the CPEs. Clearly, there are a very large number of options for the traffic distribution. The objective here, though, is not to simulate the network action for all possible traffic types but to establish a traffic load on the wireless network that is approximately representative of the actual traffic load the network will experience.

After the initial CPE geographical and traffic distributions are done, the simulation can begin with the first time step. The size of the time step can be set to be the minimum packet size at the maximum data rate for fine-grained sampling, although subsampling at much lower rates (longer time steps) can still produce statistically valid results. For a minimum packet length of 80 bytes (640 bits) at a data rate of 10 Mbps, the minimum time step would be 64 μs. At each time step in the simulation, a number of operations must take place as listed here:

1. For the downlink, the hub sectors sending packets are considered in terms of the interference they create to CPEs throughout the system. This is calculated using the information in Section 11.5.1.1 in this chapter. The resulting SINR values at the CPEs can be used to judge whether a packet intended for that CPE has been received with an adequate FER or BLER. In a system with downlink power control, the power levels for the hub sector transmissions to each CPE are adjusted to just achieve (if possible) an acceptable FER. Since increasing power to one CPE may increase interference to another, the APC adjustment is actually an iterative loop in which the APC levels are incrementally adjusted, the interference levels recalculated, and the resulting FER evaluated. If a hub sector has reached maximum power for downlink communication with a CPE and the FER is still unacceptable, then the block is classified as received in error. A tally can be made of blocks received in error as a measure of network

performance. A more sophisticated simulation would require a retransmission of the packet to mimic actual network operation. Note that for TDD systems, the potential for interference on the downlink from CPE uplink transmitters exists, so these transmitters must also be included if not synchronized on the sector (see Section 8.8.2.1 in Chapter 8). For CDMA systems, the interference calculation is done as discussed in Section 11.5.5.

2. A similar approach to step 1 can be used for the uplink with uplink traffic and interference at the hub sectors calculated as described in Section 11.5.1.1 in this chapter. The same iterative approach is used to adjust the CPE transmit APC levels to achieve acceptable FER, taking into account noise and interference from other CPEs. If maximum uplink power from a CPE does not achieve an acceptable FER, the block is received in error and tallied as a measure of uplink performance. As with the downlink, the potential for interference on the uplink from the downlink transmitters exists, so these downlink transmitters must also be included if not synchronized on the sector (see Section 8.8.2.1 in Chapter 8). For CDMA systems, the interference calculation is done as discussed in Section 11.5.5.
3. When the number of CPEs contending for access to the hub sector exceeds the data rate capacity of the resource on the hub sector, admission control and scheduling algorithms as discussed in Chapter 9 are invoked. The resulting multiple-access queuing delays or latency are an additional measure of network performance that is available with the dynamic Monte Carlo simulation. These performance metrics can be compiled for each hub sector and each CPE as another means of assessing network performance.
4. With error rate results calculated for a time step, the simulation can begin to compile several statistics. The traffic load on each hub sector is one such measure along with the error rates of the receive blocks or packets, as shown by the example of a 10-min simulation in Figure 11.16. The performance results at each individual CPE can also be obtained, although for a view of overall network performance, cumulative distribution functions (e.g. 90% of the CPEs have an FER of 10^{-2} or lower) are more useful. Geographically depicting CPE performance is also valuable using different colors or

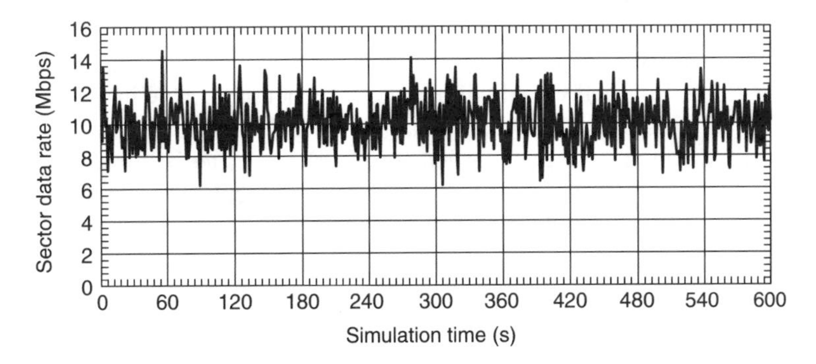

Figure 11.16 Example of the sector data rate load over 10 min from a Monte Carlo simulation of an NLOS network.

symbols to illustrate the current state or performance of that CPE. Clusters of CPEs with poor performance indicate a traffic hot spot that is underserved by the available hub sectors and the spectrum resources available on, or assigned to, those hub sectors.

5. At the next time step the traffic packet session is incremented by one time step, which means that some CPEs will start sending a packet and others will stop sending because they are in a packet spacing interval or a packet call interval. At some point the simulation also recognizes that a CPE has stopped its communication because the user has gone off to do something else. To keep the number of CPEs constant, this inactive CPE is dropped and a new one randomly added somewhere in the system using the same approach that was used for the initial CPE distribution. If desired, the total number of CPEs can also be made to fluctuate over some range by adding or dropping CPEs following some desired distribution.
6. For the new time step, simulation events 1 through 3 above are done again. This sequence is repeated for as many time steps as may be desired to achieve statistically meaningful results for the network performance. A formal determination of what is 'statistically meaningful' is difficult with a simulation like this. If a large number of CPEs are used, indicating actual CPE connections rather than just a sample, the number of steps could be set to simply represent real time; for example, the number of steps it takes to compile 30 or 60 min worth of performance results.

From the above description there are clearly a large number of calculations that must be completed for each of what may be several thousand time steps. Efficient techniques for calculating path loss and interference levels are necessary along with fast computer hardware. The benefit from this additional calculation burden is a fairly realistic view of how the wireless links in the network will perform and where the network design may be inadequate. Determining these inadequacies, and amending the network design as appropriate to correct them, is the heart of the system design process. A design that has been fine-tuned using predictions and simulation is much more likely to meet the desired service objectives when it is actually built.

11.5.3 Estimating hub sector capacity requirements

The hub sector capacity calculations follow those in Section 11.3.2, except that a direct summation of traffic from individual building CPEs is not applicable in this case. Instead, the traffic load on a sector can be found by multiplying the service area of the hub sector by the traffic density in megabits per second per square kilometer. Alternately, the number of households in the service area of a hub sector can be calculated from the census block centroid data and that tally multiplied by the expected average traffic that a household will generate.

The number of households N_H served by a hub sector i is

$$N_H(i) = \sum_{j=1}^{J}\sum_{k=1}^{K}\sum_{l=1}^{L} n_h(j,k,l) \text{ for } S(i,j,k) > S(m,j,k) \quad m = 1, M \quad m \neq i \quad (11.14)$$

where n_h is the number of households in block centroid l located in study grid tiles j, k where hub sector i has the strongest signal $S(i, j, k) > S(m, j, k)$ of all the hubs M. For the example in Figure 11.14, there are 104 862 housing units inside the service boundary spread among about 4,600 centroids according to the 2,000 Census. If a hub sector is the best server (strongest signal) for 250 of those census blocks as defined by their centroid locations, the hub is serving about 5700 households, as calculated using (11.14). Applying a 10% service penetration rate, this hub sector would be providing traffic connections to 570 households. Using the application mix in Section 9.7, and assuming that during the busiest time 20% of these households will be actively accessing the network at a given time, the maximum possible aggregate packet data rate is about 70 Mbps. Applying a hub sector capacity of 25% of the total possible aggregate data rate for reasonably low multiple-access latency and reasonably high net throughput, the capacity required on this hub sector is 17.3 Mbps. With such an estimate of the hub capacity, the capabilities of various technologies as discussed in Section 11.3.2 can be considered as candidates that can deliver the required capacity.

This example is a valid but approximate way of estimating the data rate capacity required on a hub sector. A more accurate assessment can be made including sector-specific traffic statistics using the dynamic Monte Carlo simulation described above. With the simulation, the time varying packet traffic transmission of each CPE to each hub sector, and vice versa, is tracked and complied for each sector. Some estimate of the application mix, as in Section 9.7, is still needed to determine the traffic flow to and from each CPE.

11.5.4 NLOS network performance statistics

Depending on the analysis that has been done, a wide range of different information about the system performance can be displayed. Most relevant are those data tracks that indicate revenue potential, customer satisfaction, system revenue potential, and engineering design problems. For a simple analysis using downlink signal level information, the coverage map showing signal levels and the predicted FER based on those signal levels and the predicted average interference and noise are most useful. By integrating those areas with acceptable service over the corresponding traffic density data, a measure of the data carriage potential for the system can be obtained.

While the service area maps are quite useful, other representations of the network performance can sometimes provide a concise picture of performance. A cumulative distribution graph of FER for discrete hypothetical distribution of CPEs, as shown in Figure 11.17, tells the percentage of the CPEs in the entire network that are actually achieving an acceptable FER. Similar graphs can be done on a sector-by-sector basis, which can aid in identifying network problem areas or design deficiencies. In addition, link availability cumulative distributions such as those shown in Figure 11.13 can also be compiled for NLOS PMP systems when a distribution of individual CPEs is used for system analysis.

If a Monte Carlo simulation is used, substantially more network performance statistics are available because the network state is being evaluated as a function of time. The value of this information is that it directly relates to what the customer at a CPE experiences,

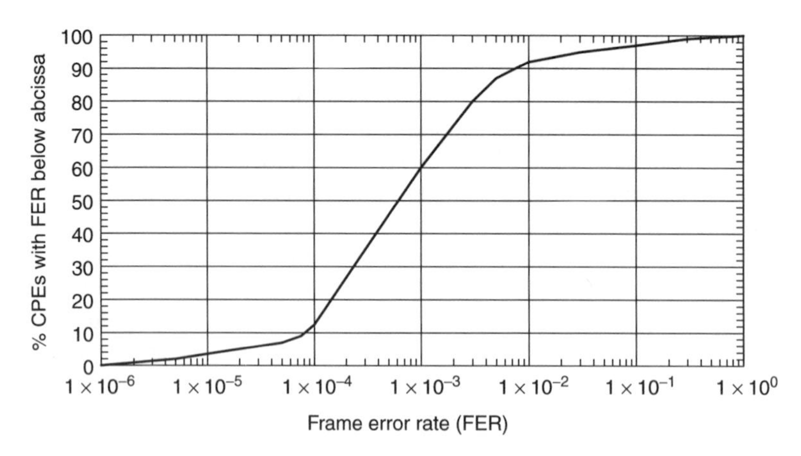

Figure 11.17 Cumulative distribution of FER for all network CPEs.

especially if a packet traffic simulation model is included along with the RF layer model. It can also help identify problems due to hot spot 'bursty' traffic loading that is particularly characteristic of Internet traffic as discussed in Chapter 9. Such time line information is not available without the time-dependent dynamic Monte Carlo simulation.

11.5.5 W-CDMA interference and capacity

The straightforward formulas to find interference power in (8.6, 11.11, and 11.13) must be taken one step further for CDMA systems because the spreading codes used mitigate the detrimental impact of the interference. Once the interference levels are calculated using these equations, the results can be employed in the downlink and the uplink interference terms I where appropriate in the equations in Sections 8.4.3 and 8.4.4. Because the summation values for I depend heavily on uplink and downlink APC settings, which, in turn, depend on the CPE locations and the traffic they are generating, fixed broadband W-CDMA systems are best analyzed using the dynamic Monte Carlo technique described earlier.

The capacity limits will occur when the system load increases to the point where the resulting traffic E_b/N_0 values fall to a level at which the FER after correction is no longer acceptable.

With these amended performance calculations, the other performance metrics can be compiled and evaluated in the same way to assess network performance and the quality of the design.

11.6 NETWORK DESIGN REVISIONS

The purpose of evaluating any design is to find out how closely it meets the objectives of the system operator. An over-designed, and therefore unnecessarily expensive, network may be as big a problem as a less expensive under-designed network that fails to provide adequate service to customers. The performance analysis results that are produced for both LOS and NLOS networks are geared toward identifying service deficiencies as well as locations where an abundance of resources have been needlessly allocated.

When a problem in the design is detected, a number of remedies may be available depending on the nature of the problem. Some of these remedies are discussed below.

11.6.1 PMP network coverage deficiencies

Where gaps or holes in the coverage area exist and traffic is expected from those locations, the reasons for the gap (propagation conditions or antenna pattern nulls) can be determined by examining individual link paths from the nearby servers into the problem areas. This will usually identify a terrain obstruction or some other obstacle. Given the position of the obstacle, a hub site can be relocated to provide better visibility into the coverage gap, or, if significant enough, a new hub site could be added.

In some cases the coverage deficiency may occur in a distant area that is normally out of range given the normal antenna gains and transmitter power being used in the system. For NLOS systems, selected cases can be addressed by using professionally installed outdoor high-gain antennas. This may be the more economical approach to providing service than relocating or adding a hub.

Future coverage deficiencies that may arise as new residential areas are built or new office parks are created can, to some extent, be anticipated by deploying network hubs in a way that is cognizant of long range planning goals and zoning. When new construction or traffic demand extends beyond what was anticipated, the conventional approach is to deploy new hub sites that are selected to best capture the new traffic.

11.6.2 High frame error rates

High FER values result from low signal level versus noise, high interference levels, or low fade margins. The obvious remedies include increasing desired signal levels and reducing the interference. In this regard, it is quite helpful to be able to identify the individual transmitters (hub sector for downlink, CPE for uplink) that are responsible for the interference levels of the problem location. Adjusting the individual channel assignments (for fixed assignments) in an LOS system with fixed antennas is one potential remedy. Another approach is to assign a CPE to a hub sector that does not necessarily provide the strongest downlink signal but would allow the CPE antenna to point in a different direction and thus sufficiently suppress the interference to achieve adequate SINR and FER.

For LOS systems, highly localized features (rooftop penthouses, equipment rooms, etc.) can be exploited by positioning antennas behind them to create obstructions on a problem interference path so that the desired link becomes viable. When this is done, the building features should be added to the building database so that they can be included in future network planning and expansion calculations.

11.6.3 High packet delay times

High latency delays observed in a dynamic Monte Carlo analysis indicates a multiple access bottleneck due to insufficient resources on the hub sector. This can be addressed by splitting the sector into two or more sectors and thus divide the traffic among the new set of sectors. Alternatively, the coverage area of the hub sector can be reduced by antenna reorientation or down-tilting, thereby forcing the excessive CPEs to load onto another hub sector. Network management software may also be able to accomplish this CPE assignment redistribution independently.

11.7 CONCLUSION

The process of designing a fixed broadband PMP network will require different engineering methods depending on whether LOS or NLOS links from the network hubs to the CPEs can be used. The available service locations for an LOS system are determined through a visibility or shadowing analysis. When an LOS path is available, a number of techniques can be used to achieve high system capacity including deploying high-gain directional antennas at the CPEs and using polarization discrimination to permit channel sharing. Although the LOS requirement restricts the number of served locations from a given set of hub sites, and the cost of the system deployment is much higher because of the CPE installation costs, the potential capacity and spectrum efficiency that can be achieved with an LOS PMP network is higher than an NLOS PMP network with user-installed CPEs with low gain antennas, even if those antennas are adaptive or if MIMO techniques are used. Chapter 12 includes a quantitative discussion of this comparison.

The paramount virtue of NLOS PMP systems and the reason they are the focus of current interest for fixed broadband network deployments to residential customers is the potential to have customer-installed CPEs with a hardware cost that is comparable to the to the cost of DSL or cable modems. The performance of NLOS systems is more difficult to model since the propagation environment factors that affect path loss, coverage, and interference are more difficult to predict. The path loss for both desired and interference paths inherently contain a greater range of variability, with building clutter, foliage, and building penetration losses playing a significant role.

The performance of either an LOS or an NLOS network design may be assessed with conventional static signal level prediction techniques or with more powerful dynamic Monte Carlo simulation methods. The Monte Carlo methods can model the changing traffic patterns on a packet-by-packet basis as well as the systematic adjustment to the uplink and the downlink APC levels in response to changing interference from transmitters throughout the system.

While the PMP network design methods described in this chapter are generally applicable to any LOS and NLOS system, the limitation on network performance and its ability to handle the desired communication traffic will depend on the spectrum resources available to the system operator and how the network makes use of those resources. These resources take the form of frequency, time, and code channels. The task of efficiently assigning channel resources throughout a network is the subject of Chapter 12.

11.8 REFERENCES

[1] IEEE Standard 802.16–2001. "IEEE Standard for Local and Metropolitan Area Networks, Part 16: Air Interface for Fixed Broadband Wireless Access Systems," IEEE, April 8, 2002.
[2] H.R. Anderson, "Simulations of channel capacity and frequency reuse in multipoint LMDS systems," *Proceedings of the 1999 IEEE Radio and Wireless Conference*, Denver, pp. 9–12, August, 1999.
[3] H.R. Anderson, et al., "Optimizing microcell base station locations using simulated annealing techniques," *Proceedings of the 44th Vehicular Technology Society Conference*, Stockholm, pp. 858–862, June, 1994.
[4] G. Xu, "Unwiring broadband: the challenges and solutions for delivering non-line-of-sight, zero-install, nomadic wireless broadband," *Proceedings of the 8th Annual WCA Annual Technical Symposium*, San Jose, CA, slideshow presentation, January, 2002.
[5] Gabriel Electronics, Inc., Scarborough, Maine, U.S.A. *http://www.gabrielnet.com.*

12

Channel assignment strategies

12.1 INTRODUCTION

The effectiveness of a wireless network will depend on the geographical layout of the physical network components [hubs, customer premise equipments (CPEs), etc.] in relation to the locations where service is required. It will also depend on how efficiently the spectrum resources available to the operator are utilized at the network nodes. The spectrum resources are usually defined in terms of a band of wireless spectrum frequencies available in a defined service area (a particular metropolitan area, for example), and permissible transmission power levels or interference prohibitions to wireless systems operating in the same or in neighboring regions.

Depending on the system type and frequency band, and country of operation, there may be administrative rules that dictate exactly how the spectrum is to be used. For example, with cellular systems, the channel width in kHz, the specific channel center frequencies, and in some cases, the permissible modulation types have all been set forth in the laws that govern how the spectrum is used. Even license-exempt bands such as the Wi-Fi band at 2.4 GHz and the U-NII band at 5.8 GHz have restrictions on the modulation and power levels of equipment that can be used. In general, licensed fixed broadband spectrum such as the multipoint, multi-channel distribution service (MMDS) and LMDS (Local Multipoint Distribution Service) bands in the United States carry no channelization or modulation type restrictions – the only limits imposed on their use are derived from interference restriction to other systems.

Even though the spectrum can be channelized in a variety of ways, and each channel can be populated with transmitters using a variety of adaptable modulation and multiple-access techniques, some specific strategies about systematically assigning channels, time slots, and codes have been developed that achieve varying degrees of efficient spectrum utilization. These strategies are not mutually exclusive and vary considerably in their complexity and implementation difficulty. All of them have the goal of accommodating as many system users as possible in the allotted spectrum while realizing the desired

Fixed Broadband Wireless System Design Harry R. Anderson
 ISBN: 0-470-84438-8

performance and availability objectives. For commercial networks, this objective will maximize the potential revenue for the systems operator. The purpose of this chapter is to explain some of these channel assignment strategies and evaluate them in terms of their effectiveness at using the spectrum efficiently.

As mentioned previously in this book, the word 'channel' is used in the broad sense to mean not only a frequency channel but also channels created through a group of time slots, or through the use of various spreading codes in a code division multiple access (CDMA) system. Fundamentally, all channel assignment discussions begin with the problem of assigning frequency channels. If there is only one channel allotted to the system operator, then the problem is trivial. In fact, it is not unusual that only a single 5-MHz channel may be allotted in a region to accommodate both uplink and downlink in some countries. The assignment challenge then becomes one of choosing time slots, directional antenna patterns and pointing angles, CDMA codes, equipment with automatic power control (APC), adaptive modulation, and any other system feature to maximize the number of users who can effectively communicate in the 5 MHz.

In general, channel assignment strategies can be divided into two types: static or fixed channel assignments (FCAs) and dynamic channel assignment (DCA). FCAs involve a channel plan of some sort in which the traffic volume predictions are used to estimate the channel resources necessary to service those users. Those resources are then permanently assigned to the hub sectors serving those users. With some older equipment, the channel assignments are fixed settings that cannot be readily or remotely changed. For point-to-point (PTP) microwave link systems in which the traffic source nodes are defined and the system is operated continuously, fixed channel assignment is the logical approach. FCAs have also been the traditional and simplest approach used in cellular systems and point-to-multipoint (PMP) fixed broadband systems, although in recent years more cellular systems have exploited the advantages of DCA. FCAs also rely on predictions of interference conflicts that are calculated, as discussed in Chapters 10 and 11. To the extent those predictions are inaccurate, the channel assignments will be similarly inaccurate, and unexpected real interference conflicts will result. This is actually one of the primary reasons for unexpected dropped and blocked calls in cellular systems.

DCA in its purest form makes no predetermined channel assignments but simply assigns resources to hub sectors, CPEs or other general link transmitters as required when they have traffic to send. The strategy may involve simply assigning or reserving a channel when a transmission or packet session is initiated by a transmitter or it may involve assigning channel resources on a packet-by-packet basis: an approach known as dynamic packet assignment (DPA) [1]. With DCA, the information needed to make an intelligent assignment decision may be difficult to obtain; this becomes one of the critical issues in deciding the effectiveness of dynamic strategies.

Like the multiple access techniques discussed in Chapter 8, the core goal of channel planning strategies is to achieve S/I ratios that provide acceptable link performance. Without link adaptation (LA) techniques, a network that provides 40 dB S/I ratios on some links in which only 12 dB is needed squanders spectrum resources. Uniform utilization for spectrum resources across the links in a wireless network is one of the challenges of channel planning.

For links using adaptation, the higher S/I ratios can potentially be exploited to boost transmission speed during those periods. LA along with other capacity enhancement techniques are discussed at the end of this chapter in the context of the capacity achieved through intelligent channel assignment methods. The basic engineering principles underlying these capacity enhancement techniques have been introduced in earlier chapters.

12.2 FREQUENCY, TIME SLOT, AND CODE PLANNING

As pointed out in Chapter 8, all wireless systems use different frequency channel assignments in some form, whether those assignments are administratively defined or created as sub-channels by the operator within some larger license frequency block using equipment with a particular channelization scheme. Within a frequency channel, different transmitters may occupy different synchronized time slot patterns to avoid interference at a *common* receiving location [time division multiple access (TDMA)]. Similarly, one or more different spreading codes may be assigned to transmitters operating on a single channel to allow successful separate detection at a common receiving location (CDMA).

In conventional channel planning, the frequency channel assignments are made at the time the system is planned or built, while time slot and code assignments are made dynamically by the system, as users require network access. In more modern network design, frequency channel assignments can also be done dynamically as users require network access. *Channel planning* has thus become synonymous with *frequency planning* because no planning prior to system operation is done for specific time slot patterns or spreading codes.

For this reason, channel planning in this chapter will primarily involve the process of choosing frequency channels to assign to links or hubs, but making those assignment choices in a way that is cognizant of the interference factors governing TDMA and CDMA networks. Where the network architecture or multiple access technique can accommodate the use of all available frequencies on all operating transmitters, the frequency planning task becomes trivial, and the capacity of the network is governed solely by the capabilities of the modulation scheme, the multiple access technique and the antenna performance. These are some of the dimensions of the spectrum space model discussed at the end of this chapter.

12.3 FIXED ASSIGNMENTS FOR POINT-TO-POINT LOS NETWORKS

The frequency channel that can be assigned to a link in a network consisting of one or more PTP links, as described in Chapter 10, is largely controlled by the gain patterns of the antennas used on the links and the susceptibility of the link modulation scheme, coding, etc. to external interference. Typically, these links are expected to operate on a continuous priority basis, so multiple access techniques such as TDMA are not available to permit frequency sharing. It is also the case that such links may be operating in a geographical region that is shared by many operators with different, perhaps competitive, networks so that cooperative frequency assignments that respond to internal network requirements may be limited.

For such situations, the basic frequency assignment process consists of obtaining the basic technical information for all the other links in the immediate and surrounding areas that are operating in the available frequency band. For each such link, the basic information includes:

- Latitude, longitude, and height [above mean sea level (AMSL)] locations for the link transmitter and receiver. The transmitter and receiver site elevation and the antenna-mounting height information should be separately specified.
- Frequency and occupied bandwidth.
- Transmit power level.
- Receiver threshold.
- Complete transmit antenna gain pattern including copolarized and cross-polarized patterns.
- Complete receiver antenna gain pattern including copolarized and cross-polarized patterns.
- Any antenna tilt angles used, and the antenna pointing azimuths, if they are not in the direction of the great circle path to the other end of the link.
- Polarization(s) used.

Other details may also be helpful, but this information is the minimum necessary. Unless locally known, this information can often be obtained from engineering services companies or regulatory agencies that maintain current and accurate databases of existing and proposed systems.

The existing links represent the current occupancy of the spectrum band. The frequency planning process is a step-by-step procedure of searching for a frequency channel with adequate bandwidth that can be used without (1) causing objectionable interference to existing or proposed links and (2) receiving objectionable interference from existing or proposed links. The most straightforward algorithm or strategy for doing this is to simply systematically search through the list of available frequency channels in the appropriate band and, at each channel, calculate the S/I ratio at existing and new link transmitters, as illustrated in Figure 12.1.

At each channel considered, the calculation must be repeated for any existing link that is a potential interferer. The calculation burden can be reduced by summarily excluding calculations to links that are using frequencies sufficiently removed from the channel being evaluated. Figure 12.2 shows an example of a criterion that can be used to exclude existing links from the interference calculation. Similarly, existing links that are sufficiently distant can also be eliminated from consideration on all evaluated channels. The information in [2] suggests a separation distance of less than 60 km between links or system service areas as the threshold when frequency coordination calculations should be used.

The antenna gains and pointing angles can also be used to quickly exclude evaluation of particular link interference combinations. For example, if the sum of the azimuth to the victim receiver from an interfering transmitter, φ_T, and reverse path azimuth, φ_R, are

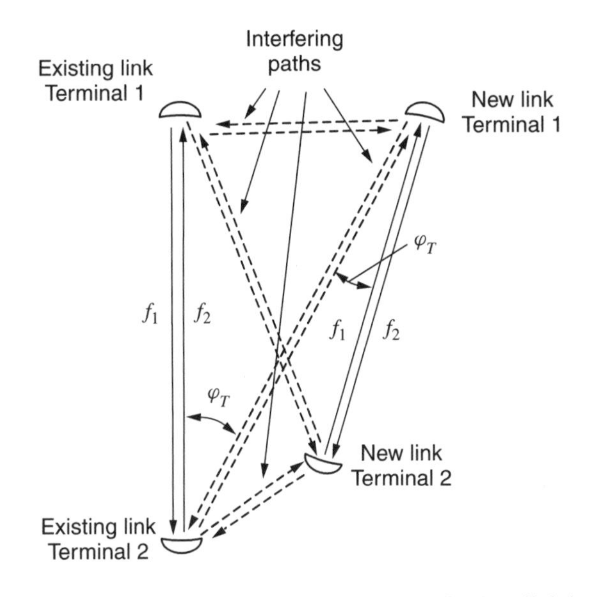

Figure 12.1 Interference paths to evaluate potential for interlink interference.

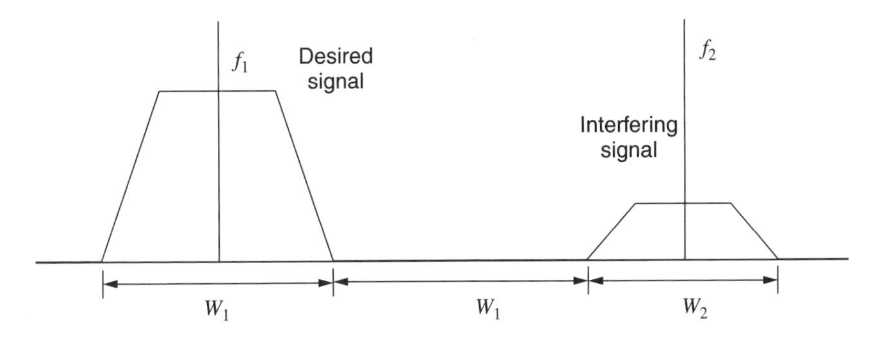

Figure 12.2 Frequency separation criterion for excluding an interferer from interference evaluation during a channel search.

greater than a threshold angle value, the antenna gain pattern attenuation will ensure that interference will not occur without explicit path analysis.

The S/I ratio is formed by calculating the strength of the desired signal and the interfering signal. The desired signal level is found, as described in Chapter 10, from the link budget. The basis equation for calculation interference levels is (8.6), which can accommodate arbitrary cochannel and adjacent channel links when used in conjunction with (8.7). For convenience, (8.6) for interfering power P_i is repeated here:

$$P_i = P_T + G_T(\varphi_T, \theta_T) + G_R(\varphi_R, \theta_R) - L_P - A_F - A_S \text{ dBW} \tag{12.1}$$

where

P_T = transmitter power in dBW

$G_T(\varphi_T, \theta_T)$ = the gain in dB of the transmitting antenna in the direction of the victim receiver where φ is horizontal plane angle (azimuth) and θ is the vertical plane angle (elevation).

$G_R(\varphi_R, \theta_R)$ = the gain in dB of the victim receiving antenna in the direction of the interferer where φ is horizontal plane angle (azimuth) and θ is the vertical plane angle (elevation).

L_p = mean pathloss in dB that includes free space loss, clutter loss, reflections, diffraction, and multipath.

A_F = the attenuation factor in dB because of the receiver filter rejecting some or all of the interfering power.

A_S = the attenuation in dB because of other system losses such as transmission line or waveguide losses, radome losses, etc.

These equations take into account the pathloss along interfering paths that may include obstacles due to terrain features or structures. In fact, it is a common approach to position link transmitters and receivers such that they are shielded from receiving or causing interference. As mentioned in Chapter 11, careful positioning of antennas behind rooftop structures, such as equipment rooms and elevator shafts, can make a frequency sharing situation viable that otherwise would not be.

The S/I ratios are assessed against the required S/I ratio for adequate fade margin and on the basis of that assessment, a determination can be made about whether that frequency can be used. The determination of whether a frequency can be successfully used may also be controlled by administrative restrictions that require the received interference power resulting from newly built links not exceeding the existing link's noise floor by more than some amount such as 1 dB, or by not degrading the existing link's fade margin by more than 1 dB, for example. In the latter case, an accurate calculation of the existing link fade margin performance is needed that both the existing link operator and the new link operator can accept.

The search algorithm proceeds until any suitable channel (with adequate S/I ratios at all receivers) is found or until the most interference-free channel is found with adequate S/I ratios at all receivers. For networks with many links, such as consecutive point and mesh networks discussed in Chapter 10, the process is repeated for each link. Of course, when an assignment is made, it must be added to the list of existing links so that it is considered in the interference calculations for subsequent new links.

12.3.1 Multiple interferers on a channel

Where there are multiple interferers using the channel being evaluated, the interference can be summed using the approximation that the sum of the interferers is noise and the noise thus can be added to the receiver thermal noise

$$I + N = \sum_{i=1}^{M} P_i + P_N \tag{12.2}$$

in which the P_i are the interfering powers from the existing links from (12.1). This approximation is discussed in more detail in Section 8.2.1.3.

12.3.2 Impact of automatic power control (APC)

The power level used in (12.1) is the nominal transmit output power. For existing links that are licensed and operated by others, the transmitter power that should be used is the licensed maximum power. However, for multiple links within the same network, as in a consecutive point network (CPN) or mesh network, the nominally adjusted clear sky (no rain attenuation) APC levels can be used. This can be found independently of frequency and interference from the link budget in Table 10.1. If the fade margin on the link with full transmit power is greater than that needed to achieve the target availability, the transmit power can be reduced by this excess amount to arrive at the nominal clear sky APC level. This APC level is then used for P_T in (12.1).

12.4 FIXED ASSIGNMENTS FOR LOS PMP NETWORKS

Channels for networks comprised of individual PTP links can be found using the systematic approach in Section 12.3; however, the assignment process is not as straightforward for PMP networks that use broad beam antennas at the hub sectors intended to provide service to unknown discrete points in a service area. This problem is somewhat analogous to the frequency assignment task for cellular networks, especially for non-line-of-sight (NLOS) broadband networks, although for line-of-sight (LOS) networks the use of directional CPE antennas has a profound impact on frequency planning. Also, since multiple CPEs will access the network via a single hub sector, the multiple accessing methods play a pivotal role in how interference is created and how it can be managed through the channel assignment process.

12.4.1 LOS networks

The discussion in Section 12.3 of interference calculations can be applied to PMP systems with the important difference that the hub-sector antennas are broad beam and therefore create more interference and receive more interference than narrow beam antennas in a way that is directly proportional to the beam width. As Figure 12.3 illustrates, only certain geometries in the network layout will have problems with uplink and downlink interference. It is clear that the narrower the CPE antenna beam width is, the less likely that a CPE will be located in an angular zone where the CPE antenna response cause S/I ratios that result in interference. The dashed zone lines in Figure 12.3 are for simplified illustration purposes only. In reality, the S/I ratio is a quantity that varies continuously as a function of azimuth and elevation angle relative to the antenna pointing angles.

At some distance, the interfering signal is sufficiently attenuated because of distance spreading. For example, the distance d_2 to interfering hub 2, may be sufficiently greater

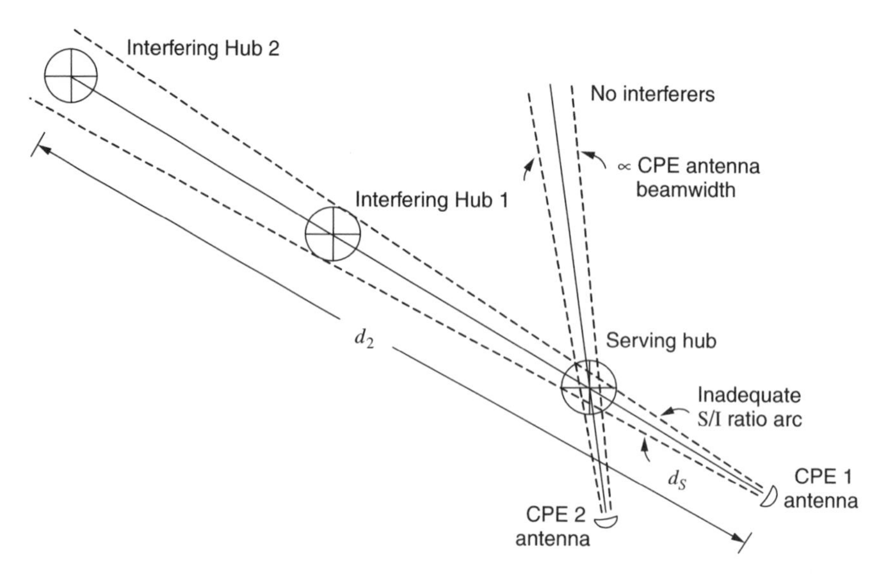

Figure 12.3 Geometry for interference in an LOS PMP system.

than the distance to the serving hub d_s so that the relative pathloss provides an adequate S/I ratio. For free space environments, this can be a very substantial distance because the interfering signal power only weakens by 6 dB for every doubling of distance. If the S/I ratio objective is 30 dB including the fade margin, the interfering hub would have to be 32 times farther away than the serving hub. For complex environments such as the urban environment used for the studies in Figure 11.11, buildings or other features may obstruct the interfering signal, increasing its effective distance significantly. From this, it is apparent that a better measure of 'distance' in the spectrum is given in terms of pathloss rather than physical distance. This subject will be taken up again at the end of this chapter.

The link geometry impacts the frequency assignment process and the CPE assignment process. If a CPE is situated such that pointing it at its nearest server results in a problem geometry as illustrated in Figure 12.3, the problem can be eliminated by directing the CPE instead at a second or third best server if one is available. The polarization in the interfering hub 2 sector may also be orthogonal to the polarization on the serving hub sector so that S/I ratio is improved by the cross polarization discrimination.

A second type of geometry that can potentially result in unacceptable S/I ratios are CPEs that are located at azimuth angles from the serving hub that are at or near the crossover azimuth of the hub-sector antenna patterns as shown in Figure 12.4. Using alternating polarizations on the sectors as in Section 11.2.2, the S/I ratio is derived directly from the polarization discrimination at this point on the pattern. Figures 11.7 and 11.8 show examples of hub antenna patterns for some standard LMDS hub-sector antennas. The cross polarization discrimination at the cross azimuth (±45 degrees) is greater than 30 dB.

These particular geometries illustrate the special situations in which interference conflicts can potentially arise in LOS PMP networks such that the frequency channel

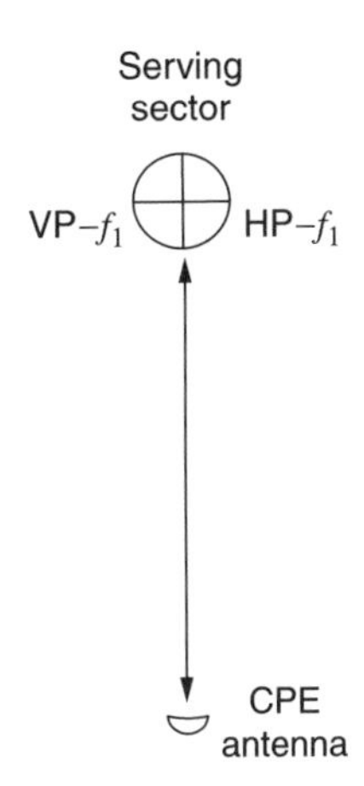

Figure 12.4 Interference to/from sectors on the same hub.

assignments need to be adjusted to accommodate them. Otherwise, it should be possible to use every frequency on every sector. Taking into account the channels that cannot be reused on adjacent hub sectors because of the conflicts illustrated in Figure 12.4, the goal of total reuse on all sectors is diminished by an amount that depends on the cross polarization pattern performance of the hub-sector antennas. From the examples Figures 11.7 and 11.8, and again using a required S/I ratio of 30 dB, the pattern transition azimuth range at all sector transitions in which the S/I ratio is inadequate is essentially 0 degrees. The resulting *frequency reuse efficiency* from sector to sector is 100%. The frequency reuse efficiency multiplied by the number of hub sectors provides the number of channels available for use on that hub.

For antennas that don't have this level of cross polarization discrimination, a common approach is to divide the available frequencies into two or more sub-bands and assign the sub-bands to sectors such that adjacent sectors are not using the same sub-band. This is illustrated in Figure 12.5 using the 16-sector hub from Figure 11.6(b) and two sub-bands.

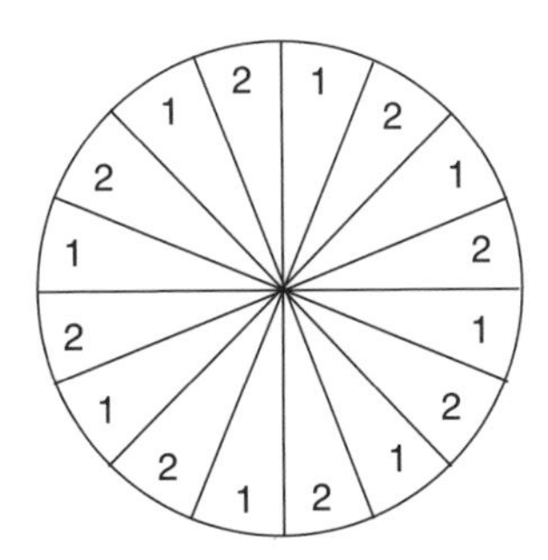

Figure 12.5 16-sector hub with two alternating sub-band channel assignments.

The frequency reuse on a hub is sometimes given in terms of the hub site frequency reuse factor, which is defined as

$$F_H = \frac{\text{Number of channels used at the hub site}}{\text{Number of channels avaliable for use in the sytem}} \tag{12.3}$$

Note that this definition is different from the frequency reuse number used to describe channel assignments in cellular systems. For cellular systems, the reuse factor is the inverse of (12.3); that is, a cellular system that uses 1/7 of the system channels on a given cell site is said to have a reuse factor of 7.

For the 4-sector hub in the above example, the reuse factor would be approximately 4 less whatever channel may be unavailable in the crossover arc. For the 16-sector alternating sub-band example in Figure 12.5, the reuse factor is 8. Because of the sub-channel divisions, the number of channels available to serve CPEs on each of the 16-sector hub is half that of the 4-sector hub. All else being equal, such as effective radiated power (ERP) and multiple access methods, the number of CPEs that can be served on each sector is also half.

It is apparent that there is a trade-off between using essentially all the frequencies on every hub sector except for those in the pattern transition arcs versus using a preplanned sub-band allocation approach. Depending on the patterns of the antennas, the former approach is much more efficient. If the 16-sector hub had all the channels assigned to every sector, with perhaps 20% restricted because of transition arc interference, it would still offer many more channels to the CPEs it serves than the sub-band approach. In a more extreme example, a rooftop hub could consist of 100 high-gain antennas, each servicing two CPEs from each sector on two channels. The reuse factor as defined above for this hub would be 50, but the number of CPEs served is only 200. The reuse factor then can be a deceptive indicator of the efficiency and frequency utilization. The frequency reuse efficiency defined above is more useful because it directly indicates the capacity that is available for serving CPEs within a hub sector.

The sub-band trade-off brings up the important concept of fixed channel assignment patterns versus a broader range of flexible channel assignment options that can be made on a case-by-case interference-limited basis. This is a theme that will recur later in this chapter when DCA is discussed. With flexible assignments on the sectors, the link-to-link interference calculation methods described in Section 12.3 can be applied to determine when a conflict exists and a channel in the transition arc is precluded from being used by a CPE in an adjacent sector. A wireless planning tool capable of these interference calculations is then used on an ongoing basis to find channels for CPEs that subscribe to the network service.

Whether or nor this assignment flexibility can be exploited depends on the capabilities of the hub-sector equipment. First generation LMDS equipment, for example, was usually shipped with specific channel assignments, or a least a fixed number of possible channels (channel cards), installed in the hardware for each sector. A fixed channel plan of some sort would be needed to know how to populate the equipment with channel cards before shipment.

12.4.2 Conventional cluster frequency planning

When multiple cell sites are considered, the problem of the CPE receiving interference from distance cells within its significant gain arc, as shown in Figure 12.3, becomes a controlling factor. As noted above, there are CPE-server assignment strategies that can reduce or eliminate such conflicts, especially when overlapping cell service areas offer the potential for multiserver options for most CPE locations. Nonetheless, from the influence of early cellular frequency planning methods, many in the fixed broadband industry have pursued a similar approach, apparently under the assumption that preassigned reuse patterns and nonoverlapping cell coverage are essential to the design process. Because of its use in the industry, this approach is discussed here.

Figure 12.6 shows a schematic of a typical example of this approach using a grid of 4×4 cells with four 90-degree sectors each [3]. This cluster pattern makes use of 8 separate sub-bands (or individual channels) and orthogonal polarizations. Reuse of the same polarization/band occurs after $4R$, in which R is the spacing between the hub sites (twice the nominal service radius of a hub). There are many variations of this pattern that can achieve the same basic interference separation result. The frequency reuse factor for

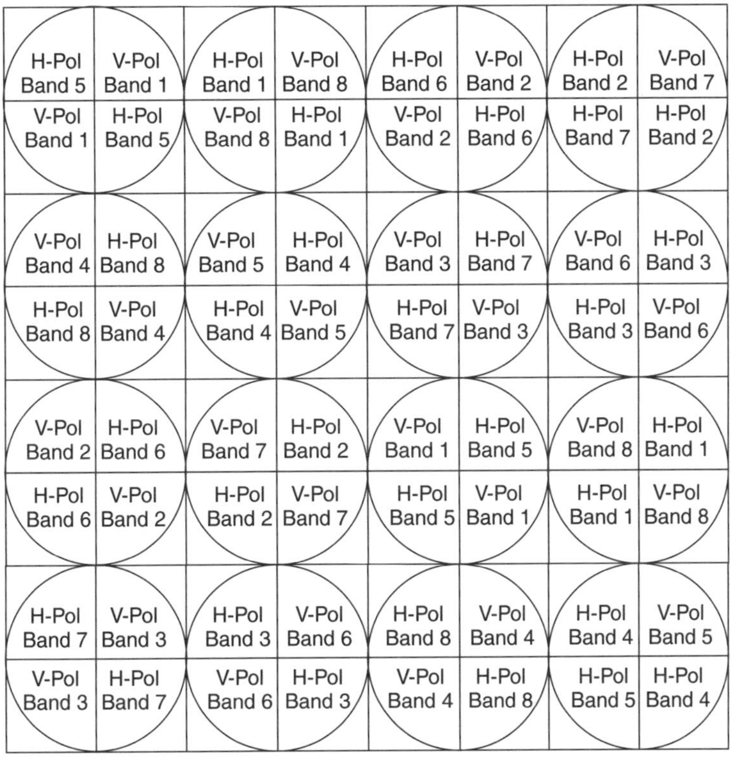

Figure 12.6 Example of a hub-sector polarization-frequency assignment pattern using 8 frequency sub-bands.

this layout is $1/2$, but the capacity available in any sector is only 1/8 of the total capacity available to the system operator.

The circular areas around the hubs in Figure 12.6 are representational only. Obviously, if these were maximum service radii, this pattern would leave large gaps in service between the cells, even if the cell hub positions were offset by one cell radius so that the gap sizes are reduced.

The primary problem with fixed sector band allocations as shown here, and it is significant, is the inflexibility that inhibits the ability of assigning spectrum resources to the system locations where they are needed. As subscribers are added to the network, a situation could easily arise in which all the channels on a given server hub sector are used but channels on other sectors remain unused. If one more subscriber wishes to have service on the fully loaded sector, the system operator would have no choice but to deny service, or overload the sector and reduce the service quality for all users on that sector. While overloading or overbooking is a common practice, offering degraded service to customers when spectrum frequency resources lie fallow elsewhere is not a strategy that maximizes revenue.

Flexibility is a general concept in managing resources of any kind, whether it is inventory in warehouses, police and fire services, or financial assets. Resources that are locked in and preassigned to only one task will inevitably lead to nonoptimum resource usage. Moreover, flexibility of bandwidth resources is one of the most significant advantages that wireless networks offer when compared with wireline communication networks in which expensive, inflexible decisions about where service is expected are unavoidable. The consequence of incorrect estimates of demand lead to unused (dark) fiber, unprofitable cable plant, and telephone cable bundles that have too few twisted pairs to serve new neighborhoods or office parks. The later case leads to the expensive and time-consuming process of digging up the street to lay new cable.

To fully exploit the flexibility that wireless has over wired systems, the channel or network resource allocation should be as flexible as possible. One method that provides much greater flexibility than fixed band allocation is described in Sections 12.4.3 and 12.8.

12.4.3 Impact of adaptive antennas in fixed LOS networks

Adaptive antennas that can steer their primary gain pattern lobes in various directions have seen relatively limited application in fixed LOS networks. An adaptive antenna of the steered beam type or smart antenna that maximized signal-to-interference + noise ratio (SINR) could be deployed at the CPE locations, however, it is unlikely to perform better than high-gain fixed pattern antenna that is correctly oriented toward the serving hub. Clearly, the fixed antenna is much less expensive than an adaptive antenna. The only tangible benefit to using an adaptive antenna at this network node is for reconfiguring the network if necessary because of network growth. If new hub sites are added to the network, the orientation of the CPE antenna may need to be changed to point at a new serving hub. An adaptive antenna could be remotely instructed to reorient its main lobe toward the new hub-sector assignment rather than requiring a technician to visit the CPE location to manually reorient the CPE antenna. An adaptive CPE antenna thus improves the scalability of the network.

At the hub sector, an adaptive antenna can offer somewhat more improvement in the network performance. Because a more directive, higher gain beam can be created and directed toward each CPE, the potential range of the system can be extended over that achieved with lower-gain 90-degree sector antennas. This difference obviously diminishes as the fixed sector antenna beam width decreases to 45 or 30 degrees. The value of range extension is most significant for hub sites on the edge of the network in which distance and noise thresholds limit link performance. Adaptive hub antennas can also be useful in lightly loaded networks in which the initial hub site deployment location may be widely spaced. However, as the network is loaded toward maximum capacity and becomes interference-limited, the appropriate strategy is to limit the range of the cells rather than extend them. This is evident from the progression in cellular system architecture in which network capacity is increased by supplementing the macro cells with micro cells or pico cells of increasingly limited power and range. As mentioned elsewhere in this book, interference management rather than service range is the most important element in maximizing the capacity and revenue potential of any high-density wireless network.

12.4.4 Demand-based fixed LOS assignments

The inflexibility of fixed-sector sub-band frequency assignments can be avoided by simply making no prior channel assignments until customers begin subscribing to network services. The channels are then assigned as needed to serve these subscribers wherever they may be located. This demand-based channel assignment strategy avoids squandering resources by reserving them for sector service areas where the service uptake is poor.

The demand-based algorithm makes direct use of the S/I ratio calculations described earlier in this chapter. As each CPE requests service and is added to the network, a separate channel search is done taking into account all the uplink and downlink channels assigned to serve currently subscribed CPEs. When a channel that offers adequate CPE S/I ratios is found, including fade margin, the channel assignment can be made. The channel allotment process makes use of S/I ratio calculations that explicitly incorporate pathloss factors such as terrain and building obstacles, which can be significant in maximizing the spectrum utilization in an area. A fixed channel band plan as described in Section 12.4.2 necessarily assumes that the propagation environment is uniform and usually flat. Because it doesn't incorporate the specific nature of the areas where the network will be built, this assumption can obviously lead to erroneous projections of sectors that will interfere with others.

The steps to finding a channel for a newly subscribed CPE using the generic demand-based channel assignment algorithm for downlink channels are

1. The received signal level and fade margin for each existing hub-CPE downlink is calculated.
2. The received signal level from each hub sector to the new CPE is calculated using the assumption that the CPE antenna is pointed toward each candidate hub. Based on these calculated signal levels, the potential servers for the CPE can be ranked as the first best, second best, third best, etc., servers. It could also be discovered at this point that

no hub sector can provide service to the new CPE location. For such circumstances, the CPE equipment [antenna gain, transmit power, Low noise amplifiers (LNAs), etc.] may need to be improved and customized compared with the standard network CPE configuration.

3. With the candidate servers ranked, a channel is first sought on the best server. To find a channel on this server, the range of channels available *to the network* are searched to determine whether a channel is available that can provide adequate SINR and fade margin. The search and channel SINR evaluation is done as described in Section 12.3 using actual clear sky APC settings for existing hub-CPE links. Both, interference caused to existing hub-CPE links and interference received from existing hub-CPE links, must be considered in evaluating the viability of the channel.
4. If a channel is found on the best server hub sector, then the channel assignment is made for this CPE and added to the list of existing CPE assignments so that it is taken into account for further CPE channel searches.
5. If the channel is not available on the best server simply because they are all assigned to other CPEs, or because of interference conflicts because of path geometries illustrated in Figures 12.3 and 12.4, then a channel is sought on the second best server using the same SINR evaluation technique as described above.
6. If no channel can be found on the second best server, the process continues with the third, fourth, etc., best servers until a viable server is found. If no viable server is found, at this point service is either denied or a channel assignment optimization process is initiated that will potentially reassign the channels for existing CPEs to make space the for the new CPE.

This algorithm clearly is biased toward the CPEs that are first to be assigned on the network. For a new network in which channel resources must be provided or planned for a known set of potential CPE locations (such as all building rooftops as in Figure 11.11), the bias can be removed by randomizing the order in which the search algorithm finds channels for CPEs. If an artificial bias is desired, such as premium customers that desire exclusive rather than shared channel access, these CPEs can be put at the top of the search order so the probability of them receiving the required resources is greatly increased.

For the uplink in frequency division duplexing (FDD) systems, the uplink channel is almost always tied directly to a paired downlink channel through a fixed duplex spacing, so assigning the downlink channel automatically assigns the uplink channel as well. In time division duplexing (TDD) systems, the downlink and uplink channels will be the same, but this actually complicates the channel search algorithm described above because interference from both hub sectors and other CPEs must be considered in all directions (see Figure 8.16) to determine whether the SINR is adequate. Other hub-CPE connections in the same or neighboring sectors that are synchronized so that they don't cause interference conflicts can be excluded from the interference conflict consideration, or included at some reduced level to take into account that the synchronization may not be perfect. With high gain fixed directional CPE antennas, the potential for CPE-CPE interference in TDD systems is small.

The demand-based algorithm was used to evaluate loading and efficiency in an LMDS system reported in [4]. Within a hypothetical system network with a uniform layout of 16 4-sector hubs and varying numbers of CPEs distributed on a regular grid, a frequency reuse efficiency approaching 90% (90% of the available network channels used on every sector) could be realized with typical CPE high-gain antennas, APC, a required SINR of 15 dB, and link availabilities of 99.99%.

A similar study done using regular hub layout but using a random rather than regularly spaced CPE distribution and full S/I ratio channel searching is shown in Figure 12.7. Simulations using the above demand-based channel assignment algorithm show results that are similar to those found in [4].

The demand-based approach requires the hub-sector radio frequency (RF) equipment to be agile enough to use any available network channel on any sector. First generation LOS PMP equipment generally lacked this flexibility since the channel cards and their setup at the factory limited channel assignments and the number of channels. With remotely programmable frequency synthesizers, there is no fundamental reason why hub sector and CPE equipment cannot provide this type of channel assignment flexibility.

The demand-based channel assignment approach is intended to find channels that are assigned on a full-time basis to CPEs. Even when the channel is shared with other CPEs through multiple access, the assumption is that the CPE will use this assigned channel and no other. The inability to jump to another channel is a limitation on flexibility that, as noted above, is detrimental to obtaining optimum use of the spectrum resources. A fully

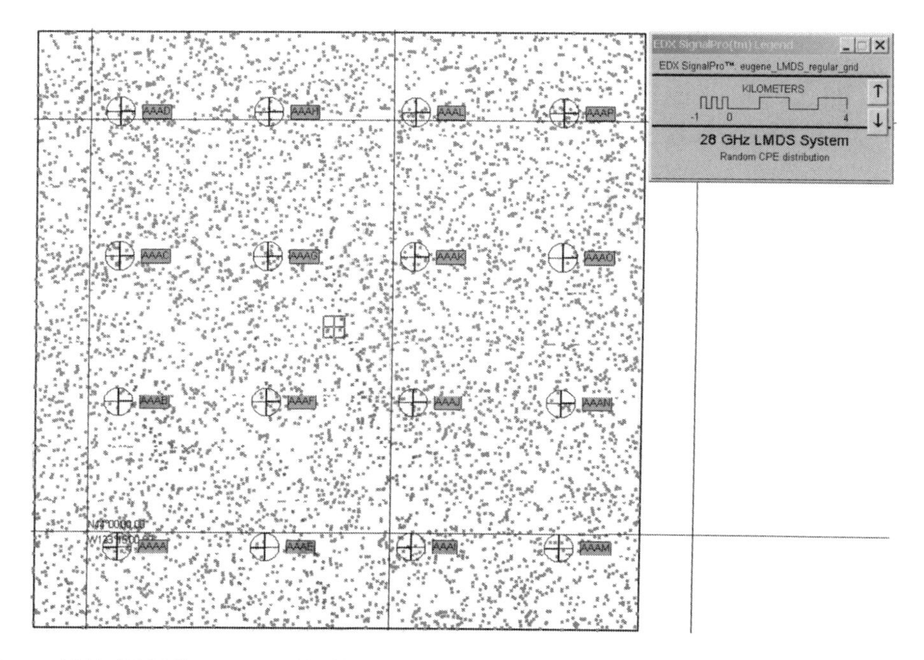

Figure 12.7 LMDS systems with 64 sectors with alternating polarizations to test frequency planning algorithms with 6400 random, uniformly distributed CPEs.

dynamic channel assignment process would allow channel, time slot, and code selection to be done, as packets are required to be transmitted. Such a fully dynamic approach is discussed in Sections 12.6 and 12.7.

12.4.5 Number of CPEs supported in fixed LOS networks

The revenue potential of a network is directly proportional to the number of CPE subscribers or customers it has. From the preceding discussion, the number of CPEs that can be served is a function of

- Number of hub sectors
- Number of channels available for use in the network
- The data rate capacity of each channel
- The multiple access efficiency
- The required quality of service (QoS) (link availability, delay, throughput)
- The channel reuse efficiency as described here

These elements can be formed into simple equations that can be used to compare one network configuration against another in terms of how many customers it can serve and what revenue it might expect from serving them. This will direct trade-off decisions over the type of applications that should be supported (is streaming video and audio commercial worthwhile?) and also service areas that will be most profitable.

12.5 FIXED ASSIGNMENTS FOR NLOS PMP NETWORKS

The approach to assigning frequency channels in NLOS networks is substantially different from that used for LOS networks. Because of the scattering in an NLOS environment, the use of high-gain antennas with fixed orientations at the CPE is not feasible. Also, because of scattering, polarizations cannot be effectively used to provide isolated channels to separate CPEs. Instead, the NLOS channel assignment process is similar to that for cellular systems that provide service to unknown user locations throughout a defined service area. The user density is established by traffic information that, along with the QoS, determines the data rate capacity the network must provide.

Because some of the basic concepts of frequency reuse and interference are the same as for cellular systems, the conventional approach to cellular channel planning will first be discussed. Following that, the particular aspects of fixed (rather than mobile) remote terminals will be incorporated along with suitable modifications that exploit the simplifications that fixed can bring versus mobile.

The traditional cell layout for cellular systems is shown in Figure 12.8. The basic hexagon grid structure for cell site layouts was adopted because it represents continuous coverage of the service area (unlike circles) and approximates a circularly symmetric service area for each cell. Since the remote units are assumed to have omni-directional

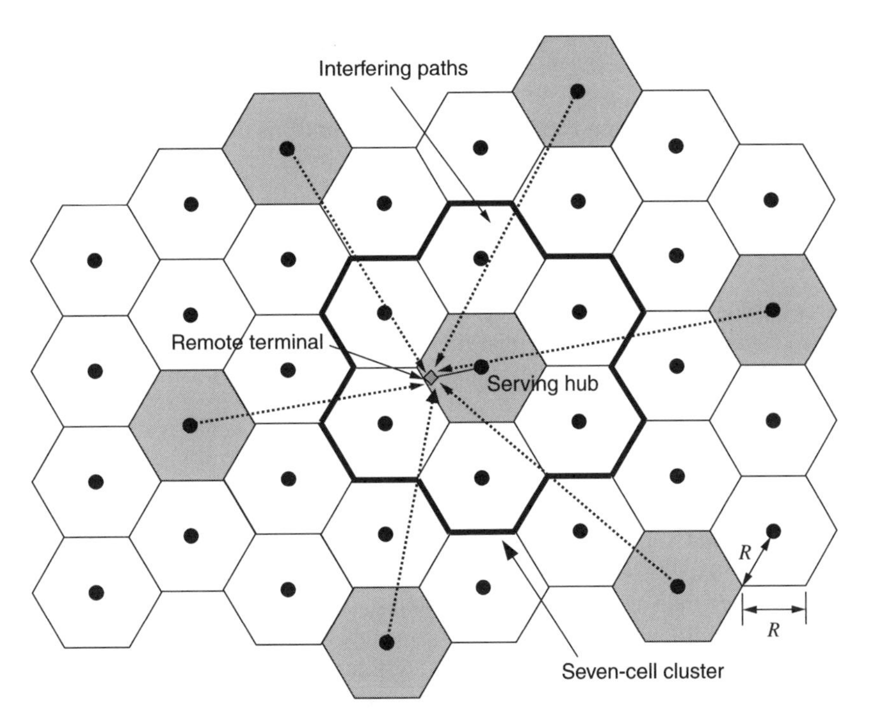

Figure 12.8 Basic hexagon fixed broadband NLOS network cell layout.

antennas, they will receive interference from all other cells in the network, with the closest neighboring cells representing the biggest interference contributors as shown by the dotted lines in Figure 12.8. The distance between cells using the same frequency (in units of hexagon cell radius R) is called the *reuse distance*. The reuse distance depends on two factors: (1) the required S/I ratio needed for acceptable service and (2) the pathloss for the desired and interference paths.

12.5.1 Target S/I ratio

The channel reuse distance depends on the S/I ratio that is required to achieve acceptable service at the CPE. Acceptable service can be described by achieving link availability for some percentage of the time (99% for example) in certain percentage of locations (95%, for example) inside the network service area. The first value is indicative of the quality of the link while the second value indicates how many locations are to be provided with adequate service. As discussed in Chapter 4, both the desired signal and the interfering signals are subject to fading effects that are generally modeled by two mechanisms:

1. Variations in the signal voltage envelope that are usually described by a Rice probability distribution.
2. Variations in the mean signal level from location to location that are not predictable using simple propagation models such as the IEEE 802.16 model or the model given

in (10.33). Shadow fading is usually described by a lognormal distribution of the mean values.

Taking lognormal fading into account, the S/I ratio in dB is described with a probability distribution given by

$$\rho = S - \left(\sum_{n=1}^{N_I} 10^{I_n/10} \right) \text{ dB} \tag{12.4}$$

where S is the lognormal-distributed desired signal and I_n are the lognormal-distributed interference signals. The probability that the link to a CPE has acceptable availability P_A is the probability that ρ is above the required S/I threshold ρ_{TH} for the required bit error rate (BER), or

$$P_A = \Pr(\rho > \rho_{\text{TH}}) \tag{12.5}$$

Normally in cellular systems the channel assignment process is done using the mean values of the desired signal and interference rather than taking into account the fading distributions. For fixed broadband networks in which the expectation of service availability is much higher than that for cellular systems, it is important to include the fading distributions in the S/I values that are used to establish channel assignments and reuse. The channel assignment algorithm for LOS systems described in Section 12.4.4 takes fading into account by including an appropriate fade margin in the target S/I ratio when searching for a usable channel.

A similar approach can be used here. If 16QAM modulation is used, the required S/I ratio for a raw BER of 10^{-3} is about 12 dB (that is FEC-correctable to 10^{-6}). If a fade margin of 16 dB is used to achieve 90% link availability with a Rician k factor of 0 dB (from Figure 4.15), the target S/I is 28 dB. If the lognormal location variability is also considered, and assuming a standard deviation of 8 dB, achieving an acceptable S/I ratio in 90% of the locations requires the S/I ratio be increased a further 10.5 dB (from Figure 4.17) to 38.5 dB. This signal-to-interference ratio (SIR) objective can be significantly reduced assuming diversity antenna can mitigate the multipath fading, reducing the Rician fading margin by 10 dB or more. The net SIR objective is thus reduced to about 28.5 dB. Because of the imposed availability requirements, this target S/I value is substantially higher than values normally used for cell cluster planning in cellular systems.

12.5.2 Frequency reuse distance

If the cell sites and CPEs are assumed to have omni-directional antennas, the only thing that attenuates an interfering signal versus the desired signal from the nearest cell is distance. In free space with pathloss proportional to $20 \log d$, the ratio of the distance to the serving hub and the distance to interfering hub has to be 6.3 to achieve an S/I ratio of 16 dB. However, since these are NLOS paths, the actual pathloss from a hub site to a CPE will be substantially higher that free space pathloss. This is usually approximated in a simple way by increasing the exponent value for the distance attenuation term in the pathloss equation from 2 (for free space) to some higher value. This is represented in dB form as

$$L \propto n 10 \log d \tag{12.6}$$

where $n > 2$ For example, in the IEEE 802.16 propagation model for MMDS frequencies discussed in Section 3.4.1, this exponent is a simple function that depends on the terrain type (A, B, or C) and the height of the hub antenna above ground. For a hub antenna height of 30 m, the values of the exponent are 4.79, 4.38, and 4.12 for terrain types A, B and C, respectively. The general assumption can be made that local foliage and building penetration loss will affect the desired and interfering signals in roughly equivalent ways, so these pathloss factors will not affect the S/I ratio.

If an exponent value of $n = 4.38$ is used in (12.6), then a 16 dB S/I ratio can be achieved with distance ratio of 2.3 instead of 6.3. However, the CPEs in a cell have six nearest neighbors using the same channel. If the assumption is made that the powers of the interfering signals add, then the interfering power on a channel is six times greater than calculation above. Taking this into account raises the reuse distance ratio to 3.5. This ratio indicates that the interfering hubs using the same channel must be approximately 3.5 times farther away from the CPE than the serving hub.

Under the worst-case conditions when the CPE is positioned at the edge of the cell service area, it is a distance from the desired hub site. Using the separation ratio of 3.5, this indicates that the cells using the same channel must be at least $3.5R$ away from the CPEs serving cell. The grayed cells in Figure 12.8 are examples of where sufficient cell separation exists so that a channel can be reused and achieve a target S/I ratio of 16 dB.

12.5.3 Cell layout and channel assignment patterns

The conventional cell system approach to channel planning assumes a grid of hexagons that can be viewed as a collection of cell clusters in which each cluster contains a given number of cells in which the channels can be used just once. The clusters make a tessellated (repeated mosaic) pattern. For this repeating pattern, there are certain cluster sizes that work as given by

$$M = i^2 + ij + j^2 \tag{12.7}$$

where i and j are nonnegative integers with $i > j$. Working through the allowed combinations of i and j yields values of $M = 1, 3, 4, 7, 12 \ldots$. Figure 12.8 has a reuse cluster size of 7; the hexagon grid can be continued infinitely by repeating this cluster and its channel assignments.

By examining the geometry of the various clusters sizes, the intercell distance D can be related to the cell radius R for a given cluster size M by

$$\frac{D}{R} = \sqrt{3M} \tag{12.8}$$

The choice of the reuse cluster size can be approximately determined by the required S/I ratio. From Section 12.5.2 and its assumptions, the S/I ratio is given by the ratio of the desired signal level to the sum of the interference signal levels, which in turn are determined by the pathloss values to the serving cell hub and the interfering hubs. All else being equal, pathloss values are proportional to the distances and the pathloss exponent n. For a CPE on the edge of cell service area at distance R from the serving hub, the signal-to-interference ratio, ρ this is then written as

$$\rho = \frac{R^{-n}}{\sum_{i=1}^{N_I} D_i^{-n}} \tag{12.9}$$

Using the assumption that all the N_I interferers are at the same distance from the CPE ($D_1 = D_2 = \cdots = D_{N_i} = D$), then SIR can be written in terms of the cluster size

$$\rho = \frac{(\sqrt{3M})^n}{N_I} \tag{12.10}$$

or

$$M = \frac{(N_I \rho)^{2/n}}{3} \tag{12.11}$$

For a cluster size of 7, and six interferers ($N_I = 6$), and a pathloss exponent of 4.38, the resulting S/I ratio from (12.10) is 21.2 dB using these assumptions.

One measure of the network capacity C is given by

$$C = \frac{N_S}{M}, \ M = \frac{N_S}{C} \tag{12.12}$$

where N_S is the number of channels available for use in the network. Substituting (12.12) into (12.11) yields

$$\frac{N_S}{C} = \frac{(N_I \rho)^{2/n}}{3} \tag{12.13}$$

$$C = \frac{3N_S}{(N_I \rho)^{2/n}} \tag{12.14}$$

An example of frequency sub-band channel assignments is shown in Figure 12.9. In this case, each of the cells has been divided into three 120-degree sectors, a typical configuration for cellular systems. Applying this channel assignment model to MMDS for example, 21 MMDS channels with 6-MHz bandwidths could be divided into 7 groups of three, and one of the three assigned to each sector α, β, or γ. This would result in 6 MHz of spectrum available for the CPEs in the service area of that sector. The 6 MHz of bandwidth may be further subdivided to suite the technologies that are being deployed.

This channel assignment assumes that the geographic distribution of the traffic is uniform. From the information in Chapter 9, it is clear that the traffic distribution is generally not uniform. Moreover, the target market for the application and services

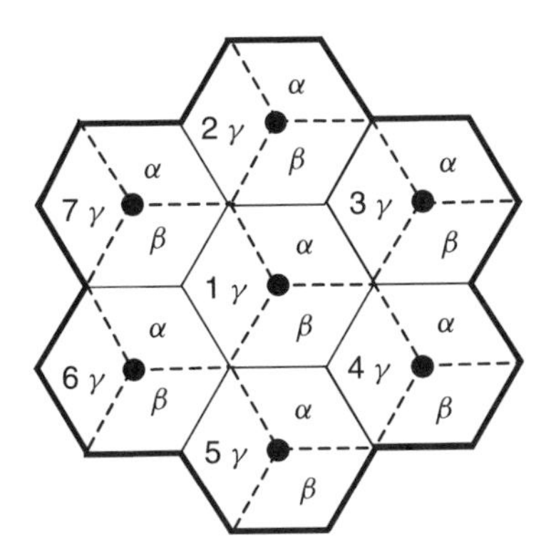

Figure 12.9 Frequency band assignments in 7-cell cluster and 3 sectors per cell.

on the network may be biased toward businesses or toward residences that will make the geographic distribution of the traffic nonuniform. In this case, it may be desirable to allocate more channels on some sectors and fewer and other to address the different loads.

The use of cell cluster patterns to allocate channels in a wireless system as described here is a fairly primitive approach that is somewhat useful for understanding the interference mechanisms involved. It is not of much value in assigning channels in a real system, although it is still used in that way. The biggest limitations of this approach are:

- The integer cluster sizes don't provide enough assignment flexibility to create an efficient channel plan for a wide range of S/I ratios. Technology improvements that can lower the target S/I by 3 or 6 dB can increase capacity significantly, but from (12.11), such an improvement may not allow a smaller cluster size (from 7 to 4, for example).
- The uniform cells don't reflect the oddly shaped and sometimes discontiguous service and interference areas of real cells.
- The regular hexagon grid layout of cells can rarely be achieved even approximately. Increasingly, the choice of cell site locations is restricted to existing wireless tower structures or the few land parcels that are zoned to allow new wireless tower construction.

Because of these limitations, the cell cluster approach to channel assignment is not a good choice for assigning channels in NLOS fixed broadband systems. Instead, an optimizing approach can be used that does not require uniform or clustered channel assignments.

12.6 OPTIMIZING CHANNEL ASSIGNMENTS IN NLOS NETWORKS

The cluster approach can be abandoned entirely in favor of a channel plan that assigns channels to hub sectors in a way that realized the traffic service goals while also achieving the target SIR throughout the service or coverage area of each hub sector. The channels do not need to be assigned in any particular pattern, which was the objective of the cluster approach above, but instead can be assigned arbitrarily on a channel-by-channel basis. Relaxing this restriction provides considerable flexibility to achieving more efficient channel assignments. This approach can also explicitly take into account nonuniform and discontiguous coverage and interference areas.

There are several algorithms for assigning channels in an optimum or quasi-optimum way. All of them require that an objective function of some kind be explicitly described. For the case of wireless systems, the objective is to provide enough spectrum resources (channels) to carry the expected traffic with the desired QoS level, and to do so while maintaining the target S/I ratio. The optimization algorithm then attempts to find a set of channel assignments for the hub sectors in the network that simultaneously meet these goals as closely as possible. In the parlance of optimization algorithms, the error or difference between the system state results and the objective function is being minimized. This error is also referred to as the *cost*.

The objective function is usually defined on a study grid that covers the network service area. The study grid approach was used for the shadow and coverage maps in Chapter 11. At each grid intersection or tile, the S/I ratio and traffic requirements are defined. The channel assignment algorithm then attempts to find a set of channel assignments for the hub sectors that meet these objectives at all grid locations. Normally, all the objectives cannot be simultaneously met, so some level of acceptable residual error must also be defined.

Among many available optimization algorithms, three approaches have emerged as applicable to the problem of finding optimum channel assignments. All of them basically provide a systematic way of guessing at a set of channel assignments, and then evaluating how well that guess achieves the traffic and S/I ratio objectives. A new guess is then made using some of the knowledge gained from the first guess so that the next guess is potentially closer to the objective. This process continues until the objective is reached within an acceptable window, or the algorithm cannot make any changes that improves the result, or the computation time becomes excessive. The three main optimization techniques are described below.

12.6.1 Steepest descent method

The steepest descent method makes trial adjustments in all the parameters that are available to determine the adjustments that have the greatest impact on lowering the error. Those adjustments that lower error the most are made and the process repeated. If the error is visualized as a multidimensional surface, this approach finds the steepest slope down the error surface to a minimum error point. The main drawback to this method is that it can get stuck in a local error minimum in which any change in assignments results in a higher error and consequently no further parameter adjustments will improve the result. The local minimum may not be the optimum minimum error, so the resulting quality of the channel assignment plan may not be optimum. Some tricks, such as changing the parameter adjustment size, can sometimes be used to jump out of a local minimum, but these in turn may leap over the optimum ideal minimum.

12.6.2 Simulated annealing method (SA)

Simulated annealing (SA) is described in [5–7]. SA starts with a random set of channel assignments s (also called a system state) at hub sectors. The error or cost $C(s)$ for this state is calculated. A new system state s' is then randomly selected and its cost $C(s')$ calculated. If $C(s') \leq C(s)$, then the new state or channel assignment set is kept and a new set of random assignments made from there. If $C(s') > C(s)$, the channel configuration is accepted with a probability given by

$$P = \min\left[1, \exp\left(\frac{-C(s') - C(s)}{T}\right)\right] \tag{12.15}$$

where T is the 'temperature'. Equation (12.15) is known as the Metropolis criteria [8].

The temperature basically determines how large a system state change can be made from one state to another. As the SA process continues, the temperature is lowered according to a schedule (the cooling schedule). At the final low temperatures, only small changes in system states are possible. The actual size of the system state change is drawn from a Gaussian distribution so the temperature is actually controlling the standard deviation of the Gaussian distribution that determines the state changes. In this way, the process is analogous to the annealing process for cooling a piece of metal in small incremental steps. By cooling it in this way, the molecules in the metal have the opportunity to migrate into a regular lattice pattern that creates a durable and malleable metal with a low internal energy (cost). In contrast, a heated metal can be 'quenched' by immersion in cold water. This results in a hard but brittle metal that more easily fractures because the molecules are not positioned in an ideal lattice structure.

The main virtue of SA for channel assignment problems is that it cannot easily get stuck in a local minimum as with the steepest descent method. This is achieved by allowing higher cost states to sometimes be accepted with probability given by (12.12). However, SA can be a calculation-intensive method, especially if a large number of hub sectors and channels are involved.

12.6.3 Genetic or evolutionary algorithm method

Genetic algorithms (GA) mimic evolutionary selection to arrive at a system state (population) that provides the minimum error in achieving the channel assignment objectives. The population is randomly modified using mechanisms such as inheritance in which a new population member is derived using most of the 'genes' from two parents in the preceding population, and mutation in which a given number of genes are randomly changed. A complete description of using GA for channel assignment problems can be found in [9]. The use of GA for channel assignments in cellular systems is described in [18,19].

Like SA, GA provide a systematic way of guessing to achieve increasingly well-adapted populations that do a better job of achieving the network goals. It also offers reasonable immunity from getting stuck in local minimums; but as with SA, though, it can be computationally intensive for large networks.

12.6.4 Channel assignments in W-CDMA systems

For W-CDMA systems, the multiple access technique is designed to accommodate an interference-rich environment in which every frequency channel is used on every hub sector. Discrimination between downlink or uplink signals is achieved through the use of different codes as described in Section 8.4. The assignment of codes for uplink and downlink signals is done in real time by the network hardware as communication links between hubs and CPEs are needed for communication. For this reason, no channel assignment plan is needed.

However, there are pseudorandom noise (PN) offset codes that are used to distinguish the spreading codes used on one hub sector from those used on another. Sometimes, conflicts in PN offset codes can arise at locations within the service areas where the signal level ratios and time delays between two hubs have the same PN offset code conflict. The equations involved with calculated areas with potential PN offset conflict for IS-95 CDMA systems can be found in [10]. Similar methods can be used for scrambling code planning in universal mobile telecommunications service (UMTS) W-CDMA. Other W-CDMA scrambling code methods are discussed in [11].

For TD-CDMA used in UTRA-TDD, there is the potential for interference from hub sector to hub sector and from CPE-to-CPE, as illustrated in Figure 8.16. Depending on the frame synchronization between adjacent cells and allocation of time slots between uplink and downlink, intercell interference can occur with UTRA-TDD, especially when the traffic distribution is highly nonuniform. This is primarily addressed using DCAs rather than network planning. Using DCA for UMTS terrestrial radio access (UTRA) TDD is discussed in Section 12.8.5.

12.7 NLOS NETWORK CAPACITY

One way of assessing the capacity of the network is to calculate the available data in bits-per-second per Hertz of bandwidth per square kilometer of service area. In the example above, 6 MHz of bandwidth was available for each sector service area. If the efficiency of the modulation scheme and coding is 3.2 bps/Hz, for example (typical of coded 16QAM), and the service radius of each cell is 10 km (104 km^2 per sector), the capacity is about 183 kbps/km^2.

From Chapter 11, the service radius for the example MMDS system was established to achieve adequate link availability with typical hub-sector power levels. However, the frequency reuse pattern is based on the ratio between cell service radius R and intercell spacing D. If the hub power is lowered so that the service radius is only 5 km, the sector service area reduces to 26.2 km^2 and the capacity increases to about 733 kbps/km^2. Of course, the trade-off is that many more cells and hub sites are needed to cover the same service area, so the network infrastructure cost increases proportionally to achieve this higher capacity.

12.8 DYNAMIC FREQUENCY CHANNEL ASSIGNMENTS

The discussion of channel assignments thus far has assumed that the channels, once assigned, are fixed assignments at that hub sector. In some cases, it can be recognized that the geographical distribution of traffic has two or three distinct patterns during the day, so two or three static channel plans can be developed and the network switched from one channel allocation scheme to another to best match the traffic load distribution. For example, in a cellular system during the day the large concentration of traffic is in business and office areas. At the end of the workday, the traffic load shifts to locations along the roads and transportations corridors used by commuters. Two channel plans that reassigned channels from the cell sectors in the business districts to the cell sectors along the transportation corridors at 5 or 6 p.m. is a basic form of DCA. In this example, the channel assignments were preplanned used estimated or measured traffic loads. True DCA assigns channels and other network resources in real time as needed to accommodate the traffic wherever it might be in the service area.

For fixed broadband wireless networks, the mobility issues associated with cellular phone systems are not present, but similar kinds of traffic redistribution during the day may occur. Shifting channel resources from business areas during the day to residential areas during the night could also be reasonably applied to fixed NLOS broadband systems. If the relative traffic load estimates are available, multiple channel assignment plans can be formulated and each automatically implemented by the network for the appropriate daily time period.

Using multiple preplanned channel assignments for the network is appealing because it does not impose any burden on the network operation other than making the channel assignment switch a few times during the day. However, it does rely on estimates of traffic load and on the quality of the propagation modeling that predicts signal level and interference. Errors in these estimates or predictions will result in channel assignments plans that are potentially a poor match to the network and, at worst, result in serious interference conflicts and the resulting service gaps. For real-time DCA, the various approaches can be divided into two categories on the basis of how decisions are made to assign channels.

12.8.1 Centralized DCA

With centralized control, all the information in the network on channel usage, interference conflicts, and traffic is brought to a central location where decisions are made to assign channels throughout the network. The hub-to-hub communication resources needed to carry the information, as well as the computations required to make the assignment decisions, can be prohibitive especially for networks with a large number of hubs. In the extreme case, every time a CPE request service (or a circuit in a circuit-switched network), the centralized DCA would have to recalculate the optimum channel assignment strategy for the entire network. The delay involved in completing this task would be added to the system delay or latency. This approach is really not feasible for circuit-switched 2G cellular systems [11]. For packet-switched systems, the computational difficulties would be worse.

The computation and communication drawbacks of fully centralized control can be mitigated to a useful extent by breaking the network up into neighboring cell clusters, a method known as distributed DCA. With this approach, the decisions about channel assignments are confined within the cell cluster rather than being made on a network-wide basis. The drawback is that most networks operate in a nonhomogenous propagation environment in which the areas of service and interference may be discontiguous. This is especially true in a cellular microcell system in urban areas where turning a street corner may initiate a handoff to a cell some distance away. In such cases, the interference conflict to be avoided through intelligent dynamic channel assignment may be coming from a cell some distance across the network and therefore outside the cell cluster in which channel assignments are being coordinated.

12.8.2 Decentralized DCA

Decentralized control essentially lets the hub sectors make their own decisions about which channels are useable and which are not at the time the hub sector has to handle traffic to/from a CPE. The general approach is to use some method for the individual hub sectors and the CPEs to sense or measure the interference in the channels available to the system. On the basis of these measurements, a choice is made as to the best channel (if any) to use for the transmission.

There are several approaches to finding the interference level on the channels. A short idle period with no transmissions can be prearranged by the serving sector so that the CPEs can detect downlink signals from the other hub sectors by scanning the channels in the band during the idle time interval. Similarly, for the uplink channel, the hub sector can scan the channels to determine those that have the lowest level of interfering power. The hub sector and CPE can share the results of these scans to assist in the channel choice. Orthogonal Frequency Division Multiplexing (OFDM) systems that used fast Fourier transform (FFT) technology and multiplier banks to correct frequency – selective fading are already equipped with this type of signal measuring feature.

In [12], simulation results are presented for DCA for conventional circuit-switched traffic using an idealized propagation model with a pathloss exponent $n = 4$ along with the lognormal fading model described in Section 4.6.2. It also uses ideal interference sensing without the conflicting assignment choices that arise because of distance-induced time delays for sectors scanning the same channels. The simulations demonstrate that DCA with and without power control can produce network capacities that exceed those of both asynchronous and synchronous CDMA. The synchronous CDMA system model assumes complete intracell interference cancellation as would be achieved with perfect joint detection (see Sections 8.4.5 and 8.4.6). Figure 12.10 shows an example of the spectrum efficiency data taken from [12] comparing capacity results for DCA with those of asynchronous and synchronous CDMA. The efficiency in Figure 12.10 is the multiple-access efficiency defined as the ratio of the number of simultaneously active remotes at

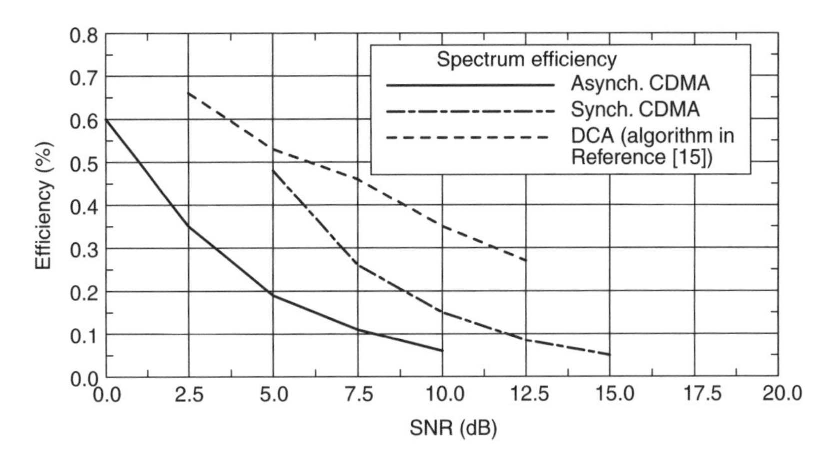

Figure 12.10 Spectrum efficiency of idealized DCA compared with CDMA. Data from Pottie (1995).

a given data rate on a sector in a semi-infinite hexagon cell structure versus the number of users at the same data rate permitted for a single sector by itself (with no intercell interference). The improvement with DCA over CDMA is in part due to interference averaging over the spreading bandwidth that occurs with CDMA, even for synchronous CDMA in which the intracell interference is eliminated. A DCA is basically injecting intelligence into the spectrum resource utilization process by choosing channels with high SIR values rather than accepting a channel with an average SIR. Because of the added intelligence and processed information, it is reasonable to expect it to more efficiently use the spectrum.

12.8.3 Channel segregation

Another type of DCA is called channel segregation [13] and, like most DCA schemes, for TDMA systems it requires that each hub sector be able to transmit in any channel allowed in the system. This method uses a training approach in which a priority list of channels is maintained for each hub sector. Initially, the channels are all considered equally desirable in terms of carrying traffic. As information is gathered on the success of using one channel versus another, the successful channel is given a higher ranking on the priority list. As the list develops, for any given transmission the choice is made to use the highest-ranking channel that is not already being used by another CPE on that hub sector.

Figure 12.11 shows a flow chart for the channel segregation algorithm. Each hub sector ranks each channel using a priority function $P(i)$ where

$$P(i) = \frac{N_s}{N_t} \tag{12.16}$$

where N_s is the number of successful accesses to the channel plus the number of accesses to the channel when it is idle but not accessible, and N_t is the total number of trials for

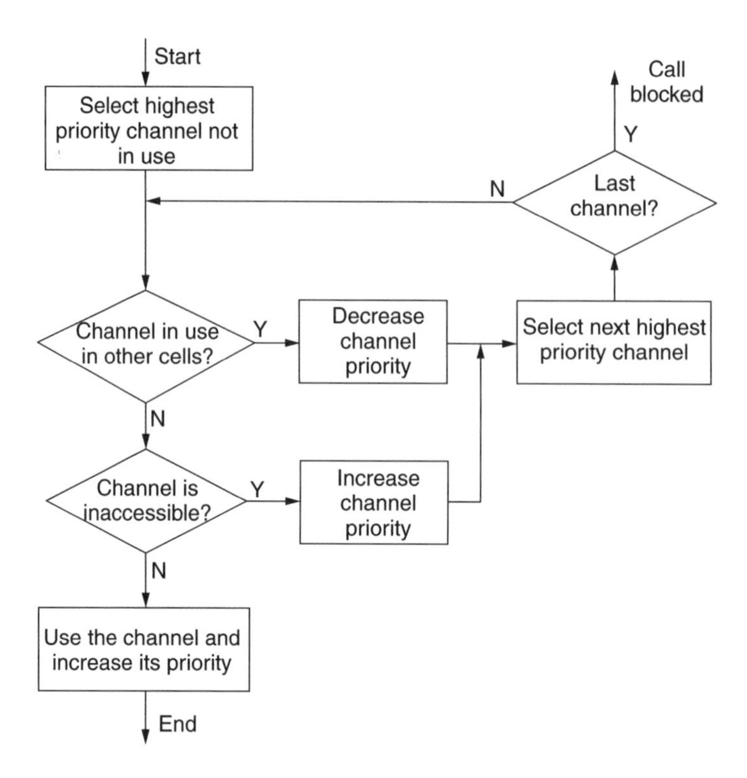

Figure 12.11 Flow chart for channel segregation DCA algorithm.

the channel. From 12.11, the hub sector selects a channel from the current channel list on the basis of the highest priority $P(i)$. If the channel is idle, it is checked for accessibility. If it is also accessible, the channel is selected for use and its priority count is increased by one. If the channel is idle but inaccessible because there are no RF transmitters available on the hub, its priority is also increased by one but the algorithm seeks a new channel from the priority list. This recursive process continues until the transmission is handled or the transmission is blocked.

The simulations in [13] show that the channel segregation algorithm can outperform FCA described earlier.

12.8.4 Dynamic packet assignment

When the channel assignment decisions are made on a packet-by-packet or packet session-by-packet session basis, DCA is called dynamic packet assignment (DPA). References [1,14] are recent publications that focus on DPA and the potential advantages it can offer.

The work in [14] addresses one drawback to the interference measurement method. When different sectors scan for a low interference channel, the same channel may be detected by two or more sectors as desirable so they all make the same conflicting channel assignment. In the simulations reported in [14], the additional condition was imposed that the interference sampling frames (idle times) be staggered among nearby cells in the same cluster to avoid this circumstance. As mentioned before, in complex propagation environments, the nearby cells may not be the closest in terms of distance. Identifying which cells should be tied together through a staggering scheme may be difficult, and may also be impossible when cells are members of multiple clusters as a result of their interference presence.

In a related issue, an interference scan at some fixed point will gradually become obsolete during the time interval until the next scan. Transmissions done on the channel determined during the initial interference scan can expect to see interference conditions on the channel deteriorate during the time interval to the next scan. In [14], the S/I ratio 1 percent of the time decreased by about 9 dB between the initial SIR scan and an SIR scan done near the end of the transmission. The transmission errors that result from the lower SIR will be a function of the length of transmission that in turn depends on the traffic statistics, as discussed in Chapter 9.

Like those in [12], the simulation results reported in [14] generally use idealized network models with a simple propagation model using a distance attenuation exponent $n = 4$. Within these idealized simulation limitations, the DPA simulation results show significant capacity gains through the use of DCA.

12.8.5 DCA for UTRA-TDD networks

As mentioned, CDMA systems in general do not require channel planning because they have been designed to accommodate intercell interference by averaging the interference power over the spreading bandwidth. For TDD CDMA systems that segment the time domain into slots occupied by a limited number of users, there is an additional interference mechanism cause by hub-to-hub and CPE-to-CPE interference on both the uplink and downlink, as illustrated in Figure 8.16. While time frames can be synchronized among neighboring cells, there remains the issue of different allocations of downlink and uplink time slots resulting in a downlink time slot occurring during an uplink time slot in the neighboring cells. Coordinating the uplink–downlink time slot allocations among cells restricts the flexibility that cells have to adjust to changes in their traffic flows.

This interference issue has been studied through simulation in [15]. Although the simulation did not take into account joint detection for incell interference reduction, it did demonstrate that the dominant interference concern is between hub sectors that can be mitigated through the use of frame synchronization, as mentioned. Network planning taking into account reuse distance separation as with traditional TDMA systems also helps, but this will reduce capacity. In [16], the use of DCA for UTRA TDD is discussed as a away of addressing the interference. In a similar fashion to DPA, it uses interference measurements during time slots to assign priorities to time slots in which the time slot with the lowest interference is being assigned the highest priority. The priority list is periodically updated. This approach should enhance spectrum utilization in a similar way to DPA.

12.9 OTHER CAPACITY ENHANCEMENT TECHNIQUES

The capacity available using one of the channel assignment strategies described in the preceding sections is limited by the intercell interference that determines how often the frequencies can be reused. The approaches used to improve the situation either: (1) find some way to suppress or remove the interference on the selected channel and/or (2) for a given channel SINR, adapt the modulation and coding schemes to achieve lower error rate communications at the highest possible data rate. Some of these techniques are discussed in the following sections. Adaptive antennas, joint detection, and link adaptation are three approaches that can be used to suppress interference.

12.9.1 Adaptive antennas

If adaptive antennas are used to suppress interference, the cell spacing for frequency reuse can be decreased and the capacity of the network increased. Interference suppression through the use of highly directional CPE antennas in LOS networks was the primary reason the capacity of such networks is much larger than NLOS systems.

This effect can be approximately accounted for by assuming that the target S/I ratio used in (12.14) for determining which sector signals conflict with others is reduced by some amount that reflects the ability of the antenna to suppress interference from other cells. As this suppression value increases, the frequency reuse can be packed more tightly and the capacity of the system increased accordingly. For example, if an adaptive antenna decreases the relative interference contribution represented by N_I in (12.14) by an adaptive antenna improvement factor $\alpha = 1/4$ (6 dB interference suppression), the increase in capacity from C to C' is

$$C' = \frac{C}{\alpha^{2/n}} \tag{12.17}$$

If the previous propagation attenuation exponent $n = 4.38$ is used, the network capacity is increased by a factor of 1.88. For smaller values of n, as with LOS systems, the improvement is greater, as might be expected, since the capacity limit is more dependent on interference when the pathloss between receiver and interference transmitters in the network is lower.

Simulation results in [1] include the effects of beam-forming at the hub sector and interference suppression from an adaptive antenna at the CPE. Using these techniques, the packet retransmission rate drops significantly when these additional enhancement techniques are employed, indicating an improvement in QoS when using adaptive antennas. This comes about as a direct consequence of interference suppression.

12.9.2 Joint detection

Joint detection was described in Section 8.4.5 in the context of uplink channels in CDMA systems. In this role, it can potentially eliminate all the intracell interference and potentially boost capacity by a factor of 2.8 (given the assumptions discussed there). It also lessens the burden of executing rapid and accurate power control adjustments.

The concept of joint or multiuser detection can be more broadly used to suppress interference from any external source for which knowledge can be obtained about the interfering signal and the channel state information (CSI) between the interferer and the victim receiver. Even incomplete interference cancellation can provide worthwhile capacity gains proportional to the extent of the suppression. Currently, UMTS TDD W-CDMA is the only technology applicable to NLOS fixed broadband network that contains joint detection as part of its specification.

12.9.3 Link adaptation

Link adaptation (LA) discussed in Section 10.5 is a technique for modifying the characteristics of the signal (modulation and coding), the transmit power level, or the parameters of an adaptive antenna (if used), to maintain link performance when the characteristics of the channel change or degrade from their nominal conditions. Such techniques can be applied to single isolated links so that they respond to changing propagation environment conditions such as rain, multipath fading, etc. LA can also be applied to links in LOS and NLOS networks in which interference is the limiting factor. For a given link in the network, LA would permit the current CSI and SINR to be exploited to the maximum extent. If a high SINR were achieved for even part of the transmission, the mode can be changed to use a more efficient modulation constellation with lower spectrum occupancy. This has the potential to achieve better capacity than when using DCA alone.

12.10 SPECTRUM VECTORS, OCCUPANCY, AND UTILIZATION

The utilization of the spectrum can be generalized into how many locations (users) have access and what the quality and value of that access is. The space or distance between users, or more specifically their access mechanisms (transmitters, receivers, signals) is a fundamental quantity that can be used in determining the potential for interference and how many users can occupy the spectrum space. This distance metric is a *spectrum vector* formed in a multidimensional spectrum space that includes the propagation environment features (distance) and signal characteristics.

The spectrum vectors can be used to formulate definitions of spectrum occupiers and occupancy that can be used to assess the degree of spectrum utilization. This section will therefore deal with the following concepts

- Spectrum vectors
- Spectrum occupancy
- Communication value
- Spectrum utilization
- Spectrum creation

Characterizing the spectrum in a more generalized form is useful to gain insight into where limitations occur and what strategies could be successful for removing capacity constraints that are normally assumed to be inevitable by regulators and other spectrum managers.

12.10.1 Spectrum vectors

The distance between two occupiers in the spectrum space can be described with two vectors; one indicating position in space and the other indicating the signal characteristics. A spectrum occupier can be a transmitter or a receiver. If the position of one wireless spectrum occupier is described by location x_1, y_1, z_1 and a second's position is described by x_2, y_2, z_2, the Euclidean distance between them is

$$|\mathbf{d}_{1,2}| = \sqrt{(x_1 - x_2)^2 + (y_1 - y_2)^2 + (z_1 - z_2)^2} \tag{12.18}$$

However, as noted in various places in this book, the real distance between two points in spectrum space is not the physical distance but the pathloss between the points $L(p_1, p_2)$. More generally, the pathloss is included in the wireless channel impulse response $h_{1,2}$ from (3.5) (time dependence suppressed)

$$|\mathbf{d}_{1,2}| = L(p_1, p_2) = |h_{1,2}| \tag{12.19}$$

The radiation/reception antenna patterns used by the spectrum occupiers also affect the separation between them in spectrum space. Along with power at the transmitter and sensitivity at the receiver, these values can be included in (12.22) as scalars so that

$$d_{1,2} = \frac{|\mathbf{d}_{1,2}|}{P_1 G_1(\phi_{1-2}, \theta_{1-2}) R_2 G_2(\phi_{2-1}, \theta_{2-1})} \tag{12.20}$$

where P_1 is the transmitter power, $G_1(\phi_{1-2}, \theta_{1-2})$ is the gain of the transmit antenna in the direction of the receiver, R_2 is the receiver sensitivity, $G_2(\phi_{2-1}, \theta_{2-1})$ is the gain of the receive antenna in the direction of the transmitter. As before, for a communicating pair of occupiers, the distance in (12.23) should be minimized. For an interfering pair of occupiers, the antenna gains are minimized toward each other (the distance $d_{1,2}$ is maximized).

The signal emanating from a transmitter, and the signal a receiver is designed to accept and decode, can be described by a vector **s**

$$\mathbf{s} = (f, t, \kappa, \nu) \tag{12.21}$$

where

f = frequency
t = time
κ = a heuristic correlation coefficient that is indicative of the uniqueness or 'distinguishability' of the signal characteristics versus others. An example is the orthogonal spreading codes used in CDMA systems that are designed to make the signals sharing the spectrum individually distinguishable.
ν = polarization that takes on one of two orthogonal states.

The distance between two signals can also be written as a Euclidean distance

$$|\mathbf{\Delta s}_{1,2}| = \sqrt{(f_2 - f_1) + (t_2 - t_1) + (\kappa_2 - \kappa_1) + (\nu_2 - \nu_1)} \tag{12.22}$$

The total spectrum distance between two occupiers can be defined as the sum of the distances from (12.20 and 12.22):

$$D = d + |\mathbf{\Delta s}| \tag{12.23}$$

Since the elements of $\mathbf{d}$ and $\mathbf{\Delta s}$ are different, the sum in (12.23) is representative of the magnitude of the spectrum distance between any two spectrum occupiers – typically a transmitter and a receiver. For the communication link between a transmitter and its intended receiver, the objective is to minimize D. Conversely, in the interfering case, the objective is to maximize D in (12.23). All of the spectrum dimensions are available to minimize or maximum spectrum separation D. For example, an intended transmitter–receiver link will have a signal difference vector $\mathbf{\Delta s} = \mathbf{0}$. The only thing separating the transmitter and receiver is the channel pathloss. To suppress interference, the frequency, time, coding and polarization differences between the interfering transmitter and the victim receiver can be maximized along with the pathloss between them.

In a frequency band with multiple occupiers, the signal received at a receiver occupier is the sum of the desired signal and other interfering signals. This can be written for one receiver occupier as

$$r_k = \sum_{j=1, j\neq k}^{M} \frac{\mathbf{s}_j}{|\mathbf{\Delta s}_{jk}|} d_{jk} + \frac{\mathbf{s}_k}{|\mathbf{\Delta s}_{kk}|} d_{kk} + N_k \tag{12.24}$$

where d_{jk} is the pathloss/antenna pattern distance from (12.20) and N_k is the Gaussian-distributed thermal noise. The signal intended for receiver k comes from transmitter k. By design, these should be synchronized in frequency, time, coding, and polarization, so $|\mathbf{\Delta s}_{kk}| = 1$. The signal-to-noise + interference ratio ρ_k at receiver k is

$$\rho_k = \frac{\mathbf{s}_k d_{kk}}{\displaystyle\sum_{j=1, j\neq k}^{M} \frac{\mathbf{s}_j d_{jk}}{|\mathbf{\Delta s}_{jk}|} + N_k} \tag{12.25}$$

For all receiver occupiers in band, (12.25) can be written in matrix form in the usual way

$$\mathbf{R} = \mathbf{S}\mathbf{D}^T + \mathbf{N} \tag{12.26}$$

Assuming that each receiver has an associated transmitter (although it may be the same transmitter as in a broadcast situation), the spectrum distance matrix $\mathbf{D}$ in (12.26) is a square matrix and $\mathbf{R}$, $\mathbf{S}$, and $\mathbf{N}$ are column vectors.

12.10.2 Spectrum occupancy

A conventional view of spectrum occupancy considers the power, directional antenna, and bandwidth of a transmitter and formulates an occupied spectrum volume [17]. The occupied spectrum is regarded as the volume in which the use of a frequency or the presence of a receiver is precluded. The particular work in [17] was done at a time when the advanced CDMA, space division multiple access (SDMA) and other technologies were not part of the commercial or wireless communication lexicon.

For the discussion here, a different approach to describing spectrum occupancy will be used. The occupied spectrum is defined in terms of the spectrum space distance to other occupiers. If a transmitter is in a band by itself (no other transmitters or receivers to which a separation distances can be calculated), then the spectrum occupied by that transmitter is zero.

$$O_k = \frac{1}{\sum_{j=1, j \neq k}^{N} D_{j,k}} \tag{12.27}$$

where M is the number of spectrum occupiers in the band. As the spectrum distance $D_{j,k}$ increases, the spectrum occupied k decreases. The spectrum distance can be increased by increasing the pathloss d, the signal difference $\mathbf{\Delta s}$ or by decreasing antenna gains along the path between the two occupiers. This model of spectrum occupancy recognizes the decrease in occupancy through the use of technologies that differentiate signals used by occupiers with frequencies, time slots, coding and polarization. An occupier that was using a signal that was undetectable by all other occupiers (except the intended one) would have zero occupancy. License-exempt, ultra wideband systems are an example of technologies whose impact to existing occupiers is minimized through large spreading codes and very low power spectral density.

Inserting hypothetical occupiers in the spectrum space and calculating a surface in the multidimensional spectrum volume can calculate a generic assessment of occupancy. When evaluating the volume at various fixed dimension points (same frequency, same time, same code, for example) the volume precluded to other occupiers in the remaining dimensions can be found. Using (12.20 and 12.23) in (12.24), it is clear that occupancy is proportional to transmitter power level and receiver sensitivity.

12.10.3 Communication value

Spectrum occupiers make use of frequency bands of various sizes depending on administrative restrictions. The number of occupiers and the 'value' of their communication within a band is one measure of spectrum utilization. The value of the communication can be measured in many nontechnical ways. For example, the value of emergency police and fire communication is regarded as higher than normal commercial or personal communications. While such value metrics can be applied to assessing spectrum utilization, for the purposes here, no value judgment is assigned to the nature of the communication. Instead, the value V of the communication is simply given by the data throughput multiplied by the distance between the information source and the information destination and divided by the communication transit time:

$$V = \frac{B \cdot x_{s-d}}{t_T} \tag{12.28}$$

where B is the data throughput, x_{s-d} is the distance between the source and destination, and t_T is the transit time. The spectral efficiency e_s is then defined as the communication valued achieved for a given spectrum occupancy from (12.28)

$$e_k = \frac{V}{O_k} \tag{12.29}$$

From (12.29), as $O_k \to 0$, $e_k \to \infty$.

12.10.4 Spectrum utilization

From Chapters 10 and 11, there are several network topologies for achieving communications between two points, including point-to-point, consecutive point, multipoint, and mesh networks. The various topologies can be compared in terms of the occupancy required to achieve the same communication value. For example, consider two methods illustrated in Figure 12.12 for getting a message from the source to the destination. One method uses a single wireless link, while the second method uses multiple links that relay the message from one occupier to another until it reaches its destinations. The latter is typical of consecutive, mesh and generalized packet radio networks.

For the single link, a certain transmitter power will be needed to achieve successful communication. Occupancy is proportional to the power level. For the multilink case, the power on each link is lower because the path is shorter (in pathloss terms), but the message must use multiple time slots, one each time the message is relayed. Assuming the signaling techniques are the same, the occupancy in the single link case is proportional to the transmit power required

$$O_S = \frac{P_S}{P_m} O_m \tag{12.30}$$

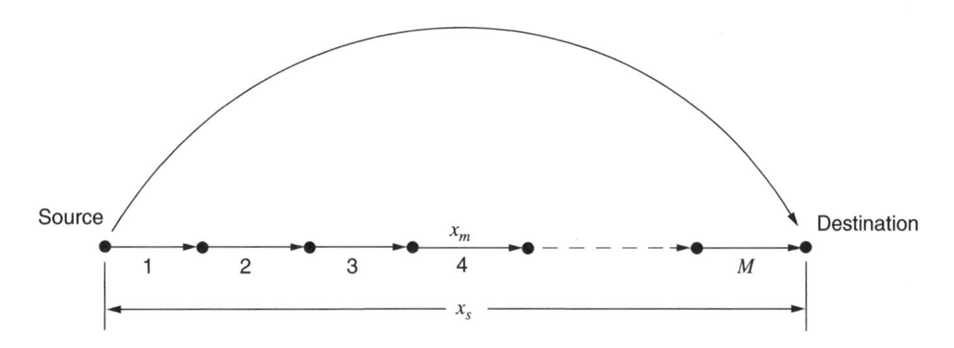

Figure 12.12 Single link and multilink methods to get a message from the source to the destination.

where P_S is the power of the single link transmitter transmitting over distance x_S, P_m is the power of each of M individual transmitters in the multiple link case, each of which operates over a distance x_m.

$$x_m = \frac{x_S}{M} \rightarrow P_m = \frac{P_S}{M^2} \tag{12.31}$$

Equation (12.28) assumes free space propagation attenuation. The actual pathloss exponent may be or higher than two, or stated another way, the shorter links are less likely to a have unknown path obstacles and associated attenuation (because they are shorter). Each of the powers of the multiple transmitter case may actually be less than the single transmitter power by an exponent factor that is greater than two.

The multiple transmitter approach uses less power at each transmitter, but the message must be handled M times consuming M time slots, time slots being used here in the generic rather than TDMA sense. Comparing the occupancy of the two approaches

$$o_M = \frac{o_S}{M} \tag{12.32}$$

The multilink case has $1/M$ of the occupancy of the single link case. This also implies that the occupancy can be made arbitrarily small by increasing the number of relays – a system with a dense network of closely spaced intercommunication nodes. However, the realized communication value and spectrum efficiency is reduced because several short signal relays will introduce delay. This is especially true if the intermediate nodes involved are also handling other messages (not dedicated to the single message flow), introducing a multiple access delay. There will also be other delays associated with calculating the most efficient message routing that is usually nontrivial.

12.10.5 Spectrum creation

With the expression for spectrum occupancy in (12.27), the capacity of the spectrum is found as the number of occupiers that can fit in a given spectrum volume multiplied by the communication value for each of those occupiers from (12.28). The challenge is to determine the volume of the multidimensional spectrum space that necessarily includes the code space available to each signaling scheme. This calculation is beyond the scope of this book. However, the basic approach for characterizing the spectrum occupiers, distances and communication efficiencies described here can be used as an initial framework for pursing a calculation of maximum spectrum capacity that would be analogous to channel capacity.

If spectrum is regarded generally as that entity that supports wireless communication, the multilink example above suggests the concept that spectrum is *created* through the use of more nodes. This is somewhat counterintuitive to the common notion that more nodes consume more spectrum space.

12.11 CONCLUSIONS

Channel assignment is the process of choosing how best to use the wireless network resources in the form of frequency, time slots, and codes to carry the data traffic between the network nodes. While the network hardware (transmitters, receivers, antennas, etc.) is the enabling technology, the communicating value added to the network comes from the efficient utilization of the resources through intelligent and flexible channel assignment strategies.

For LOS systems using fixed high-gain antennas at the CPEs and reasonable levels of sectorization at the hubs with alternating polarizations, it is possible to achieve frequency reuse factors of essentially one (every frequency can be used on every hub sector). Adaptive antennas used in LOS systems can improve link margins by directing gain toward the communicating CPEs and providing flexibility to reconfigure the network as it scales up by adding more hubs to accommodate more users.

NLOS systems can employ channel assignment strategies that have been used for cellular networks, including FCA and DCA schemes that use network resources to carry traffic where and when it is needed. Several DCA schemes have been devised with various performance trade-offs that are primarily intended for circuit-switched systems. For current and future packet-switched data networks, DPA can achieve very efficient spectrum utilization on the basis of recent simulations.

Since the capacity of a high-density wireless network is interference-limited, the capacity achieved with any channel assignment strategy can be enhanced through several interference suppression methods. Adaptive antennas, joint detection, and link adaptation are methods for suppressing interference and otherwise improving individual link performance during temporary degradations because of fading or interference bursts.

Channel assignment strategy is really part of the larger subject of how wireless spectrum is created and managed. To better understand and formulate strategies that could aid in the management process, a generic multidimension spectrum space structure using vector distance metrics has been introduced in this chapter. This structure provides a framework for quantitatively analyzing the relative merits of various technological innovations and network topologies that not only enable wireless communications but also demonstrate that new technology actually creates spectrum. Rather than being a limited natural resource, in fact, the available spectrum is only bound by the current state of the applied technology and methodologies for using it. As technology advances, and the densities of deployed fixed wireless access points intensify, new spectrum is created. This is a more accurate and constructive view of how spectrum is evolved and utilized than the conventional limited resource spectrum model currently used by regulators.

12.12 REFERENCES

[1] J.C.-I. Chuang and N.R. Sollenberger, "Beyond 3G: wideband wireless data access based on OFDM and dynamic packet assignment," *IEEE Communications Magazine*, vol. 38, no. 7, pp. 78–87, July, 2000.

[2] IEEE Standard 802.16.2-2001. "IEEE Standard for Local and Metropolitan Area Networks: Coexistence of Fixed Broadband Wireless Access Systems," IEEE, April 8, 2002.

[3] R. Foster, et al., "Radio frequency deployment considerations for TDD BWA systems," *Proceedings of the 7th Annual WCA Technical Symposium*, San Jose, CD-ROM, January, 2001.

[4] H.R. Anderson, "Simulations of channel capacity and frequency reuse in multipoint LMDS systems," *Proceedings of the 1999 IEEE Radio and Wireless Conference*," Denver, pp. 9–12, August, 1999.

[5] E. Aarts and J. Korst. *Simulated Annealing and Boltzmann Machines*. New York: John Wiley & Sons, 1989.

[6] M. Duque-Anton, D. Kunz, and B. Ruber. "Channel assignment for cellular radio using simulated annealing," *IEEE Transactions on Vehicular Technology*, vol. VT-42, no. 1, pp 14–21, Feb. 1993.

[7] H.R. Anderson, et al., "Optimizing microcell base station locations using simulating annealing techniques," *Proceedings of the 44th Vehicular Technology Society Conference*, Stockholm, pp. 858–862, June 1994.

[8] N. Metropolis, A. Rosenbaum, M. Rosenbluth, A. Teller, and E. Teller, "Equation of state calculations by fast computing machines," *J. Chem. Phys.* vol. 21, pp. 1087–1092, 1953.

[9] M. Cuppini, "A genetic algorithm for channel assignment problems," *European Transactions On Telecommunications and Related Technologies*, vol. 5, no. 2, pp. 285–294, March-April, 1996.

[10] Samuel C. Yang. *CDMA RF System Engineering*. Boston: Artech House Publishers, 1998, pp. 165–174.

[11] Young-Ho Jung, Y.H. Lee, "Scrambling code planning for 3GPP W-CDMA systems," *Proceedings of the 51st Vehicular Technology Society Conference*, Athens, pp. 2431–2434, May, 2001.

[12] G.J. Pottie, "System design choices in personal communications," *IEEE Personal Communications*, vol. 2, pp. 50–67, October, 1995.

[13] Y. Akaiwa and H. Andoh, "Channel segregation – a self-organized dynamic channel allocation method: application to TDMA/FDMA microcellular system," *IEEE Journal on Selected Areas in Communications*, vol. 11, no. 6, pp. 949–954, August, 1993.

[14] J.C.-I. Chuang and N.R. Sollenberger, "Spectrum resource allocation for wireless packet access with application to advanced cellular internet service," *IEEE Journal on Selected Areas in Communications*, vol. 16, no. 6, pp. 820–829, August, 1998.

[15] H. Holma, S. Heikkinen, O.-A. Lehtinen, and A. Toskala, "Interference considerations for the time division duplex mode of the UMTS terrestrial radio access," *IEEE Journal on Selected Areas in Communications*, vol. 18, no. 8, pp. 1386–1393, August, 2000.

[16] M. Haardt, et al., "The TD-CDMA based UTRA TDD Mode," *IEEE Journal on Selected Areas in Communications*, vol. 18, no. 8, pp. 1375–1385, August, 2000.

[17] L. Berry, "Spectrum metrics and spectrum efficiency: proposed definitions," *IEEE Transactions of Electromagnetic Compatibility*, vol. EMC-19, no. 3, pp. 254–260, August, 1977.

[18] W.K. Lai and G.G. Coghill, "Channel assignment through evolutionary optimization," *IEEE Transactions on Vehicular Technology*, vol. 45, no. 1, pp. 91–96, February, 1996.

[19] C.Y. Ngo and V.O.K. Li, "Fixed channel assignment in cellular radio networks using a modified genetic algorithm," *IEEE Transactions on Vehicular Technology*, vol. 47, no. 1, pp. 163–172, February, 1998.

Appendix A

Atmospheric and rain data

The fading models described in Chapter 4 and used for link availability calculations in Chapter 10 rely on two types of data. The first type is atmospheric refractivity data that shows either the statistics of the refractivity gradient or an empirical atmospheric factor that is indicative of the variability of radio multipath. The atmospheric conditions described by this data affect link path clearance and multipath fading probabilities. There are two methods primarily used for atmospheric multipath fade outage predictions:

- *Vigants–Barnett Multipath Fade Method*: The maps in Figures A.1 and A.2 show the appropriate C factor that is used in (4.4).
- *ITU-R Multipath Fade Method*: The maps in Figures A.3 through A.6 show the percentage of time that the refractivity gradient is less than -100 N-units/km. These values are those used for p_L in (4.8).

The second type of data describes rain events. The rain rate tables show the percentage of time that rain rates in millimeter per hour occur for various regions in the world. The maps associated with the rain rate tables show where various rain regions are located. There are two methods used in rain fade outage predictions:

- *Crane Rain Fade Outage Method*: The map in Figure A.7 shows the 1996 Crane rain regions throughout the world designated by letters. Table A.1 shows the percentage of time that the rain rates in millimeter per hour occur in each of the defined regions.
- *ITU-R Rain Fade Outage Method*: The map in Figure A.8 shows the ITU-R rain regions throughout the world designated by letters. Table A.2 shows the percentage of time that the rain rates in millimeter per hour occur in each of the defined regions.

Fixed Broadband Wireless System Design Harry R. Anderson
 ISBN: 0-470-84438-8

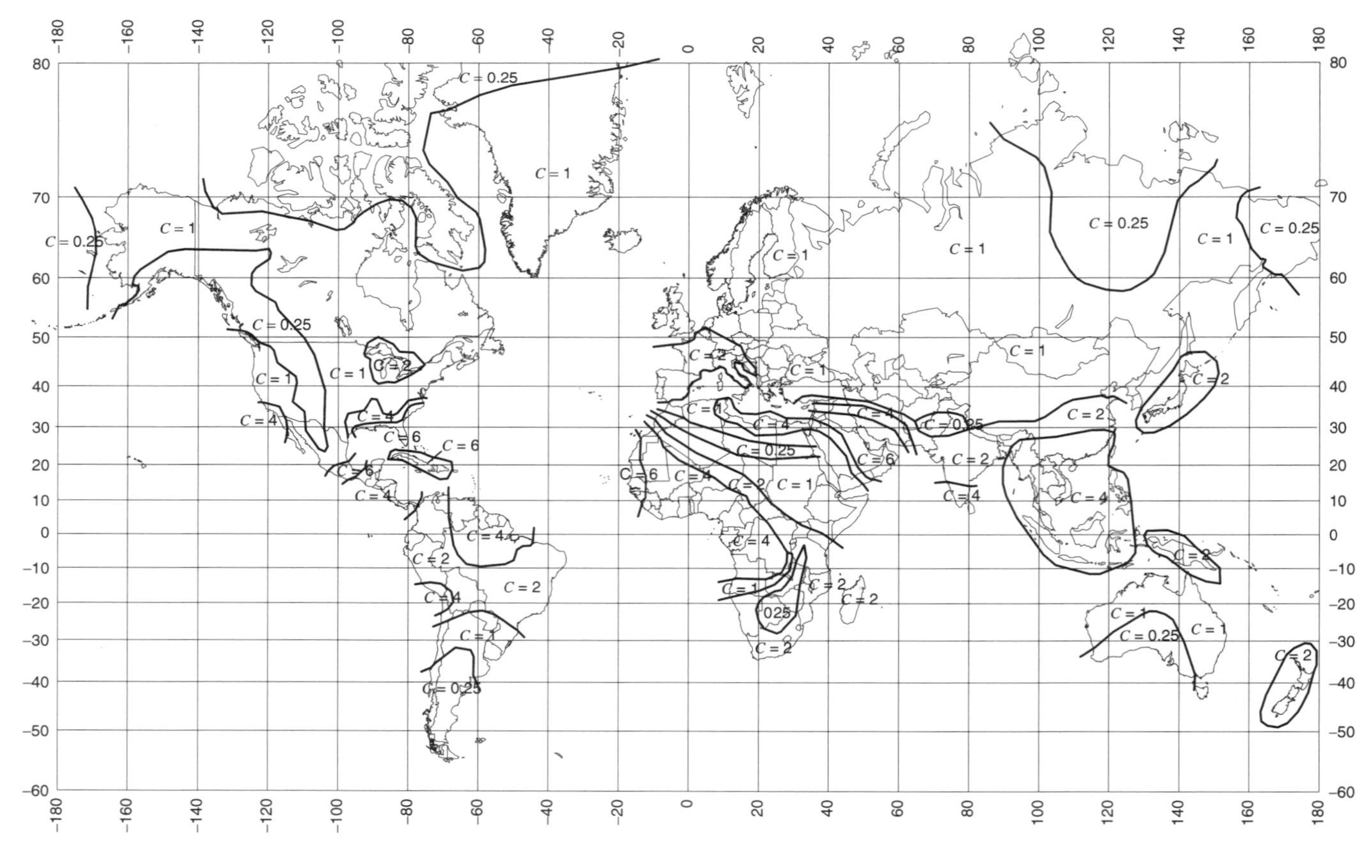

Figure A.1 Vigants–Barnett atmospheric conditions map for the United States.

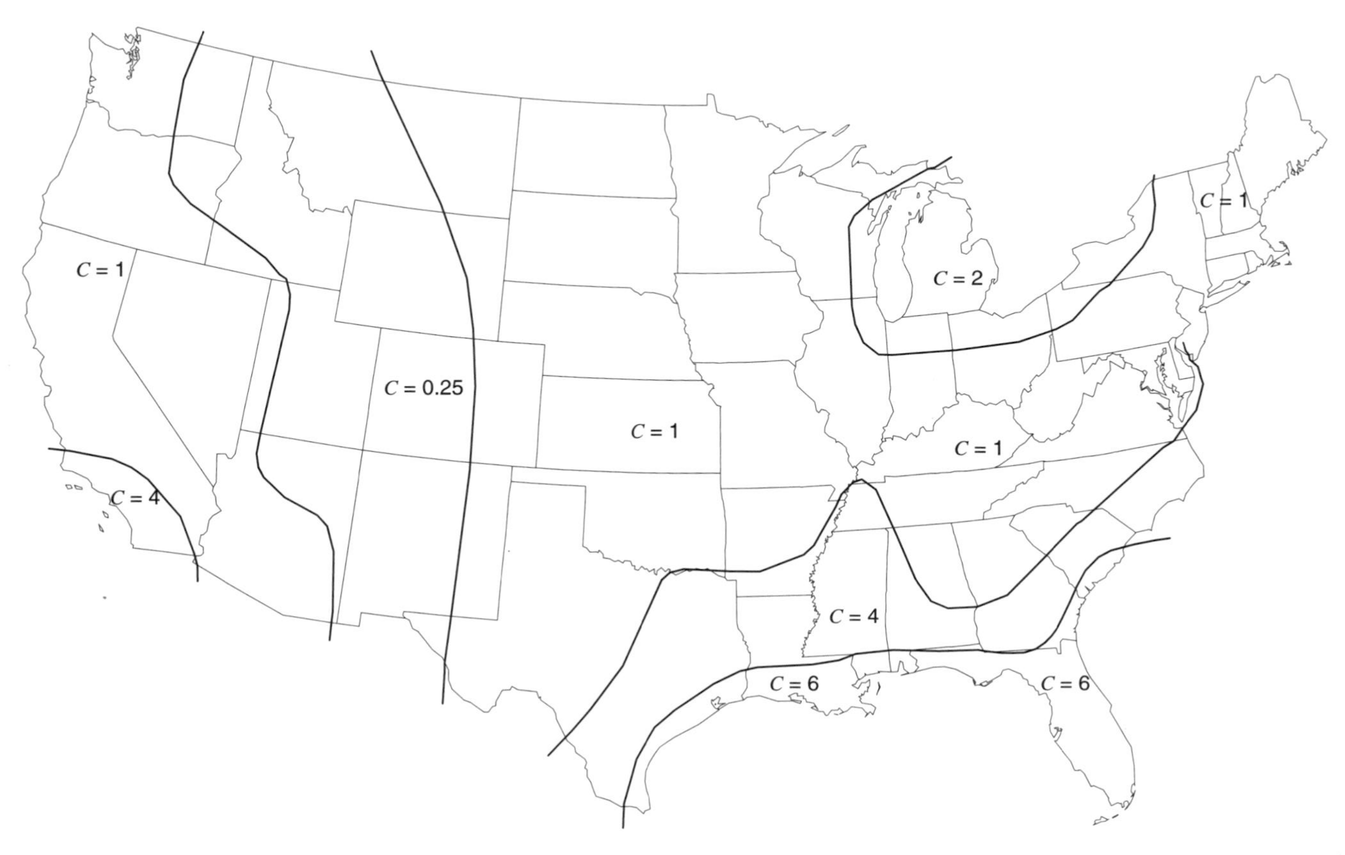

Figure A.2 Vigants–Barnett atmospheric conditions map for the United States.

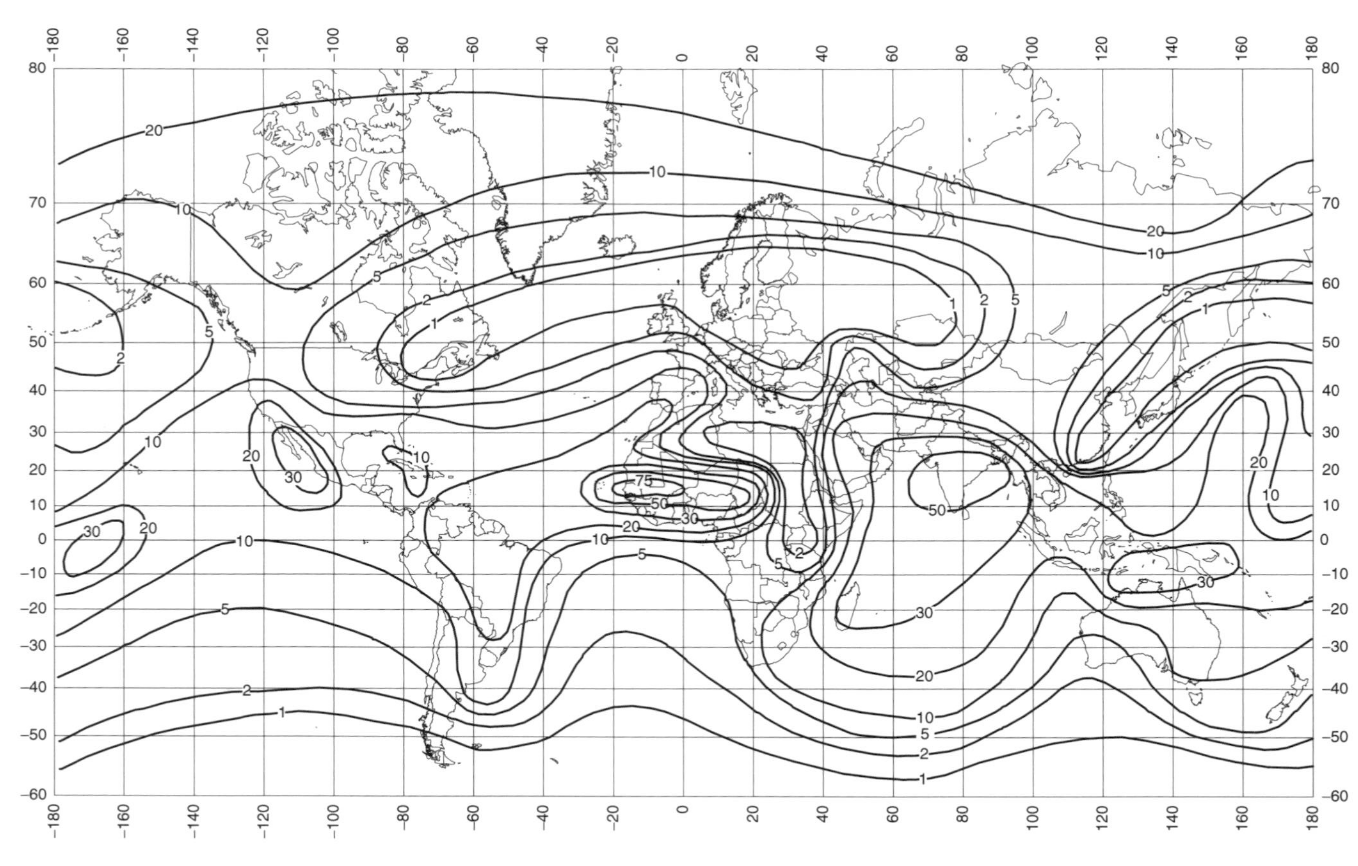

Figure A.3 ITU-R map of the percentage of time that the refractivity gradient is <-100 N-units/km for the month of February.

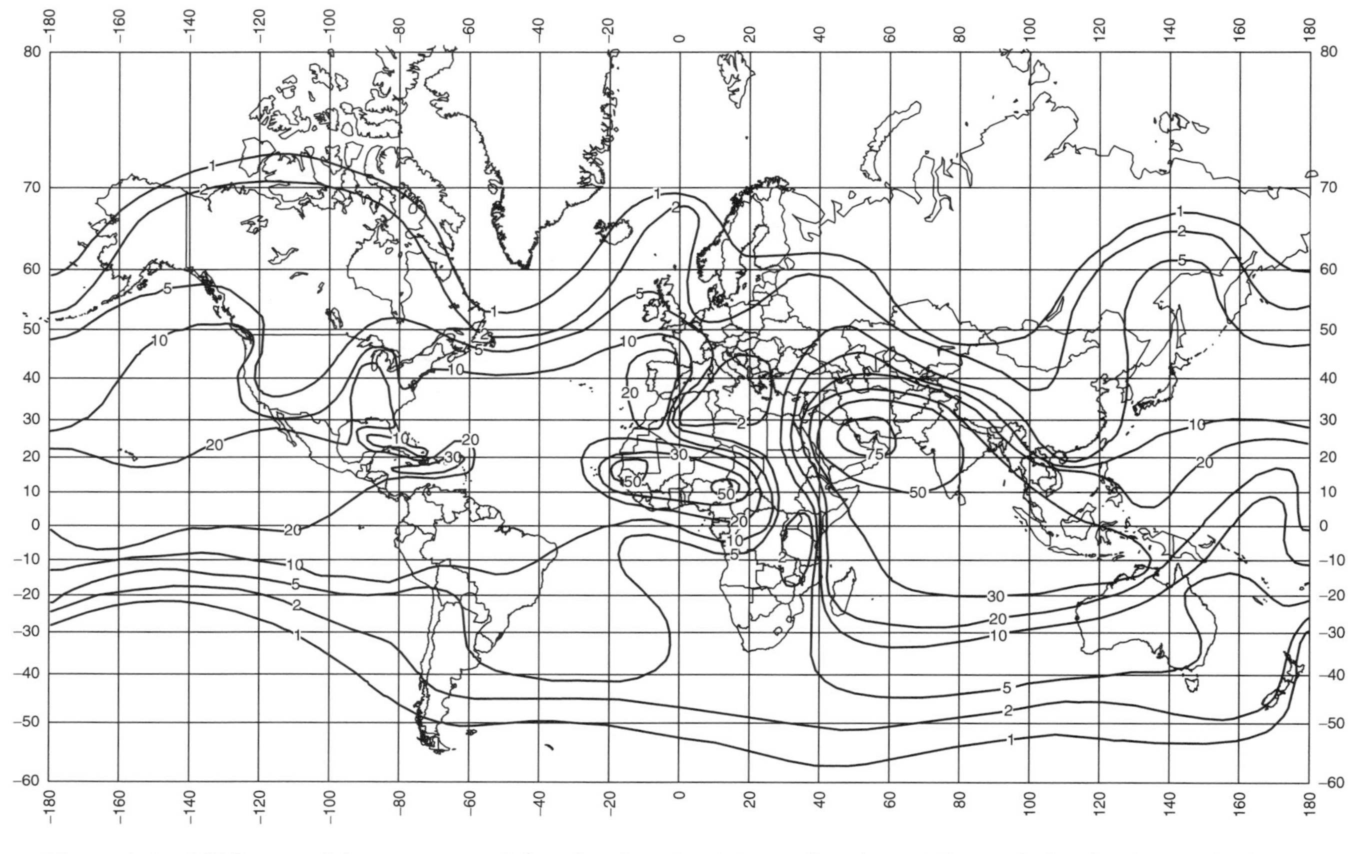

Figure A.4 ITU-R map of the percentage of time that the refractivity gradient is < -100 N-units/km for the month of May.

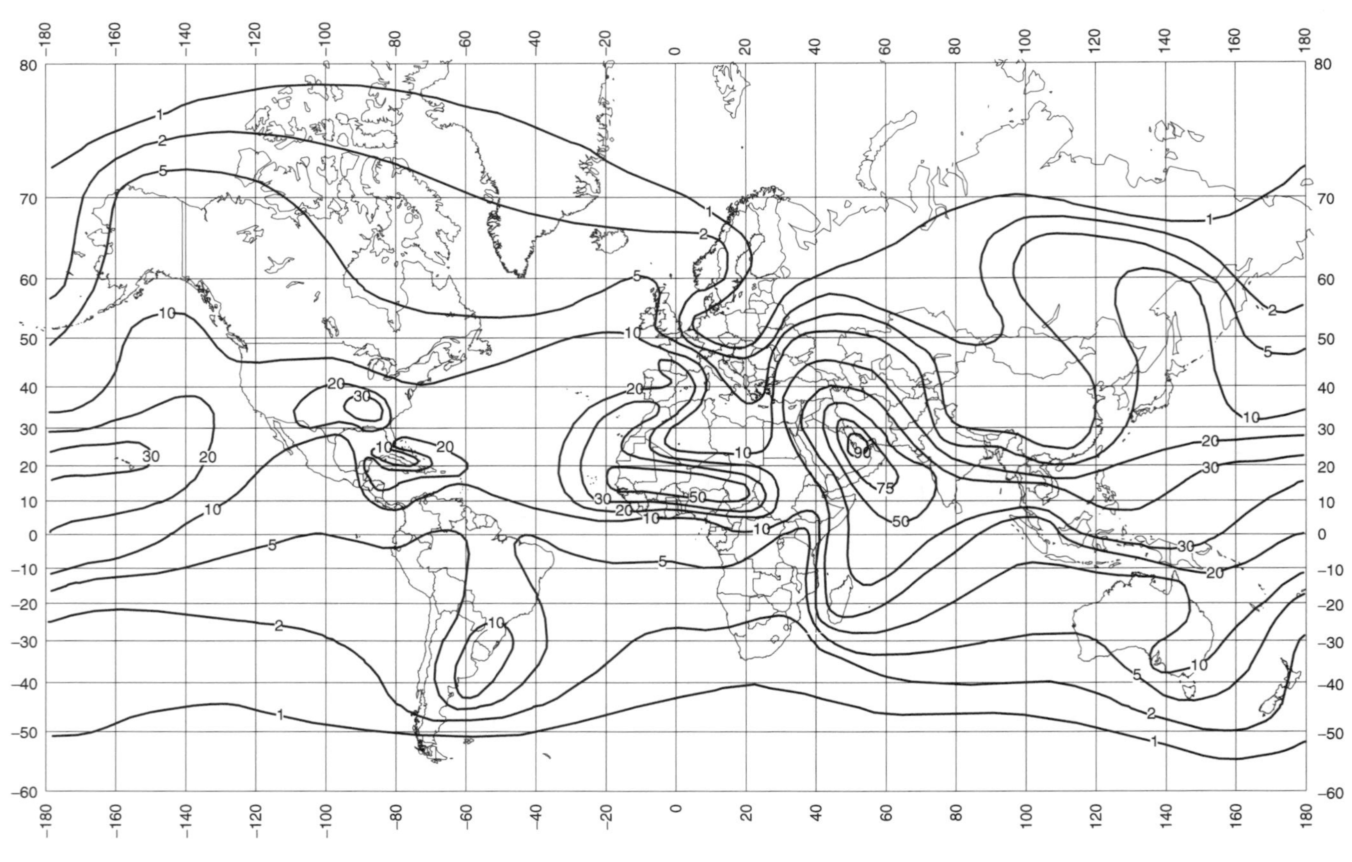

Figure A.5 ITU-R map of the percentage of time that the refractivity gradient is < -100 N-units/km for the month of August.

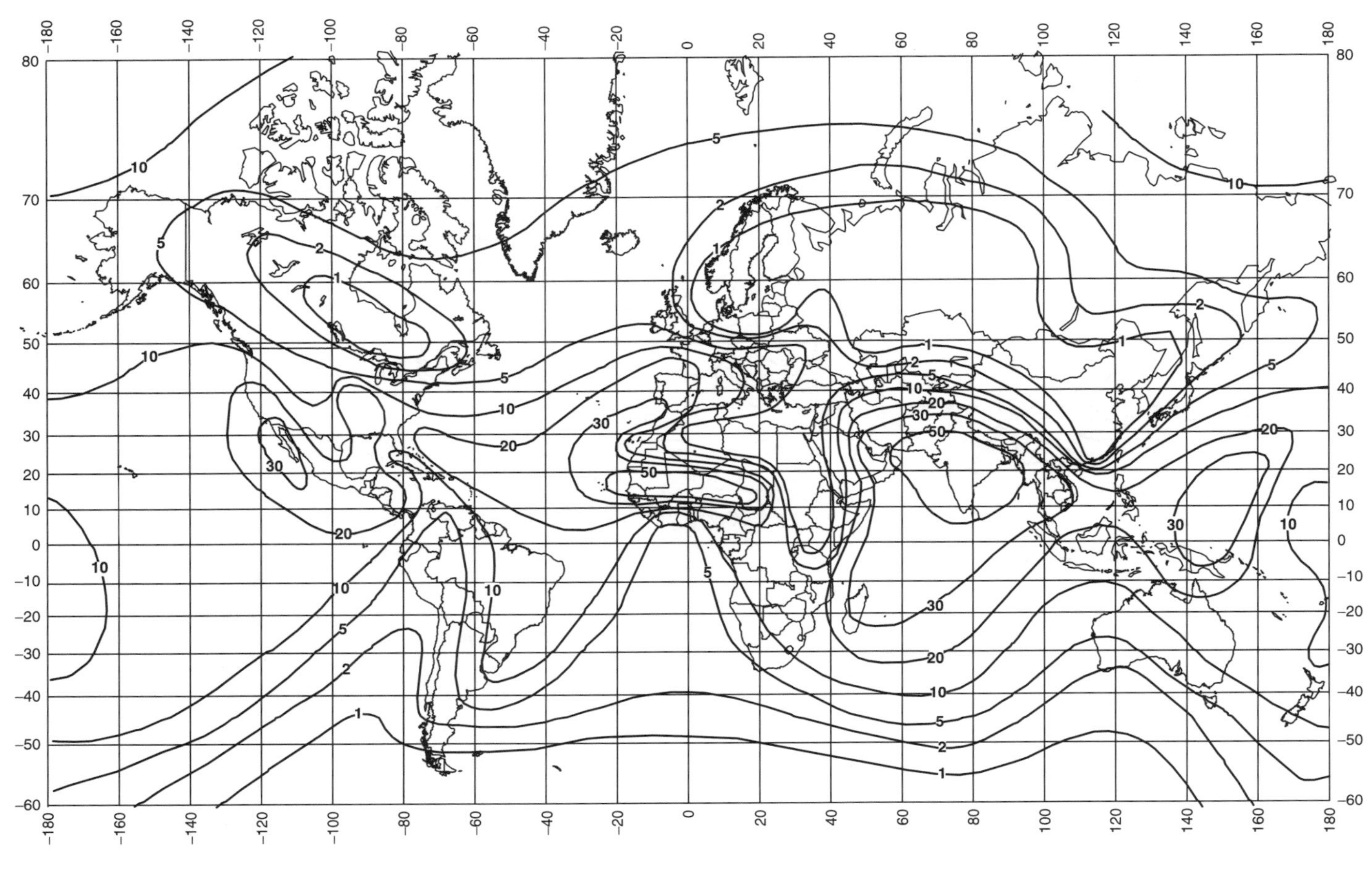

Figure A.6 ITU-R map of the percentage of time that the refractivity gradient is <-100 N-units/km for the month of November.

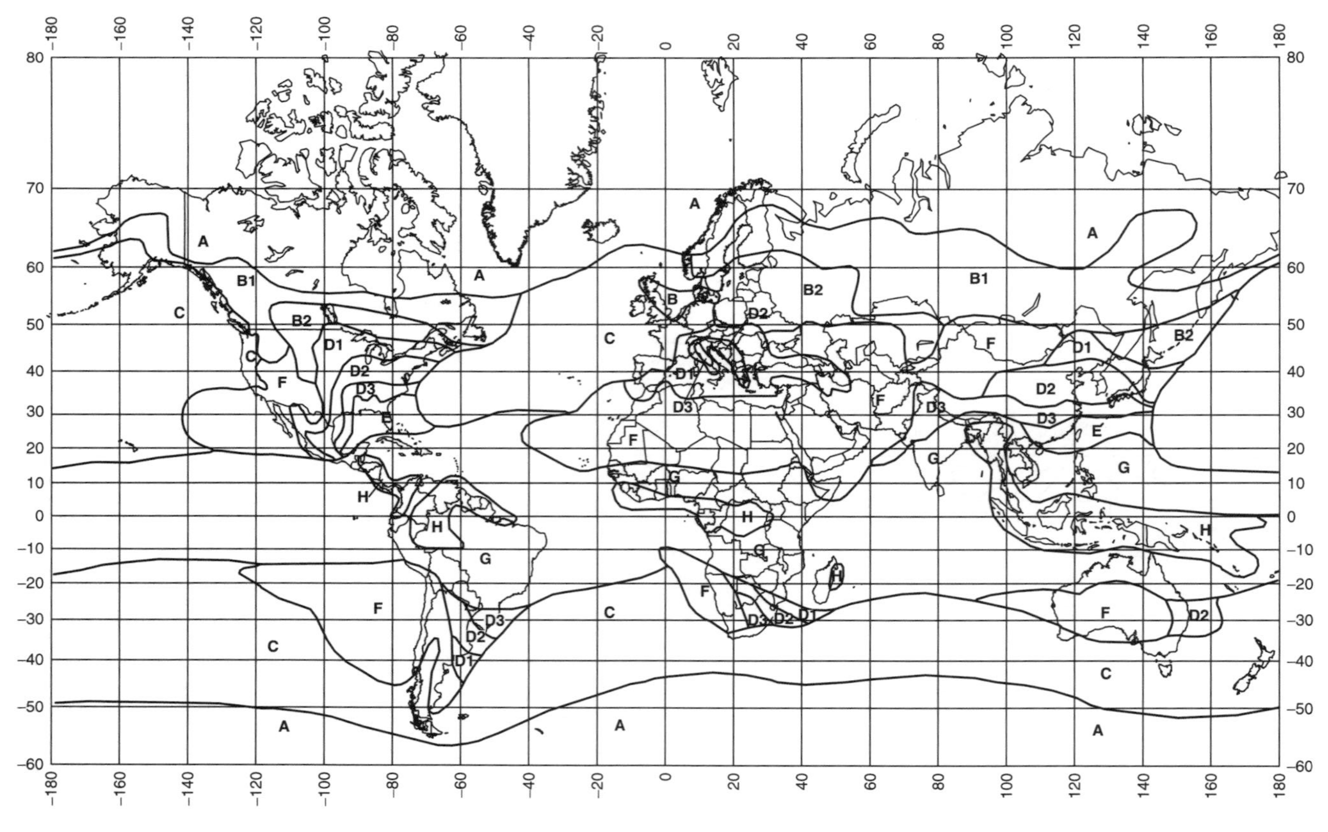

Figure A.7 Worldwide map of 1996 Crane rain regions.

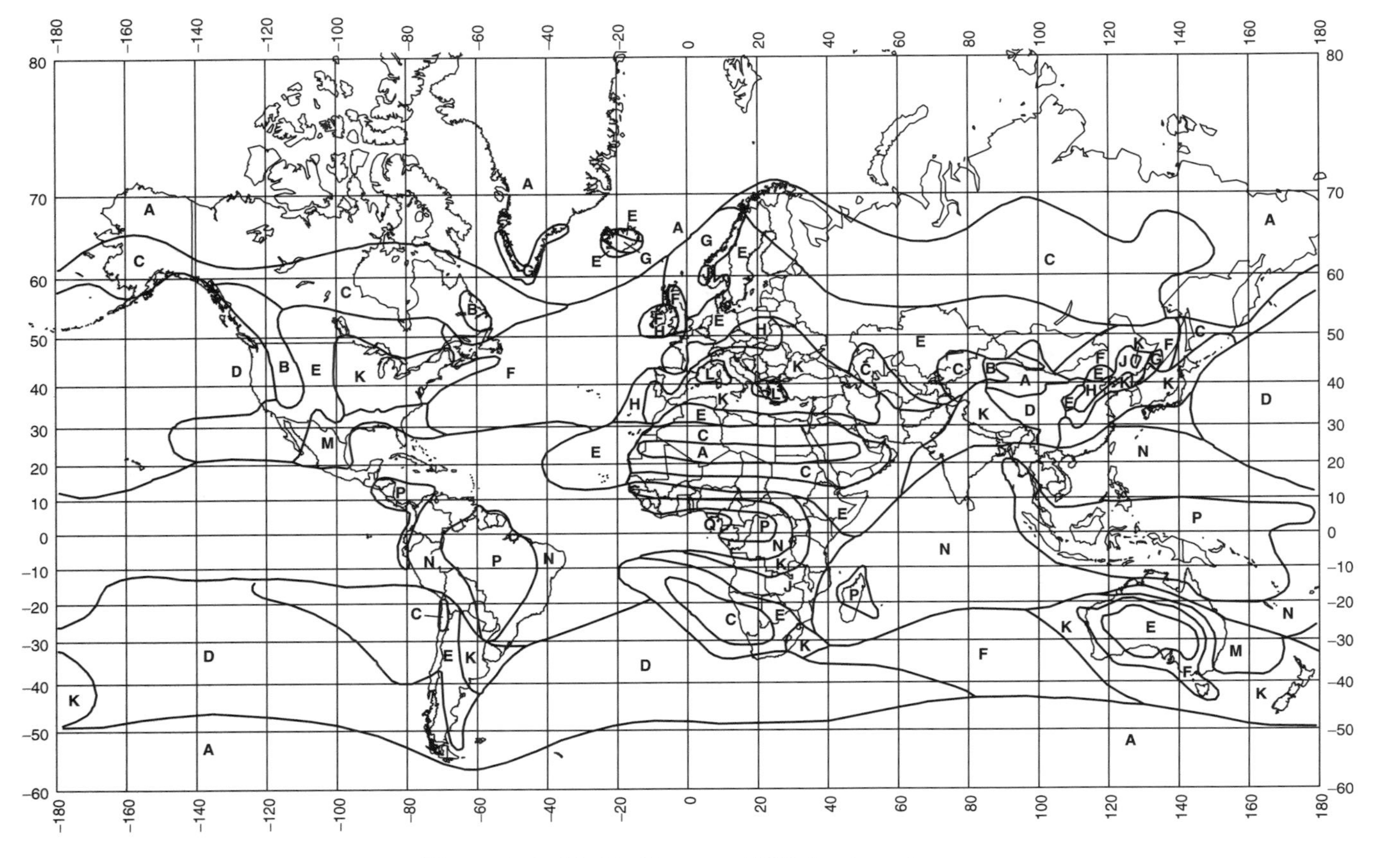

Figure A.8 Worldwide map of ITU-R rain regions.

Table A.1 1996 Crane rain rates for map regions A to H

% of year exceeded	A	B	B1	B2	C	D1	D2	D3	E	F	G	H
0.001	28.1	52.1	42.6	63.8	71.6	86.6	114.1	133.2	176.0	70.7	197.0	542.6
0.002	20.9	41.7	32.7	50.9	58.9	69.0	88.3	106.6	145.4	50.4	159.6	413.9
0.003	17.5	36.1	27.8	43.8	50.6	60.4	75.6	93.5	130.0	41.4	140.8	350.3
0.005	13.8	29.2	22.3	35.7	41.4	49.2	62.1	78.7	112.0	31.9	118.0	283.4
0.01	9.9	21.1	16.1	25.8	29.5	36.2	46.8	61.6	91.5	22.2	90.2	209.3
0.02	6.9	14.6	11.3	17.6	19.9	25.4	34.7	47.0	72.2	15.0	66.8	152.4
0.03	5.5	11.6	9.0	13.9	15.6	20.3	28.6	39.9	62.4	11.8	55.8	125.9
0.05	4.0	8.6	6.8	10.3	11.5	15.3	22.2	31.6	50.4	8.5	43.8	97.2
0.1	2.5	5.7	4.5	6.8	7.7	10.3	15.1	22.4	36.2	5.3	31.3	66.5
0.2	1.5	3.8	2.9	4.4	5.2	6.8	9.9	15.2	24.1	3.1	22.0	43.5
0.3	1.1	2.9	2.2	3.4	4.1	5.3	7.6	11.8	18.4	2.2	17.7	33.1
0.5	0.5	2.0	1.5	2.4	2.9	3.8	5.3	8.2	12.6	1.4	13.2	22.6
1	0.2	1.2	0.8	1.4	1.8	2.2	3.0	4.6	7.0	0.6	8.4	12.4
2	0.1	0.5	0.4	0.7	1.1	1.2	1.5	2.0	3.3	0.2	5.0	5.8
3	0.0	0.3	0.2	0.4	0.6	0.6	0.9	0.8	1.8	0.1	3.4	3.3
5	0.0	0.2	0.1	0.2	0.2	0.2	0.3	0.0	0.2	0.1	1.8	1.1

Table A.2 ITU-R rain rates for map regions A to Q

% of year	ITU-R rain regions. Rainfall rate exceeded (mm/h)														
exceeded	A	B	C	D	E	F	G	H	J	K	L	M	N	P	Q
0.001	22	32	42	42	70	78	65	83	55	100	150	120	180	250	170
0.003	14	21	26	29	41	54	45	55	45	70	105	95	140	200	142
0.01	8	12	15	19	22	28	30	32	35	42	60	63	95	145	115
0.03	5	6	9	13	12	15	20	18	28	23	33	40	65	105	96
0.1	2	3	5	8	6	8	12	10	20	12	15	22	35	65	75
0.3	0.8	2	2.8	4.5	2.4	4.5	7	4	13	4.2	7	11	15	34	49
1	<0.1	0.5	0.7	2.1	0.6	1.7	3	2	8	1.5	2	4	5	12	24

Appendix B

PDF of a signal with interference and noise

B.1 INTRODUCTION

A standard vector approach can be used to find the probability density function (pdf) of a signal vector perturbed by sine-wave interference and noise as shown in Figure B.1. The quadrature noise vectors are shown as n_x and n_y with the sine-wave interference as a vector with amplitude b. The objective is to find the pdf of the phase and the amplitude of this ensemble of vectors.

The general problem of determining the pdf of the phase and the amplitude of the vector with additive white gaussian noise (AWGN) was treated in [1]. For the pdf of the phase only [for determining error rate performance of phase shift keying (PSK) systems] both AWGN and cochannel sine wave were addressed by Rosenbaum [2]. The approach followed by Rosenbaum was also used in [3] for PSK error analysis with non-Gaussian atmospheric noise. Here a more generalized treatment is presented in which arbitrary transmitted phase and amplitudes are included. The derived pdf results will be given for the amplitude and the phase of the received signal vector to be used for those modulation types such as *M-ary* PSK, amplitude phase shift keying (APSK), and so on, in which these symbol parameters are used to convey information. In addition, derived pdf results for the x, y components of the received signal vector will also be presented, which can be used for modulation constellations such as *M-ary* quadrature amplitude modulation (QAM).

The following derivation follows the notation and the geometry in Figure B.1. The transmitted signal vector is shown with amplitude A and phase α. These are known quantities. An interference vector with amplitude b and phase ϕ is added to the transmitted vector. The amplitude b is known and the phase is assumed to be a random variable uniformly distributed from 0 to 2π. The composite received signal vector also includes

Fixed Broadband Wireless System Design Harry R. Anderson
 ISBN: 0-470-84438-8

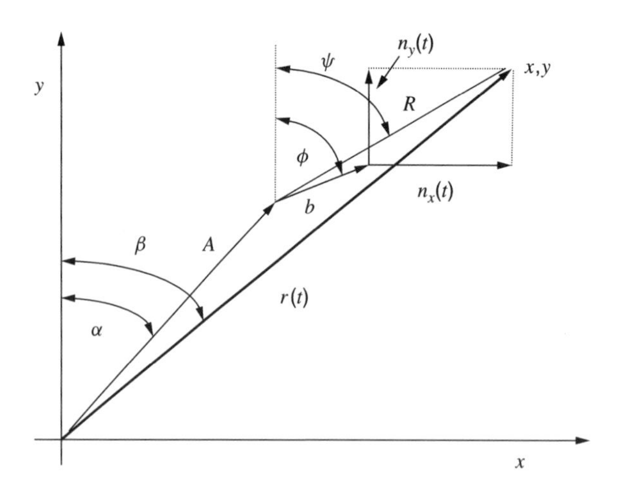

Figure B.1 Signal, interference, and noise vectors for a sine-wave interference case.

the AWGN resolved into two quadrature components n_x and n_y, each with zero mean and variances of σ^2. The resultant vector $r(t)$ has orthogonal components x, y, which have values given by

$$\overline{x} = A\sin\alpha + b\sin\phi\overline{y} = A\cos\alpha + b\sin\phi \tag{B.1}$$

Since x and y are independent, the joint distribution of x and y conditioned on ϕ is then

$$p(x, y\,|\,\phi) = \frac{1}{2\pi\sigma^2}\exp\left(-\left\{\frac{[x-(A\sin\alpha + b\sin\phi)]^2 + [y-(A\cos\alpha + b\sin\phi)]^2}{2\sigma^2}\right\}\right) \tag{B.2}$$

The ϕ can be removed by integrating over its range, namely,

$$p(x, y) = \int_0^{2\pi} p(x, y\,|\,\phi)\,d\phi \tag{B.3}$$

In order to perform this integration, the terms of the integrand that depend on ϕ must be isolated as shown in the following operations on the argument of the exponential function in (B.2).

$$\begin{aligned}[x-(A\sin\alpha + b\sin\phi)]^2 &= x^2 - 2Ax\sin\alpha + A^2\sin^2\alpha \\ &\quad + (-2bx\sin\phi + 2Ab\sin\phi\sin\alpha + b^2\sin^2\phi)\end{aligned} \tag{B.4}$$

$$\begin{aligned}[y-(A\cos\alpha + b\cos\phi)]^2 &= y^2 - 2Ay\cos\alpha + A^2\cos^2\alpha \\ &\quad + (-2by\cos\phi + 2Ab\cos\phi\cos\alpha + b^2\cos^2\phi)\end{aligned} \tag{B.5}$$

Collecting the terms in brackets in (B.4) and (B.5), which depend on ϕ, yields

$$2b(-x\sin\phi + A\sin\phi\sin\alpha - y\cos\phi + A\cos\phi\cos\alpha) + b^2(\sin^2\phi + \cos^2\phi) \tag{B.6}$$

Defining vector R and angle ψ as shown in Figure B.1,

$$R = \sqrt{(x - A\sin\alpha)^2 + (y - A\cos\alpha)^2} \quad \psi = \tan^{-1}\left[\frac{(x - A\sin\alpha)}{(y - A\cos\alpha)}\right] \tag{B.7}$$

Multiplying and dividing (B.6) by R results in

$$-2bR\left[\sin\phi\left(\frac{(x - A\sin\alpha)}{R}\right) + \cos\phi\left(\frac{(y - A\cos\alpha)}{R}\right)\right] + b^2 \tag{B.8}$$

but

$$\frac{x - A\sin\alpha}{R} = \sin\psi \text{ and } \frac{y - A\cos\alpha}{R} = \cos\psi \tag{B.9}$$

Using (B.9), (B.8) becomes

$$-2bR(\sin\phi\sin\psi + \cos\phi\cos\psi) + b^2 \tag{B.10}$$

Using the trigonometric identity $\cos\alpha\cos\beta + \sin\alpha\sin\beta = \cos(\alpha - \beta)$, (B.10) becomes

$$-2bR\cos(\phi - \psi) + b^2 \tag{B.11}$$

Collecting terms and rewriting (B.3) results in

$$p(x, y) = \frac{1}{(2\pi\sigma)^2}\exp\left(\frac{x^2 + y^2 + A^2 + b^2 - 2Ax\sin\alpha - 2Ay\cos\alpha}{2\sigma^2}\right) \times \int_0^{2\pi}\exp\left(\frac{b}{\sigma^2}R\cos(\phi + \psi)\right)d\phi \tag{B.12}$$

Since ψ is not a function of ϕ, we can make use of the following identity [4]

$$I_0(z) = \frac{1}{2\pi}\int_0^{2\pi}\exp(z\cos\theta)\,d\theta \tag{B.13}$$

where $I_0(\cdot)$ is the modified Bessel function of the first kind and order 0. Equation (B.12) then becomes (with the condition on A and b explicitly noted)

$$p(x, y \mid A, b) = \frac{1}{2\pi\sigma^2}\exp\left[-\frac{(x - A\sin\alpha)^2 + (y - A\cos\alpha)^2 + b^2}{2\sigma^2}\right] \times I_0\left[\frac{b\sqrt{(x - A\sin\alpha)^2 + (y - A\cos\alpha)^2}}{\sigma^2}\right] \tag{B.14}$$

To find the pdf of x or y alone, it is necessary to integrate (B.14) over one or the other random variable:

$$p(x) = \int_{-\infty}^{\infty} p(x, y \mid A, b)\, dy \quad p(y) = \int_{-\infty}^{\infty} p(x, y \mid A, b)\, dx \tag{B.15}$$

The pdf in (B.13) does not easily yield a closed-form expression for the probability of error. However, it can readily be integrated by numerical methods to find the volume of the pdf in the error region for any particular modulation scheme, especially those where the error boundaries between modulation states are rectilinear in x and y. As examples, in Chapter 7 this pdf is applied to finding error rates for 16QAM and quadrature phase shift keying (QPSK) when perturbed by noise and interference.

B.2 REFERENCES

[1] J.G. Proakis. *Digital Communications*. New York: McGraw-Hill. 2nd Edition. 1989. p. 262.
[2] A.S. Rosenbaum, "PSK error performance with gaussian noise and interference," *Bell System Technical. Journal*, Vol. 48, pp. 413–442. Feb. 1969.
[3] H.R. Anderson, "Theoretical performance of binary coherent PSK on AM SCA channels," *IEEE Transactions on Broadcasting*, vol. BC-29, no. 4, pp. 113–120, December, 1983.
[4] M. Abramowitz and I.A. Stegun. *Handbook of Mathematical Functions*. Washington: National Bureau of Standards, Tenth printing, December, 1972. Chapter 9.

Index